$r(t)$	traffic intensity at time t
S	service time
$S(n)$	service time of the n'th customer to arrive
$V[\ ]$	variance of enclosed random variable
W_q	expected time in queue
$\overline{W}_q$	average time in queue among a data set
$W_q(n)$	time in queue, for the n'th customer to arrive
$\overline{\overline{W}}_q$	total time in queue among all customers
W_s	expected time in system
$\overline{W}_s$	average time in system among a data set
$W_s(n)$	time in system, for the n'th customer to arrive
β	value of time per customer
$\overline{\lambda}$	arrival rate
$\lambda(t)$	arrival rate at time t
λ	average arrival rate
$\Lambda(t)$	expected arrivals by time t
μ	service rate per server
$\mu(t)$	service rate per server at time t
$\overline{\mu}$	average service rate per server
ρ	absolute utilization $= \overline{\lambda}/\mu$
$\rho(t)$	absolute utilization at time t $= \lambda(t)/\mu(t)$
$\tilde{\rho}$	proportional utilization $= \overline{\lambda}/m\mu = \overline{\lambda}/c$
$\tilde{\rho}(t)$	proportional utilization at time t $= \lambda(t)/c(t)$
ω	departure rate
$\omega(t)$	departure rate at time t

Queueing Methods

PRENTICE HALL INTERNATIONAL SERIES
IN INDUSTRIAL AND SYSTEMS ENGINEERING

W. J. Fabrycky and J. H. Mize, Editors

ALEXANDER *The Practice and Management of Industrial Ergonomics*
AMOS AND SARCHET *Management for Engineers*
ASFHAL *Industrial Safety and Health Management* 2/E
BABCOCK *Managing Engineering and Technology*
BANKS AND CARSON *Discrete-Event System Simulation*
BEIGHTLER, PHILLIPS, AND WILDE *Foundations of Optimization*, 2/E
BANKS AND FABRYCKY *Procurement and Inventory Systems Analysis*
BLANCHARD *Logistics Engineering and Management*, 3/E
BLANCHARD AND FABRYCKY *Systems Engineering and Analysis*, 2/E
BROWN *Systems Analysis and Design for Safety*
BUSSEY *The Economic Analysis of Industrial Projects*
CANADA/SULLIVAN *Economic and Multi-Attribute Evaluation of Advanced Manufacturing Systems*
CHANG AND WYSK *An Introduction to Automated Process Planning Systems*
CLYMER *Systems Analysis Using Simulation and Markov Models*
ELSAYED AND BOUCHER *Analysis and Control of Production Systems*
FABRYCKY, GHARE, AND TORGERSEN *Applied Operations Research and Management Science*
FRANCIS AND WHITE *Facility Layout and Location: An Analytical Approach*
GIBSON *Modern Management of the High Technology Enterprise*
HALL *Queueing Methods: For Services and Manufacturing*
HAMMER *Occupational Safety Management and Engineering*, 4/E
HUTCHINSON *An Integrated Approach to Logistics Management*
IGNIZIO *Linear Programming in Single- and Multiple-Objective Systems*
KUSIAK *Intelligent Manufacturing Systems*
MUNDEL *Improving Productivity and Effectiveness*
MUNDEL *Motion and Time Study: Improving Productivity*, 6/E
OSTWALD *Cost Estimating*, 2/E
PHILLIPS AND GARCIA-DIAZ *Fundamentals of Network Analysis*
SANDQUIST *Introduction to System Science*
SMALLEY *Hospital Management Engineering*
TAHA *Simulation Modeling and SIMNET*
THUESEN AND FABRYCKY *Engineering Economy*, 7/E
TURNER, MIZE, AND CASE *Introduction and Systems Engineering*, 2/E
WHITEHOUSE *Systems Analysis and Design Using Network Techniques*
WOLFF *Stochastic Modeling and the Theory of Queues*

Queueing Methods

For Services and Manufacturing

Randolph W. Hall

University of California at Berkeley

PRENTICE HALL
Englewood Cliffs, NJ 07632

Library of Congress Cataloging-in-Publication Data

HALL, RANDOLPH W., (date)
Queueing methods : for services and manufacturing / by Randolph W. Hall.

Includes bibliographical references.
ISBN 0-13-744756-6
1. Service industries—Management—Mathematical models.
2. Manufactures—Management—Mathematical models. 3. Queueing theory. I. Title.
HD9980.5.H35 1991
658.4'034—dc20 90-6877
CIP

PRENTICE HALL INTERNATIONAL SERIES
IN INDUSTRIAL AND SYSTEMS ENGINEERING

W. J. Fabrycky and J. H. Mize, Editors

Editorial/production supervision and
interior design: Marianne Peters
Cover design: Lundgren Graphics, Ltd.
Prepress buyer: Margaret Rizzi
Manufacturing buyer: David Dickey

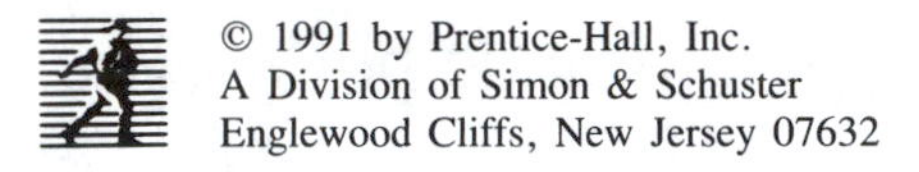

Printed in the United States of America
10 9 8 7 6 5 4 3 2 1

ISBN 0-13-744756-6

Prentice-Hall International (UK) Limited, *London*
Prentice-Hall of Australia Pty. Limited, *Sydney*
Prentice-Hall Canada Inc., *Toronto*
Prentice-Hall Hispanoamericana, S.A., *Mexico*
Prentice-Hall of India Private Limited, *New Delhi*
Prentice-Hall of Japan, Inc., *Tokyo*
Simon & Schuster Asia Pte. Ltd., *Singapore*
Editora Prentice-Hall do Brasil, Ltda., *Rio de Janeiro*

To my parents

Contents

Preface

I began this work with a computer search of the University of California library catalog for entries containing the word ''queues'' or ''queueing.'' After compiling a list of 100 or so books, I was surprised to find one written by Pablo Picasso. Little had I known that a man known for abstract art was also an author—let alone an author on queues. To my chagrin, I discovered that Picasso's book, titled *Le Desire Attrape par la Queue* translated to *Desire Caught by the Tail*. I had forgotten that the original meaning of ''queue'' is ''tail of a beast.''

Lest there be confusion, this book is not about tails. It is about another type of queue, the type of queue known in the United States as ''a line or file of people or vehicles awaiting turn, as at a ticket window.''

Although my interest is in queueing lines, I like to think of beast tails whenever I see the word queue, for queues are, in a way, part of a beast. More often than not, queues are the most visible symptom of a problem. This problem is not the tail itself but the beast at the front of the tail—the person, company, or object serving the queue. If the beast is incapable of serving its customers, the queue grows. But queueing problems are not solved by ''chopping off the tail''; they are solved by looking directly at the heart of the beast.

It is unfortunate that many societies view queues as our modern-day manifest destiny, ranked right behind death and taxes. We expect to wait when we see our doctors, we expect to wait when we commute to work, and we expect to wait when we visit the post office. Our sense of hopelessness is heightened by the literature on queueing. Invariably, pessimists will state that delay can only be reduced at the expense of providing more servers, or that it is inefficient for a system *not* to have queues from time to time. Thus, queueing appears to be the unfortunate consequence of an economic compromise between service costs and delay costs.

Despite what has been written, long queues and long waits are not inevitable. There are ways to eliminate queues without adding servers. And reducing delay does not always have to cost money.

Though anyone who has waited in line probably recognizes that queueing is an important problem, that is not my sole motivation in writing this book. The operations research (OR) profession (to which I belong) has been criticized in recent years for doing work that is either unrealistic or irrelevant. And the first place that critics point to seems to be queueing research. I find this criticism ironic because queueing is one of the areas of operations research concerned with how physical systems behave. Most of operations research is technique oriented (for example, linear programming, nonlinear programming, network optimization), without any specific application.

In 1984, I wrote an article titled "What's So Scientific about MS/OR" in which I argued that the OR profession could and should be more "scientific" and less "mathematical." By this, I meant that the profession should be more concerned with how systems behave and less concerned with abstract symbol manipulation. I recognized then that queueing was one of the best opportunities to shift in this direction. Even with the preponderance of research on queueing mathematics, there is still an impressive amount of literature on "queueing science." To take two well known examples, a paper by Edie on queueing at toll booths and another by Oliver and Samuel on letter delays in post offices both won the Lanchester prize, awarded annually by the Operations Research Society of America to the outstanding work written each year in operations research.

So my aim is not simply to write a book on "applied queueing theory" (although I do provide a number of practical suggestions), but to write a book on the science of queueing. I believe that the best way to solve queueing problems is to understand how a queue operates. Understanding does not follow directly from the mathematics of queueing, but from the observation and measurement of queueing systems. The idea is not to take queueing models and apply them to real problems, but to look at real problems, then develop appropriate models.

The objective of this book is to provide the knowledge needed to diagnose and correct the problems that are creating queues, whether they be queues of shoppers at a market, queues of work in a factory, or queues of checks at a service center. The book will also provide an understanding of the general principles of how queues operate and how to use models to analyze queueing phenomena and develop queueing solutions.

But even if I fail to achieve these objectives, be sure to think of tails when you see the word *queue*. Soon after I was reminded of this meaning I read in our student newspaper about queues snaking hundreds of feet out the doors of the campus administration building as students waited to register for classes. Perhaps you too can have a good laugh imagining what kind of beast must have lurked inside to have such a tremendous tail!

ACKNOWLEDGMENTS

I owe my deepest gratitude to the students of IEOR 164 for making this book possible. Steve Bender, Inwook Cho, Carty Chock, Toby Choy, Charles Harris, Vivienne Ip, Mei Po Kwan, Sophia Lam, Hsueh Lee, Joel Lehrer, May Ma, Craig Matsumoto, Kathleen McDivitt, John O'Neal, Karen Rawson, Scott Rosenthal, Sol Solorzano, Ok Kyung Suk, and Jeh-Wuan Tan all contributed to this book. Special thanks go to Michael Grivett for his thorough reading—he was an enormous help.

I am grateful to the Literary Executor of the late Sir Ronald A. Fisher, F. R. S., to Dr. Frank Yates, F. R. S., and the Longman Group Ltd, London, for permission to reprint Table III from their book *Statistical Tables for Biological, Agricultural, and Medical Research* (6th Edition, 1974).

My gratitude also goes to Gordon Newell whose comments and inspiration greatly influenced the content of this book. Thanks also go to Carlos Daganzo, Yafeng Du, Bob Foote, Bill Jordan, Richard Larson, Julia Lin, Michael Racer, and Bart Stuck, as well as many anonymous reviewers, for their constructive suggestions.

Most of all, my appreciation goes to Janice Partyka, for her love and support over the last three years.

NOTE TO EDUCATORS

Queueing Methods was written to teach students and professionals how to use basic concepts of queueing theory to solve queueing problems. At the University of California at Berkeley, *Queueing Methods* has been used in a junior/senior level Industrial Engineering course in Queues and Inventories. However, the book could also be used to teach queueing courses in Business Administration, Transportation Engineering, and Applied Operations Research. The book could also serve as a supplemental text for courses in Simulation or Applied Statistics.

The prerequisites for the book are an introductory course in Probability and Statistics and a semester of Calculus. It does not require background in transformation methods or advanced concepts of probability. Though much of the material is quantitative, the book is written in simple language and encourages students to use mathematics and statistics to solve problems. The techniques presented are not frightening or incomprehensible and all important equations are illustrated with graphs.

The book has two main parts. The first (Chapters 1–6) emphasizes modeling and the second (Chapters 7–11) emphasizes solving queueing problems. The modeling section shows how to build a model from the ground up—beginning from data collection and observation. The steps are repeated in a series of exercises at the ends of Chapters 2–5, which are recommended in teaching students how to create stochastic models. In the second part of the book, the concepts developed in the first six chapters are used to solve queueing problems. Case studies are provided here that describe the operations of real queueing systems, such as fast-food restaurants, health clinics, or telephone directory assistance. The case studies should be used in class for discussion on the causes and remedies for queueing delays. Problem sets are included in all chapters, which can help readers master queueing techniques.

SUPPLEMENTAL CASE STUDIES

In addition to the studies provided in this book, the following cases are excellent for in class discussion.

Browne, J. 1984. *Management and Analysis of Service Operations*. New York: North-Holland. (contains several case studies drawn from the Port Authority of New York; can be used in Chapters 7, 11)

Edie, L. C. 1954. "Traffic Delays at Toll Booths," *Journal of the Operations Research Society*, 2, 107–138. (Chapters 3, 7)

Green, L. and P. Kolesar. 1989. "Testing the Validity of a Queueing Model of Police Patrol," *Management Science*, 35, 127–148. (Chapter 3)

Horonjeff, R. 1969. "Analyses of Passenger and Baggage Flows in Airport Terminal Buildings," *Journal of Aircraft*, 6, 446–451. (Chapter 6)

Ignall, E. J., P. Kolesar, A. J. Swersey, W. E. Walker, E. H. Blum, and G. Carter. 1975. "Improving the Deployment of New York City Fire Companies," *Interfaces*, 5:2, 48–61. (Chapter 11)

Kolesar, P. 1984. "Stalking the Endangered CAT: A Queueing Analysis of Congestion at Automatic Teller Machines," *Interfaces*, 14:6, 16–26.

Kolesar, P. J., K. L. Rider, T. B. Crabill, and W. E. Walker. 1975. "A Queueing-Linear Programming Approach to Scheduling Police Patrol Cars," *Operations Research*, 23, 1045–1062. (Chapter 7)

Linder, R. W. 1969. "The Development of Manpower and Facilities Planning Methods for Airline Telephone Reservations Offices," *Operational Research Quarterly*, 20, 3–21. (Chapter 7)

Oliver, R. M. and A. H. Samuel. 1967. "Reducing Letter Delays in Post Offices," *Operations Research*, 10, 839–892. (Chapter 10)

Vickrey, W. S. 1955. "Revising New York's Subway Fare Structure," *Journal of the Operations Research Society of America*, 2, 38–68. (Chapter 8)

Welch, N. and J. Gussow. 1986. "Expansion of Canadian National Railway's Line Capacity," *Interfaces*, 16:1, 51–64. (Chapter 10)

chapter 1

Introduction

Think back over your week and consider how much time you spent waiting in line—at the supermarket, in traffic, at the post office. Every moment you spent waiting for some type of service you were part of a queue. But queues are not always this obvious and they do not have to involve people. A suit waiting to be dry-cleaned is part of a queue; a memo waiting to be typed is part of a queue; and a lawsuit waiting to be heard in court is part of a queue. A queue is a group of people, tasks, or objects waiting to be served. Waiting is the essence of queueing.

Though we are most aware of our own waiting time, queueing is foremost a problem for industry and government. The success of any organization depends on maximizing the utilization of its resources. Every minute that an employee spends waiting for another department and every minute that a job spends waiting to be processed is money wasted. The success of any organization also depends on attracting and keeping customers. Every minute that a customer spends waiting to be served translates into lost business and lost revenue.

Continued growth in the service industries in the United States has heightened the need to manage and control queueing effectively. Though queueing is also an important concern to manufacturers, it is felt especially strongly in the service sector because of the heavy reliance on customer interaction. In their article ''Will Services Follow Manufacturing into Decline?'' James Quinn and Christopher Gagnon highlight the need to keep service industries competitive:

> Although they probably know better, many executives still think of the service sector in terms of people making hamburgers or shining shoes. These images belie the complexity, power, technical sophistication, and economic value of activities that now account for more than 68% of the nation's GNP and 71% of its employment. Worse, they help perpetuate a set of myths about service industries that lead managers and policymakers to ignore their full potential. Worse still, such inattention and complacency threaten to undermine the competitive ability of these industries at a time when their importance to the national economy has never been greater.[1]

Though not always recognized as such, queueing problems in the service sector constantly appear in the headlines of major newspapers. Note these examples:

Airline Delays

> Operating the Eastern shuttle on time is one of the toughest jobs in the industry. The New York–Washington corridor traverses some of the most congested airspace in the world. Shuttle pilots must dodge a multitude of other planes, especially near New York's three major airports, La Guardia, Kennedy and Newark. . . . Congestion in the sky can quickly cause a frustrating backup of 20 planes waiting to take off at either La Guardia or at National airport in Washington.[2]

Medical Transplants

> "Modern Technology has provided us with a double-edged sword," said Dr. Gary Friedlander, chief of orthopedics at Yale University. "On the one hand, organ transplants are miracles that will save many lives. On the other hand, we are faced with thousands of people whose lives potentially could be saved with a transplant but who will die because they are not available." The hard truth is there are nowhere near enough donor organs and probably never will be.[3]

Court Congestion

> New York City's criminal-justice system is in a state of crisis, just barely able to cope with a growing flood of new drug cases generated by the Police Department's drive against crack. . . . So desperate is the situation . . . that without a major new infusion of money, manpower, new courtrooms and more jail space, the system will be swamped by this summer. . . . As a result, judges and district attorneys say, more and more accused drug traffickers and users will simply be returned to the streets, their cases dismissed or the charges greatly reduced.[4]

Fortunately, queues are not problems without solutions. Queueing problems go unsolved only because the organizations causing the problems do not directly pay the price. The server does not bear the cost and aggravation of waiting. And though long waits likely translate into lost business, organizations do not always see the cause and effect. Lost business can be blamed on any number of things—competition, weather, or other "acts of God"—which have nothing to do with queueing. Casting blame elsewhere perpetuates the problem.

In *Quality Is Free*, Philip Crosby challenges the notion that product quality depends on a trade-off with cost. He writes: "Quality is free. . . . What costs money are the unquality things—all the actions that involve not doing jobs right the first time."[5] Just as no one benefits from poor quality, no one directly benefits from queueing. The time customers spend waiting does not benefit the organization providing the service, and it surely does not benefit the customer. Time spent queueing is time wasted forever.

The point to recognize is that queueing is an important problem that affects us all both directly and indirectly. But just as importantly queueing is a solvable problem. The solution may not be simple, but with some understanding and an ability to put itself in the customer's shoes, an organization can substantially reduce waiting times at little or no expense.

1.1 HISTORY OF QUEUEING

Since the day that humans gathered into societies, there have been queues. History books are replete with occurrences of citizens waiting for the dole, courtiers waiting for an audience with the king, or the unemployed waiting for a bowl of soup. Though queueing is by no means new, the *study* of queues is modern, dating only to the beginning of the twentieth century and the work of A. K. Erlang, a man who spent years investigating a then cutting-edge technology—telephones (see Brockmeyer et al. 1948).

Theoretical analysis of queueing systems grew considerably with the advent of operations research in the late 1940s and early 1950s. The first textbook on the subject, *Queues, Inventories, and Maintenance*, was written in 1958 by Morse. Saaty wrote his famous *Elements of Queueing Theory with Applications* in 1961, and Kleinrock completed his *Queueing Systems* in 1976. Today, more than 40 books have been written on the subject.

Over the last 30 years the mathematics of queueing systems has advanced tremendously. Journals such as *Operations Research* and *Naval Research Logistics Quarterly* frequently contain contributions on the subject, and the total number of papers exceeds 1000. Less has been written on the scientific aspects of queueing systems—that is, on how real queues actually behave. Most of the work in this area is found in applications-oriented journals, such as *Interfaces*, *Transportation Science*, and *International Journal of Production Research*.

1.2 ELEMENTS OF QUEUEING SYSTEMS

It may seem that a queue of lawsuits waiting for trial has little in common with a queue of shoppers waiting at the checkout counter. In fact, both possess the same basic elements. They are both processes by which "customers" wait for service. This concept may seem straightforward, but the terms *customer*, *server*, and *queue* do require elaboration if they are to be applied to disparate systems.

The *customer* is the person or thing that waits for service. In the case of a queue of people waiting to buy stamps, the ‘‘customer’’ is a customer in the ordinary sense of the word. But identifying the customer is not always trivial. At a supermarket checkout counter, the customer may be either the shopper in line or the items being purchased, depending on one’s point of view. Both are waiting for service. *A customer does not have to be a person.*

The *server* is the person or thing providing the service. Like the customer, identifying the server is sometimes obvious and sometimes not. For the stamp queue, the server might simply be the postal clerk. But if one takes a broader perspective, the server might encompass the clerk, a cash register, and other machines that serve the customer in concert. All are necessary to perform the service and all are part of the server.

The *queue* is the group of customers waiting to be served. The queue does not have to be an orderly line. It does not even have to be visible. The queue is simply the group of customers that have requested, but have not yet received, service. The request for service can take many different forms: the physical arrival of a person at the back of a line, a phone call, a submission of a piece of work, or the like.

Thus, a queueing system has just three basic elements: customer, server, and queue. Despite the apparent simplicity, identifying the basic elements is not always easy and is almost always open to debate. There are even times when the roles of customer and server are interchanged (an idle server might be viewed as a ‘‘customer’’ waiting for a true customer to serve).

1.3 QUEUEING CHARACTERISTICS

The best way to understand how a queue operates is to examine the characteristics of the basic queueing elements. Here are some of the things to consider.

1.3.1 Customer

The ***arrival process*** depicts the timing of customer arrivals at the queue. Do customers arrive independently of each other, or do they arrive in groups? Do customers arrive at a fairly constant rate, or is there some pattern to their arrivals (for example, a ‘‘rush hour’’)? Is the arrival process predictable or random? Do different types of customers arrive at different times of the day?

In addition to understanding customer arrivals, one should also understand customer reneging, balking, and jockeying. ***Reneging*** is the act of leaving a queue before being served; ***balking*** is the act of not joining a queue upon arrival. Conceptually, the two concepts are the same. The only difference is the timing of when the customer leaves (either immediately in the case of balking or later in the case of reneging). For this reason, both actions will be called *reneging* in this book. ***Jockeying*** is the act of switching from one queue to another. Reneging, balking, and jockeying are three of the most difficult aspects of a queueing system to measure because the customer may never be recorded by the system. How long must a queue be before customers renege? Are some types of customers more likely to renege than others? Will a customer that reneges come back or never return? If the customer returns, when will that be?

1.3.2 Server

The ***service process*** represents the time taken to serve customers, commonly referred to as the *service time*. Is the service time constant, or does it vary from customer to customer? Are customers served in bulk, as at a traffic signal? Does the service time depend on the type of customer, and can it be predicted in advance? Does the service time depend on the server or the time of day? Is there any way to shorten the service time by improving server efficiency?

A second important aspect of the server is its *configuration*. How many servers are there? Which servers work in parallel, performing identical tasks, and which perform in series, performing different tasks? What are the rules governing how customers move from server to server?

1.3.3 Queue

The ***queue discipline*** specifies the order in which customers in the queue are served. Are customers served on a first-come, first-served (FCFS) basis, or perhaps on a last-come, first-served (LCFS) basis? Do different customers receive different priority, and if so, what is the system for assigning priority? Is service for a customer ever interrupted to serve another with higher priority?

A second important characteristic of the queue is its general *organization*. Is there a single queue that feeds all servers, separate queues at each server, or some variation of the two? Is there a limit on the total number of customers in the queue? What is the system for keeping track of customers and ensuring orderliness?

These are some of the key questions to ask in assessing a queueing system. Many of these aspects can be changed in order to improve the performance of the system; others are beyond control. Regardless, understanding these aspects is the first step toward improving a queueing system's performance.

1.4 EXAMPLES OF QUEUEING SYSTEMS

To give you a better feel for the basic queue elements, this section describes three familiar queueing systems: bank, supermarket, and motor vehicles department. These systems represent three common configurations: parallel servers fed by a single queue; parallel servers, each fed by its own queue; and a combination of parallel and serial servers.

1.4.1 Bank

The primary functions of commercial banks are borrowing and lending money. Borrowers are served by loan officers, and lenders (that is, depositors) are served by account officers and cashiers. Depending on the transaction, a bank client might visit any of several places inside the bank (Fig. 1.1).

Let us concentrate on the depositor who wishes to make a withdrawal. The depositor has two choices for service: an ordinary (human) teller or an automated (computer) teller. For the first case, the client first walks to a desk and fills out a withdrawal form, then joins a first-come, first-served queue to wait for the first available teller. In the second case, the client does not fill out a withdrawal form, but instead joins a single queue waiting for the automated teller.

The two types of tellers are represented by the queue diagram in Fig. 1.2. The upside-down triangle is the symbol for a queue, and the circle is the symbol for a server. The arrows represent the movement of customers from queue to server. The ordinary tellers work in parallel and are fed by a single queue. The automated teller works alone and is also fed by a single queue.

What Is the Customer? Either the depositor or the transaction can be viewed as the customer. If a depositor wishes to perform several transactions, each transaction might be considered a separate customer.

What Is the Server? The server is either an automated teller or the human teller (perhaps in concert with a computer console and other machines, some of which may be shared with other tellers). Less directly, a centralized computer also serves the customer.

The configuration, queue discipline, and queue organization have been outlined. Consider on your own the other key characteristics of the system: What is the arrival process—Are there times of the day or week when customers arrive at an especially fast rate? What is the reneging process—Do customers renege from the ordinary teller queue to join the automated queue? What is the service process—Do some transactions take longer than others, and why? In light of these characteristics, consider why the system is designed the way it is.

1.4.2 Supermarket

Supermarkets sell a wide variety of products ranging from produce to packaged food to stationery. A supermarket operates on a self-service basis. Shoppers take a cart as they enter the store and walk up and down the aisles, retrieving goods along the way. After selecting their goods, shoppers join the queue for one of the checkout counters and wait for a checker to compute the bill, accept payment, and bag the groceries.

Figure 1.3 shows the checkout configuration for a market with eight checkout counters. Each counter has a cash register, a scale (for weighing produce), and a laser scanner (for reading bar codes from packaged goods). Two of these counters can be used for quick-check (12 items or less). Thus, customers making small purchases have some

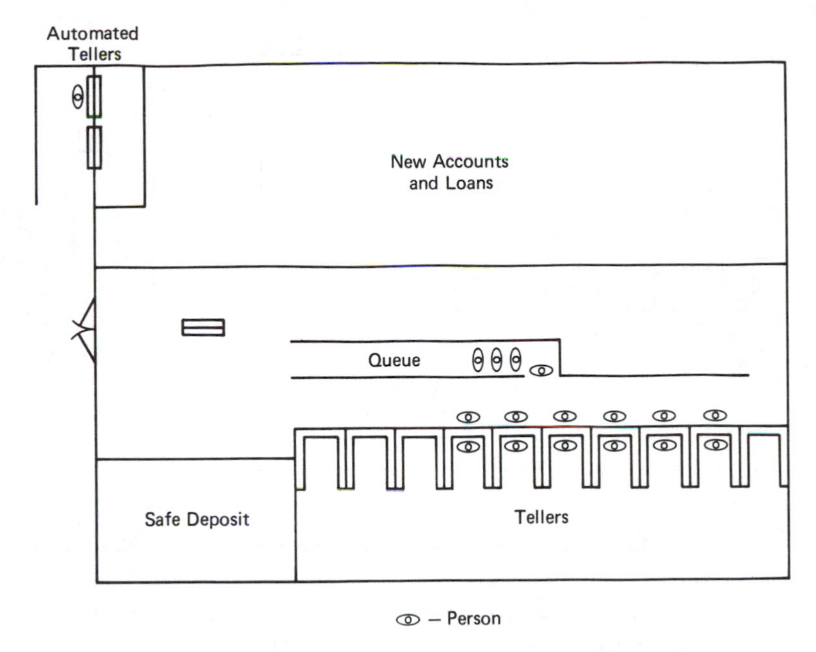

Figure 1.1 Bank layout showing parallel servers fed by a single queue.

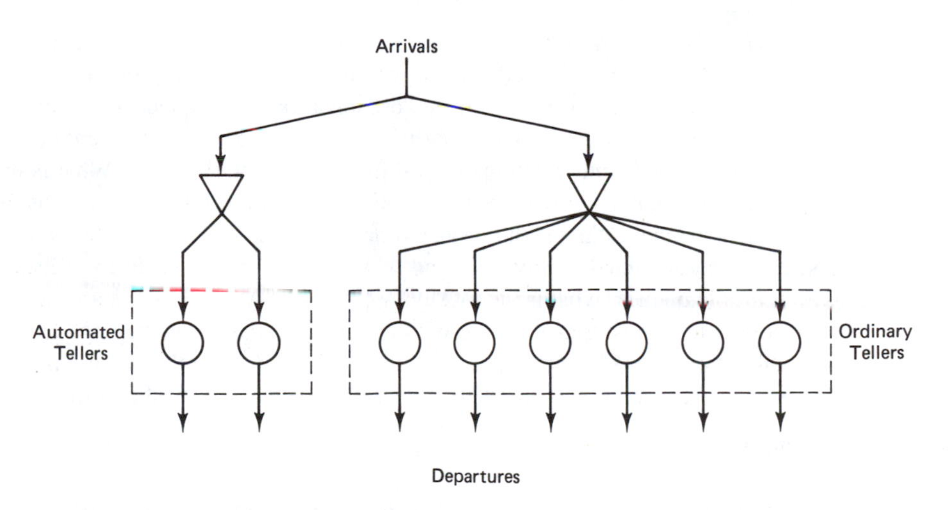

Figure 1.2 Queue diagram of bank.

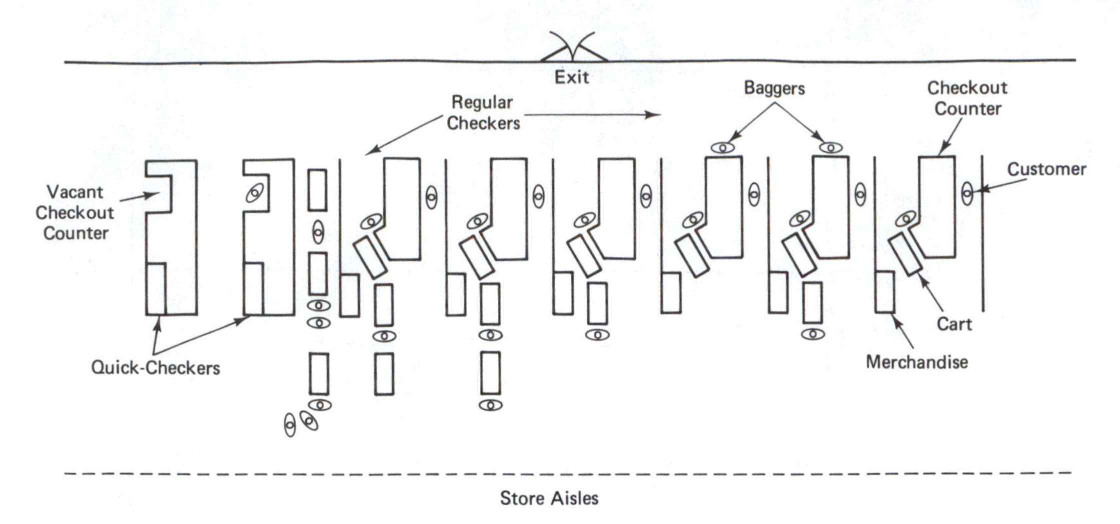

Figure 1.3 Grocery store layout showing parallel servers fed by separate queues.

priority over other customers. The number of counters in operation varies with demand. Each counter can be served by one (checker only) or two (checker and bagger) people, also depending on demand. These features are illustrated by the queue diagram in Fig. 1.4, which shows a system with parallel servers, each fed by its own queue.

What Is the Customer? The customer is either the shopper or the set of items purchased (along with the payment transaction). In the latter case, the queue diagram might be more detailed and account for queues occurring at the various pieces of equipment (for example, scale versus laser scanner).

What Is the Server? The server is the checker along with the cash register, scanner, and scale. When used, the bagger is also part of the server.

Again, the configuration, queue organization, and queue discipline have been outlined. Consider on your own: What is the arrival process—Are there certain times of the day, week, or month when the arrival rate is particularly large? What is the reneging process—Will a customer renege after walking in the door, after selecting the products, or some time between? What is the service process—Does service time vary among customers, can it be predicted in advance, and what can the customer do to influence it? And, with multiple queues, what is the jockeying process—At what points are customers likely to jockey over to a shorter line? How do grocery carts affect jockeying? In light of these characteristics, consider why queues are organized differently at supermarkets than at banks, and why some supermarkets are organized differently than others.

1.4.3 Motor Vehicles Department

The California Motor Vehicles Department issues driver's licenses, identification cards, and vehicle registrations. To be issued a license, first-time California drivers must pass a road test as well as a written test. A license is renewed by passing the written test alone.

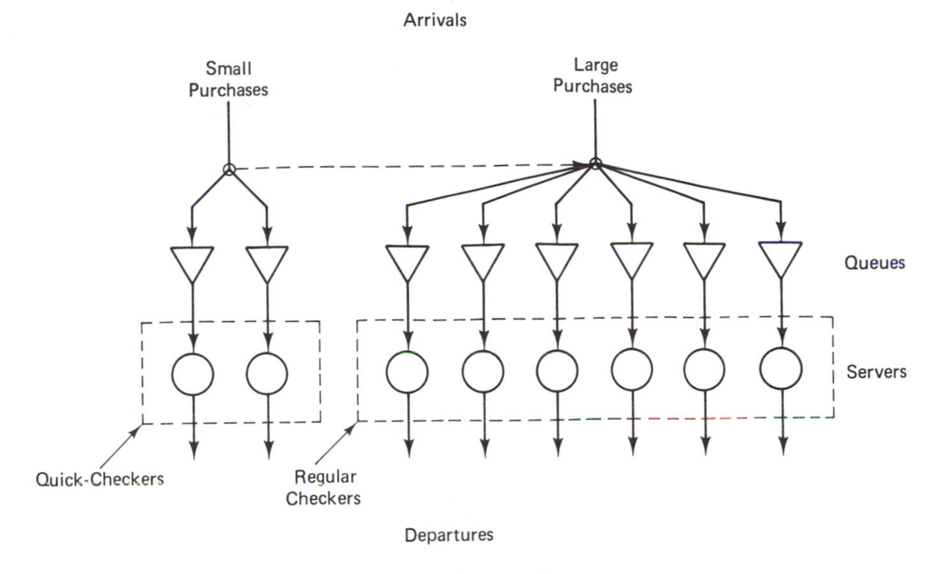

Figure 1.4 Queue diagram of grocery store.

Vehicle registrations are normally handled through the mail, though late registrations must be completed at the motor vehicles office.

A driver wishing to renew a license first fills out a form, then joins an FCFS queue to wait for the first available clerk to type the information. The clerk provides the driver with a copy of the written test and directs him or her to a desk to complete the test. Once finished, the driver joins a second FCFS queue to wait for the first available grader to score the test and check the driver's eyesight. If the driver passes the test, he or she then joins a third queue to have his or her picture taken. The grader also takes the photograph, but a single camera is shared among all graders. The actual license arrives two to three weeks later in the mail. If a driver is new to California, then he or she must also complete a road test subsequent to passing the written test. Figure 1.5 shows the layout for a motor vehicles office. Figure 1.6 is the flow diagram and shows that the system has a combination of serial and parallel servers.

What Is the Customer? The driver is the customer, and all customers are nearly identical. However, for the last queue, the grader and the driver wait for the camera together. That is, both are, in a sense, the customer.

What Is the Server? Three servers perform three distinct tasks: clerk/typewriter types the form, grader scores test and administers eye exam, and camera takes picture. Though the grader is needed to take the picture, it is the camera that provides the service.

Again, the server configuration, queue configuration, and queue discipline have already been outlined. Consider on your own: What is the arrival process—Does the department of motor vehicles have any information that can help predict, or control, arrivals? What is the reneging process—If a person reneges, will he or she come back

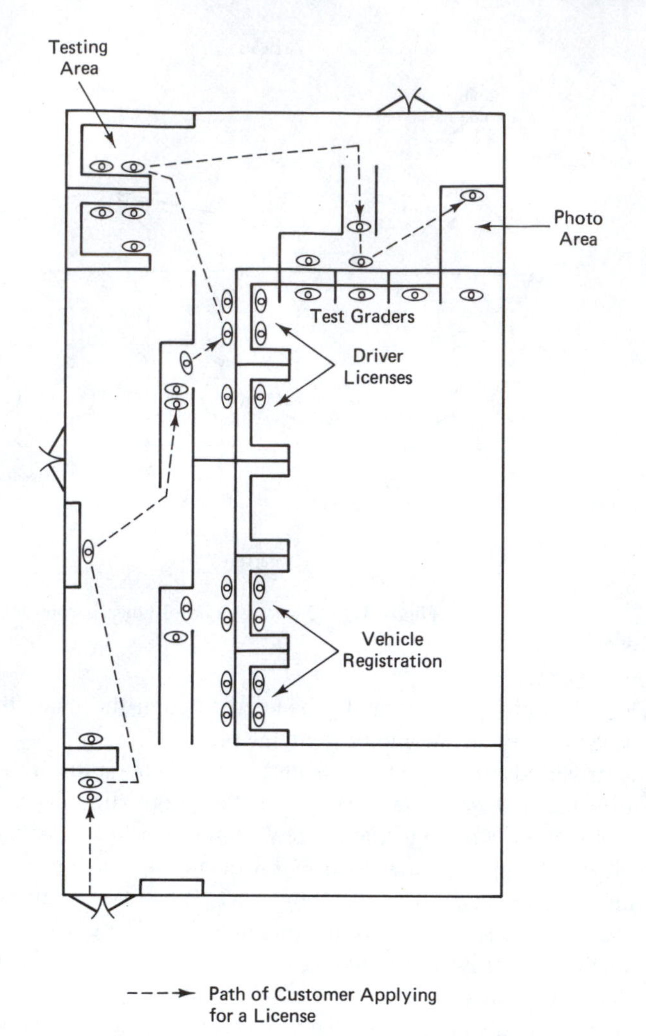

Figure 1.5 Layout showing serial servers in motor vehicles department.

later? What is the service process—Does service time vary from customer to customer? In light of these characteristics, consider why servers are assigned distinct tasks rather than identical tasks.

1.4.4 Other Queues

The bank, supermarket, and motor vehicles department examples were chosen for their familiarity, not because they are in any way unique. Other queues are not so familiar. For example, income tax returns pass through several queues as they are processed (cashing

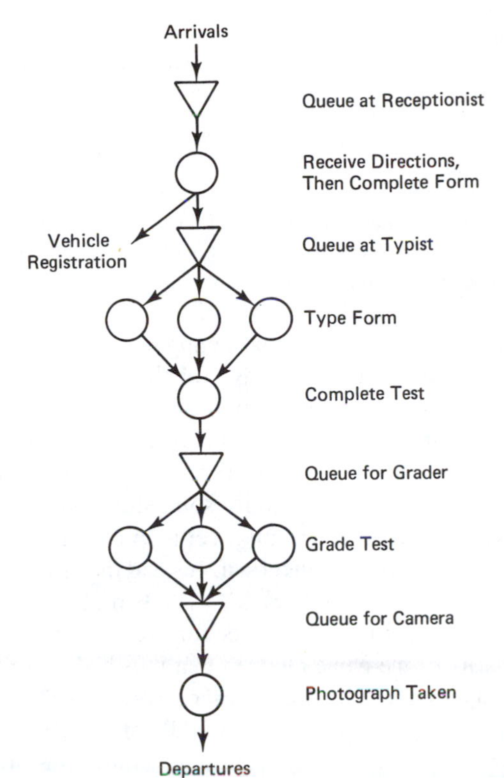

Figure 1.6 Queue diagram of motor vehicles department.

check, recording information, verifying information, and so on). Phone calls are transformed into data packets that queue at switching centers. Silicon wafers form queues at fabrication machines inside manufacturing plants. Other queues may be familiar but are more difficult to classify. For example, congestion forms on freeways when the lanes do not have sufficient capacity to accommodate all of the vehicles. The server is not a machine or a person but a restriction that impedes the flow of vehicles, the service time being defined by the minimum spacing between vehicles.

1.5 ATTITUDES TOWARD QUEUEING

An East German visits a friend in Moscow. A little girl answers the door. The East German asks her:

— Where is your father?
— He is not home.
— When will he be at home?
— At eight hours, forty minutes and twenty-three seconds.

— Where is he?
— He is going around the world thirty-three and a half times.
— What about your mother? Is she at home?
— No, she isn't.
— When is she expected?
— I don't have any idea.
— How come you know the hour, minute, and second when your father will return, but you have no idea when your mother will return? Where is she?
— She is at the market in the line for meat.[6]

One does not have to live in Eastern Europe to appreciate the humor in this story. Dislike (some would even say hatred) of queues is probably universal. Though there is much that different cultures hold in common, behavior in queues is not always the same. Adherence to the principle of first-come, first-served is one example. In the words of the sociologist Edward Hall, "To Americans, to be first is to be more deserving. If an American has been sitting at a table in a restaurant for some time and a latecomer is served before he is, his blood pressure will rise noticeably. Yet in most places outside of Europe [FCFS] ordering in situations of this type is unknown. Instead the laws of selection apply; that is, service is dependent upon a person's rank" (1959, p. 157).

Leon Mann's study of queues to buy tickets for Australian football matches is indicative. The great popularity of the game dictated that fans had to wait overnight, and even up to six days, to be assured a seat. Mann describes the fans' efforts to maintain order in the absence of strict control: "People in the middle of the queue worked together to erect barricades from material left in the park" as constraints against would-be infiltrators (1969, p. 347). He also notes that "at times of maximum danger, and in the hour before the ticket windows opened, there was a visible bunching together, or shrinkage, in the physical length of the queue, literally a closing of the ranks" to prevent infiltration. In addition to "vociferous catcalls and jeering," Mann observes that "the most extreme constraint was physical force. During the early hours of August 15, five men were taken to hospitals after four separate brawls broke out in ticket lines."

Though first-come, first-served is the norm for queues of people waiting for service, such is not the case for all queues. In business, it is considered proper to give priority to the employee with the highest rank (for example, the department manager gets her memos typed first). Other times, priority is given to the work with the most urgent deadline. Favors can also influence priorities (such as tipping the maître d'hôtel to obtain preferential seating at a restaurant). And whereas people tend to queue in orderly lines for tellers or ticketing (Fig. 1.7), they do not in all cases, such as waiting to enter an elevator or cross a street (Fig. 1.8). In extreme situations, lack of control can lead to panic, as in stampedes to enter stadiums for rock-and-roll concerts and soccer matches.

Maintaining a pleasant waiting environment can also have a tremendous impact on our attitudes. Systems for maintaining order, such as roped barriers and signs, could have eliminated the anxiety witnessed in the Australian football queues. They can also prevent crowding, as when customers bunch together to prevent infiltration of the line. Adequate

Figure 1.7 Customers waiting in orderly queue to purchase tickets.
Source: Courtesy of Project for Public Spaces, Inc., New York.

lighting, ventilation, and sound control, as well as smoking prohibitions, are also important. Better yet is a system in which a patron can drop off a piece of work (the work being the customer) and retrieve it later, without ever having to be present in the queue.

The attitude of the server toward the customer is also significant. Servers should be aware that the act of waiting automatically puts the customer in a subordinate position. Thus, lack of regard for customer needs not only delays the customer but angers him or her. Monopolies, both private and public, are notorious for an indifferent attitude.

Figure 1.8 Queue at an intersection is not FCFS.
Source: Courtesy of Project for Public Spaces, Inc., New York.

1.6 SYSTEMS APPROACH

The ideas presented in this book fit within a larger scheme known as the systems approach (see Churchman 1968; de Neufville and Stafford 1971). The end objective of the approach is to decide the best way to operate a system—in the case of this book, the system being a queueing system. This objective is achieved with these four steps:

1. Formulation
2. Modeling
3. Evaluation
4. Decision

1.6.1 Formulation

The purpose of the formulation step is to assess the characteristics of the situation at hand and then isolate and define a ''problem.'' One might think of this as putting a box around a part of the system. Anything inside the box is a matter of interest; anything outside is taken as given. In the three queueing examples (bank, supermarket, and motor vehicles department), the system was defined by isolating the activities that occur at a given location and ignoring things that happen elsewhere. Problem definition is an essential first step toward problem solution.

The mathematician G. Polya (1945) emphasizes answering these questions in defining a problem:

What is the unknown?
What are the data?
What is the condition?

The unknown are those things that can be controlled. In the parlance of operations research, the unknown are the decision variables. The number of servers, the server configuration, and the queue organization are three examples. The data are the quantitative characteristics of the situation—perhaps a history of customer arrivals and service times. These data are the facts that can be used to arrive at the best possible decision. The condition is a word description of the situation, such as the type of service requested, restrictions on server configurations, and so on.

1.6.2 Modeling

The purpose of modeling is to develop a representation of the system that provides a better understanding of how the system operates. A model might be physical (for example, a scale model of a bank), conceptual (a word description of how a bank operates), or mathematical (an equation showing average waiting time as a function of customer arrival rate). Of these three types, mathematical models are most applicable to queues, and this is where the book concentrates. However, physical and conceptual models will also be mentioned.

Creating good models is by no means a simple task. It requires a mixture of scientific talent and a good dose of creativity. Fundamentally, the two main considerations are that the model should be realistic and the model should be meaningful. The model should also be suited to the problem at hand and possess just the right amount of detail. If the model is not realistic, it cannot be trusted, and if the model is not meaningful, it cannot be understood. Unfortunately, these two qualifications usually conflict. One can go to great effort to create a very realistic model, only to find that it provides no insights into how the system operates. The important message is that a balance should be achieved.

Creating good models is not a matter of creating the ''right'' models, for there is rarely one right answer. We have already seen that the customer can be defined in any of several ways. How the customer is represented is but the first point in the modeling process, and the first point where the modeler's judgment must be exercised. The same type of judgment must be exercised throughout the process.

1.6.3. Evaluation

After the model is created, the next step is to use it to evaluate the system. The evaluation step can be divided into two parts: generation of alternatives and evaluation of alternatives.

The purpose of the first step is to find alternative ways to run the system. The model might provide insights into alternative configurations—more servers, serial servers, or the like. Optimization techniques might also be applied to a simplified model to identify how many servers to operate or server schedules. Or the alternative might come from one's own imagination. The best ideas often result from one's own observations of the system.

After the alternatives are identified, they can be analyzed more rigorously with a detailed model. This usually involves predicting how well the alternatives will work according to quantitative ***measures of performance*** (**MOP**). What is the cost of the system? What is the average waiting time? How many customers will renege? These are some of the possible MOPs. A measure of performance should, first of all, be measurable—something that can actually be recorded by observing the queue. And the measure of performance should provide an important indicator of how well the alternative meets major system objectives, such as maximizing return on investment.

1.6.4 Decision

The decision phase is the least structured of the steps in systems analysis. The main objective here is to consider the information obtained from the evaluation phase and select the best alternative. Because objectives may be conflicting, and some alternatives may score well according to some measures of performance and not according to others, considerable judgment must be exercised. The most important thing to recognize is that the best alternative is not an automatic by-product of the evaluation process. The model does not pick the best alternative. Rather, the model is used to provide the information needed for a *person* to select the best alternative. There is no simple prescription.

1.6.5 Systems Analysis in Queueing

This book does not cover all the steps of the systems analysis process as it applies to queues. The emphasis here is on the observation and modeling aspects. Other pertinent subjects, such as evaluation techniques and decision making, are covered in books on operations research, decision analysis, and systems analysis (references are provided at the end of this chapter). Anyone who wishes to develop and analyze queueing models should also be exposed to these areas.

Despite the appeal of the systems approach, do not feel compelled to follow it whenever a problem is confronted. With a little understanding of how queues behave, the solution might become obvious. If you already know how to fix a problem, there is no point in studying it.

1.7 BOOK ORGANIZATION

This book is written with the intention of developing the understanding needed to diagnose and correct queueing problems. It will provide awareness of what solutions can be applied to a given situation and exposure to various modeling techniques (simulation, fluid approximations, stochastic models, and so on). The book is written in a cumulative fashion, each chapter dependent on the material in the preceding chapter.

The first half of the book, Chaps. 2 to 6, explores the behavior of simple queueing systems. Understanding how a queue operates must always begin with observation and measurement, the subject of Chap. 2. This is followed by a group of four chapters on modeling queueing systems. The first of these, Chap. 3, introduces the Poisson process as a model for customer arrivals. Chapter 4 explains the philosophy behind simulation and describes how to use simulation to model a random queueing process. In Chap. 5, the behavior of queueing systems that operate in steady state is described. And Chap. 6 examines queueing systems where the customer arrival rate is not constant but varies over time.

Chapters 7 through 9 concentrate on solving queueing problems, and are somewhat less mathematical than the previous four. Chapter 7 provides methods for reducing waiting time through changes in the service process. Chapter 8 also provides suggestions for reducing waiting time, but through changes in the arrival process. Chapter 9 examines the queue discipline and ways to reduce the cost of delay by adjusting the order in which customers are served.

Chapter 10 covers queueing systems with multiple servers, examining queueing behavior and solutions to queueing problems. Included in Chap. 10 is a description of some of the computer packages available for simulating the performance of queueing networks.

Finally, Chap. 11 looks at the design aspects of queueing systems. It includes such topics as keeping track of customers, queueing layout, queueing location, and queueing environment.

Chapters 2 to 5 include a series of exercises in which students can observe a queueing system, measure its performance, and create a queueing model. The aim here is to provide some experience in model development. Chapters 7 to 10 consider a series of case studies, describing the operation of a variety of real queueing systems. The studies serve to illustrate the concepts presented in each chapter. In particular, they can provide the focus for class discussion on ways to eliminate queueing problems. Chapter 11 includes a design exercise addressed at improving all aspects of queueing performance.

FURTHER READING

BROCKMEYER, E., H. A. HALSTROM, AND A. JENSEN. 1948. *The Life and Works of A. K. Erlang*. Copenhagen: Danish Academy of Technical Science.

CHURCHMAN, C. W. 1968. *The Systems Approach*. New York: Dell.

———, R. L. ACKOFF, AND E. L. ARNOFF. 1957. *Introduction to Operations Research*. New York: John Wiley.

DE NEUFVILLE, R., AND J. H. STAFFORD. 1971. *Systems Analysis for Engineers and Managers*. New York: McGraw-Hill.

EDIE, L. C. 1954. "Traffic Delays at Toll Booths," *Journal of the Operations Research Society of America*, 2, 107–138.

HALL, E. T. 1959. *The Silent Language*. Garden City, N.Y.: Doubleday.

HALL, R. W. 1985. "What's So Scientific about MS/OR?" *Interfaces*, 15, 40–45.

KLEINROCK, L. 1975, 1976. *Queueing Systems*. New York: John Wiley.

MANN, L. 1969. "Queue Culture: The Waiting Line as a Social System," *American Journal of Sociology*, 75, 340–354.

MORSE, P. 1958. *Queues, Inventories and Maintenance*. New York: John Wiley.

OLIVER, R. M., AND A. H. SAMUEL. 1967. "Reducing Letter Delays in Post Offices," *Operations Research*, 10, 839–892.

POLYA, G. 1945. *How to Solve It*. Princeton, N.J.: Princeton University Press.

RAIFFA, H. 1968. *Decision Analysis, Introductory Lectures on Choices under Uncertainty*. Menlo Park, Calif.: Addison-Wesley.

SAATY, T. L. 1961. *Elements of Queueing Theory with Applications*. New York: McGraw-Hill.

SCHWARTZ, B. 1975. *Queueing and Waiting, Studies in the Social Organization of Access and Delay*. Chicago: University of Chicago Press.

NOTES

1. Reprinted by permission of the *Harvard Business Review*. Excerpt from "Will Services Follow Manufacturing into Decline?" by James Brian Quinn and Christopher E. Gagnon (November/December 1986), p. 95.

2. W. Carley, "A Ride on the Eastern Shuttle Is Rush Hour in Crowded Skies," *The Wall Street Journal*, September 24, 1985, p. 31. Reprinted by permission of *The Wall Street Journal*, © Dow Jones and Company, Inc., 1985. All rights reserved.

3. R. Kotulak, "Transplant Gift in Short Supply," *Chicago Tribune*, July 16, 1986, p. 1. © Copyrighted, Chicago Tribune Company. All rights reserved. Used with permission.

4. D. E. Pitt, "Drug Cases Clog New York Courts," *The New York Times*, April 4, 1989. © Copyright 1989 by The New York Times Company. Reprinted by permission.

5. P. B. Crosby, *Quality Is Free* (New York: McGraw-Hill, 1979), p. 1.

6. From *First Prize: Fifteen Years!* by C. Banc and A. Dundes. Fairleigh Dickinson University Press, Rutherford, N.J., 1986, p. 58. Reprinted by permission of Associated University Presses.

PROBLEMS

1. Based on your understanding of each of the following queueing systems, describe the server, customer, queue organization, queue discipline, and reneging and jockeying processes.

- **a.** Toll booths on a highway
- **b.** Traffic intersection controlled by a traffic light
- **c.** Taxi service at a hotel
- **d.** Elevators in a large building
- **e.** Copying service at a copy shop
- **f.** Mail sorting by the postal service
- **g.** Medical service at a health center
- **h.** Airplanes landing at an airport
- **i.** Emergency telephone lines
- **j.** A sit-down restaurant
- **k.** Time-shared computer system

2. Describe a situation (or situations) in which waiting or delay is harmful to productivity. Discuss how the situation might be improved.

EXERCISE: QUEUEING TOUR

Select five queueing systems to visit and observe (your instructor may provide suggestions). Answer the following questions for each site. Be succinct.

1. Describe the server(s) and service process. How many servers are there?

2. Describe the customer(s).

3. Diagram the server configuration and the queue organization.

4. Record the service time for five successive customers. Describe the factors that influence the service time.

5. Count and record the number of customers that arrive over five successive 1-minute intervals. Describe the factors that influence the arrival process.

6. If you observe any reneging or jockeying, describe the process.

7. Describe the queue discipline.

8. Record the time of day and the date.

chapter 2

Observation and Measurement

A doctor examining a patient looks for symptoms of whether the patient is healthy or ill. Blood pressure, body temperature, and pulse are just a few of the things to check. An analyst examining a queueing system also looks for symptoms of health or illness. These symptoms are known as measures of performance (MOP).

Observing the queueing system is the key to the formulation stage of systems analysis. It directly answers the questions: What are the data? and What are the conditions? Understanding how the system currently operates should also be the basis for choosing a new form of operation. Observation should provide insights to help answer the question: What is the unknown? That is, it should help identify, define, and isolate the problem.

Chapter 1 describes how to assess the characteristics of a queueing system. This chapter describes how to measure the performance of the system. It details which queueing symptoms to examine and how they should be measured. It begins by listing some of the key MOPs that apply to almost any queueing system. Next, a major topic that will be used throughout the book is introduced: cumulative arrival and departure diagrams. These diagrams are used to display and analyze the most important MOPs. The diagrams are followed by a discussion of how to calculate averages and standard deviations of the performance measures and how to create an empirical probability distribution. Little's formula is introduced. The chapter concludes with a review of techniques for measuring and observing queues. Chapter 2 does not address the question of how long to observe the system, nor does it consider how to use observations to predict future queue performance, subjects that are covered later in the book.

2.1 MEASURES OF PERFORMANCE

One should consider two perspectives in measuring the performance of a queueing system. From the customer's perspective, the major concern is the quality of the service provided. From the server's perspective (or the server's employer's perspective), the major concerns are the cost of providing the service and the impact of service quality on business (that is, revenue). Measures of performance allow these concerns to be quantified.

The following sections divide the MOPs along the lines of customer concerns and server concerns. But before discussing the measures of performance, a few basic terms must be defined:

Definitions 2.1

Arrival time The time that the customer arrives at the queue

Departure time The time that the customer completes service and leaves the queueing system

Departure time from queue The time that the customer leaves the queue to enter service

Time in queue Departure time *from queue* minus the arrival time

Service time Departure time minus departure time *from queue*

Time in system Departure time minus the arrival time = time in queue plus service time

As an example, suppose that a customer joined a bank queue at 2:00, stepped up to the teller window at 2:03, and finished the transaction at 2:05. Thus, 2:00 would be the arrival time, 2:03 the departure time from queue, and 2:05 the departure time from the system. The time in queue would be 3 minutes, service time 2 minutes, and the time in system 5 minutes. Therefore, *a customer is "in the system" when it is either in the queue or in service.*

2.1.1. Customer Measures of Performance

In most situations, the major concerns of the customer are the length of time spent in the queue and in service and the cost associated with this time. These are reflected in the following measures of performance:

Time in queue In essence, a shorter wait is better than a longer wait.

Service time The time in service might be treated separately from the time in queue if customers find their time waiting in queue more costly than their time being served (or vice versa). If they are perceived more or less the same, *time in system* might substitute for service time and time in queue.

Waiting cost Waiting time might be more costly for some customers than others, in which case some weighting scheme might be used to combine the various waiting times. The combined waiting time could represent cost.

Proportion of work completed on time Customers may have deadlines to meet. If they are not served by a deadline, a penalty is incurred.

Tardiness If a job does not meet the deadline, it is better to be late by a small amount than a long amount. The tardiness equals zero if the job is completed on time and equals the departure time minus the deadline time if the job is late.

Example 1

A job is received at 3:00, begins processing at 4:00, and is finished at 4:30. The deadline for the job is 4:45. The time in queue is 60 minutes, the service time is 30 minutes, and the tardiness is zero (job completed on time).

Example 2

Another job is received at 3:15, begins processing at 4:30, and is finished at 5:00. The deadline for the job is also 4:45. The time in queue is 75 minutes, the service time is 30 minutes, and the tardiness is 15 minutes (job completed late).

Because many MOPs are random variables, they should be defined more precisely. It is not enough simply to say that the MOP is time in queue. The MOP should be specified as *average* time in queue, *standard deviation* of the time in queue, or *maximum* and *minimum* time in queue. The MOPs might also vary in a predictable (nonrandom) fashion with time of day, day of week, or the like. For example, the waiting time at a toll plaza might be very long every day from 7:00 A.M. to 8:00 A.M. and short the remainder of the day. If this is the case, the MOPs should be calculated for distinct time periods.

In addition to the quantitative measures, such qualitative factors, as the waiting environment should be examined. Is the customer idle while waiting? Do customers receive adequate information about how long they will have to wait? Can customers sit? Is the room crowded? Is there adequate ventilation? All of these factors are important to queue performance.

2.1.2 Server Measures of Performance

The server is concerned with minimizing the cost of providing service and attracting and keeping customers. The cost of providing the service is reflected in the service time and server utilization; attracting and keeping customers is reflected in arrivals and reneging and balking. The MOPs are listed below:

Service time A shorter service time is indicative of a more efficient operation.

Proportional utilization This is the proportion of time that servers are busy serving customers. The proportional utilization is a number between zero and one, and the closer the number is to one the more efficient the operation is.

Throughput The rate at which customers are served. In some systems (such as toll plazas), the throughput increases as queues become longer because customers are better prepared before they reach the server; in others, service deteriorates when queues become long due to customer inattention or server fatigue.

Arrival rate Arrival rate is a measure of the amount of business. A large arrival rate is desirable because it means more revenue.

Reneging proportion Each customer who reneges may translate into lost business. Reneging might be measured as the proportion of customers who renege after joining the queue, or as the proportion who decide not to join the queue after seeing its length. Reneging is the most difficult MOP to measure because the customer who reneges may never be recorded.

Queue length A long queue is more costly to accommodate than a short queue because of the space required.

Qualitative aspects include the following: Do servers have something to do when they are not serving customers? Are servers fully productive? Are there problem servers? Are servers courteous? Are servers ever idle when customers are waiting in the queue?

2.1.3 Relation of Measures of Performance to Organization Goals

It is important to be aware of how the measures of performance fit within the organization's overall goals. If the organization is a private business, that goal is almost certainly to maximize long-term profitability. In government, the goal is more nebulous, but probably has something to do with maximizing the welfare of its constituents. Ideally, when evaluating queueing alternatives, one would like to measure directly the impact of the alternative on the overall goal. Rarely is this possible. Instead, one looks at the measures of performance as indicators of the potential impact on profits and the potential impact on welfare. As illustrated in Fig. 2.1, both time in system and queue length affect the arrival rate and reneging, which in turn affects revenue. The utilization, queue length, and service time affect cost. Combined, cost and revenue determine profitability. The measures of performance are proxies for the overall goal, which, in practice, cannot be directly measured.

2.2 CUMULATIVE ARRIVAL AND DEPARTURE DIAGRAMS

The word *cumulative* means ''resulting from accumulation.'' Thus, a *cumulative diagram* indicates how many customers have accumulated up to some point in time from an initial starting time. A *cumulative arrival diagram* indicates how many customers have arrived

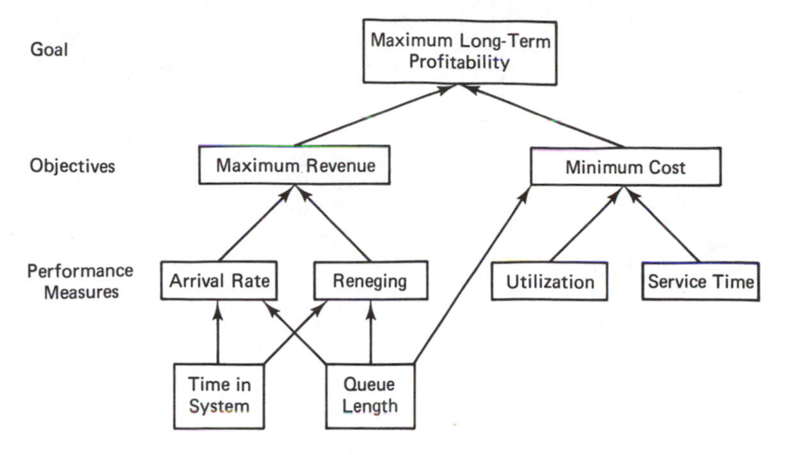

Figure 2.1 Performance measures indicate whether system objectives and goals are achieved.

up to a point in time, and a *cumulative departure diagram* indicates how many customers have departed up to a point in time.

Cumulative diagrams are important because they provide many of the measures of performance in one simple picture. Creating a cumulative diagram is the easiest and simplest way to assess the health of a queueing system. (Remember, though, cumulative diagrams are not the only way to evaluate a queueing system and should not be the sole tool used.)

Definitions 2.2

$A(t)$ = *cumulative arrivals* from time 0 to time t

$D_s(t)$ = *cumulative departures from the system* from time 0 to time t

$D_q(t)$ = *cumulative departures from the queue* from time 0 to time t

The starting time, time 0, can be set at any time that is convenient to the analysis. For example, if a store opens at 9:30 A.M., time 0 would be 9:30 and $A(t)$ would be the number of customers who arrived between 9:30 and time t.

Consider the following data:

Customer	Arrival time	Departure from queue	Departure from system
1	9:36	9:36	9:40
2	9:37	9:40	9:44
3	9:38	9:44	9:48
4	9:40	9:48	9:52
5	9:45	9:52	9:56

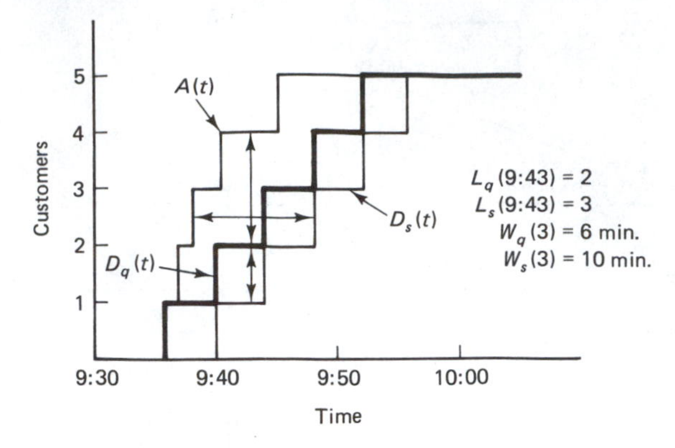

Figure 2.2 Cumulative arrival and departure diagram. Queue lengths are determined by vertical separation between curves. Waiting times are determined by horizontal separation.

The first customer arrives at 9:36, finds no other customer in the system, and goes directly into service. The second customer arrives at 9:37, before the first customer completes service, and joins the queue. The fact that this customer did not enter service immediately indicates that the system has a single server. As soon as customer 1 completes service (9:40), customer 2 departs from the queue and begins service. Each subsequent customer enters service at the time that the previous customer completes service, indicating that the queue operates on a first-come, first-served basis.

The data presented above are plotted in Fig. 2.2 in the form of cumulative arrival and departure diagrams, with time 0 set at 9:30. Each step has height of one and represents one customer. The placement of the step on the horizontal axis corresponds to an "event" (arrival, departure from queue, or departure from the system). The steps in the arrival diagram correspond to the times that customers arrived; the steps in the departure diagram correspond to the times that customers departed.

At any point in time, the height of each curve represents the number of customers that have passed a certain point in the system. At time 9:43, four customers have arrived, two have departed from the queue, and one has departed from the system. One can imagine that these values represent counts made by three observers, stationed at the entrance to the queue, the exit from the queue, and the exit from the system. The vertical distance between the arrival curve and the departure curves indicates how many customers are in the queue and in the system:

Definitions 2.3

$$
\begin{aligned}
L_q(t) &= \text{number of customers in the queue at time } t \\
&= A(t) - D_q(t) \\
L_s(t) &= \text{number of customers in the system at time } t \\
&= A(t) - D_s(t)
\end{aligned}
$$

At time 9:43, $A(t) = 4$ and $D_q(t) = 2$. Therefore, two customers remain in the queue ($L_q(t) = 2$). At the same time, only one customer so far has left the system ($D_s(t) =$

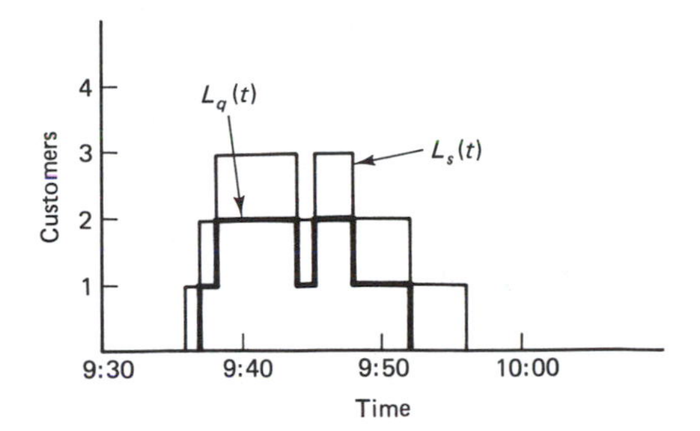

Figure 2.3 Customers in the system and in the queue versus time.

1), so the number of customers in the system is three ($L_s(t) = 3$). Figure 2.2 shows that the queue eventually vanishes at 9:52 and the last customer leaves the system at 9:56. For comparison, Fig. 2.3 plots $L_q(t)$ and $L_s(t)$. Note that Fig. 2.3 provides no information that is not already shown in Fig. 2.2. And, as will be discussed later in the book, the cumulative diagram provides information that cannot be shown on the graphs of $L_q(t)$ and $L_s(t)$.

A customer cannot leave the queue before it arrives, and it cannot leave the system before it leaves the queue. Therefore

$$A(t) \geqslant D_q(t) \geqslant D_s(t) \qquad t \geqslant 0$$

which guarantees that both $L_q(t)$ and $L_s(t)$ are greater than or equal to zero.

2.2.1 Waiting Time

When Fig. 2.2 is read vertically, the queue size and number of customers in the system are identified. Reading Fig. 2.2 horizontally reveals the time in queue and the time in system.

Definitions 2.4

$$A^{-1}(n) = \text{time of the } n\text{th arrival}$$

$$D_q^{-1}(n) = \text{time of the } n\text{th departure from queue}$$

$$D_s^{-1}(n) = \text{time of the } n\text{th departure from system}$$

Whereas $A(t)$, $D_q(t)$, and $D_s(t)$ convert a time into a customer number, $A^{-1}(n)$, $D_q^{-1}(n)$ and $D_s^{-1}(n)$ take a customer number and convert it to a time. They correspond exactly to the data provided before:

n	$A^{-1}(n)$	$D_q^{-1}(n)$	$D_s^{-1}(n)$
1	9:36	9:36	9:40
2	9:37	9:40	9:44
3	9:38	9:44	9:48
4	9:40	9:48	9:52
5	9:45	9:52	9:56

Definitions 2.5

$W_q(n)$ = time in queue, for nth customer to arrive

$W_s(n)$ = time in system, for nth customer to arrive

When the discipline is FCFS, the waiting times, $W_q(n)$ and $W_s(n)$, are found by computing the horizontal distance between the steps in Fig. 2.2:

FCFS Waiting Time

$$W_q(n) = D_q^{-1}(n) - A^{-1}(n) \tag{2.1}$$

$$W_s(n) = D_s^{-1}(n) - A^{-1}(n) \tag{2.2}$$

For example, the steps for customer 3 occur at 9:38, 9:44, and 9:48. Therefore, $W_q(3) =$ 6 minutes and $W_s(3) =$ 10 minutes. Equations (2.1) and (2.2) do not apply to non-FCFS disciplines because the customers do not necessarily depart in the same order that they arrive. An alternative representation for non-FCFS systems is presented later in the chapter.

2.2.2 Other Measures of Performance

Server utilization (the proportion of time that the server is busy) is also found on cumulative diagrams. The number of servers that are busy at any time equals $D_q(t) - D_s(t)$, the number of customers that have entered service but have not yet completed service. The ***absolute utilization*** is the average number of busy servers (over time), and the ***proportional utilization*** is the absolute utilization divided by the total number of available servers. In Fig. 2.2, the single server is busy for 20 of the 30 minutes, so both utilizations are 2/3.

The cumulative diagram can also be examined to see whether servers are ever idle when customers are waiting in the queue. When this occurs, $D_q(t) - D_s(t)$ is less than the number of servers, and $A(t) - D_q(t) = L_q(t)$ is greater than zero. This is an obvious sign of inefficiency, for all servers ought to be busy when customers are waiting. Although this situation is not illustrated in Fig. 2.2, it frequently occurs in practice.

Reneging can be presented on a cumulative diagram, but it should be treated separately from ordinary service, as in Fig. 2.4. A major difficulty in creating a reneging diagram is identifying exactly which customers reneged. Reneges do not occur in a strict FCFS order; one customer might remain in the queue 5 seconds before reneging, another 5 minutes. So it is difficult to correlate the time a customer reneged with the time the customer arrived. In Fig. 2.4, four customers are shown to renege between 9:43 and 9:46.

If both an arrival/departure diagram and a reneging diagram are created, the queue size equals the combined number of customers in both diagrams. The waiting time should be based on the arrival/departure diagram alone (this is the waiting time for the customers

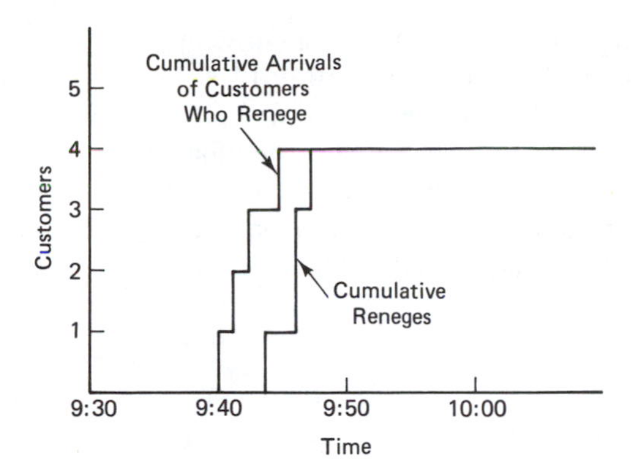

Figure 2.4 Cumulative diagram of reneges.

served). However, in presenting the data, the percentage of customers that reneged, along with their average time in queue, should be noted.

If for some reason reneging was not measured, then the cumulative arrival/departure diagram will contain a discrepancy between the number of customers that arrived and the number that were served, which will lead to overestimation of queue lengths and waiting times. If the discrepancy is small, this problem might be resolved by scaling the departure curve upward by a slight percentage. However, there is no accurate way to resolve a large discrepancy other than measuring the data again by a different method.

2.2.3 Average and Standard Deviation of Waiting Time

The average waiting time per customer equals the total waiting time divided by the total number of customers served. To be most meaningful, it should be computed over a period of time that begins and ends with no customers in the system. If this is not possible, only those customers that arrived before the "oldest" customer in the queue (that is, the customer that has been in the queue the longest time) should be included in the average. Keep in mind that the average waiting time may be different for the customers remaining in the queue, particularly if the discipline is not FCFS.

Definitions 2.6

$$W_q = \text{average waiting time in queue}$$

$$= \frac{\sum_{n=1}^{N} W_q(n)}{N} \tag{2.3}$$

$$W_s = \text{average waiting time in system}$$

$$= \frac{\sum_{n=1}^{N} W_s(n)}{N} \tag{2.4}$$

where N is the number of data points in the sample. For our sample data, $W_q = (0 + 3 + 6 + 8 + 7)/5 = 4.8$ minutes, and $W_s = (4 + 7 + 10 + 12 + 11)/5 = 8.8$ minutes. The difference between W_q and W_s is the average service time, 4 minutes.

The standard deviation is an indicator of the variation in data. If the standard deviation is small, then the time in queue and time in system did not vary much between customers; if the standard deviation is large, the variation was large. Because a predictable wait is generally preferable to an unpredictable wait, the standard deviation is an important measure of performance.

Definitions 2.7

$$\sigma_{W_q} = \text{standard deviation of time in queue}$$

$$= \sqrt{\frac{\sum_{n=1}^{N} [W_q(n) - W_q]^2}{N}} \tag{2.5}$$

$$\sigma_{W_s} = \text{standard deviation of time in system}$$

$$= \sqrt{\frac{\sum_{n=1}^{N} [W_s(n) - W_s]^2}{N}} \tag{2.6}$$

For our sample problem, $\sigma_{W_q} = \sigma_{W_s} = \sqrt{(4.8^2 + 1.8^2 + 1.2^2 + 3.2^2 + 2.2^2)/5} = 2.9$ minutes. The standard deviations are the same for the problem because the service time is the same for all customers.

It is important to recognize that the average and standard deviation of the waiting time depend on when the data are recorded and on how much data are recorded. Waiting times tend to be highly correlated between successive customers. That is, if a customer has a long wait, the next customer is likely to have a long wait as well. The average and standard deviation are foremost descriptors of what happened in a given time interval. One should not necessarily infer that the values are indicative of what should always occur.

2.2.4 Average and Standard Deviation of Queue Length

One can think of the average waiting time as an average that is computed vertically across Fig. 2.2. The number in the denominator of Eqs. (2.3) and (2.4) is the number of customers that arrived. The average queue length is computed horizontally, and the number in the denominator is an elapsed time. Hence, the average queue length is interpreted as the average number of customers in the queue over some interval of time. Suppose that a = start time of interval and b = end time of interval. Then

Definition 2.8

$$L_q = \text{average queue length (customers)}$$

$$= \frac{\int_a^b L_q(t)dt}{b - a} \tag{2.7}$$

The numerator is defined by an integral rather than a summation. The reason for this is simple. Whereas customers are discrete entities, time is continuous. An integral is the continuous analog of a summation.

The integral of $L_q(t)$ equals the area between $A(t)$ and $D_q(t)$ over the interval from a to b. The integral does not have to be calculated in the customary form of an equation. It can be calculated geometrically by determining the length of time that one customer is in the queue, the length of time that two customers are in the queue, and so on. (The appendix at the end of this chapter expands on this concept.) For our sample data:

Time interval	$L_q(t)$
9:30 – 9:37	0
9:37 – 9:38	1
9:38 – 9:44	2
9:44 – 9:45	1
9:45 – 9:48	2
9:48 – 9:52	1
9:52 –10:00	0

$L_q(t)$ equals zero for 15 minutes, one for 6 minutes, and two for 9 minutes. Therefore, the integral of $L_q(t)$ equals $1 \times 6 + 2 \times 9 = 24$ *customer-minutes*. The average queue length over the half-hour interval is 24 customer-minutes/30 minutes = .8 customer. Whereas waiting time is measured in terms of time (minutes, hours), queue length is measured in terms of customers.

The standard deviation of the queue length, like the average queue length, is defined over a time interval:

Definition 2.9

$$\sigma_{L_q} = \text{standard deviation of queue length}$$

$$= \sqrt{\frac{\int_a^b [L_q(t) - L_q]^2 dt}{b - a}} \tag{2.8}$$

For our sample data, the numerator of Eq. (2.8) is $(.8^2 \times 15 + .2^2 \times 6 + 1.2^2 \times 9) = 22.8$ customer^2-minutes. Dividing through by 30 minutes and taking the square root, the standard deviation of $L_q(t)$ is .87 customer.

The average and standard deviation of the number of customers in the system are calculated in similar fashion:

Definitions 2.10

$$L_s = \text{average number of customers in the system}$$

$$= \frac{\int_a^b L_s(t) dt}{b - a} \tag{2.9}$$

$$\sigma_{L_s} = \text{standard deviation of customers in the system}$$

$$= \sqrt{\frac{\int_a^b [L_s(t) - L_s]^2 dt}{b - a}} \tag{2.10}$$

For our sample problem, the integral in Eq. (2.9) equals 44 customer-minutes, so the average number of customers in the system is 44/30 = 1.47 customers. The standard deviation is 1.23 customers. Even though the service time is the same for all customers, the standard deviation of $L_s(t)$ does not equal the standard deviation of $L_q(t)$.

2.3 LITTLE'S FORMULA

In the calculations for average queue length, a coincidence occurred. The numerator of the equation for L_q equaled the numerator of the equation for W_q, and the numerator of the equation for L_s equaled the numerator of the equation for W_s. Actually, this is no coincidence at all. It is a simple implication of one of the most important results of queueing theory: Little's formula (Little 1961).

So why does the following occur?

$$\sum_{n=1}^{N} W_q(n) = \int_a^b L_q(t)dt \tag{2.11}$$

Both the left and right sides of Eq. (2.11) equal the area between the curves $A(t)$ and $D_q(t)$, *whether or not* the discipline is FCFS (an exception is mentioned below). The left side calculates the area with a vertical summation and the right side calculates the area with a horizontal integral (Fig. 2.5). The two methods measure the exact same quantity in two different ways and must, therefore, arrive at the same result. This area can be interpreted as the total waiting time for all customers:

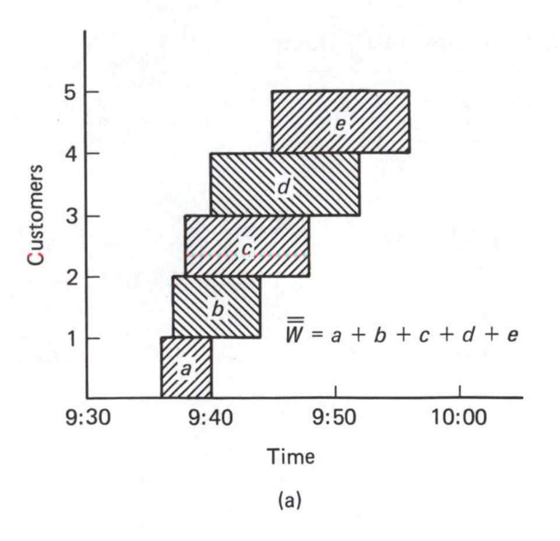

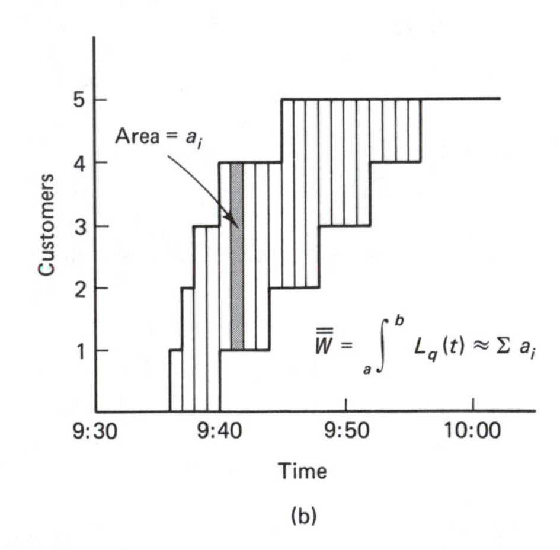

Figure 2.5 Total waiting time can be calculated by (a) summing individual waiting times or (b) summing queue lengths over short time intervals.

Definition 2.11

$$\bar{\bar{W}}_q = \text{total time in queue for all customers}$$

$\bar{\bar{W}}_q$ has the dimension customers × time. For example, $\bar{\bar{W}}_q$ might equal 100 customer-minutes, meaning that 100 customers might have waited 1 minute each, or 50 customers might have waited 2 minutes each, or any combination whose product is 100 customer-minutes.

One caveat: The areas are only equal if the system begins and ends the time interval with no customers in the system. Otherwise, a few customers will be left dangling and there will be a slight discrepancy in the calculations.

Little's formula, which is quite general and applies to any queue discipline, specifies how to convert the average waiting time into the average time in queue, and vice versa. Based on the definition of $\bar{\bar{W}}_q$, the average time in queue and average customers in queue can be defined as follows:

$$W_q = \frac{\bar{\bar{W}}_q}{N} \qquad L_q = \frac{\bar{\bar{W}}_q}{b - a} \tag{2.12}$$

Combining these two equations provides the following:

$$NW_q = (b - a)L_q \tag{2.13}$$

Definition 2.12

$$\lambda = \text{average arrival rate over an interval } [a,b]$$
$$= N/(b - a)$$

The arrival rate, λ, is interpreted as the average number of customers that arrive per unit time. Substitution of λ in Eq. (2.13) yields

Little's Formula I

$$L_q = \lambda W_q \tag{2.14}$$

The same result applies to time in system:

Little's Formula II

$$L_s = \lambda W_s \tag{2.15}$$

The only restriction on Little's formula is that the time interval begins and ends with no customers in the system. However, even if there are customers in the system, Little's formula is approximately correct. It is correct even if there is reneging. However, the waiting time must account for the waiting time of customers who left the system without receiving service.

Returning to our sample problem, recall that $W_q = 4.8$ minutes and that $\lambda = 5/30 = .167$ customer-minute. Thus, L_q must be $.167 \times 4.8 = .80$ customer, the value calculated earlier. Also recall that $W_s = 8.8$ minutes, so L_s must be $.167 \times 8.8 = 1.47$ customers, also the same value calculated previously.

Little's formula is particularly significant with respect to data measurement. Because average waiting time can be calculated from average queue length, and vice versa, one does not have to measure both queue lengths and waiting times to estimate the averages of both; only one of the two must be measured.

Example

A large machine shop has records on the number of jobs in the work queue, by hourly interval, for the month of March. It also has records on the number of jobs submitted each month. A queueing analyst calculated the average queue length to be 16 jobs. The number of jobs submitted during the month was 253. From Little's formula, average waiting time in queue is approximately $16/253 = .063$ month (1.9 days).

No simple relation holds between the standard deviation of waiting time and the standard deviation of queue length. Depending on the queue discipline, it is possible to have a very small deviation in queue length yet at the same time have very large deviations in waiting time. This is especially true if the discipline is not first-come, first-served.

2.4 PARALLEL SERVERS

Cumulative diagrams can be used when servers work in parallel. Consider the following data:

n	$A^{-1}(n)$	$D_q^{-1}(n)$	$D_s^{-1}(n)$
1	9:36	9:36	9:40
2	9:37	9:37	9:41
3	9:38	9:40	9:44
4	9:40	9:41	9:45
5	9:45	9:45	9:49

The arrival data are the same as before, but the departure data are different. There are now two servers working in parallel. Customer 2 is immediately served by the second server, and customer 3 begins service as soon as customer 1 (not customer 2) completes service. Figure 2.6 is the cumulative diagram. The separation between $D_q(t)$ and $D_s(t)$ now equals two (the number of servers) from 9:37 to 9:44. Between 9:36 and 9:37 and between 9:44 and 9:49 only one of the two servers is busy and the separation equals one. In general, cumulative diagrams can be used for any number of parallel servers and, as will be shown in later chapters, any number of serial servers.

Time in queue, queue length, customers in the system, and utilization are calculated in the exact same manner as for a single server. Time in system is also calculated in the same way, with an additional caveat. It is possible for a parallel server queue to follow a first-come, first-*served* discipline without following a first-come, first-*out* (FCFO) disci-

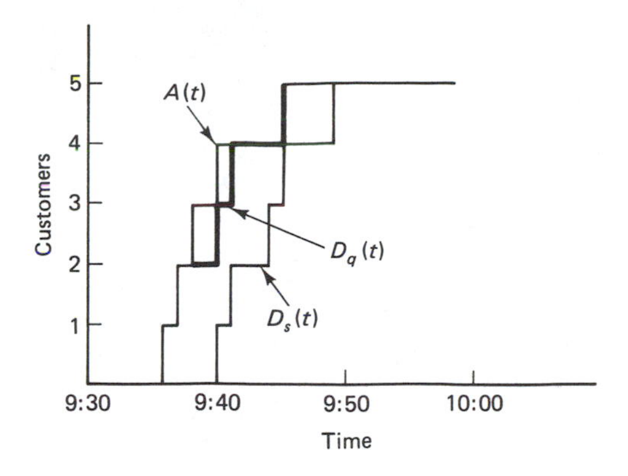

Figure 2.6 Cumulative arrival and departure diagram with two parallel servers.

pline. For example, the first customer might begin service at 2:00, ahead of the second customer, which begins at 2:02. But because the service time is longer for customer 1 than customer 2, the first does not complete service until 2:10 while the second finishes at 2:07. Thus, the discipline is FCFS but not FCFO. Time in system only equal $D_s^{-1}(n) - A^{-1}(n)$ if the discipline is FCFO.

2.5 REPRESENTATION OF NON-FCFS SYSTEMS

As already noted, caution must be exercised in interpreting the cumulative graphs when the discipline is not FCFS (or not FCFO for parallel servers), because the *n*th customer to arrive is not necessarily the *n*th customer to depart. Non-FCFS disciplines present no problem with respect to measuring queue lengths. They also present no problem with respect to measuring average waiting times. But they do present problems with respect to measuring waiting times of individual customers and calculating the waiting time standard deviation.

Consider the following data:

n	$A^{-1}(n)$	Departure from Queue	Departure from System
1	9:36	9:36	9:40
2	9:37	9:52	9:56
3	9:38	9:44	9:48
4	9:40	9:40	9:44
5	9:45	9:48	9:52

The arrival data are the same as before, but customers are served according to the last-come, first-served discipline. When the first customer arrives, no one is in the system, and the customer begins service immediately. When customer 1 completes service, customers 2, 3, and 4 are in the queue, and customer 4 is served first because it arrived last.

Figure 2.7 is a ***Gantt chart*** representation of the data. One bar represents each

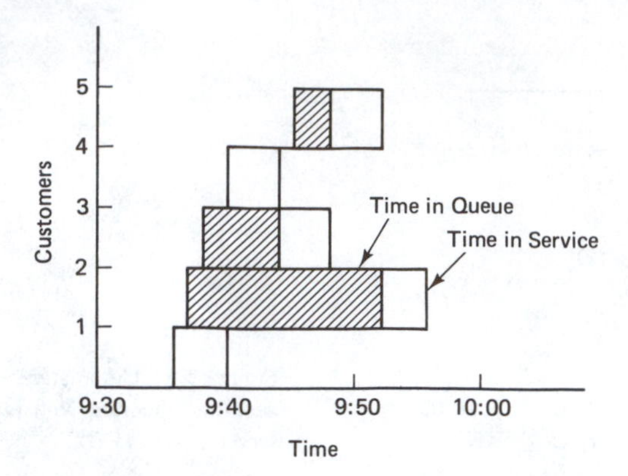

Figure 2.7 Gantt chart showing last-come, first-served (LCFS) discipline.

customer, and the bar is divided into two segments, representing time in queue and time in service. The entire length of the bar is time in system. By comparing with Fig. 2.2, we see that the LCFS discipline allows customers 4 to 5 to leave earlier, forces customer 2 to leave much later, and has no impact on customers 1 and 3. The average time in queue (and time in system) for LCFS is no different from FCFS (4.8 minutes and 8.8 minutes). However, the standard deviation is now much larger (5.9 minutes). Thus, the major impact of the LCFS discipline is an increase in the *variation* of the waiting times, not an increase in the average waiting time.

Gantt charts are also useful when customers have deadlines, whether or not the discipline is FCFS. Vertical lines indicating customer deadlines can be compared against the times service was actually completed. The separation between the deadline and the departure time from the system reveals the lateness.

Note that the Gantt chart is not a good way to show queue lengths. If this is the purpose, then the cumulative arrival and departure curves should be used (independent of the discipline).

2.6 EMPIRICAL PROBABILITY DISTRIBUTIONS

We have already seen that it is important to calculate the average and standard deviation of the queue length and waiting time. These statistics reduce a large set of data into a few measures of performance. Sometimes probabilistic data should be presented in greater detail, in the form of a diagram.

A probability distribution function represents the probability that a random variable occurs at or below a set value. For example, suppose that a coin is flipped five times. The total number of heads recorded has a binomial distribution with parameters $n = 5$ (number of flips) and $p = .5$ (probability of heads). The binomial distribution function indicates the probability of observing no heads, the probability of having one or fewer heads, the probability of having two or fewer heads, and so on, in the sample of five flips.

Unlike flipping coins, there may be no reason to believe that the data from a queueing system conform to any of the familiar theoretical distributions. This does not mean that there is no explanation for the underlying process generating the data, only that the process is something with which we are not familiar.

Whether or not the process conforms to a theoretical distribution, the observations can be displayed as an ***empirical distribution***. The empirical (*empirical* means "relying on observation") distribution indicates the proportion of data points *in a sample* that falls at or below a set value. The empirical probability distribution is drawn in much the same manner that a cumulative arrival and departure diagram is drawn (even though the cumulative diagrams are not directly concerned with probability).

Suppose that the following data are the service times (in seconds) recorded for 20 consecutive customers:

69	55	49	75	36	50	57	77	99	11
94	86	30	60	33	14	75	79	23	3

The first step in creating the empirical distribution is to sort the data in ascending order:

3	11	14	23	30	33	36	49	50	55
57	60	69	75	75	77	79	86	94	99

The second step is to plot the data in ascending order as a step curve. Each step has a height of one and is placed on the horizontal axis at a point corresponding to one of the service times. The graph is completed when the vertical axis is rescaled from zero to one, as in Fig. 2.8. The empirical distribution, $F(x)$, indicates the proportion of data points that have values at or below the value x. For example, 60 percent of the data points are less than or equal to 65.

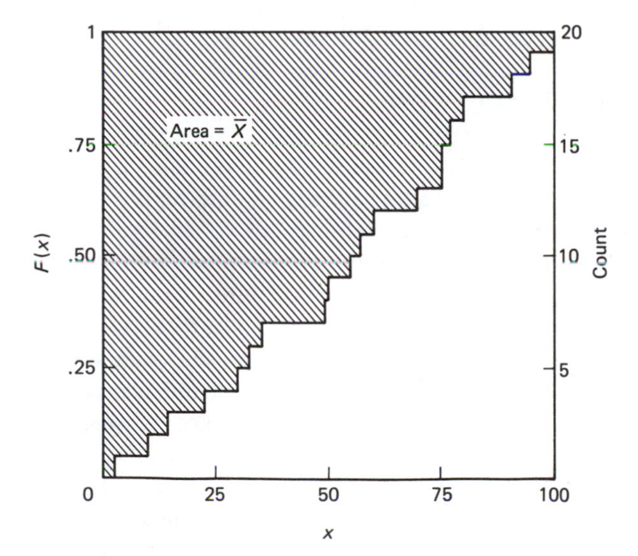

Figure 2.8 Empirical probability distribution showing calculation of mean.

If all of the data points are greater than or equal to zero, the cumulative diagram can also be used to find the average of the data points. This average is defined by

$$\bar{X} = \int_0^\infty [1 - F(x)]dx \qquad F(0) = 0 \tag{2.16}$$

That is, the average equals the area between $F(x)$ and the horizontal line 1. (Equation (2.16) also applies to theoretical distribution functions.)

2.7 MEASUREMENT TECHNIQUES

A fundamental principle in performing any type of experiment is that the measurement technique should not interfere with the process observed. Unfortunately, this goal cannot always be achieved. From the Heisenberg uncertainty principle, we know that it is impossible to measure both the position and motion of a wave with unlimited accuracy. Though no hard-and-fast rules apply to queueing systems, there are times when something cannot be measured without disturbing the system. Sometimes the mere presence of an observer will change customer behavior. The only exceptions seem to be automated systems, such as time-shared computers and telephone information lines. These systems can continuously monitor and record queueing information and produce reports, as shown in Fig. 2.9.

The task of determining cumulative arrivals and departures can be accomplished in any of several ways:

1. Record the time that each customer arrives, departs from queue, and departs from system.

2. Record the time that each customer departs from the system. Periodically count the number of customers in the queue and the number of customers in service.

3. Record the time that each customer departs from the system. Periodically time how long a customer spends in queue and spends in service.

2.7.1 Arrival and Departure Data

Departures from the system are the easiest to measure. The server can record departures with a counting device (Fig. 2.10) or perhaps directly from a cash register. Arrivals are usually much more difficult to measure.

2.7.1.1 Arrivals. Generally, measuring arrivals requires the added cost of stationing an observer (either human or an automated counting device) at a point beyond the end of the queue. Difficulties arise if the ''free-flow'' time for the customer to travel from the measurement point to the server is long. Vehicle queues at toll plazas sometimes stretch a mile or more, meaning that the vehicle arrives at the end of the queue at least a minute before it would have arrived at the server if there had been no queue. If a vehicle spends 5 minutes in the queue, eliminating the delay can at most reduce waiting time by 4 minutes—the 1-minute travel time cannot be eliminated. This 1-minute difference should

```
                              ANSWERED-CALL PROFILE
                                TRUNK GROUP WATS
                               DECEMBER 14, 1985

   TIME      NO. OF  % OF CALLS ANSW'D     /       AVG. TIME-BEFORE-ANSWER
    OF       CALLS     WITHIN X SECONDS    /              (SECONDS)
   DAY       ANSW'D  90 180 270 360 450 540/      0   90  180  270  360  450  540
   ----      ------ --- --- --- --- --- ---/      +----+----+----+----+----+----+

                                           /      +

07:30-08:00      18  72  83 100 100 100 100/  75 +xx

08:00-08:30      74  14  31  62  66  88 100/ 258 +xxxxxxxxxxxxx

08:30-09:00      86   0   1  44  57  80  98/ 326 +xxxxxxxxxxxxxxxxx

09:00-09:30     116   0  23  89 100 100 100/ 214 +xxxxxxxxxxxx

09:30-10:00     103   0  32  68  94 100 100/ 228 +xxxxxxxxxxxxx

10:00-10:30      62   0  23  31  47  65  87/ 360 +xxxxxxxxxxxxxxxxxxx

10:30-11:00     136   4  65  71  91  93  97/ 200 +xxxxxxxxxxx

11:00-11:30     110  13  68  96 100 100 100/ 150 +xxxxxxxx

11:30-12:00     125  18  66  82  95 100 100/ 171 +xxxxxxxxx

12:00-12:30     112  16  94 100 100 100 100/ 123 +xxxxxxx

12:30-13:00     102  14  40  72 100 100 100/ 198 +xxxxxxxxxxx

13:00-13:30      71   0   0   4  37  99 100/ 367 +xxxxxxxxxxxxxxxxxxxxx

13:30-14:00      75   0   0  24  53  64  79/ 387 +xxxxxxxxxxxxxxxxxxxxxx

14:00-14:30     116   0  34  67  88 100 100/ 232 +xxxxxxxxxxxx

14:30-15:00      92   0  14  53  98 100 100/ 261 +xxxxxxxxxxxxxx

15:00-15:30      82   0   0  23  71  89 100/ 328 +xxxxxxxxxxxxxxxxx

15:30-16:00     135  19  70  92  99 100 100/ 150 +xxxxxxxx

16:00-16:30      95   5  25  58 100 100 100/ 238 +xxxxxxxxxxx

16:30-17:00      81   0   0  21  81  98 100/ 311 +xxxxxxxxxxxxxxxx

17:00-17:30      12  50  67  75 100 100 100/ 120 +xxxxxx

                                           /      +

                                           /      +----------------------------

             ------ --- --- --- --- --- ---/

TOTAL          1803   7  38  64  86  95  98/ 233
```

(a)

Figure 2.9 Output generated by call sequencing system.

Source: Reprinted by permission of IBM Corporation (Rolm Systems).

SEQUENCE # 53782

ABANDONED-CALL PROFILE
TRUNK GROUP WATS
DECEMBER 13, 1985

TIME OF DAY	NO. OF CALLS ABAN'D	% OF CALLS ABAN'D WITHIN X SECONDS 10	20	30	40	50	60	AVERAGE TIME-BEFORE-ABANDONMENT (SECONDS)	0 10 20 30 40 50 60
----	------	--	---	---	---	---	---		+----+----+----+----+----+----+
07:30-08:00	4	25	50	75	75	100	100	21	+xxxxxxxxxxx
08:00-08:30	14	43	57	57	57	64	79	29	+xxxxxxxxxxxxxx
08:30-09:00	10	20	70	90	90	100	100	18	+xxxxxxxx
09:00-09:30	18	56	67	83	89	100	100	15	+xxxxxxx
09:30-10:00	24	50	67	75	83	92	100	17	+xxxxxxxx
10:00-10:30	24	13	29	50	67	83	96	30	+xxxxxxxxxxxxxxx
10:30-11:00	12	67	83	83	83	83	92	16	+xxxxxxxx
11:00-11:30	22	41	91	100	100	100	100	14	+xxxxxxx
11:30-12:00	19	53	84	95	95	95	95	15	+xxxxxxxx
									+----+----+----+----+----+----+
	------	--	---	---	---	---	---		
TOTAL	147	41	67	78	83	90	96	19	

(b)

Figure 2.9 (*continued*)

be compensated for when drawing the cumulative arrival diagram, as in Fig. 2.11. (The large arrival rate has smoothed out the steps in the arrival and departure curves.)

As another example, a patient visiting a medical clinic might have to check in at a receptionist before waiting to see a doctor. Depending on the line for the receptionist, the patient might have several minutes of delay before his or her arrival is recorded. These minutes may never be accounted for in measuring how long patients wait to see doctors.

Another problem when the discipline is not FCFS is correlating a specific arrival with a specific departure. A counting device does not indicate which customer passed a given point, only that some customer passed the point. To measure variation in waiting time, one must be able to identify exactly which customer passed each point. When the arrival rate is large, this task is tedious at best. If vehicles are involved, the license plate can be recorded. Time-lapse photography can be used when people are being counted.

Figure 2.10 Device for counting customers.

Source: Courtesy of Timelapse, Inc., St. Petersburg, Florida.

Measuring arrivals is not as difficult when the customers are work orders or things rather than people. Computer systems routinely record the times that jobs are submitted, begin processing, and complete processing. For packages and objects, bar codes and automated reading devices can greatly simplify the task of recording arrival and departure data. Even handwritten work slips might contain the necessary information.

2.7.1.2 Departures from Queue. Departures from queue can be derived from the data on departures from the system and arrivals. When all servers are busy, one departure from queue coincides with one departure from system. When servers are idle, a

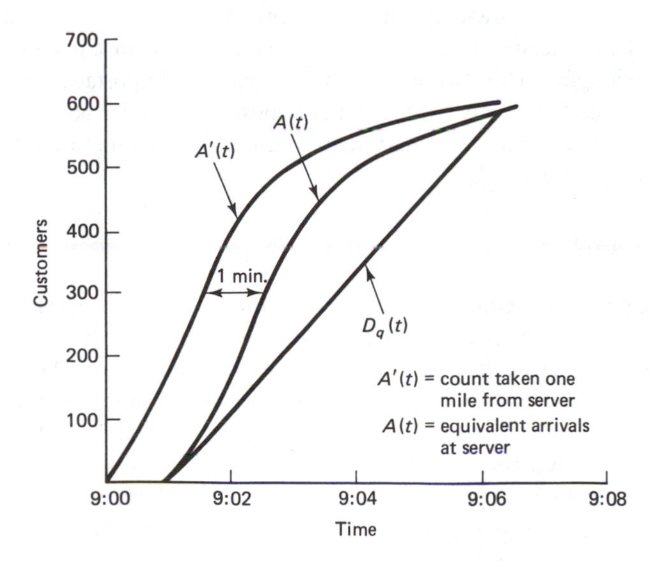

Figure 2.11 When customers are counted upstream from the server, the cumulative count should be transformed to obtain the arrival curve.

departure from queue coincides with an arrival. This at least is how a queue performs in theory. In reality, there is often a lag between the time a customer completes service and the time the next customer begins service. This lag may represent two things: a reaction time and a movement time.

Example

> George Waites arrived at his bank at 9:00, at which time he immediately joined a teller queue. Because the line was long, George began reading his newspaper. At 9:03, George reached the front of the queue, and at 9:04 a customer completed service. By this time, George was enraptured in the comics and did not notice that a teller was free. Five seconds later, seeing that George did not move, the teller announced, ''Next in line.'' But George did not hear the teller. Another 5 seconds later, the customer behind George poked him in the back, and George realized a teller was free. It then took George another 4 seconds to walk up to the teller. George completed service 30 seconds later.
>
> For the purposes of analysis, George departed from the queue at 9:04, even though he did not reach the teller until 14 seconds later. George departed from the system at 9:04.44 and the service time is 44 seconds, the total time devoted by the server to George. Of the 44 seconds, reaction took 10 seconds and movement took 4 seconds.

When a queue exists, a departure from queue occurs when a server becomes available, not when the customer realizes the server is available. Depending on the purpose of the analysis, one may wish to record the reaction and movement times in addition to the total service time. Such information may be important if changes in the queue's design are considered (see Chap. 11). Keep in mind that the reaction time and movement time may be different when a queue exists than when a queue does not exist. This may lead to slightly different service times.

2.7.1.3 Busy Systems. For systems such as highways, telephone lines, and large cafeterias, the number of customers served can be so large that recording each customer's arrival and departure time is unwieldy. Interval counts should be used instead. An interval count would tell how many customers arrived and departed in each time interval (each minute, for example). This simplification is especially important when arrivals and departures are recorded manually, because it is impossible to write down the time of each event. In making these counts, one may wish to use a metronome to emit a sound at the end of each time interval.

2.7.2 Queue Lengths

Provided that there is no reneging, the cumulative departures from the queue equal the cumulative departures from the system plus the number of customers in service; the cumulative arrivals equal the cumulative departures from the system plus the number of customers in the system. Therefore, measurement of departures from the system, customers in the system, and customers in service provides the necessary information.

Queue lengths must be measured periodically, at time intervals that are sufficiently short to detect any major changes. In most cases, some type of recording device is needed to count customers. For example, the queue might be photographed periodically and

Figure 2.12 Time-lapse photography records the formation of a queue.

counted afterward (Fig. 2.12). Or markers might be positioned according to different queue lengths so that an observer would only have to note which marker best approximates the queue length.

2.7.3 Waiting Times

The cumulative arrival diagram and the cumulative departure from queue diagram can also be developed by periodically measuring customer waiting times. The time in service

PLEASE IMPRINT OR PRINT

DATE OF SERVICE | LOCATION | STATION

LAST NAME FIRST NAME INITIAL

BIRTHDATE MONTH DAY YEAR | HEALTH INSURANCE CLAIM NO.

MEDICAL RECORD NUMBER

SEX | COVERAGE | GROUP NUMBER | ACCOUNT NO.

Page ______ of ______

DAY OF WEEK - Circle one

Su M Tu W Th F Sa

OBSERVED BY:

TIME APP'T. SCHEDULED

TIME OF ARRIVAL

OFFICE VISIT TIME RECORD

PHYSICIAN'S NAME M.D.

ACTIVITY/AREA DESCRIPTION	HOUR : MIN.
1.	:
2.	:
3.	:
4.	:
5.	:
6.	:
7.	:
8.	:
9.	:
10.	:
11.	:
Last Description Does Not Count as an Act. - Record Time Only	:

ELAPSED TIME HOURS : MIN.	ACTIVITY/AREA DESCRIP. NO. & CODE
:	1.
:	2.
:	3.
:	4.
:	5.
:	6.
:	7.
:	8.
:	9.
:	10.
:	11.

=

IF MORE EXITS AND RETURNS OCCUR,

☐ CHECK HERE AND ENTER ON BACK. SUM :

NOTES

02977 (1-83)

Figure 2.13 Sample sheet for recording customer data.

Source: Courtesy of Kaiser Permanente, Oakland, California.

is subtracted from departures from the system to obtain departures from the queue; time in system is subtracted from departures from the system to obtain arrivals.

Waiting times are measured by sending test customers through the system. This technique is used in communication systems with test data and sometimes on road networks with test vehicles. In addition to being potentially expensive, one drawback of this approach is that the test interferes with the queue. The added customers increase overall delay.

Another approach is to observe real customers and record their departure and service times. Figure 2.13 is a sample record sheet that can be used when customers are processed at several different servers.

2.7.4 Reneging

Unless the queueing system is totally automated, as in an airline telephone reservation line, reneging is bound to be difficult to measure.

Example

> Diners place their names on a waiting list as soon as they arrive at a busy restaurant. As tables becomes available, the maître d'hôtel calls off names in the order that customers arrived. Some of the diners have reneged before their names are called.

The waiting list allows the restaurant to determine when customers arrive and when they depart from queue. It also allows them to determine which customers reneged. It does not tell the restaurant when the customers reneged.

Example

> An observer counts customers as they join a ticket queue for a popular movie, and the cashier records ticket sales. Before they reach the front of the line, some of the customers decide that the wait is too long and leave to attend another movie.

The system here provides arrival and departure data. But because customers renege from the middle of the queue (not the end or the beginning), they may not be detected until the queue finally vanishes.

Example

> A company uses work tickets to keep track of customer orders. The times the order is received, begun, completed, and delivered are recorded. Work tickets are discarded if an order is canceled.

Again, the system provides data on arrivals and departures but not on reneges.

Unfortunately, measuring reneging often requires interfering with the queue's operation. For example, customers might be required to sign out before departing. Or barriers might be erected that prevent customers from leaving the middle of the line. For the work orders, a system must be instituted to collect data that the company would otherwise prefer to discard.

2.7.5 Detailed Service Data

Often large queues are the product of inefficient servers, and the best way to eliminate the queue is not to add more servers but to encourage the existing servers to work more productively. The only way to tackle this problem is first to observe the servers in action, but this in itself raises a host of problems, especially if the server is a human. It is a fair assumption that workers do not like to be watched, timed, or measured in any way, nor are they likely to behave in the same manner when they are observed as when they are not observed. Needless to say, the task is difficult. Entire books are written on the subjects of time-motion study and work methods, and it would not be appropriate to cover this topic in any detail here. Refer to the books listed at the end of this chapter.

2.8 CHAPTER SUMMARY

Queueing systems should be examined for symptoms of health and illness. These symptoms are the measures of performance (MOP), the most important of which are:

1. Time in system and time in queue
2. Customers in queue and in service
3. Service time
4. Utilization
5. Arrival rate
6. Reneging

A poor score on any one of these MOPs indicates that something is wrong, and further examination is needed.

The measures of performance should be presented through cumulative diagrams and Gantt charts, which show how the queues evolve over time. Average waiting time and average queue length should also be calculated. From Little's formula, average waiting time can be converted into average queue length with a simple calculation. The standard deviation of the waiting time and queue length provide an indication of how much these two factors vary between customers and over time, respectively. Finally, empirical probability distributions for the service time and interarrival time provide a more detailed picture of how the factors vary.

Data measurement may involve recording arrivals, departures, queue lengths, and/or waiting times. Counters, photographs, and test customers are some of the tools used in data measurement. Particular caution should be exercised when measuring arrivals and reneging. No matter what tool is used, the measurement technique should be as unobtrusive as possible.

Though data are important in themselves, they are no substitute for observing the system with your own eyes. A simple record of what happened does not necessarily reveal *why* it happened. Explanations for unusually long service times (server attitude, mechanical breakdown, problem customer) may never appear in the data. Although quantitative data are needed in evaluating the merits of an alternative, qualitative information is essential to defining and isolating the problem. It is also essential in creating models, discussion of which begins in the following chapter.

FURTHER READING

BARNES, R. M. 1980. *Motion and Time Study, Design and Measurement of Work*. New York: John Wiley.

BROWNLEE, K. A. 1965. *Statistical Theory and Methodology in Science and Engineering*. New York: John Wiley.

FREEDMAN, D., R. PISANI, and R. PURVES. 1978. *Statistics*. New York: W. W. Norton.

LITTLE, J. D. C. 1961. "A Proof for the Queueing Formula: $L = \lambda W$," *Operations Research*, 9, 383–387.

MAXWELL, E. A. 1983. *Introduction to Statistical Thinking*. Englewood Cliffs, N.J.: Prentice Hall.

MUNDEL, M. E. 1985. *Motion and Time Study, Improving Productivity*. Englewood Cliffs, N.J.: Prentice Hall.

NEWELL, G. F. 1982. *Applications of Queueing Theory*. London: Chapman and Hall.

NIEBEL, B. W. 1982. *Motion and Time Study*. Homewood, Ill.: Richard D. Irwin.

WONNACOTT, T. H., and R. J. WONNACOTT. 1977. *Introductory Statistics*. New York: John Wiley.

PROBLEMS

1. The data below represent jobs that arrived at a printing shop between the opening time, 9:00 A.M., and 12:00 noon. Calculate (a) average and standard deviation of time in queue; (b) average and standard deviation of time in system; and (c) average and standard deviation of tardiness.

Customer	Arrival time	Depart queue	Depart system	Due
1	9:03	9:03	9:23	10:00
2	9:10	9:23	9:55	10:00
3	9:40	9:55	10:35	11:00
4	10:05	10:35	10:50	11:00
5	10:15	10:50	11:20	11:00
6	10:40	11:20	11:35	12:00
7	11:50	11:50	12:20	1:00

2. For the data in Prob. 1, plot the cumulative arrivals, cumulative departures from queue, and cumulative departures from system. For the time interval from 9:00 to 12:00, determine (a) average and standard deviation of customers in queue and (b) average and standard deviation of customers in system.

3. From your answers to Probs. 1 and 2, verify Little's formula for customers in queue. Also, check Little's formula for customers in system. Explain why the formula is not verified for customers in system and give a remedy.

4. Repeat Probs. 1–3 for the data set below.

Customer	Arrival time	Depart queue	Depart system	Due
1	9:05	9:05	9:30	12:00
2	9:10	9:30	10:05	12:00
3	9:20	10:05	10:35	12:00
4	10:35	10:35	11:10	12:00
5	11:20	11:20	11:45	12:00
6	11:40	11:45	12:15	12:00

5. Determine the throughput and utilization for the data in Prob. 1 over the 9:00–12:00 period.

* **6.** The following are arrival and service times for work that was in a small machine shop during January. The system has parallel servers.

Customer	Arrival time	Depart queue	Depart system
1*			Jan. 4
2*			Jan. 6
3	Jan. 1	Jan. 1	Jan. 15
4	Jan. 3	Jan. 4	Jan. 14
5	Jan. 7	Jan. 7	Jan. 18
6	Jan. 12	Jan. 14	Jan. 29
7	Jan. 16	Jan. 16	Feb. 2
8	Jan. 16	Jan. 18	Jan. 27
9	Jan. 22	Jan. 27	Feb. 7
10	Jan. 24	Jan. 29	Feb. 20
11	Jan. 26	Feb. 2	Feb. 15

*Arrived in December.

(a) How many servers are there?

(b) Plot cumulative arrivals, cumulative departures from queue, and cumulative departures from system.

(c) Customers are served FCFS, but the system is not first-in, first-out. Explain why.

(d) Among jobs that arrived in January, calculate average time in queue and average time in system. From Little's formula, estimate average customers in queue and average customers in system.

(e) Will your estimate for average customers in queue from part d equal the average customers in queue over January? Why or why not?

7. The printer in Prob. 1 has decided to shift to an LCFS discipline. As before, one server operates. Assume that arrival and service times are the same as in Prob. 1.

(a) Draw a Gantt diagram for the customers, indicating arrival time, departure time, and due date.

(b) Plot cumulative arrivals, cumulative departures from system, and cumulative departures from queue.

(c) Calculate average time in system and the standard deviation of time in system.

8. The printer in Prob. 1 would now like to serve customers with the shortest service time first. Using this discipline, repeat parts a–c in Prob. 7. Can you see any advantages or disadvantages of this discipline over LCFS? Over FCFS?

*Difficult problem

9. These service times were recorded at a fast-food counter (in seconds):

9 57 63 30 18 7 76 97 53 59 37 54 88 39 94

(a) Draw the empirical probability distribution function.

(b) Using Eq. (2.16), along with one of the methods in the appendix, calculate the average service time.

(c) Confirm your answer to part b by summing the service times and dividing by 15.

***10.** The following data set includes some reneges. Plot the cumulative arrival/departure diagram, along with a cumulative reneging diagram. Determine all meaningful measures of performance.

Customer	Arrival time	Depart queue	Depart system	Renege time
1	9:05	9:05	9:30	
2	9:10	9:30	10:05	
3	9:20	10:05	10:35	
4	9:25			9:25
5	9:40			9:50
6	9:49			9:50
7	10:35	10:35	11:10	
8	11:20	11:20	11:45	
9	11:40	11:45	12:15	

11. Suppose that the data in Prob. 1 contain a reaction/movement time in addition to a service time. If the reaction/movement time is 30 seconds when a customer is already in the system and zero otherwise, what is the average time that the customer is actually being served (not counting the reaction/movement time)?

12. A telephone system began the day with no calls being served. The data below give the number of calls received and departed in each of the next 8 minutes.

	1	2	3	4	5	6	7	8
Arrived	10	15	12	8	16	15	11	10
Departed queue	10	13	12	10	13	13	12	13
Departed system	3	13	12	12	11	13	12	13

(a) Draw the cumulative diagrams and derive estimates for average time in queue and in system.

(b) Can anything be said about the standard deviation of these values? Explain.

(c) How many servers are operating?

13. Describe three ways for estimating the expected time in queue for the following queueing systems. In each case, describe what values you would measure and how you would measure them.

(a) Patients in a doctor's office

(b) Customers at a delicatessen

(c) Computer users on a time-shared system

(d) Autos at a traffic signal

(e) Checks waiting to be cleared by a bank

14. How would you modify your answers to Prob. 13 to estimate expected queue length?

15. How would you modify your answers to Prob. 13 to estimate standard deviation of time in queue?

*Difficult problem

16. How would you modify your answers to Prob. 13 to estimate reneges/hour?
17. Before patrons can check out books at a library, they must fill out a charge card for each book. Unfortunately, some of the patrons forget to complete the cards before reaching the clerk. In these cases, the clerk asks the patron to step aside and fill out the card, after which the patron receives priority over other customers waiting in line. Describe how you would measure the performance of this queueing system. Also, discuss how you might modify your cumulative diagram and Gantt chart.

EXERCISE: QUEUE OBSERVATION AND MEASUREMENT

The purpose of this exercise is to observe and record the behavior of a queueing system. These recordings will be used in later chapters as the basis for queue models.

In a two-person team, observe a queue (your instructor will provide direction) for *exactly one hour* and record the following (to the second):

1. Time that each person arrived at the queue
2. Time that each person began service
3. Time that each person completed service
4. Time that each person reneged (if applicable)

In addition, record the number of people in the queue and the number of people in service at the beginning of the hour and at the end of the hour.

Based on the recordings, each team should turn in the following:

1. A graph showing cumulative arrivals, cumulative departures from queue, and cumulative departures from system
2. A graph showing the empirical probability distribution for the interarrival times
3. If applicable, a cumulative graph for reneging customers
4. A graph showing the empirical probability distribution for the service times
5. Numerical estimates of the following (show work):
   - **(a)** Average and standard deviation of queue length and average number of customers in the system
   - **(b)** Average and standard deviation of time in queue and time in system
   - **(c)** Average and standard deviation of the interarrival time
   - **(d)** Average and standard deviation of the service time
   - **(e)** Average arrival rate.
6. Verify the equations $L_q = \lambda W_q$ and $L_s = \lambda W_s$.

APPENDIX—DEFINITE INTEGRALS

The *definite integral* of a non-negative single-valued function

$$\int_a^b f(x)\, dx \qquad (2.A1)$$

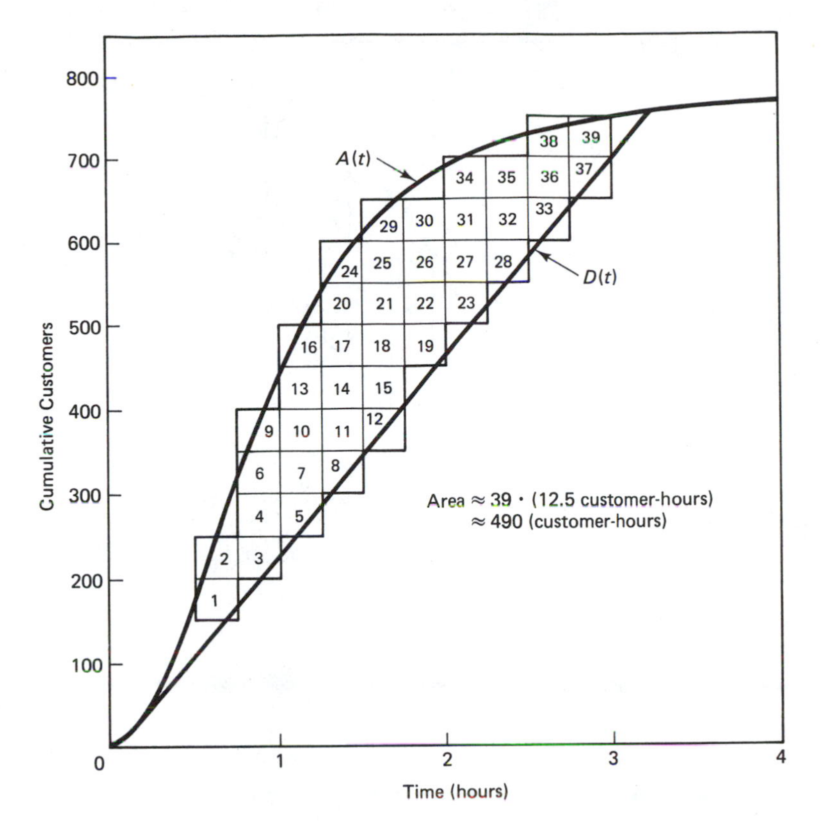

Figure 2.14 Approximation of definite integral by counting squares.

equals the area between the x-axis and $f(x)$ over the range from a to b. When $f(x)$ is expressed as an equation, a definite integral can by calculated by one of the methods of numerical integration found in any calculus text. For some functions, the definite integral can also be determined by evaluating the indefinite integral of $f(x)$ at b and a and taking the difference.

A definite integral can also be calculated when a function is not expressed as an equation but is plotted on a piece of graph paper.

Method 1: Weights. This is the simplest method but requires special equipment. From the piece of graph paper, cut out the section of paper bounded by the vertical lines through points a and b, the x-axis, and the function. Weigh this section of paper on an accurate scale (W_f). Cut out a large square, of known dimensions x by y, from the remaining section of graph paper and weigh it (W_s). The integral is approximately

$$\int_a^b f(x)dx \approx \frac{W_f}{W_s}\ (x \cdot y) \tag{2.A2}$$

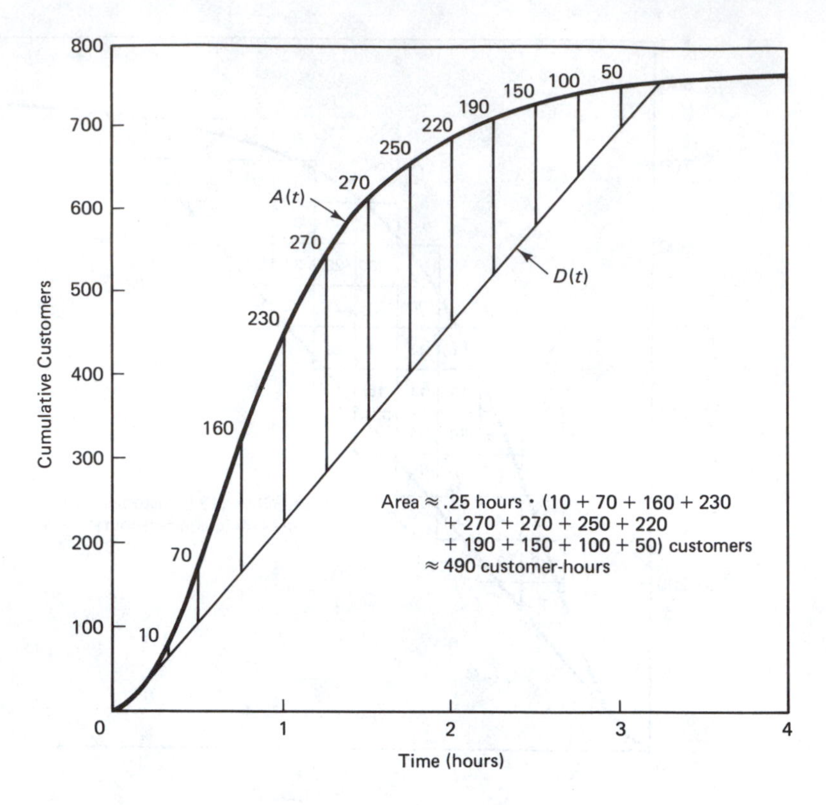

Figure 2.15 Approximation of definite integral by measuring lines.

Method 2: Squares. This approach is simple but a bit tedious. Determine the size of a single square on the graph paper (A_s). If the majority of a square's area falls in the desired region, shade it a light color. Count the number of shaded squares (N). The integral is approximately (see Fig. 2.14):

$$\int_a^b f(x)dx \approx N \cdot A_s \tag{2.A3}$$

The smaller the square is, the more accurate the approximation will be. Of course, this requires more effort.

Method 3: Lines. This approach requires more than just counting but can be more accurate than method 2. On a piece of graph paper, draw a series of equidistant vertical lines (spaced d_x apart) over the range from a to b. Calculate the height of each line, then sum the data to obtain a total height, h_t. As shown in Fig. 2.15, the integral is approximately

$$\int_a^b f(x)dx \approx d_x \cdot h_t \tag{2.A4}$$

As with method 2, accuracy increases when the spacing decreases.

Method 4: Geometry. This approach is less general. Some definite integrals represent the calculation of ordinary geometric shapes, such as rectangles, triangles, and trapezoids. This is true of cumulative diagrams, because of the discrete nature of customer counts. Partition the area of interest into rectangular, triangular, and trapezoidal regions. Calculate the area of each region and then sum the areas. Provided that the areas can be accurately measured, the result should be the exact integral. An example is shown in Fig. 2.5a.

chapter 3

The Arrival Process

Chapter 2 describes the first step in evaluating a queueing system: observation and measurement. Observation is a key part of the formulation stage of systems analysis. It answers the questions: What is the unknown? What are the data? What is the condition? The second stage of systems analysis is modeling. Whereas observation determines what happened, modeling *explains* what happened. It should reveal the underlying process that created the data. Thomas Kuhn, writing on the discovery of oxygen, states:

> Though undoubtedly correct, the sentence, ''Oxygen was discovered,'' misleads by suggesting that discovering something is a single simple act assimilable to our usual (and also questionable) concept of seeing . . . discovering a new sort of phenomenon is necessarily a complex event, one which involves recognizing *that* something is and *what* it is.[1]

Recognizing *what* it is—the act of explanation—is the role of modeling.

Modeling is the subject of this chapter, as well as Chaps. 4 to 6. This chapter concentrates on the customer arrival process, with emphasis on a particular model known as the Poisson process. As early as 1910, with Erlang's paper ''The Theory of Probabilities and Telephone Conversations,'' the Poisson process had been used as a model for customer arrivals. However, it was developed much earlier, in the early nineteenth century, by the French mathematician Denis Poisson. In addition to being a good representation of arrival processes, it turns out that the Poisson process is also a good representation of many physical systems, such as the position of molecules within gases. (See Haight 1967 for more examples.) The Poisson process is a rare example of a model that fulfills the dual objectives of realism and simplicity.

This chapter begins with a description of the conditions that create the Poisson process. Next, it covers the relationship between the Poisson process and the Poisson probability distribution. It then discusses the properties of the Poisson process. The chapter concludes with an examination of some of the ways to determine whether or not an arrival process is Poisson. One should understand random variables, probability distributions, and basic statistics before beginning this chapter.

3.1 CREATION OF THE POISSON PROCESS

The Poisson process is an example of a broader class of stochastic (that is, random) processes known as counting processes. Suppose that events occur at various times, in some random fashion. A ***counting process*** is the function representing the cumulative number of events that have occurred up to any point in time. We have already seen three examples of counting processes—$A(t)$, $D_q(t)$, and $D_s(t)$.

The Poisson process is a type of counting process that applies to customer arrivals, $A(t)$. The definition of the Poisson process will depend on the following:

Definitions 3.1

A counting process has **independent increments** if the numbers of events in any pair of disjoint time intervals are statistically independent.

A counting process has **stationary increments** if the distribution of the number of events in any time interval depends only on the length of the time interval. It does not depend on when the interval occurred.

Formally, the Poisson process is defined as follows:

Definitions 3.2

A counting process $N(t)$ is a *Poisson process* with rate λ if

A. The process has independent increments.

B. The process has stationary increments. And

C.

$$Pr\{[N(t + dt) - N(t)] \begin{Bmatrix} = 0 \\ = 1 \\ > 1 \end{Bmatrix} \begin{matrix} = 1 - \lambda dt \\ = \lambda dt \\ = 0 \end{matrix}$$

where dt is a differential (that is, very small) sized time interval.

The rate λ represents the *expected* number of customers to arrive per unit time. If $\lambda = 10$ customers per hour, then the expected number of customers to arrive in a 60-minute period is 10 and the expected number to arrive in a 30-minute period is 5. The number of customers that actually arrive in any 60-minute period can be very different from 10, and the spacing between customer arrivals does not have to be constant. The significance of condition C is that customers arrive one at a time, and the probability that an arrival occurs in a very short (differential length) time interval equals the arrival rate multiplied by the size of the interval. For example, if $\lambda = 10$ customers per hour, then the probability of one customer arriving within any 1-second interval is approximately $10/3600 = .00278$.

In words, the Poisson process can be summarized as follows:

A. The probability that a customer arrives at any time does not depend on when other customers arrived.

B. The probability that a customer arrives at any time does not depend on the time.

C. Customers arrive one at a time.

The word *description* suggests why the Poisson process is important. Consider a queue for tellers at a bank. Is there any reason to believe that past arrivals influence future arrivals? Does the probability of an arrival vary over time? Do customers arrive in groups? For most customers, the answer to the first and third questions is almost certainly no. Customers do tend to arrive independently of each other, one at a time. There may be exceptions—two people who decide to go to the bank together—but that is what they are, exceptions. The model does not have to be perfect to be useful.

With regard to the second question, the answer is more of a maybe. The arrivals may be fairly constant over short time intervals, but most likely vary over the day and week. For the moment, think of the standard Poisson process as representing what happens over time intervals when the arrival rate is nearly constant. Later on, in Chap. 6, a nonstationary version of the Poisson process is presented that accounts for this variation.

The Poisson process is sometimes viewed in terms of its ***calling population*** the group of *potential* customers. If the calling population is large, customers arrive independently, and the probability that any particular customer arrives during any small time interval is small and constant, then the arrival process will be Poisson. As the size of the calling population declines, the arrival process will look less and less like a Poisson process, in which case an alternative model may be called for (as is discussed in Chap. 5).

3.2 POISSON DISTRIBUTION

The Poisson distribution is a probability distribution that arises from the Poisson process. It is a *discrete distribution*, meaning that its random variable is limited to a set of distinct values. A Poisson random variable is limited to the set of non-negative integers. The Poisson distribution gets its name because it is the probability distribution for the number of arrivals within any time period of a Poisson process.

Derivation. Consider the random variable $N(t)$, representing the number of arrivals over the interval $[0,t]$. Suppose that the interval $[0,t]$ is divided into I segments, where I is a very large number. Consider just one of these segments. From condition C of

Definition 3.2 of the Poisson process, either one customer arrives during the segment or no customers arrive, but never more than one. The probability that exactly one customer arrives is $\lambda\,(t/I)$. The total number of arrivals over the entire interval equals the sum of I Bernoulli (0,1) random variables, each representing whether or not a customer arrived during a segment. Thus $N(t)$ is a binomial random variable with parameters I and $\lambda(t/I)$:

$$P[N(t) = n] = \lim_{I\to\infty} \binom{I}{n} p^n (1 - p)^{I-n} \qquad n = 0, 1, \ldots, \mathrm{I} \tag{3.1}$$

where $P[\ \ ]$ denotes the probability of an event, and $p = \lambda t/I$.

Through a series of calculations, Eq. 3.1 can be reduced to a very simple expression. First, by substituting $\lambda t/I$ for p, and expanding the permutation, we can rewrite Eq. 3.3 as the following:

$$P[N(t) = n] = \lim_{I\to\infty} \left[\frac{I(I-1)\cdot\ldots\cdot(I-n+1)}{I^n}\right] \left[\frac{(\lambda t)^n}{n!}\right]\left[\frac{(1-\lambda t/I)^I}{(1-\lambda t/I)^n}\right] \tag{3.2}$$

In the limit, as I approaches infinity, the numerator of the first term equals I^n. Thus, the first term approaches one. Also in the limit, the denominator of the third term equals 1^n, or just one. This leaves the following:

$$P[N(t) = n] = \lim_{I\to\infty} \frac{(\lambda t)^n}{n!} (1 - \lambda t/I)^I \tag{3.3}$$

From the definition of the number e, the last term in Eq. 3.3 is $e^{-\lambda t}$. Making this substitution allows Eq. 3.1 to be expressed in final form:

$$P[N(t) = n] = \frac{(\lambda t)^n}{n!} e^{-\lambda t} \qquad n = 0, 1, \ldots \tag{3.4}$$

For a *discrete* random variable, the ***probability function***, $f(x)$, specifies the probability that the random variable *equals* a set value x. Equation 3.4 is the probability function for a Poisson random variable with mean λt. Hence, *the probability distribution for the number of events in any time interval is Poisson.*

Definition 3.3: Poisson Probability Function, Mean λt

$$f(x) = \frac{(\lambda t)^x}{x!} e^{-\lambda t} \qquad x = 0, 1, 2, \ldots \tag{3.5}$$

The ***probability distribution function***, $F(x)$, specifies the probability that a random variable occurs at or below a set value x.

Definition 3.4: Poisson Probability Distribution Function, Mean λt

$$F(x) = \sum_{n=0}^{x} f(n) = \sum_{n=0}^{x} \frac{(\lambda t)^n}{n!} e^{-\lambda t} \qquad x = 0, 1, 2, \ldots \tag{3.6}$$

Example

From past experience, customers are known to arrive at a service station at the rate of $\lambda = 15$ customers/hour. The owner would like to know the probability that more than one customer will arrive during an employee's 5-minute coffee break.

Solution The expected number of customers to arrive during 5 minutes is λt, or 15 customers/hour $\times$ (5/60) hours = 1.25 customers. The probabilities of zero customers arriving and 1 customer arriving are calculated from Eq.(3.4):

$$f(0) = \frac{1.25^0}{0!} e^{-1.25} = e^{-1.25} = .29$$

$$f(1) = \frac{1.25^1}{1!} e^{-1.25} = 1.25\, e^{-1.25} = .36$$

The probability that one or less customers arrive, $F(1)$, equals $f(0) + f(1) = .65$. Therefore, the probability that more than one arrives equals $1 - F(1) = .35$.

Given that the number of arrivals in a time interval has a Poisson distribution, the Poisson process can be defined in the following alternative form:

Definition 3.5

The *Poisson process* is a counting process with the properties:

A. The process has independent increments.

B. Number of events in any time interval of length t has a Poisson distribution with mean λt.

Condition B substitutes for both conditions B and C in Definition 3.2. Therefore, it implies stationarity.

3.3 PROPERTIES OF THE POISSON PROCESS

The Poisson process has a number of important properties. So far, we have seen the following:

Property 1. The distribution for the number of events in any time interval of length t has the Poisson distribution with mean λt.

Here are some of the important characteristics of the Poisson distribution:

Characteristics of Poisson Distribution

1. The Poisson distribution is discrete and defined over the set of non-negative integers.
2. $E[N(t)] = \lambda t$
3. $V[N(t)] = \lambda t$
4. $P[N(t) = 0] = e^{-\lambda t}$

$E[\ ]$ represents the *expectation* of the enclosed random variable, and $V[\ ]$ represents the *variance* of the enclosed random variable.

The Poisson distribution has the distinguishing characteristic that its variance equals its mean (thus, its standard deviation equals the square root of its mean).

Example

Residents of a small city are known to place telephone calls by a Poisson process with rate 1000 per hour. The expected number of phone calls made after a time t is $1000t$, and the standard deviation of the number of phone calls made after a time t is $\sqrt{1000t}$. The *coefficient of variation* (ratio of standard deviation to mean) of the number of calls is $\sqrt{1000t}/1000t = 1/\sqrt{1000t}$, which declines as t increases.

3.3.1 Interarrival Time

The Poisson process is closely related to the exponential probability distribution. The exponential distribution is a ***continuous distribution***, meaning that exponential random variables are not limited to a set of discrete values. Instead of having a probability function, a continuous random variable has a ***probability density function***. The density function, $f(x)$, does not equal the probability that the random variable equals x. Rather, $f(x)$, multiplied by the differential dx, equals the probability that the random variable is contained in an interval of width dx centered around the point x.

In the following characteristics of the exponential distribution, X represents the outcome of a random variable.

Characteristics of the Exponential Distribution

1. The exponential distribution is continuous and defined over the set of non-negative real numbers.
2. The probability density function, $f(x)$, is

$$f(x) = \lambda e^{-\lambda x} \qquad x \geq 0$$

3. The probability distribution function, $F(x)$, is

$$F(x) = \int_0^x \lambda e^{-\lambda t}\, dt = 1 - e^{-\lambda x} \qquad x \geq 0$$

4. $E(X) = 1/\lambda$
5. $V(X) = 1/\lambda^2 = E^2(X)$

Example

Calculate the probability that an exponential random variable, mean .5, is contained in [.95, 1.05].

Approximate Solution $\lambda = 1/.5 = 2$, and $dx = 1.05 - .95 = .1$. The desired probability is approximately $f(1)dx = 2e^{-2\times 1}(.1) = .027$.

Exact Solution The approximate result is confirmed from the probability distribution function. The probability that $X \leq 1.05$ equals $F(1.05) = .878$ and the probability that $X \leq .95$ equals $F(.95) = .850$. The difference rounds off to .027.

The exponential distribution function is a special case of the gamma distribution, which is discussed in greater detail later. In Fig. 3.1b, the exponential distribution is the gamma distribution with a coefficient of variation of 1. Unlike the Poisson distribution function (Fig. 3.1d), the exponential distribution function is smooth, reflecting the fact that an exponential random variable is not restricted to a discrete set of values. Put another way, time (the exponential variable) is continuous, but counts (the Poisson variable) are discrete.

The relationship between the Poisson process and the exponential distribution is revealed by the third characteristic of the exponential distribution:

$$P(X \geq t) = 1 - F(t) = e^{-\lambda t} = P[N(t) = 0] \tag{3.7}$$

Thus, the probability that the random variable X is greater than or equal to some value t is identical to the probability that no events occur over an interval of length t. It should be apparent that these are two ways of stating the same thing. Furthermore, the independent increments property guarantees that interarrival times are mutually independent. Hence, we have another property of the Poisson process:

Property 2. The interarrival times for a Poisson process with rate λ are independent exponential random variables with mean $1/\lambda$.

Example

Customers are known to arrive at a medical clinic at the rate of 8 per hour. The receptionist is called away from his desk at 10:00, immediately after a customer arrived. How long can he stay away from his desk if he is willing to take a 50 percent chance of being away when the next customer arrives?

Solution The time until the next arrival has an exponential distribution with mean 1/8 hour. The receptionist would like to determine the value of x for which:

$$F(x) = 1 - e^{-(1/8)x} = .5$$

Inverting $F(x)$, this is equivalent to finding the value of x for which

$$x = \frac{-1}{8}\ ln(.5) = .087 \text{ hr} = 5.2 \text{ min}$$

Thus, if the receptionist is gone for 5.2 minutes, there is a 50 percent chance that a customer will arrive before he returns.

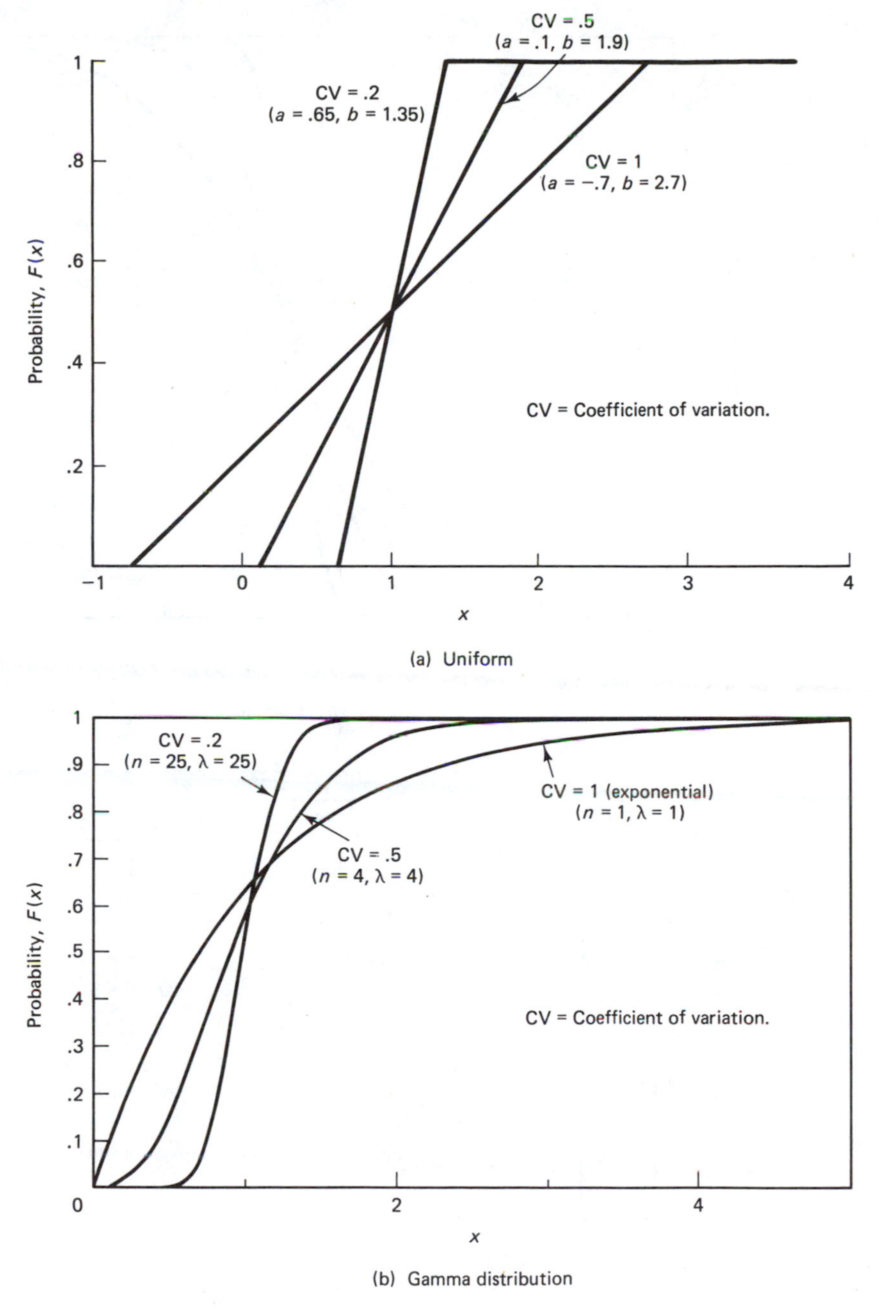

Figure 3.1 Comparison of probability distribution functions: (a) uniform, (b) gamma, (c) normal, and (d) Poisson.

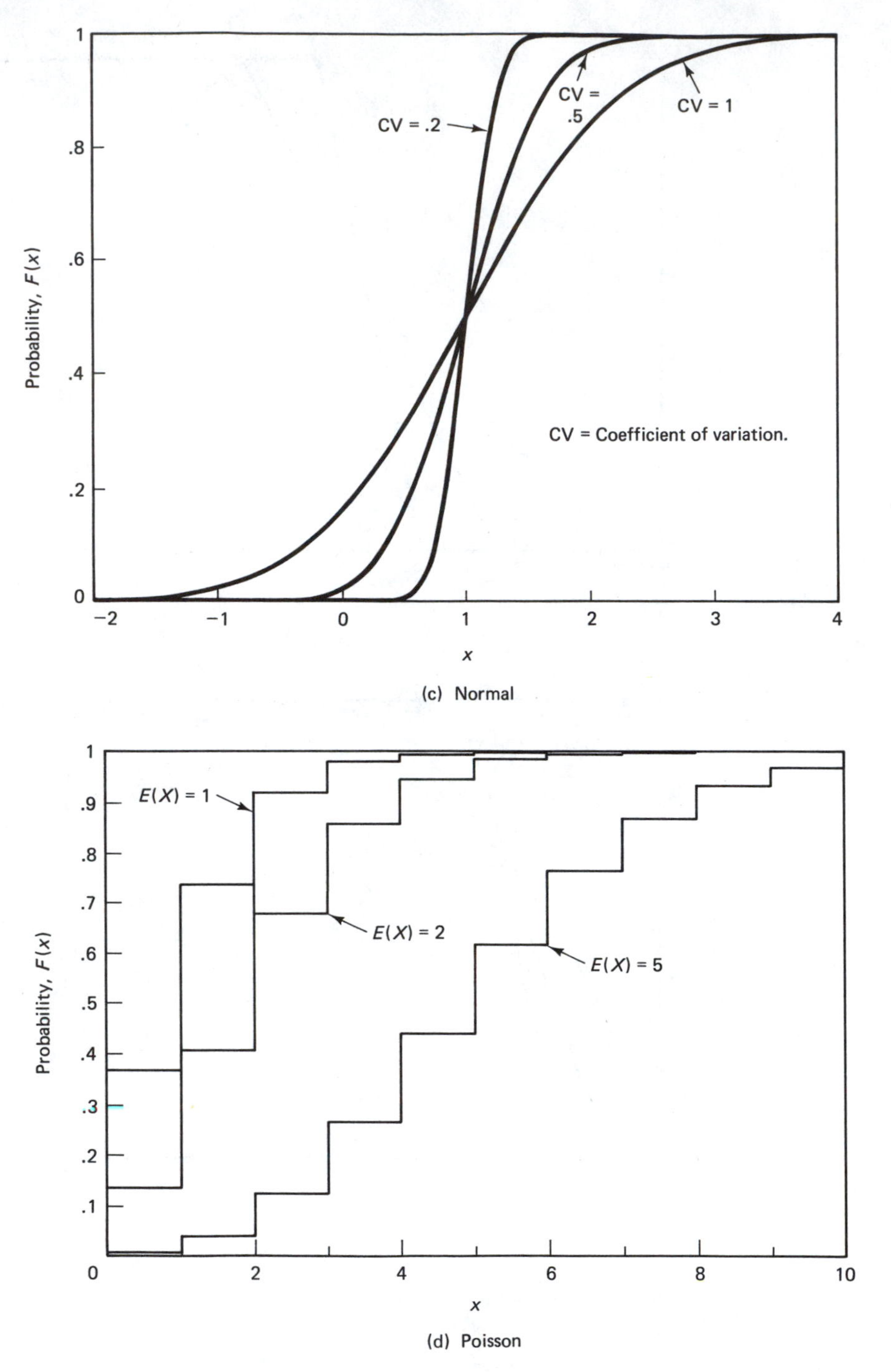

Figure 3.1 *(continued)*

3.3.2 Memoryless Property

A random variable is said to be memoryless if the time until the next event does not depend on how much time has already elapsed since the last event:

Definition 3.6

A random variable is ***memoryless*** if:

$$P\{X > s + t \mid X > t\} = P\{X > s\} \qquad \text{for all } s, t \geqslant 0$$

In the context of the Poisson process, the memoryless property applies to the interarrival time. Surely this is a logical consequence of independent increments and stationarity. What consequence should past arrivals have on future arrivals?

The memoryless property is easily proved for the exponential distribution (the distribution governing the time until the next arrival). From the law of conditional probability.

$$P\{X > s + t \mid X > t\} = \frac{P\{X > s + t \text{ and } X > t\}}{P\{X > t\}} = \frac{P\{X > s + t\}}{P\{X > t\}} \tag{3.8}$$

If the random variable X is greater than s plus t, it must also be greater than t. Substituting the exponential distribution function in Eq. (3.8), we obtain the following:

$$P\{X > s + t \mid X > t\} = \frac{e^{-\lambda(s+t)}}{e^{-\lambda t}} = e^{-\lambda s} = P(X > s) \tag{3.9}$$

as required. Thus

Property 3. The time until the next event does not depend on the elapsed time since the last event.

Example

Suppose that the receptionist in the previous example left his desk at 10:05, 5 minutes after the last arrival. Then, if he leaves his desk for 5.2 minutes, the probability that no one arrives is still .5, the same as if he had left at 10:00.

3.3.3 Time Until *n*th Event

So far, we know the probability distribution for the number of events within a set time interval, and we know the probability distribution for the length of time until the next event. What about the probability distribution for the time until the second, third, fourth, . . . event?

Definitions 3.7

X_n = the time between the n − 1st event and the nth event

Y_n = the time of the nth event

$= X_1 + X_2 + \ldots + X_n$

Derivation. The probability that Y_2 occurs before some time t equals the probability that $X_1 + X_2$ is less than t. Suppose that X_1 equals s, where s is some non-negative number less than t. Then Y_2 occurs before t if X_2 is less than $t - s$. Considering all the possible values of s between 0 and t:

$$\begin{aligned} P(Y_2 \leq t) &= \int_0^t P(X_1 + X_2 \leq t \mid X_1 = s) P(X_1 = s) ds \\ &= \int_0^t [1 - e^{-\lambda(t-s)}] \lambda e^{-\lambda s} ds \\ &= \int_0^t \lambda e^{-\lambda s} ds - \int_0^t \lambda e^{-\lambda t} ds \\ &= (1 - e^{-\lambda t}) - (\lambda t e^{-\lambda t}) \end{aligned} \tag{3.10}$$

Equation (3.10) gives the probability distribution function for Y_2. The probability density function, $f_{Y_2}(t)$, is the derivative of Eq. (3.10), $F_{Y_2}(t)$, with respect to t, and equals

$$f_{Y_2}(t) = \frac{dF_{Y_2}(t)}{dt} = \lambda^2 t e^{-\lambda t} \qquad t \geq 0 \tag{3.11}$$

Equation (3.11) is recognized as the density function of the ***gamma distribution*** with parameters $(2,\lambda)$. In general, the distribution for the sum of n successive interarrival times has the following property:

Property 4. The sum of n independent exponential random variables, each with mean $1/\lambda$, has the gamma distribution with parameters (n,λ).

The probability density function for the gamma distribution is the following:

Gamma Probability Density Function

$$f_{Y_n}(t) = \frac{\lambda^n}{\Gamma(n)} t^{n-1} e^{-\lambda t} \qquad t \geq 0 \tag{3.12}$$

where $\Gamma(n)$ is the ***gamma function*** (the gamma function is *not* a probability distribution function). When n is an integer (as will be the case in queueing analysis), the gamma function is defined as

$$\Gamma(n) = (n - 1)! \qquad n \geq 1 \text{ and integer} \tag{3.13}$$

The gamma distribution is continuous and defined over the domain $t \geq 0$. The mean and variance of the gamma distribution are

$$E(T) = \frac{n}{\lambda} \qquad V(T) = \frac{n}{\lambda^2} \tag{3.14}$$

The gamma distribution function is found by integrating Eq. (3.12):

Gamma Probability Distribution Function

$$F_{Y_n}(t) = 1 - \sum_{j=0}^{n-1} \frac{e^{-\lambda t}(\lambda t)^j}{j!} \qquad n \geqslant 1 \text{ and integer} \tag{3.15}$$

For noninteger values of n, the integration must be performed numerically. $F_{Y_n}(t)$ is shown in Fig. 3.1b, for $E(T) = 1$, and coefficients of variation of .2, .5, and 1 (the last also being an exponential distribution). Note that the distribution loses symmetry as the coefficient of variation increases.

Although the gamma distribution may appear complicated, it is a natural consequence of the conditions underlying the Poisson process—conditions that are quite plausible.

3.3.4 Event Times Within a Time Interval

The Poisson process has one more property that deserves mention. Suppose that n events are known to have occurred within a time interval. What is the probability distribution for the time of each event?

First, suppose that just one event occurred over the interval $[0,t]$, and let the time of this event be represented by the random variable Y_1. Then the following two statements, representing probabilities of joint events, are equivalent:

$$P\{[Y_1 < s] \text{ and } [N(t) = 1]\} = P\{[1 \text{ event} \in [0,s)] \text{ and } [0 \text{ events} \in [s,t]]\} \tag{3.16}$$

From the law of conditional probability, the first statement can also be expressed as

$$P\{[Y_1 < s] \text{ and } [N(t) = 1]\} = P[Y_1 < s \mid N(t) = 1] \cdot P[N(t) = 1] \tag{3.17}$$

Combining the expressions and substituting the Poisson probability distribution yield the following:

$$P[Y_1 < s \mid N(t) = 1] = \frac{(\lambda s e^{-\lambda s})(e^{-\lambda(t-s)})}{\lambda t e^{-\lambda t}} = \frac{s\, e^{-\lambda t}}{t e^{-\lambda t}} = \frac{s}{t} \tag{3.18}$$

Equation (3.18) is the *uniform* $[0,t]$ probability distribution function, meaning that the event is equally likely to occur at any time during the interval. This should come as no surprise, given the memoryless property of the Poisson process. More generally,

Property 5. If $N(t) = n$, the unordered event times are defined by $N(t)$ independent uniform $[0,t]$ random variables.

The uniform distribution has the following characteristics:

Characteristics of Uniform Distribution Defined over $[a,b]$

1. The uniform distribution is continuous. Depending on a and b, uniform random variables can be either negative or positive.
2. $f(x) = \begin{cases} 1/(b - a) & x \in [a, b] \\ 0 & \text{elsewhere} \end{cases}$
3. $F(x) = \begin{cases} 0, & x < a \\ (x - a)/(b - a) & x \in [a, b] \\ 1, & x > b \end{cases}$
4. $E(X) = \dfrac{b + a}{2}$
5. $V(X) = \dfrac{(b - a)^2}{12}$

Example distribution functions are shown in Fig. 3.1a. As with the gamma figure, the functions have a mean of 1 and coefficient of variations of .2, .5, and 1. Note that the uniform distribution is always symmetric about its mean and that the slope of the distribution function is constant over $[a,b]$.

Example

Three events are known to have occurred over a 2-hour period of a Poisson process. The probability that any one event occurred in the first half hour is $(1/2)/2 = .25$. The probability that all three events occurred in the first half hour is $.25^3 = 1/64$. The probability that none of the events occurred in the first half hour is $(1 - .25)^3 = 27/64$.

3.3.5 Summary

In this section we have seen that a few plausible assumptions (independence, stationarity, and arriving one at a time) lead to a number of important consequences:

1. The number of events within an interval of length t has a Poisson distribution with mean λt.
2. The time until the next event has the exponential distribution.
3. The time until the next event is independent of the elapsed time since the last event (memoryless property).
4. The time until the nth event has a gamma distribution with parameters (n,λ).

5. If $N(t) = n$, the unordered event times within the interval $[0,t]$ are defined by n independent uniform $[0,t]$ random variables.

These properties are used in the next section to check whether an observed process is indeed Poisson, and in Chap. 4 as the basis for simulating a queue.

3.4 GOODNESS OF FIT

''Goodness of fit'' is a term that describes how well a model fits the behavior of a system. Occasionally, a model precisely matches the behavior of a system, as in the binomial distribution matching flipped coins. But, more commonly, there are some differences. Differences are acceptable if the model does not differ appreciably from reality. The amount of difference that is acceptable is a matter of judgment and varies from situation to situation, depending on the needs for accuracy and simplicity.

Above all else, and independent of whatever data are recorded, goodness of fit should be judged according to whether or not the model's assumptions are plausible for the system studied. This standard is best appreciated by considering situations that do *not* conform to the Poisson process.

Example 1

An observer is stationed at the end of the runway at Newark International Airport. Over the 1-hour period from 5:00 to 6:00 P.M., the observer records the time that the front wheels of each airplane touch the runway.

Airplanes certainly land one at a time. And the probability that an airplane arrives at any particular time does not vary appreciably over the hour. But arrivals are not independent of each other. Safety dictates a minimum spacing between planes. If a plane landed at 5:00, the next plane would not land until at least 90 seconds later. The process does not possess independent increments and it is not Poisson.

Example 2

Job candidates are scheduled for interviews with a personnel department. One candidate is scheduled for each hour of the day, beginning at 8:00. An observer records the time that each candidate arrives.

The job candidates most likely arrive independently of each other, one at a time. But the process is not stationary. Arrival times depend on appointment times. The likelihood that a candidate arrives at 7:55 is much greater than the likelihood that a candidate arrives at 8:15. Again, the process is not Poisson. (If the spacing between appointments is small, and arrival times are somewhat random, the process would behave *locally* like a Poisson process. That is, the interarrival times would be approximately exponential, but the number of arrivals over long time intervals would not be Poisson. See Newell 1982.)

Example 3

An observer records the time that visitors arrive at the San Francisco zoo during the period 10:00 to 11:00 on a weekday morning. Most of the visitors are members of school groups.

If it takes no longer to serve a group of visitors than an individual visitor, then the group would be considered the customer and the arrival process might be Poisson. However, if visitors are served individually, the arrival process would not be Poisson because visitors do not arrive one at a time.

Example 4

An observer records the arrivals of bank patrons at an automated teller machine over a 1-hour period from 10:00 to 11:00 on a weekday morning.

Bank patrons tend to arrive one at a time, and arrivals tend to be mutually independent. The rate at which they arrive may vary over the day. However, the arrival rate may be fairly constant over a 1-hour period from 10:00 to 11:00. The Poisson process is *plausible*.

The bank is the only one of the four examples for which the Poisson process is plausible. But plausibility does not imply that the process is definitely Poisson. It only means that the process seems to be Poisson. If there is any doubt, final verdict should not be passed until quantitative tests are performed on the data, as discussed in the following section.

If the Poisson process is not plausible, all is not lost. There are many other models for the arrival process, any of which might be appropriate for a given situation. Some of these are described in Chap. 6. The Poisson process is emphasized here because it is a reasonable model for many systems and because its simplicity facilitates evaluation of queueing system performance.

3.5 QUANTITATIVE GOODNESS OF FIT TESTS

Suppose that the Poisson process is a plausible model for customer arrivals, but you are not positive that it is the right model. Your next step is to test the arrival data to see how well they conform to the Poisson process. Quantitative testing seeks to answer two key questions:

1. Whether or not the model is correct, and
2. If the model is not correct, why is it not correct?

The answer to the latter question provides guidance in creating an alternative model that more closely matches the data.

Tests can be performed for any of the properties of the Poisson process. However, because some of the properties are implied by others, there is no need to test them all. For example, if the interarrival times are found to be independent exponential random variables, pairs of interarrival times do not have to be tested to see if they are gamma random variables.

The tests are divided into two categories: graphical tests and statistical tests. Graphical tests are less rigorous than statistical tests. Yet they are very effective at identifying patterns in the data. Statistical tests impose standards for whether or not the data conform to the properties of the Poisson process. In both cases, the tests merely check to see whether the data conform with the properties of the Poisson process. Whether the conditions will recur in the future is a matter of inference.

3.5.1 Graphical Tests

Perhaps the best way to determine whether or not the arrival process is Poisson is to plot and examine the data for patterns. For the graphs described here, absence of pattern generally is an indication that the process is Poisson; presence of pattern generally indicates that the process is not Poisson. If a pattern is found, a cause should be sought. If the process is not stationary, why is it not stationary? If the interarrival times are not exponential, why are they not exponential? If the interarrival times are not independent, why are they not independent? Understanding the process creating the data is by far the most effective means for deciding whether the model is correct or whether to design an alternative model.

Stationarity: Cumulative. Plot the cumulative arrival curve. Draw a straight line connecting the points $A(0)$ and $A(T)$, where T is the end of the time interval observed. There should be no visible pattern to the deviations of $A(\mathrm{T})$ around the straight line, and the difference between $A(T)$ and the straight line should be small for all values of t.

Stationarity/Independence: Interarrival Times. Plot the interarrival times in serial order on an (x,y) graph. The points should be scattered randomly about the line $y = \bar{X}$, where $\bar{X}$ is the average interarrival time. There should be no cyclic patterns.

Independence. Plot the Points $\{(X_n, X_{n-l}), n = 2, 3, \ldots\}$ on graph paper, where X_n represents the nth interarrival time. It should be impossible to approximate the data by a straight line or any other regular curve.

Exponential Interarrival Distribution. Plot the empirical probability distribution for the interarrival times. On the same piece of paper, plot the exponential distribution with $\lambda = A(T)/T$, where T is the length of the time interval observed. There should be no visible pattern of the deviations between the distributions.

Demonstration. The date in Table 3.1 were collected at the automatic teller machine for a large bank branch in Berkeley, California. The cumulative arrival curve is shown in Fig. 3.2. The arrival curve varies above and below the straight line, in a fairly random fashion. The interarrival times in Fig. 3.3 also show no time varying pattern. Figure 3.4 plots (X_n,X_{n-1}) for 97 data points (number of arrivals minus 1). The downward slope of the *boundary* of the data points makes the data appear to have a negative correlation. However, the data points themselves do not slope downward, and again there is no visible pattern. Finally, the empirical probability distribution for the interarrival time

TABLE 3.1 ARRIVAL, DEPARTURE, AND WAITING TIMES AT AUTOMATIC TELLER MACHINE (MINUTES, TIME 0 = 9:30)

n	$A^{-1}(n)$	$S(n)$	$D_q^{-1}(n)$	$D_s^{-1}(n)$	$W_q(n)$	$W_s(n)$
1	0.61	0.73	0.61	1.34	0.00	0.73
2	0.86	0.85	1.34	2.19	0.48	1.33
3	1.09	1.75	2.19	3.94	1.10	2.85
4	5.61	0.95	5.61	6.56	0.00	0.95
5	6.59	0.77	6.59	7.36	0.00	0.77
6	6.90	0.82	7.36	8.17	0.46	1.27
7	7.71	0.70	8.17	8.87	0.46	1.16
8	7.93	0.72	8.87	9.59	0.94	1.66
9	7.93	0.73	9.59	10.32	1.66	2.39
10	8.39	0.87	10.32	11.19	1.93	2.80
11	9.22	0.73	11.19	11.92	1.97	2.70
12	9.29	0.60	11.92	12.52	2.63	3.23
13	12.92	0.63	12.92	13.55	0.00	0.63
14	14.57	0.53	14.57	15.10	0.00	0.53
15	17.67	0.95	17.67	18.62	0.00	0.95
16	20.15	0.80	20.15	20.95	0.00	0.80
17	20.39	0.87	20.95	21.82	0.56	1.43
18	20.83	0.75	21.82	22.57	0.99	1.74
19	20.99	1.12	22.57	23.68	1.58	2.69
20	21.31	1.88	23.68	25.57	2.37	4.26
21	21.39	0.88	25.57	26.45	4.18	5.06
22	22.96	0.78	26.45	27.23	3.49	4.27
23	24.03	0.68	27.23	27.92	3.20	3.89
24	24.15	0.48	27.92	28.40	3.77	4.25
25	24.32	0.82	28.40	29.22	4.08	4.90
26	25.40	0.90	29.22	30.12	3.82	4.72
27	25.77	0.72	30.12	30.83	4.35	5.06
28	30.39	0.52	30.83	31.35	0.44	0.96
29	30.91	1.30	31.35	32.65	0.44	1.74
30	34.43	0.73	34.43	35.16	0.00	0.73
31	34.54	1.15	35.16	36.31	0.62	1.77
32	35.27	0.62	36.31	36.93	1.04	1.66
33	35.39	0.47	36.93	37.40	1.54	2.01
34	36.38	0.77	37.40	38.16	1.02	1.78
35	40.39	0.73	40.39	41.12	0.00	0.73
36	41.61	0.83	41.61	42.44	0.00	0.83
37	44.77	0.53	44.77	45.30	0.00	0.53
38	46.09	0.62	46.09	46.71	0.00	0.62
39	48.85	0.67	48.85	49.52	0.00	0.67
40	51.40	1.45	51.40	52.85	0.00	1.45
41	52.87	0.63	52.87	53.50	0.00	0.63
42	53.84	0.77	53.84	54.61	0.00	0.77
43	55.09	1.08	55.09	56.17	0.00	1.08
44	55.64	0.92	56.17	57.09	0.53	1.45
45	55.83	0.57	57.09	57.66	1.26	1.83
46	57.56	0.85	57.66	58.51	0.10	0.95
47	58.65	0.63	58.65	59.28	0.00	0.63
48	59.83	0.72	59.83	60.55	0.00	0.72
49	61.91	0.57	61.91	62.48	0.00	0.57

TABLE 3.1 *(continued)*

n	$A^{-1}(n)$	$S(n)$	$D_q^{-1}(n)$	$D_s^{-1}(n)$	$W_q(n)$	$W_s(n)$
50	64.71	0.57	64.71	65.28	0.00	0.57
51	64.95	0.95	65.28	66.23	0.33	1.28
52	70.20	1.37	70.20	71.57	0.00	1.37
53	71.18	1.48	71.57	73.05	0.39	1.87
54	72.15	0.68	73.05	73.73	0.90	1.58
55	72.72	0.72	73.73	74.45	1.01	1.73
56	72.94	0.65	74.45	75.10	1.51	2.16
57	74.74	0.63	75.10	75.73	0.36	0.99
58	75.12	0.58	75.73	76.32	0.61	1.20
59	75.64	0.65	76.32	76.97	0.68	1.33
60	75.77	0.68	76.97	77.65	1.20	1.88
61	76.95	0.72	77.65	78.37	0.70	1.42
62	78.36	0.85	78.37	79.22	0.01	0.86
63	79.98	0.80	79.98	80.78	0.00	0.80
64	80.21	0.87	80.78	81.65	0.57	1.44
65	81.93	0.67	81.93	82.60	0.00	0.67
66	82.75	0.73	82.75	83.48	0.00	0.73
67	82.95	0.60	83.48	84.08	0.53	1.13
68	83.35	0.65	84.08	84.73	0.73	1.38
69	84.55	0.68	84.73	85.42	0.18	0.87
70	84.72	0.70	85.42	86.12	0.70	1.40
71	85.28	1.00	86.12	87.12	0.84	1.84
72	86.17	0.60	87.12	87.72	0.95	1.55
73	87.58	0.87	87.72	88.58	0.14	1.00
74	87.85	1.25	88.58	89.83	0.73	1.98
75	88.33	0.45	89.83	90.28	1.50	1.95
76	88.42	1.00	90.28	91.28	1.86	2.86
77	92.17	0.67	92.17	92.84	0.00	0.67
78	92.97	0.85	92.97	93.82	0.00	0.85
79	93.63	0.83	93.82	94.65	0.19	1.02
80	96.34	0.58	96.34	96.92	0.00	0.58
81	96.53	1.85	96.92	98.77	0.39	2.24
82	97.39	1.00	98.77	99.77	1.38	2.38
83	97.88	0.77	99.77	100.54	1.89	2.66
84	99.64	0.90	100.54	101.44	0.90	1.80
85	103.82	0.77	103.82	104.59	0.00	0.77
86	104.88	0.65	104.88	105.53	0.00	0.65
87	105.12	1.35	105.53	106.88	0.41	1.76
88	105.66	0.67	106.88	107.55	1.22	1.89
89	106.92	0.93	107.55	108.48	0.63	1.56
90	107.55	1.00	108.48	109.48	0.93	1.93
91	110.02	0.90	110.02	110.92	0.00	0.90
92	113.14	2.35	113.14	115.49	0.00	2.35
93	114.12	0.73	115.49	116.22	1.37	2.10
94	115.01	0.48	116.22	116.71	1.21	1.70
95	116.32	0.58	116.71	117.29	0.39	0.97
96	117.50	0.55	117.50	118.05	0.00	0.55
97	119.18	0.55	119.18	119.73	0.00	0.55
98	119.71	0.85	119.73	120.58	0.02	0.87

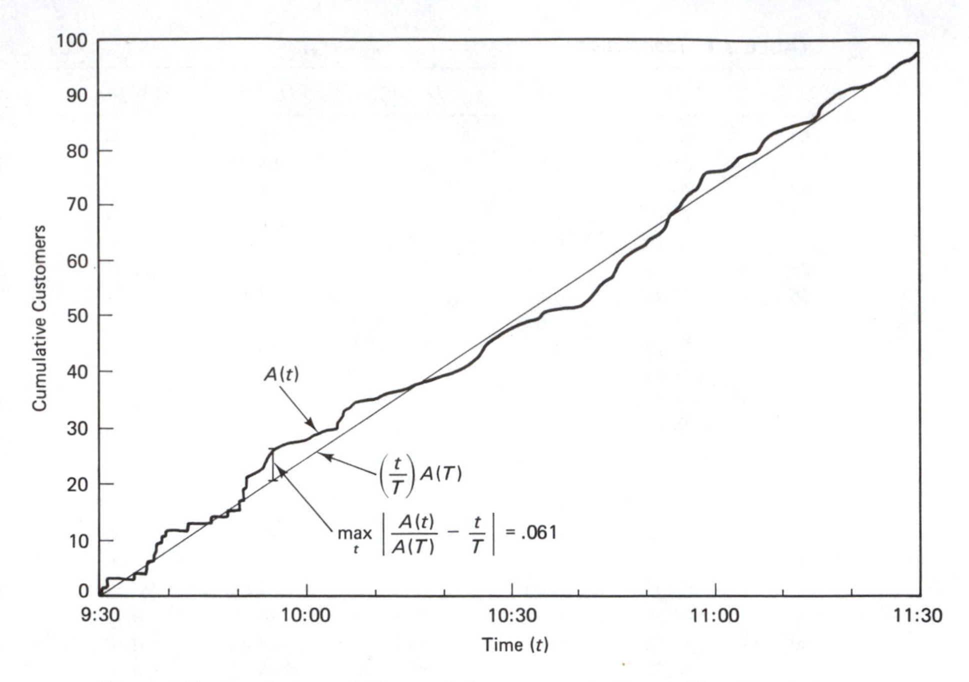

Figure 3.2 Cumulative arrivals recorded at automated teller machine. If arrivals are stationary, the deviation from the diagonal line should be small (see Kolmogorov-Smirnov test), as shown.

is shown in Fig. 3.5. The deviations are small and again show no pattern. Based on these four graphs, there is no reason to believe that the process is not Poisson.

Consider a second example, which is not Poisson. Table 3.2 provides data on arrival times of elevators at the lobby of a high-rise building on the University of California campus. The plot of the interarrival time distribution in Fig. 3.6 appears to be exponential. Though not shown, the cumulative arrival plot also appears to be stationary. However, the plot of paired interarrival times (Fig. 3.7) reveals a subtle dependency. It appears that short interarrival times tend to be followed by long interarrival times, and vice versa (note that fewer points are concentrated near the origin and more points are further out near the axes). There is a negative correlation between (X_n, X_{n-1}).

Careful observation of the arrival process should reveal the cause of the pattern. There is a common phenomenon in many modes of transportation (buses and elevators, for example) known as "bunching." The number of people who board a vehicle depends on how much time has elapsed since the previous vehicle arrived (the longer the time, the more people). If this separation happens to decline, fewer people will board the second vehicle, and the vehicle will begin to travel faster, further reducing the separation. Eventually, the second vehicle will catch up with the first vehicle, and from then on the two will travel in a pair, perhaps "leap frogging" from stop to stop. Bunching is common in elevators and explains why the interarrival times are not independent. Hence, the arrival process is not Poisson, and a different model is called for.

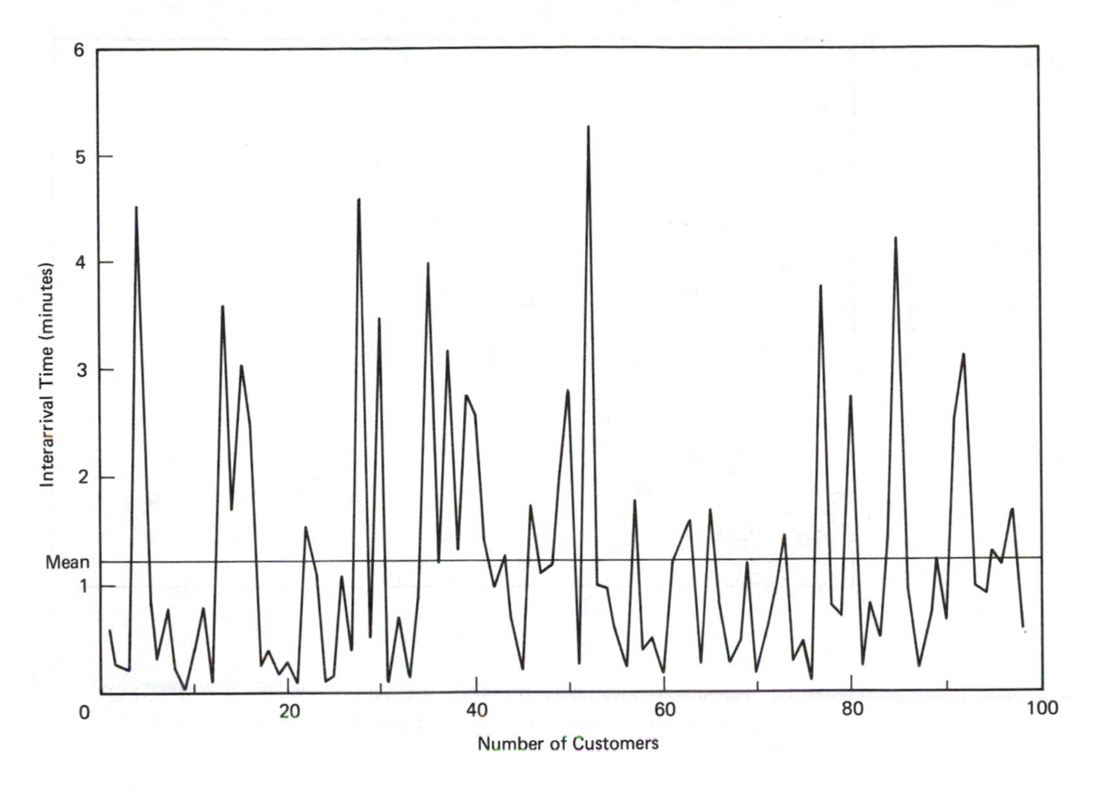

Figure 3.3 Interarrival times at automated teller machine. If arrival process possesses independent increments, there should be no cyclic pattern to the data, as shown.

The point to recognize is that an unexpected pattern is reason for further investigation. The system should be observed to identify what factors are causing the pattern. If the factors are significant, they should be incorporated into the model.

3.5.2 Statistical Tests

Graphs alone may leave doubt as to whether the process is Poisson. Perhaps a small deviation was found, but you have no idea what caused it. You may wonder how much deviation is acceptable. Where is the line drawn between what is a Poisson process and what is not? Statistical tests help in these situations.

Definition 3.8

A ***statistic*** is a function of the data. Like the data, a statistic is itself a random variable. The sample average, sample standard deviation, and maximum and minimum are all examples of statistics.

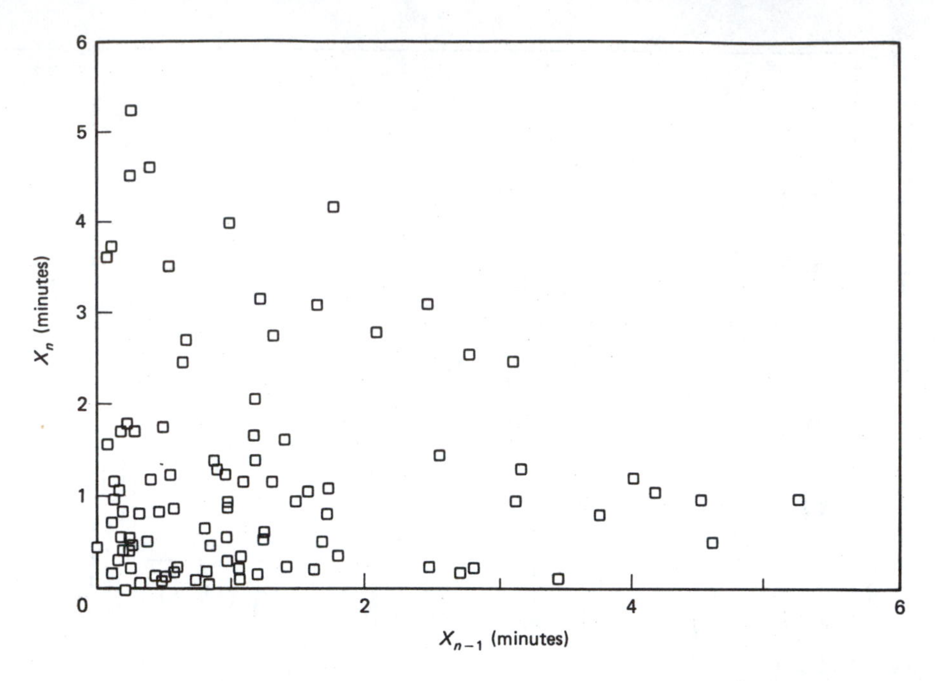

Figure 3.4 Paired successive interarrival times at automated teller machine. If arrival process possesses independent increments, the data cannot be approximated by a straight line, as shown.

Statistical tests are phrased in the form of a ***hypothesis*** (denoted by the letter H). For example, suppose that you have made a bet with a friend. Each time she flips a coin and it turns up heads, she wins a dollar. Each time it turns up tails, you win a dollar. Now suppose that after ten flips the coin has turned up heads nine times. You may then wonder whether the coin she flipped is fair. That is, you may wonder whether the following hypothesis, H, is true:

H: Probability of heads $= .5$

or whether antihypothesis, A, is true instead;

A: Probability of heads $> .5$

The statistical test only suggests whether or not the hypothesis is true; it does not provide conclusive evidence. It does this by *calculating the probability of obtaining the observed data, given that the hypothesis is true*. In the case of the coin flip, you would be interested in the probability of obtaining one or fewer heads in ten trials, given the probability of .5. From the binomial distribution, this probability can be expressed as follows:

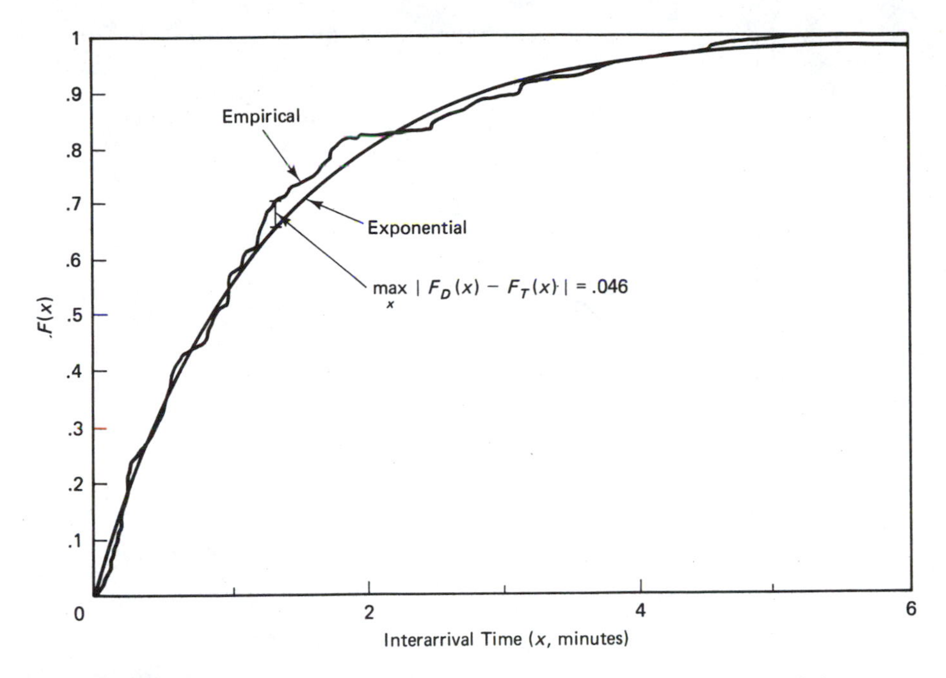

Figure 3.5 Interarrival time distribution at automated teller machine. For Poisson process, deviation from the exponential distribution should be small, as shown.

$$\begin{aligned} P(\text{data} \mid H \text{ is true}) &= P(1 \text{ or fewer heads in 10 trials} \mid p = .5) \\ &= \binom{10}{0} .5^0 (1 - .5)^{10} + \binom{10}{1} .5^1 (1 - .5)^9 \\ &= .011 \end{aligned}$$

The probability of seeing one or fewer heads is only .011, so there is reason to be suspicious that the coin (and perhaps your friend) is unfair. Nevertheless, because the probability is greater than zero, one cannot say unequivocally that the coin is unfair.

In statistics, the $P(\text{data} \mid H \text{ is true})$ is expressed as a ***significance level***, which is merely a way of rounding off the probability. In the example, the probability would be rounded up to .02, and one would say that "the hypothesis is rejected at the 2% significance level." This statement does not imply that the hypothesis is absolutely

TABLE 3.2 ARRIVAL TIMES OF ELEVATORS (MINUTES)

.8	1.2	2.8	3.2	4.1	7.5	8.7	9.7	10.2	11.1	13.5
15.0	16.2	16.2	19.1	21.9	22.1	23.5	24.1	26.0	27.0	27.1
27.9	20.7	31.0	31.4	33.3	33.8	36.2	36.3	40.0	40.2	41.1
42.6	44.8	45.1	45.6	45.8	49.0	51.2	51.8	53.5	53.8	54.5
55.0	56.5	56.8								

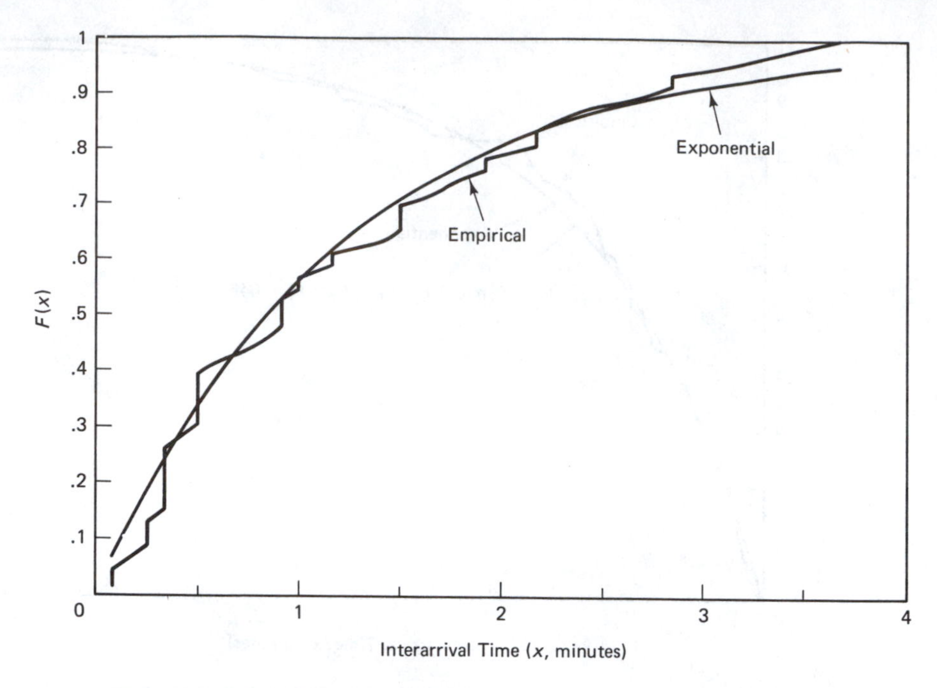

Figure 3.6 Interarrival time distribution for elevators appears to be exponential.

rejected. It only means that the data have a likelihood of less than 2%. As a matter of custom, $P(\text{data} \mid H \text{ is true})$ is compared to a significance level of 1%, 2%, or 5%. If the probability falls below .05, then it is rounded up to the next highest significance level and the hypothesis is rejected at that level. If the probability falls above .05, then we would say that "the hypothesis is not rejected at the 5% significance level." Ordinarily, one never states that a hypothesis is accepted, because one can never be completely sure.

Statistical tests are effective at identifying when something is amiss but not at finding solutions. They offer little guidance as to what is right. Therefore, they are no substitute for plotting the data and checking to see how the data actually behave.

Statistical tests are provided for the following hypotheses, all of which must be true if the arrival process is Poisson:

H_1: The interarrival times have an exponential probability distribution.

H_2: The interarrival times are independent.

H_3: The unordered arrival times have a uniform distribution over the period of observation.

If none of the three hypothesis is rejected *and* the graphical tests show no abnormal pattern *and* the assumptions of the Poisson process are plausible, then it is safe to assume that the observed arrival process was indeed Poisson. However, just because a data set does not

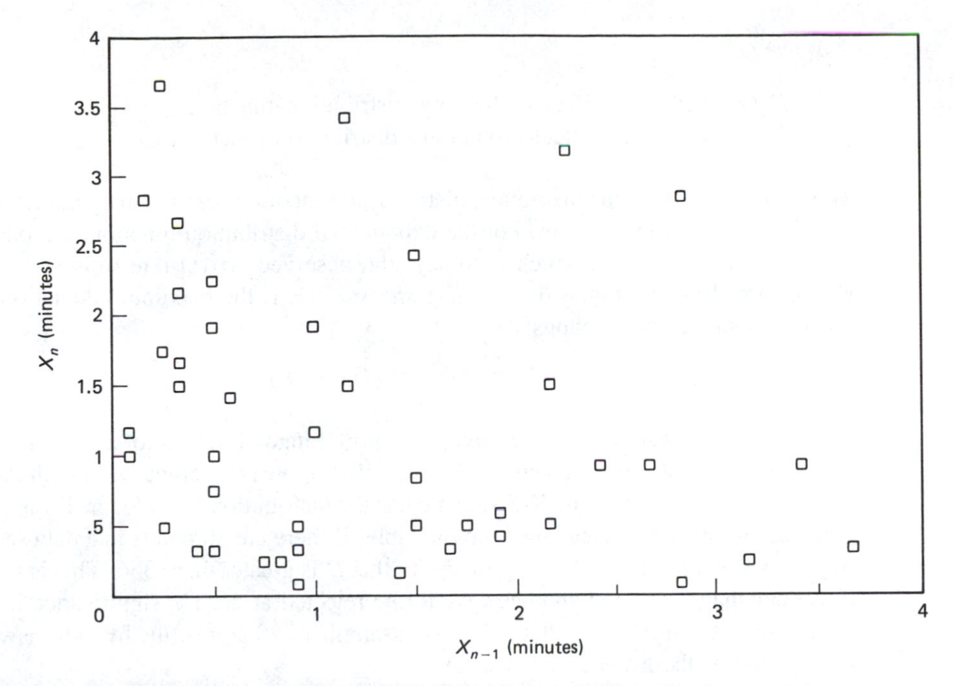

Figure 3.7 Paired successive interarrival times of elevators. Negative correlation revealed by points along axes.

conform exactly to the assumptions of the Poisson process does not imply that the Poisson process is not a reasonable model. The simplicity of the Poisson model may outweigh any gain in accuracy from using a more complicated model.

3.5.2.1 H_1: Interarrival Time Distribution. The two most common tests for whether a data set conforms to a theoretical probability distribution are the Kolmogorov-Smirnov (K-S) test and the chi-squared test. The K-S test is based on deviations between the empirical probability distribution function and the theoretical distribution function. The chi-squared test is based on deviations between a data histogram and the theoretical density function. Because the K-S test is the more powerful of the two (and also the simpler), it will be the only test presented here. Books listed at the end of this chapter provide further information on the chi-squared and other distribution tests.

The Kolmogorov-Smirnov test can be applied to any set of continuous, independent, random variables. It is based on the maximum deviation between the empirical distribution function and a theoretical distribution function. If the theoretical distribution is correct, this deviation should be small. The K-S test is an example of a *one-tail test* because a large deviation suggests that the model is incorrect and a small deviation suggests that the model is correct—in contrast to a *two-tail test* in which either a small or large value would cause concern.

Let

$F_D(x)$ be the empirical probability distribution function.
$F_T(x)$ be the theoretical probability distribution function.

The specific shape of the theoretical distribution function is defined by parameters, which may be derived from the data. For the exponential distribution function, the parameter λ can set equal to $A(T)/T$, which is to say, the observed arrival rate (this is discussed in greater detail in a section 3.6). The *K-S statistic, D*, is the maximum deviation between the two distribution functions:

$$D = \max_x | F_D(x) - F_T(x) | \tag{3.19}$$

The maximum must be computed over the entire range of values for x, including $x = 0$. Also, because $F_D(x)$ is a step curve, $F_D(x) - F_T(x)$ must be computed at both the top and the bottom of each step. The K-S statistic has the distribution provided in Table A.4 in the appendix to this book. Reading from the table, if there are 20 points in a data set, and the hypothesis is true, there is a .01 probability that D is greater than .356. This is to say, if D is greater than .356 the hypothesis would be rejected at the 1% significance level. Also from Table A.4, if D is less than .294 for a sample of 20 points, the hypothesis would not be rejected at the 5% significance level.

Example

The maximum deviation between the empirical distribution function and the theoretical distribution function in Fig. 3.5 is .046 (when $x = 1.3$). There are 97 interarrival times in the data set. From Table A.4 in the appendix, .046 is below the critical value .138 ($1.36/\sqrt{97}$) and the hypothesis is not rejected at the 5% significance level. The data are not unusual for a Poisson process.

3.5.2.2 H_2: Interarrival Time Independence. Independence is one of the most difficult characteristics to test because it has such broad implications. It is possible for an interarrival time, X_n, to be independent of X_{n-1} but not independent of X_{n-2}, X_{n-3}, ... Thus, each statement X_n is independent of X_{n-1} ($n = 2, 3, \ldots$), X_n is independent of X_{n-2} ($n = 3, 4, \ldots$), and so on, might require a separate test. There is really no limit to the types of interdependencies that might be checked. However, unless there is some reason to believe that a complicated form of dependency exists, the test for independence is usually limited to checking whether X_n is independent of X_{n-1}.

Just as there is a chi-squared test for distributions, there is a chi-squared test for independence. The test is based on categorizing the data into a contingency table and comparing the number of observations in each category to the expected number of observations. The numbers should be similar if the hypothesis is correct. The unfortunate part of this approach is that diffusing the data into categories necessitates that a large data set be collected for the test to be powerful.

An alternative to the chi-squared test for independence is the *t-test*. This checks the following weaker hypothesis: *the correlation coefficient between X_n and X_{n-1} equals* *zero*. Independent random variables must have a correlation coefficient of zero, but the reverse is not necessarily true. Figure 3.8 plots data points that are not independent, yet still have zero correlation. However, such a pattern would be most unusual in observing queues, and even if it did materialize, it would be easily identified by plotting the data. Hence, if it can be shown that the correlation coefficient is zero between pairs of successive interarrival times, then the data are likely independent.

The correlation coefficient between two data sets, $\{X_1, X_2, \ldots, X_N\}$ and $\{Y_1, Y_2, \ldots, Y_N\}$, is the ratio of the covariance to the product of the standard deviations.

Definition 3.9

r_{xy} is the *sample correlation coefficient*

$$= \frac{\sum_{n=1}^{N} (X_n - \bar{X})(Y_n - \bar{Y})}{(N - 1)s_x s_y} \tag{3.20}$$

where $\bar{X}$ and $\bar{Y}$ are sample averages, and s_x and s_y are *sample standard deviations*:

$$s_x = \sqrt{\frac{\sum_{n=1}^{N} (X_n - \bar{X})^2}{N - 1}} \tag{3.21a}$$

$$s_y = \sqrt{\frac{\sum_{n=1}^{N} (Y_n - \bar{Y})^2}{N - 1}} \tag{3.21b}$$

$N - 1$ is used instead of N in the denominator to obtain an unbiased estimate of the standard deviation. That is, the expectation of s_x and s_y equals σ_x and σ_y. The correlation coeffficent can be any value between -1 and $+1$.

The statistical test for correlation is not performed directly on r_{xy}. Rather, it is performed on the transformation of r_{xy} into the t statistic, as shown below (Note: this value of t has nothing to do with the value of t used in Sec. 3.3):

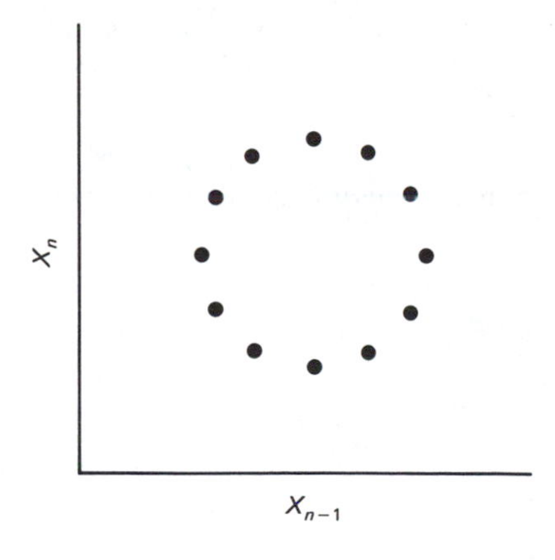

Figure 3.8 Dependent random variables might not be correlated, as shown here.

$$t = \frac{r_{xy}\sqrt{N - 2}}{\sqrt{1 - r_{xy}^2}} \tag{3.22}$$

If the random variables X and Y are independent and have normal distributions, then the statistic t should have the ***t* probability distribution** with $N - 2$ degrees of freedom (see Table A.3 in the appendix to this book). In the case of the Poisson process, X and Y have the exponential distribution, so the t-test is only approximate.

The t-test is a two-tail test because either positive correlation (r_{xy} close to 1) or negative correlation (r_{xy} close to -1) is cause for concern. The significance level is defined by the probability that $|t|$ is greater than or equal to the observed value of t, given that the hypothesis is true.

Example

The correlation coefficient (X_n, X_{n-1}) for the teller data set was calculated as $-.032$. There are 96 *pairs* of interarrival times in the data set (one less than the number of interarrival times), so $N = 96$. Substitution of $-.032$ for r_{xy} and 96 for N in Eq. (3.22) gives $t = -.31$.

Rounding N to 120 in Table A.3, $P(t < 1.98)$ equals .975. Because the t distribution is symmetric, the $P(|t| > 1.98)$ equals .05. Because $|-.31|$ is well below 1.98, the hypothesis that the correlation coefficient equals zero *cannot be rejected* at the 5% significance level.

When the data set has 50 or more data points, the normal distribution (Table A.2) can be used in place of the t distribution. The normal distribution is identical to the last line of Table A.3 for the t distribution ($N = \infty$).

3.5.2.3 Arrival Time Uniformity. If the arrival process is stationary, the unordered arrival times should conform to the uniform distribution. That is, the empirical probability distribution should be approximately a straight line passing through the points (0,0) and $(T,1)$, where T is the length of the time period observed. Though it was not introduced as such, the empirical probability distribution for arrival times is simply the cumulative arrival diagram (Fig. 3.2), with the vertical axis rescaled from 0 to 1.

The Kolmogorov-Smirnov test can be used again. The K-S statistic, D, is calculated as follows:

$$D = \max_t \left| \frac{A(t)}{A(T)} - t/T \right| \tag{3.23}$$

A large value of D would suggest that the arrival process is not stationary.

Example

For the sample teller data, $D = .061$, which occurs at time 9:55. There are 98 arrival times in the data set. Because .061 is less than the critical value, .137 ($1.36/\sqrt{98}$), the hypothesis cannot be rejected at the 5% significance level.

3.5.2.4 A Quick Test. As mentioned earlier, the coefficient of variation (ratio of standard deviation to mean) for the exponential distribution is 1. One of the easiest checks for whether an arrival process is Poisson is to calculate this value and compare it to 1.

Example

The standard deviation of the interarrival times is 1.18 and the average is 1.22. Therefore, the coefficient of variation is .97, which is nearly 1.

3.5.2.5 Interpretation of Statistical Tests. Models are rarely perfect representations of reality. Occasional disturbances might disrupt the usual arrival pattern—some customers might arrive in pairs or the arrival rate might fluctuate slightly. Small imperfections such as these are difficult to detect in small data samples, yet will certainly arise if the sample is large enough. This means that a slightly imperfect model will invariably be statistically rejected if the data sample is sufficiently large.

For practical purposes, one should not discard slightly imperfect models, only grossly imperfect models. Just because a model is statistically rejected does not mean it should not be used, particularly if the data sample is large. It does mean though, that the cause of the imperfection should be identified and considered for inclusion in the model. If the gain in accuracy does not justify the added complexity, then the cause should not be included. Like many aspects of modeling, this is a matter of judgment.

3.6 PARAMETER ESTIMATION

Section 3.5 discusses how to determine whether the Poisson process is a good model for an observed arrival process. Suppose that the arrival process passes the test of plausibility, passes the graphical tests, and passes the statistical tests, as do the teller data. There is still one more task to undertake before the model is complete: ***parameter estimation***. The Poisson process is not a single model but actually a family of models, defined by different values of the parameter λ. The model is complete when the value of λ is estimated.

Parameter estimation is somewhat different from assessing goodness of fit because it is not a matter of answering a simple yes or no question. λ can be any real positive number, and there usually is no a priori reason to believe that λ should be any particular value. So parameter estimation normally does not appeal to one's knowledge of the underlying arrival process. Rather, it depends almost exclusively on analysis of the data. Along these lines, the accuracy of a parameter estimate is a matter of degree rather than simply true or false. If the parameter estimate is 10, the true value (the value of λ that created the data) might be 10.1, 9.5, or even 13.6. This is because the parameter estimate, or the ***estimator***, is itself a random variable.

Parameter estimation is the process of determining the best estimate of a distribution's parameter or parameters, given the observed data. Usually, this means that the parameter estimate should be precise (that is, the standard deviation of the estimator is small) and the parameter estimate is ***unbiased*** (that is, the expected value of the estimator equals the true vale of the parameter).

The simplest estimation technique is called the ***method of moments*** (***MOM***), which amounts to equating sample moments to population moments, such as $E(X)$ or $E(X^2)$. The method of moments was used in the previous section to estimate λ, the arrival rate. An estimator is denoted by the caret symbol. Thus, $\hat{\lambda}$ is the estimate for the true value of λ. For the method of moments

$$\hat{\lambda} = \frac{A(T)}{T} \tag{3.24}$$

For the teller data, $\hat{\lambda}$ equals $98/120 = .817$. This estimator is the guess for the true value of λ, the value of λ that actually created the data.

A more rigorous technique for estimating parameters is the ***maximum likelihood method***. This approach determines the parameter values that are most likely to generate the observed data sample. The maximum likelihood estimator (MLE) tends to be more precise than the method of moments estimator. However, determining the MLE may be more difficult, partly because it depends on the probability distribution for the random variable.

There are two ways to determine the MLE for λ. The first is based on the Poisson distribution for the number of events, and the second is based on the exponential distribution for the interarrival times.

Definition 3.10

The ***likelihood function,*** $\boldsymbol{L}$**(data |** $\boldsymbol{\lambda}$**)**, specifies the probability density function for the observed data, given that the parameter equals λ.

Suppose that $N = A(T)$ arrivals are observed over a time interval of length T. Then the likelihood of observing these data is defined by the Poisson probability function as follows:

$$L(N \text{ arrivals} \mid \lambda) = \frac{(\lambda T)^N}{N!} e^{-\lambda T} \tag{3.25}$$

The MLE is the value of λ that maximizes the likelihood function in Eq. (3.25). For the particular expression above, the MLE is found by determining the point where the derivative *with respect to the parameter* λ equals 0:

$$\frac{\partial L}{\partial \lambda} = N \frac{\lambda^{N-1} T^N}{N!} e^{-\lambda T} - T \frac{(\lambda T)^N}{N!} e^{-\lambda T} = 0$$

$$[N/\lambda - T] \frac{(\lambda T)^N}{N!} e^{-\lambda T} = 0$$

$$\hat{\lambda} = \frac{N}{T} = \frac{A(T)}{T} \tag{3.26}$$

For the Poisson process, the method of moments estimator and the Poisson distribution MLE estimator happen to be the same. This will not be the case for all estimators, such as the estimator for the standard deviation of a normal distribution.

The second likelihood function, based on the exponential distribution, requires more data than the first, because it depends on the interarrival times, not just the number of events. It also differs because X_n is a continuous, rather than a discrete, random variable. Because the likelihood function is a probability density function, it is not a true probability for a continuous random variable. As before, let X_n represent the nth interarrival time. Then

$$L(X_1, \ldots, X_n \mid \lambda) = \left[\prod_{n=1}^{N} \lambda e^{-\lambda X_n} \right] e^{-\lambda(T - \Sigma X_n)} \tag{3.27}$$

The first term is the likelihood associated with the first N interarrival times. The second term is the likelihood associated with the last interval, length $T - \Sigma X_n$, during which no arrival occurred. Equation (3.27) can be simplified to

$$L(X_1, \ldots, X_N \mid \lambda) = \lambda^N e^{-\lambda T} \tag{3.28}$$

Surprisingly, the likelihood reduces to a function of just two observations, N and T, not the entire set of interarrival times. As before, the estimator is found by taking the derivative *with respect to the parameter* λ. It should come as no surprise that the MLE estimator is the same as the Poisson distribution MLE: N/T.

Both the method of moments and the maximum likelihood method are general techniques for estimating distribution parameters from a data set and can be used for virtually any probability distribution (not just Poisson or exponential). For the Poisson process, the two techniques happen to yield the same result. In general, the MLE is guaranteed to be the more precise estimator for large data sets. However, in choosing an estimator, the added precision (which tends to be small) must be weighed against the effort needed to determine the estimator.

3.6.1 Confidence Intervals

The accuracy of a parameter estimate can be measured by way of a confidence interval. While the parameter estimate is the best guess for λ, the ***confidence interval*** specifies a range of plausible values for λ. If this range is large, then one has little confidence that the estimate is correct. This situation can only be corrected by collecting more data and/or changing the method for collecting the data.

The confidence interval is based on the ***standard error*** of the parameter being estimated, which is just another name for the standard deviation of the estimator. The standard error for the mean of a sample of N *independent* random variables ($\bar{X}$) equals

$$\sigma_{\bar{X}} = \frac{\sigma}{\sqrt{N}} \tag{3.29}$$

where σ is the standard deviation of the random variable.

The standard error for the statistic $\bar{X}$ is smaller than the standard deviation for the random variable, X, because averaging reduces fluctuations. As N becomes large, fluctuations in the average become smaller and $\sigma_{\bar{X}}$ approaches zero.

Just as the sample mean has an estimator, so does the standard deviation. Usually this is defined by the following:

$$\hat{\sigma} = \sqrt{\frac{\sum_{n=1}^{N} (X_n - \bar{X})^2}{N - 1}} \tag{3.30}$$

To provide an unbiased estimate of σ, $N - 1$, not N, appears in the denominator (note that this in not a MOM estimator).

Recall that a property of the exponential distribution is that the mean equals the standard deviation. For the exponential distribution, it turns out that the best estimator (that is, the MLE) for σ is not the sample standard deviation; the best estimator for σ is the sample mean (T/N):

Exponential Random Variable

$$\hat{\sigma}_{\bar{X}} = \frac{\hat{\sigma}}{\sqrt{N}} = \frac{\mathrm{T/N}}{\sqrt{N}} = T \cdot N^{-1.5} \tag{3.31}$$

The confidence interval is always specified for a given level of confidence. For example, a 95% confidence interval means that there is a 95% probability that the true value of the parameter falls inside the confidence interval. To determine the confidence interval, one usually refers to the normal probability distribution. From the ***central limit theorem***, we know that probability distribution for a sum of n independent, identically distributed random variables will approach the normal distribution as n becomes large. And because the limiting distribution for the sum is normal, the limiting distribution for the ***sample mean*** (that is, average) must also be normal.

In terms of the arrival process, the confidence interval for $1/\lambda$ (mean of exponential distribution) can be derived from tables of the normal distribution (see Table A.2 in the appendix to this book) if the sample is large (greater than 50 data points). The normal distribution does not apply to λ directly because it is not a sample mean.

$$P(\widehat{1/\lambda} - 1.96\hat{\sigma}_{\bar{X}} \leq 1/\lambda \leq \widehat{1/\lambda} + 1.96\hat{\sigma}_{\bar{X}}) = .95 \tag{3.32}$$

$$P(\widehat{1/\lambda} - 2.58\hat{\sigma}_{\bar{X}} \leq 1/\lambda \leq \widehat{1/\lambda} + 2.58\hat{\sigma}_{\bar{X}}) = .99 \tag{3.33}$$

The symbol $1/\lambda$ represents the true mean of the random variable; $\widehat{1/\lambda}$ and $\hat{\sigma}_{\bar{X}}$ represent the estimators for the mean and mean standard error, based on the observed data.

The first line gives the 95% confidence region and the second gives the 99% confidence region. Both regions are symmetric around $\widehat{1/\lambda}$, indicating that $\widehat{1/\lambda}$ could be either be larger or smaller than the true value. More specifically, the confidence region for $1/\lambda$ in a Poisson process can be derived from the exponential distribution:

$$P[T/N - 1.96(T \cdot N^{-1.5}) \leq 1/\lambda \leq T/N + 1.96(T \cdot N^{-1.5})] = .95 \quad (3.34)$$

$$P[T/N - 2.58(T \cdot N^{-1.5}) \leq 1/\lambda \leq T/N + 2.58(T \cdot N^{-1.5})] = .99 \quad (3.35)$$

Example

For the automatic teller machine data in Table 3.1, $T/N = 120/98 = 1.22$ minutes and $N = 98$. Thus, the confidence intervals are:

95% confidence $1.22 - 1.96\ (1.22/\sqrt{98}) \leq 1/\lambda \leq 1.22 + 1.96\ (1.22/\sqrt{98})$

$$.98 \leq 1/\lambda \leq 1.46$$

99% confidence $1.22 - 2.58\ (1.22/\sqrt{98}) \leq 1/\lambda \leq 1.22 + 2.58\ (1.22/\sqrt{98})$

$$.90 \leq 1/\lambda \leq 1.54$$

Based on the data sample, there is a 95% probability that the true value of $1/\lambda$ (the value that created the data) is between .98 and 1.46 and a 99% probability that the true value of $1/\lambda$ is between .90 and 1.54. The 95% and 99% confidence intervals for λ are found by inverting these bounds, (1.02, .68) and (1.11, .65), respectively. These confidence intervals are not particularly small, which suggests that more data need to be collected to obtain a more precise estimate.

3.6.2 Sample Size

One final issue needs to be addressed: sample size. The sample should be sufficiently large to provide a precise estimate of the model parameters. But data collection is expensive, and no more data should be collected than necessary.

A reasonable approach for selecting the sample size is to begin by specifying a desired level of accuracy and then calculate how long the process must be observed in order to obtain the desired level of accuracy. For the teller example, one might specify that the 95% confidence interval should have a width of no more than .2 minute. This means that

$$P(|\widehat{1/\lambda} - 1/\lambda| \leq .1) \geq .95 \quad (3.36)$$

The width of the 95% confidence interval is 2×1.96 multiplied by the standard error for $\widehat{1/\lambda}$ (the number 2 accounts for the two tails of the distribution). This quantity must be less than or equal to the desired width, .2 minute.

$$2\ [1.96\ 1/(\lambda\sqrt{N})] \leq .2 \quad (3.37)$$

Simplifying the equation leads to

$$N \geq \left[\frac{2 \cdot 1.96}{.2\lambda}\right]^2 = \frac{384}{\lambda^2} \quad (3.38)$$

From the teller data in Table 3.1, $1/\lambda$ is estimated to be 1.22 minutes. Substituting this value in Eq. (3.38) yields 572 arrivals (11.6 hours). This is approximately how many observations would be needed to obtain a 95% confidence interval of width .2. Fortunately, the arrival rate is very easy to measure, and only involves counting arrivals. So observing the process for 11.6 hours may not be difficult. In general, if

$$w = \text{desired width of confidence interval (in time units)}$$

the desired accuracy requires a sample size of

95% Confidence	**99% Confidence**	
$N \geq \left[\dfrac{3.92}{w\lambda}\right]^2$	$N \geq \left[\dfrac{5.16}{w\lambda}\right]^2$	(3.39)

Of course, λ is not known until the sample is taken. But one usually can roughly guess the arrival rate beforehand. This prior guess can be the basis for the sample size calculation. The calculation might also be based on a ‘‘presample’’—a sample over a short time interval used to obtain a preliminary guess for the parameter. In either case, determining the sample size does not require absolute precision.

Example

Twelve arrivals are observed over a 1-hour period. The sample size that would produce a 99% confidence interval of width .05 hour is desired.

Solution Substituting the value .05 hour for w and 12/hour for λ, we get a sample size of 74 arrivals. This translates into 6.2 hours of observation.

There is no right way to set the width of the confidence interval or the confidence level. Both the cost of collecting the data and the need for precision affect the answer.

One final comment: The sample size calculations presume that it is possible to observe the arrival process for the required length of time. But what if the process is short-lived, lasting only one or two hours? It would then be impossible to observe the process long enough to obtain the desired precision. But if the process is observed over its entire lifetime, then there is no need to collect more data. The data already collected should accurately depict what happened, and since the process has ended, there is no need to infer what will happen in the future.

3.7 CHAPTER SUMMARY

A model is a way to explain the underlying process creating the data. Sometimes the model is a perfect representation of the underlying process, as in the binomial distribution representing flipped coins. More often, the model is a plausible representation that closely matches observations.

The Poisson process represents the customer arrival process of systems for which

1. The probability of a customer arriving at any time does not depend on when other customers arrived.

2. The probability that a customer arrives at any time does not depend on the time.
3. Customers arrive one at a time.

The Poisson process is a particularly important model because it accurately represents many real arrival processes and because it is fairly simple to analyze. The Poisson process possesses a number of unique properties summarized at the end of Sec. 3.3.

Determining whether or not an arrival process is Poisson always begins with an assessment of plausibility. The most important question is whether the conditions underlying the Poisson process accurately represent the situation. If they do, then further quantitative tests—graphical and statistical—can be performed to see whether the data displays the properties of the model. In the case of the Poisson process, those properties included stationarity, exponential interarrival times, and independent interarrival times. However, only a few of the many possible statistical tests were provided. For a more complete survey, consult Bhat (1978), Green and Kolesar (1989), or one of the statistics texts cited at the end of this chapter.

If the model passes the plausibility test and quantitative tests, then the next step is to determine the exact shape of the model through parameter estimation. Once this is done, a confidence interval can be calculated to estimate the precision of the parameter estimator. If the precision is not sufficiently accurate, further data should be collected. The exact amount of data to collect can be calculated with the method provided for determining sample size in Sec. 3.6.2.

Keep in mind that just because a process is Poisson with rate λ today does not mean that it will be Poisson with rate λ tomorrow, or at any other time. Predicting the future must always rely on inference—inference as to whether the conditions that created the historical data will recur in the future. The subject of prediction is addressed in the following chapter.

Although this chapter focuses on the Poisson process, the general steps of

1. Assessing the conditions creating the data
2. Gauging model plausibility
3. Testing data for goodness of fit
4. Estimating parameters and confidence intervals

apply to most any situation. The model does not have to be Poisson, but it does have to be a reasonable representation of reality. Just what this representation should be varies from situation to situation.

FURTHER READING

Allen, A. O. 1979. *Probability, Statistics and Queueing Theory*, New York: Academic Press.

Bhat, U. N. 1978. "Theory of Queues." in *Handbook of Operations Research, Foundations and Fundamentals*, ed. J. J. Moder and S. E. Elmaghraby. New York: Van Nostrand Reinhold.

GREEN, L., and P. KOLESAR. 1989. "Testing the Validity of a Queueing Model of Police Patrol," *Management Science*, 35, 127–148.

GREENSHIELDS, B. D., and F. M. WEIDA. 1978. *Statistics with Applications to Highway Traffic Analysis*. Westport, Connecticut: Eno Foundation for Transportation.

HAIGHT, F. A. 1967. *Handbook of the Poisson Distribution*, New York: John Wiley.

HAYS, W. L., and R. L. WINKLER. 1971. *Statistics, Probability, Inference and Decision*, New York: Holt, Rinehart, Winston.

MOOD, A. M., F. A. GRAYBILL, and D. C. BOES. 1974. *Introduction to the Theory of Statistics*, New York: McGraw-Hill.

NEWELL, G. F. 1982. *Applied Queueing Theory*, London: Chapman and Hall.

ROSS, S. M. 1972. *Introduction to Probability Models*, New York: Academic Press.

TUFTE, E. R. 1988. *The Visual Display of Quantitative Information*, Cheshire, Conn.: Graphics Press.

WONNACOTT, T. H., and R. J. WONNACOTT. 1969. *Introductory Statistics*, New York: John Wiley.

NOTE

1. KUHN, T. THOMAS, *The Structure of Scientific Revolutions* (Chicago: University of Chicago Press, 1970), p. 55.

PROBLEMS

Probability Distributions

1. In each of 10 minutes, the probability of exactly one arrival equals .1 and the probability of no arrivals equals .9.

(a) Using the binomial distribution, calculate the probability of 0 arrivals over 10 minutes. Repeat for 1, 2, 3, and 4 arrivals.

(b) Now assume that arrivals occur by a Poisson process at the rate of .1 per minute. Repeat your calculations for part a, based on the Poisson distribution. Why are your results similar, or different?

***2.** Suppose that an arrival comprises either a pair of customers or a single customer. The probability that any arrival has 2 customers equals .5 and the probability that any arrival has 1 customer equals .5. In each of 10 minutes, the probability of exactly one arrival equals .0667 and the probability of no arrivals equals .9333.

(a) Calculate the probability that 0 customers arrive over 10 minutes. Repeat for 1, 2, and 3 customers. (Hint: For any number of customers n, sum the probabilities of n customers given 1 arrival, n customers given 2 arrivals, ...)

(b) Why are your results similar, or different, from those in part b of Prob. 1?

3. Telephone calls are known to arrive by a Poisson process with rate 20 per hour between 1:00 and 3:00 P.M. Determine the following:

(a) The probability function for the number of arrivals in a 5-minute interval (for up to 5 arrivals).

*Difficult problem

(b) Probability that no customer arrives over a 10-minute interval.

(c) Probability that the second arrival after 1:00 occurs before 1:10.

(d) Given that three arrivals occurred between 1:00 and 1:30, the probability that the second arrival occurred after 1:15.

4. The chancellor of a university has determined that student complaints arrive by a Poisson process, with a rate of 25 per year. Determine the following:

(a) Probability that one month passes with no more than one complaint received (assume that a month is 1/12 of a year).

(b) The probability that the first complaint of the year occurs during the second month.

(c) The probability that the third complaint of the year occurs during March (use the gamma distribution).

5. Based on Definition 3.2 for the Poisson process, prove that the variance for the Poisson distribution equals λt. (Hint: Derive the result from the variance of a binomial random variable.)

***6.** Individual customers arrive by a Poisson process with rate of .0333 per minute, and pairs of customers also arrive by a Poisson process, with the same rate. Determine the probability function for the number of customers arriving during a 10-minute period, for 0 to 3 customers. Compare your result to the calculations in Prob. 2. Explain why your answer is the same, or different.

7. The Poisson distribution is discrete, but the gamma and exponential distributions are continuous. Is this inconsistent, given that all three define properties of the Poisson process? Explain.

8. A desperate gambler has flown to Lake Tahoe to win his fortune. He has chosen a $1 slot machine, which will pay a prize of $1 million, with a probability of 1/2,000,000 (success is independent among tries).

(a) Assuming that there is only one possible prize, what is the probability of winning before 500,000 tries? (Approximate from Poisson process.)

(b) Explain why your answer to part a is *not* 500,000/2,000,000.

(c) Suppose that the gambler knows that the machine has not paid out in the last 2 million tries. Will this information affect your answer to part a? Explain.

9. From any familiar application, give examples of three continuous random variables and three discrete random variables. Based on your intuition alone, plot the probability distribution function for each example. Which, if any, of your examples conforms to a familiar theoretical distribution?

10. Collect ten observations of any continuous random, then ten observations of any discrete random variable. Plot the empirical distribution function for each. By examining these distribution functions, can you tell which random variable was discrete and which was continuous? (If possible, show these functions to a classmate, and see whether he or she can tell which is discrete and which is continuous.)

Goodness of Fit

***11.** The data below represent the times that people stopped to deliver letters at a mailbox over a 255-minute period.

3.1	48.6	117.0	177.5
6.2	50.8	123.3	183.0
6.5	72.2	128.8	195.0

*Difficult problem

10.1	76.0	131.1	200.0
13.9	78.2	145.7	204.2
29.3	90.3	147.7	207.6
34.4	91.5	150.7	207.9
35.2	104.2	156.2	239.7
39.2	113.1	162.3	240.2
45.0	114.4	169.1	251.6

(a) Is it plausible that the arrival process of people is Poisson?
(b) Is it plausible that the arrival process of letters is Poisson?
(c) Derive an MOM estimate for the arrival rate of people. Calculate a 95% confidence interval for your estimate. Is your estimate also a maximum likelihood estimator? Briefly discuss.
(d) Plot the empirical probability distribution function for the interarrival times. On the same graph, plot the distribution function for an exponential distribution with mean defined by part c. Perform the K-S goodness of fit test. Do you believe the interarrival times are exponential random variables?
(e) Plot successive interarrival times, as in Fig. 3.4. Then perform the t-test for correlation. Do you believe that the interarrival times are independent?
(f) Plot the cumulative arrival curve. On the same graph, plot a straight line connecting $A(T)$ and $A(0)$. Next, perform the K-S test for goodness of fit. Do you believe that the arrival process is stationary?
(g) Based on all the evidence collected, do you believe the arrival process is Poisson?

***12.** The data below are the arrival times of westbound BART trains at the Montgomery Street station in San Francisco, between time 6.85 A.M. and time 8.85 A.M. (in hours).

6.861	6.926	7.063	7.159	7.208	7.271	7.356	7.415	7.484	7.532	7.613
7.665	7.739	7.793	7.860	7.905	7.965	8.060	8.110	8.155	8.237	8.296
8.359	8.403	8.489	8.569	8.640	8.732	8.818				

(a) Is it plausible that the arrival of trains is a Poisson process?
(b) Is it plausible that the arrival of people is a Poisson process?
(c–g) Repeat parts c–g from Prob. 11 for the new data set.

13. BART trains are scheduled to arrive every 3.75 minutes at the Montgomery Street Station, beginning at time 7.35 hours and ending at time 8.48 hours. Based on this information and the data given in Prob. 12, develop a new model for the arrival process (other than the Poisson process). That is, define a probability distribution function for the arrival time of the nth train.

14. For each of the three examples in Sec. 3.4 that do not conform to the Poisson process, describe a mathematical model that represents the arrival process.

15. Using the data from Prob. 11, estimate the minimum time that the system would have to be observed to obtain a 95% confidence interval for $1/\lambda$ of width no more than 30 seconds.

*Difficult problem

EXERCISE: GOODNESS OF FIT

The purpose of this exercise is to test the data recorded in Chap. 2 to see whether a Poisson process was observed.

1. Discuss whether the arrival process matches the conditions underlying the Poisson process, referring to the three properties:
   (a) Stationarity
   (b) Independent increments
   (c) Customers arrive one at a time.

   If the process does not precisely match the Poisson process, discuss whether the differences are appreciable.
2. As a quick check, calculate the coefficient of variation for the interarrival times. Does the arrival process seem to be Poisson?
3. On a graph of cumulative arrivals, draw a straight line connecting $A(T)$ to $A(0)$. Is there any pattern in the difference between the two curves? Discuss whether the process appears to be stationary.
4. On a graph for the empirical probability distribution for interarrival times, drawn an exponential distribution function with identical mean. Is there any pattern in the difference between the two curves? Discuss whether the interarrival times appear to be exponential.
5. Plot the paired intervals (X_n, X_{n-1}) on graph paper. Is there any pattern to the data? Discuss whether the data appear to be independent.
6. Perform the following statistical tests at the 5% significance level:
   (a) Kolmogorov-Smirnov test for stationarity of arrivals
   (b) Kolmogorov-Smirnov test for an exponential interarrival distribution
   (c) Correlation t-test between X_n and X_{n-1}

   Can you conclude that the arrival process was Poisson?
7. Calculate 95% and 99% confidence intervals for $1/\lambda$. Convert these intervals into 95% and 99% confidence intervals for λ.
8. Determine how long (in minutes) the process would have to be observed to obtain a 95% confidence interval for $1/\lambda$ of width $(1/\lambda)/20$. (Remember that the confidence interval extends on both sides of $1/\lambda$).

chapter 4

Simulation

There are two reasons why a data set might not conform to a model. The first, addressed in the previous chapter, is structural: The data were not created by the process underlying the model. The second reason is not structural but, rather, simple randomness. From time to time, a process may create unusual data that appear to have been produced by an entirely different process. This does not mean that the model is incorrect. It only means that something is out of the ordinary.

How much variability is to be expected if the model is correct? This is the issue addressed in this chapter. A technique is presented for creating ''pure'' data—pure in the sense that the data have the exact distribution assumed by the model. The pure data are compared to the real data (arrivals at automatic teller machines) to demonstrate how much variability is ordinary for an arrival process that is truly Poisson.

Simulation is the technique that creates the data. According to Webster's dictionary, a simulation is the act of imitation. It is a word used in many contexts. NASA uses chambers to simulate zero gravity for astronauts. The Army Corp of Engineers uses a warehouse-size model to simulate water flows inside the San Francisco Bay. United Airlines uses computer programs to simulate jet flight. Though the applications are different, all three reproduce the *dynamic* behavior of a real environment. Dynamic, time varying behavior is a key characteristic of simulation.

In the domain of operations research, simulation is a powerful technique for *evaluating* the behavior of a model (simulation itself is not a model). Simulation is a way to learn about a system by experimenting on a copy rather than the system itself—the copy being the model of the system (the Poisson process, for example). Simulation is a powerful tool because it is usually much quicker and easier to experiment on a copy of the system than the real system.

The type of simulation used in operations research has a more specific name: ***Monte Carlo, discrete event, computer simulation***. The city of Monte Carlo, well known for its gambling, symbolizes the element of chance incorporated in the simulation. Monte Carlo simulation tests the performance of the system under randomized scenarios. As an example, suppose that the Federal Aviation Administration would like to assess the impact on passenger delay of allowing more flights to operate out of Dulles Airport. The randomized scenario might comprise flight delays en route to the airport, weather conditions, passenger boarding times, and other pertinent factors. The simulation would calculate the delay under each scenario.

Simulation is most effective at studying complicated models that cannot be evaluated by other means. The properties of the Poisson process are well known, so simulation is not really needed to evaluate it. However, more complicated models, such as models of nonstationary multiserver queueing systems, are not so easily evaluated. Simulation may then be used (as a last resort) to estimate the system's performance.

This chapter describes how to create a simple Monte Carlo simulation of a queueing system. But, more importantly, it is also concerned with understanding the relationship between the simulation and the model and the relationship between the simulation and the system. The chapter begins by discussing the philosophy of simulation. Next, a general technique is presented for simulating a random variable. Specific equations are provided for simulating a few common probability distributions. At this point, a sample simulation for an arrival process is compared to the observed arrival process. The chapter concludes with a demonstration of how to simulate a queueing process.

4.1 SIMULATION PHILOSOPHY

Before the mechanics of Monte Carlo simulation are presented, we should first consider a point of philosophical debate that dates to at least the eighteenth century. That debate concerns the existence and meaning of probability. *A Philosophical Essay on Probability*, by the French mathematician Pierre Simon de Laplace, posits that probability is no more than an ''expression of man's ignorance.'' That is, something is random when we do not possess sufficient information to predict its outcome. To take an example, we may feel that the outcome from the spin of a roulette wheel is totally random. Yet, with sufficient knowledge of the force exerted on the wheel and the properties of the wheel itself, the outcome should be predictable with certainty.

When it comes to the behavior of real systems, queueing systems included, we indeed are very ignorant. We do not know when customers will arrive and we do not know how long it will take to serve them. And it is hard to imagine a day when we do possess the knowledge to predict this sort of human behavior. We may be able to predict how a system behaves on average, but any prediction as to the exact queue length or exact waiting time at any particular time of day would be no more than speculation.

When it comes to the behavior of simulated systems, we are much less ignorant. Most (though not all) Monte Carlo simulations employ a computer to create simulated random variables. But computers possess that most unusual characteristic of total predictability. Every time we run a computer program we expect the exact same result. One may question whether a predictable device such as a computer is capable of simulating random variables.

Even though computer simulations do produce the exact same result every time, they can still replicate the *properties* of a random variable. Computer simulations create what are known as ***pseudo-random numbers***. Beginning with a seed number specified by the user, the program performs a series of operations that create numbers that appear to have a uniform [0,1] probability distribution. If 1000 numbers are generated by the computer, then the empirical distribution of the numbers should closely conform to the theoretical uniform [0,1] distribution. Yet these numbers are not random in the sense that, should one know the internal formula, they are entirely predictable. Nevertheless, for practical purposes, the variables can be considered random because they behave as though they are random. This is the goal of simulation.

4.2 MECHANISM FOR SIMULATING A RANDOM VARIABLE

Most random variables can be simulated by one basic mechanism. That mechanism is driven by the simulation of pseudo-random $U[0,1]$ (uniform over the region, [0,1]) random variables. Once the uniform variables are created, they can be transformed into any desired univariate random variable defined by a probability distribution function.

4.2.1 Creation of Uniform [0,1] Variables

Before considering how a computer might create a $U[0,1]$ random variable, think of a way for you to create a $U[0,1]$ random variable. Can you think of an experiment that will create such a random variable? Or do you know of a list of numbers that possesses the uniform distribution (try looking in the phone book)? Computer simulation may be quicker, but there is no reason why simulation cannot also be performed by hand.

Computers create pseudo-random number sequences through use of the modulus operation, which provides the remainder from a division. The initial seed for the sequence, Y_0, is specified by the user. This is multiplied by parameter a and added to another parameter b. The result is divided by a third parameter, which is a very large number, m. The remainder of the division is then multiplied by m to obtain the next number in the sequence. The process is repeated using the previous result as the seed for the next calculation. These calculations are summarized in the equation

$$Y_i = (aY_{i-1} + b) - \left[\frac{aY_{i-1} + b}{m} \right]^{-} \times m \tag{4.1}$$

where

m = the maximum value of Y_i
a,b = user-specified parameters
Y_0 = seed value for simulation

The $[\ ]^-$ symbol denotes the largest integer less than or equal to the enclosed quantity. (See Fishman 1978 on how to select the parameters a, b, and m.

Once Y_i is obtained, a $U[0,1]$ random variable is derived by dividing Y_i by m:

$$U_i = \frac{Y_i}{m^-} \quad (4.2)$$

Besides being pseudo-random, U_i is not truly continuous, because it is limited to m discrete outcomes. However, if m is a very large number, this small discrepancy presents no practical problems.

Example

Suppose that $m = 10{,}000$, $a = 101$, and $b = 7$. The random number seed is 4567. Then the first five pseudo-random $U[0,1]$ random variables are

i	Y_{i-1}	Y_i	U_i
1	4567	$461274 - [461274/10000]^- \times 10000 = 1274$	.1274
2	1274	$128681 - [128681/10000]^- \times 10000 = 8681$	.8681
3	8681	$876788 - [876788/10000]^- \times 10000 = 6788$	.6788
4	6788	$685595 - [685595/10000]^- \times 10000 = 5595$	.5595

4.2.2 Transformation into Desired Distribution

A desired random variable can be simulated through a simple transformation of the $U[0,1]$ random variable. This transformation applies to ***univariate*** random variables that are defined by a probability distribution function. Univariate means that the distribution pertains to single independent random variables, rather than pairs, triplets, or the like, of dependent random variables, as is the case of ***multivariate*** distributions.

The transformation is a simple consequence of the properties of probability distribution functions. All probability distribution functions must be nondecreasing and have the range [0,1]. The lower bound on the range is the probability that the random variable is less than a very small value x, and the upper bound is the probability that the random variable is less than a very large value x. A probability cannot be less than zero or greater than one. Now consider a possible outcome of the random variable X. The value $F(X)$ (the distribution function evaluated at X) is itself a random variable, just as any function of a random variable would be. By the definition of $F(x)$, the probability that $F(X)$ is less than or equal to any value y must be

$$P[F(X) \leq y] = y \qquad 0 \leq y \leq 1 \quad (4.3)$$

That is, $F(X)$ has a uniform [0,1] distribution. Therefore, if the $U[0,1]$ random variable, U, represents $F(X)$, $F(X)$ can be inverted to obtain the desired random variable X:

$$X = F^{-1}[U] \tag{4.4}$$

The technique is best appreciated through an illustration. Figure 4.1 shows how a uniform random variable is transformed into the desired distribution of a continuous random variable. The $U[0,1]$ variable is generated first and plotted on the vertical axis of the probability distribution function. Then a horizontal line is drawn over to the distribution function and a vertical line is drawn down to the value X. Thus, $F(X) = .7$ is transformed into the value $X = 2.6$. Note that $F(X)$ does not have the shape of any familiar distribution function. The technique only requires that the distribution function be defined, not that it be written in a standard form. The distribution function does not even have to be defined by an equation—a figure will suffice (however, if a computer is used to simulate the random variable, the picture must be translated into computer-readable data).

The same basic technique applies to discrete distribution functions, as shown in Fig. 4.2. However, because $F(x)$ is a step function, different values of $F(X)$ are transformed into the same value of X. Generating a random variable amounts to categorizing the outcome of the $U[0,1]$ variable. For the function shown, the categorization is as follows:

U	***X***
[0,.2]	0
(.2,.5]	1
(.5,.7]	2
(.7,.9]	3
(.9,1]	3.5

Note that discrete random variables do not have to be integers. They merely have to be limited to distinct values.

If there is no reason to believe that the random variables are created by one of the standard probability distributions, the simulation can be based entirely on the empirical distribution function. Figure 4.3 shows such a function for a sample of service times. Conceptually, the simulation is performed in the exact same manner as for a theoretical distribution function, by plotting U on the vertical axis, drawing a horizontal line over to $F(X)$, and drawing a vertical line down to X. However, because each step in $F(X)$ represents one data point, there is a simpler way to perform the simulation: Simply pick the data points at random (with replacement). If there are 20 data points, a value of U between 0 and .05 amounts to drawing the first point, a value of U between .05 and .1 amounts to drawing the second point, and so on.

A potential weakness of this last approach is that the outcomes of the simulation are limited to the set of observations in the data sample. Potential outcomes of the random variable will never appear in the simulation if they were not observed in the data sample. If the random variable is continuous (not if it is discrete), a compromise approach would be to base the simulation on a smooth approximation to the empirical distribution function, as shown in Fig. 4.4. This approximation would create a greater range of outcomes, which may be more realistic.

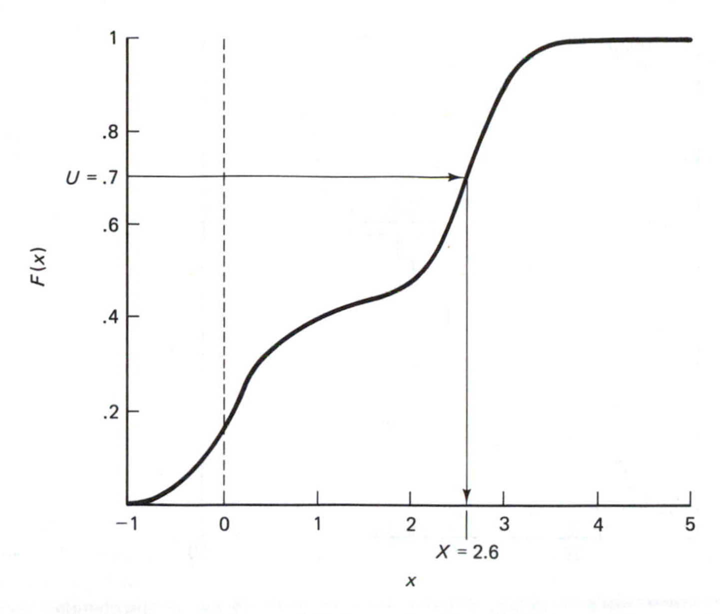

Figure 4.1 Continuous random variable is simulated by substituting a uniform random variable in the inverse of the probability distribution function.

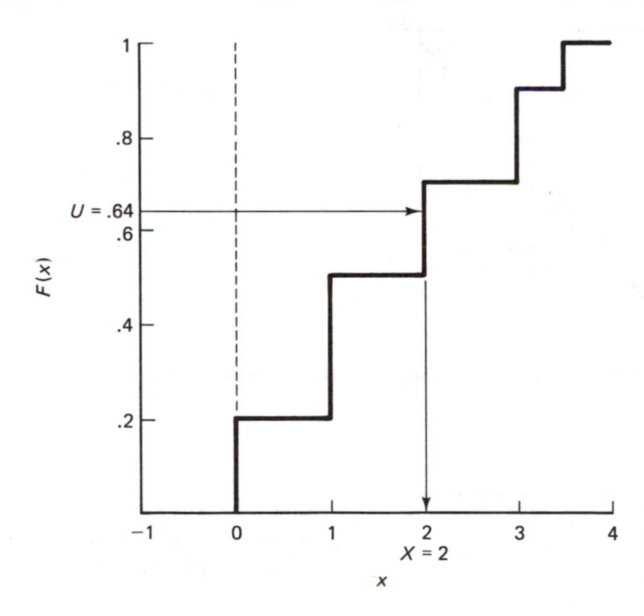

Figure 4.2 Discrete random variable can be simulated in the same manner as a continuous random variable.

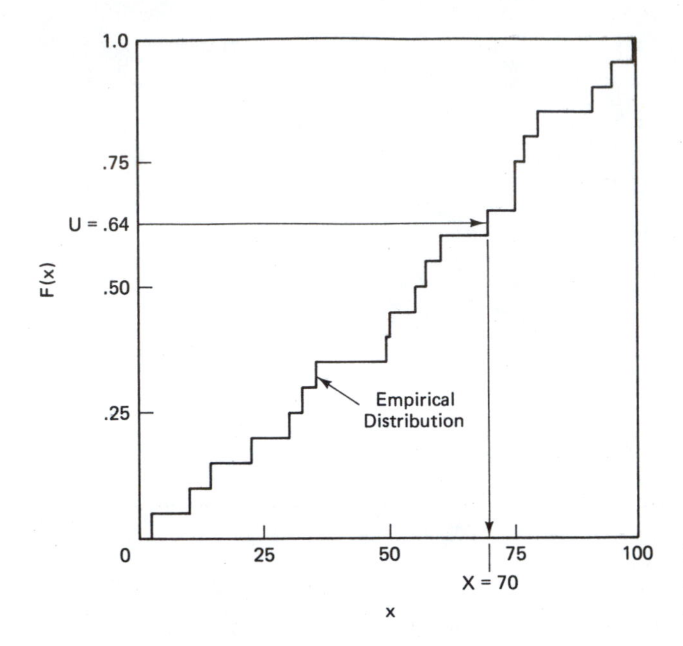

Figure 4.3 Random variable can also be simulated from an empirical probability distribution.

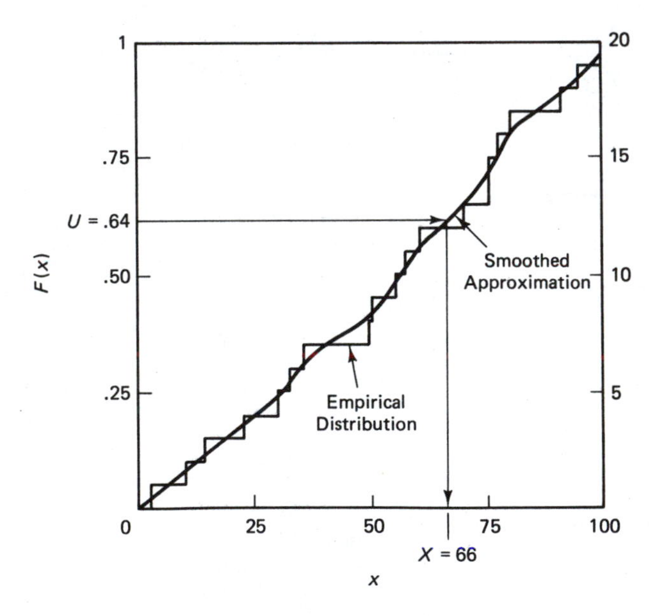

Figure 4.4 Simulation of a random variable from a smoothed approximation to an empirical probability distribution.

4.3 TECHNIQUES FOR GENERATING SPECIFIC RANDOM VARIABLES

Random variables are particularly easy to generate if the inverse of the probability distribution function can be written as a closed form equation. Two examples follow.

Uniform [*a*,*b*]. The random variable is uniformly distributed over the closed interval $[a,b]$. The distribution function, $F(x)$, is the following:

$$F(x) = \frac{x - a}{b - a} \qquad a \leq x \leq b \tag{4.5}$$

The inverse of $F(x)$ is found by solving

$$U = \frac{x - a}{b - a} \rightarrow U \cdot (b - a) = x - a \rightarrow a + U \cdot (b - a) = x \tag{4.6}$$

Example

A $U[5,20]$ random variable is to be simulated. The $U[0,1]$ random equals .6. The corresponding $U[5,20]$ random variable is $5 + .6(15) = 14$.

Exponential. As shown in Chap. 3, the exponential distribution function is

$$F(x) = 1 - e^{-\lambda x} \qquad x \geq 0 \tag{4.7}$$

The inverse of $F(x)$ is found by solving

$$U = 1 - e^{-\lambda x} \rightarrow \ln(1 - U) = \ln[e^{-\lambda x}] \rightarrow \frac{\ln(1 - U)}{-\lambda} = x \tag{4.8}$$

If U has a $U[0,1]$ distribution, so must $1-U$. Therefore, the simulation can be simplified by substituting U for $1-U$ in Eq. (4.8):

$$x = \frac{\ln(U)}{-\lambda} \tag{4.9}$$

Throughout this book, Eq. (4-9) will be used to simulate exponential random variables.

Example

An exponential random variable with mean 3 is to be simulated. The $U[0,1]$ random variable equals .6. The associated exponential random variable equals $ln(.6)/(-1/3) = 1.53$.

If the inverse of the distribution function cannot be written as a closed form equation (for example, if the random variable has the normal or gamma distribution), then simulation is more difficult. Conceptually, such a random variable can be generated in the exact same manner as any other univariate random variable. However, either $F(x)$ must be plotted and the simulation performed manually, or the inverse of $F(x)$ must be tabulated, as in a normal distribution table. These options make computation more difficult. Alternative techniques for simulating normal and gamma random variables follow.

Normal. Box and Muller (1958) developed a technique for generating a pair of independent normal (0,1) random variables from a pair of independent $U[0,1]$ random variables. The transformation works as follows, with cosine and sine calculated from radians:

$$X_1 = \sqrt{-2 \ln(U_1)} \cos(2\pi U_2) \tag{4.10}$$

$$X_2 = \sqrt{-2 \ln(U_1)} \sin(2\pi U_2) \tag{4.11}$$

To create independent random variables Y_1 and Y_2 with a more general normal (μ,σ) distribution, the following transformation is performed:

$$Y_1 = \mu + \sigma X_1 \tag{4.12}$$

$$Y_2 = \mu + \sigma X_2 \tag{4.13}$$

Example

A normal random variable with mean 10 and standard deviation 15 is to be simulated. The $U[0,1]$ random variables are .6 and .3. The corresponding normal (0,1) and normal (10,15) random variables are

$$X = \sqrt{-2 \ln(.6)} \cos(2\pi .3) = 1.01$$

$$Y = 10 + 15(1.01) \qquad = 25.2$$

Gamma. Recall that a random variable has the gamma distribution if it is the sum of independent exponential random variables. Hence, a random variable with the gamma (n,λ) distribution (with n integer) can be simulated by summing n simulated exponential (λ) random variables.

Example

A gamma (3,5) random variable is to be simulated. The $U[0,1]$ random variables .3, .6, and .78 are generated and transformed into the exponential random variables with $\lambda = 5$: .241, .102, and .050. The gamma random variable is the sum of these values: .393.

4.4 SIMULATION OF A POISSON PROCESS

Simulating a stochastic process is not the same as simulating a random variable. A stochastic process represents a composition of random variables, not individual random variables. The Poisson process is one of the easiest stochastic processes to simulate, due to its independence and stationarity properties (outlined in Chap. 3). There are several ways to carry out the simulation, the easiest of which follow:

1. Simulate exponential random variables $X_1, X_2, \ldots$, representing interarrival times. Set $Y_1 = X_1$, $Y_2 = Y_1 + X_2$, $Y_3 = Y_2 + X_3, \ldots$.
2. Simulate a Poisson random variable, $N(T)$, representing the number of arrivals over a time interval length T. Then simulate $N(T)$ uniform $[0, T]$ random variables,

representing arrival times. Sort the arrival times in ascending order to obtain Y_1, Y_2, ...

As you can see, these two ways of simulating the Poisson process exploit two different properties of the process: that interarrival times are exponential random variables and that the number of events in a time interval is a Poisson random variable and the event times are uniform random variables.

A third technique is to exploit the property that $P\{N(t + dt) - N(t) = 1\} = \lambda dt$, for differential dt. This technique is presented in chapter 6, when a nonstationary version of the Poisson process is discussed.

Example 1

A Poisson process with rate 4 customers/hour is to be simulated over the period 9:00 to 11:00 A.M. by method 1. First, suppose that $U[0,1]$ random variables are generated from Eq. (4.1). Next, the $U[0,1]$ random variables are transformed into exponential random variables, $\lambda = 4$, with Eq. (4.9). These exponential variables are summed, and the process terminates when an arrival time exceeds the 2-hour limit. The three steps are illustrated in the table below:

n	$U[0,1]$	X_n	Y_n
1	.0702	.664	9.664
2	.3117	.291	9.955
3	.1257	.518	10.473
4	.2397	.357	10.830
5	.5522	.148	10.978
6	.8537	.040	—

Example 2

The same Poisson process is to be simulated by method 2, using the same $U[0,1]$ random variables as in Example 1. $N(2)$ is found as follows:
The $U[0,1]$ random variable .0702 is compared to the Poisson distribution:

$$P[N(2) \leq 0] = \frac{(\lambda t)^0}{0!} e^{-\lambda t} = \frac{8^0}{0!} e^{-8} = .00034$$

$$P[N(2) \leq 1] = \frac{(8)^1}{1!} e^{-8} + P[N(2) \leq 0] = .0027 + .00034 = .0030$$

$$P[N(2) \leq 2] = \frac{(8)^2}{2!} e^{-8} + .003 = .0137$$

$$P[N(2) \leq 3] = .0286 + .0137 = .0423$$

$$P[N(2) \leq 4] = .0573 + .0423 = .0996 \leftarrow \text{Round up to 4}$$

If the cumulative probability distribution for $N(2)$ were plotted on graph paper, as in Fig. 4.2, the $U[0,1]$ random variable .0702 would correspond to four arrivals. That is, the range (.0423,.0996) represents four arrivals.

In the second step of the simulation, $N(2) = 4$ arrival times are generated from the $U[9,11]$ distribution:

n	$U[0,1]$	$U[9,11]$
1	.3117	9.623
2	.1257	9.251
3	.2397	9.479
4	.5522	10.104

In the final step, the arrival times are sorted in ascending order:

n	Y_n
1	9.251
2	9.479
3	9.623
4	10.104

In the example, both alternatives used the same set of pseudo-random numbers, yet they arrived at different results. This should be no surprise because the different methods transform the uniform variables in different ways.

4.4.1 Simulation of Renewal Processes

Method 1 can be used to simulate a broader class of stochastic processes, known as ***renewal processes***. A renewal process is a counting process for which the interarrival times are independent and identically distributed (this does not imply that the process has *independent increments*). The Poisson process is a renewal process, but so are other processes.

Example

A busy freeway lane was observed over a period of 2 hours. Each time the front of a vehicle passed a set marker, an arrival was recorded. The interarrival times were found to be independent, but the *arrival* times were not independent. The minimum separation between arrivals was 1 second. Beyond this minimum, the probability distribution function for the separation had the shape of an exponential distribution function. The average separation was 2 seconds.

From the information provided, the separation, H, has a *shifted exponential distribution*:

$$P(H \leq h) = \begin{cases} 0 & h < a \\ 1 - e^{-\lambda(h - a)} & h \geq a \end{cases} \tag{4.14}$$

The parameter a represents the minimum possible separation (1 second). $a + 1/\lambda$ equals the mean separation, so $1/\lambda$ also equals 1 second.

The arrival process is simulated by summing shifted exponential random variables:

n	U_n	$X_n = a + \frac{\ln(U_n)}{-\lambda}$	Y_n
1	.392	1.94	1.94
2	.418	1.87	3.81
3	.956	1.04	4.85
4	.987	1.01	5.86

A small caveat: The shifted exponential distribution does not possess the memoryless property. This means that X_1 depends on when the last customer arrived prior to the start of the simulation. The simulation above assumes that the last arrival was at time 0. Unless the arrival rate is very small, this assumption should have negligible impact on the simulation.

4.4.2 Goodness of Fit

For practical purposes, there is no need to test the goodness of fit of pure data. A hypothesis does not have to be tested if it is known to be true. Nevertheless, it is instructive to see how a pure data set—one that was, without a doubt, created by the Poisson process—would score. It is also instructive to compare the pure data set to a real (observed) data set to see the differences.

Table 4.1 provides a list of 92 customer arrival times generated from a Poisson process with rate $\lambda = 98/120$, the exact same rate observed at the automated teller machines (see Chap. 3), and Figs. 4.5 through 4.8 show the cumulative arrivals, interarrivals, successive interarrivals, and the interarrival distribution. Although the specific data points are different from before, the general pattern is nearly the same. This pattern is distinctive of the Poisson process.

How does the data score according to the statistical goodness of fit tests? First, consider the Kolmogorov-Smirnov (K-S) test for the interarrival distribution. Note that the theoretical interarrival distribution in Fig. 4.8 is based on the arrival rate of 98/120 customers/minute, not the rate of 92/120 observed in the simulation. Plotting the data in this way changes the hypothesis test somewhat, from

H_1: Interarrival times have exponential distribution, rate $\lambda = A(T)/T = 92/120$

to

H_1': Interarrival times have exponential distribution, rate $\lambda = 98/120$

Thus, the test is based on the theoretical value of λ rather than the sample estimate of λ. Even so, the K-S statistic equals .087 (for $x = .42$ minute), and the hypothesis cannot be rejected at the 5% significance level. The correlation coefficient for the simulated data is

TABLE 4.1 SIMULATED SERVICE TIMES AT AUTOMATIC TELLER MACHINE (MINUTES)

n	$A^{-1}(n)$	n	$A^{-1}(n)$	n	$A^{-1}(n)$	n	$A^{-1}(n)$
1	0.13	24	31.84	47	65.98	70	83.86
2	2.94	25	34.54	48	66.50	71	85.51
3	3.22	26	36.55	49	67.27	72	87.80
4	4.20	27	38.26	50	67.86	73	89.08
5	7.35	28	38.72	51	69.61	74	90.57
6	9.49	29	39.20	52	70.36	75	92.88
7	11.07	30	39.41	53	72.06	76	93.53
8	11.19	31	41.22	54	72.90	77	93.90
9	11.80	32	43.39	55	73.11	78	101.15
10	12.51	33	45.17	56	73.24	79	101.61
11	13.46	34	45.70	57	73.48	80	102.99
12	14.74	35	46.48	58	73.96	81	105.84
13	18.91	36	48.56	59	74.87	82	107.32
14	19.60	37	50.56	60	75.37	83	107.59
15	22.70	38	51.71	61	75.89	84	108.39
16	24.91	39	58.24	62	75.97	85	109.27
17	26.51	40	58.62	63	76.40	86	109.76
18	26.69	41	60.25	64	78.08	87	110.04
19	28.63	42	60.53	65	78.53	88	111.91
20	29.34	43	64.37	66	79.00	89	113.32
21	30.32	44	65.03	67	79.81	90	116.86
22	31.59	45	65.31	68	81.63	91	118.59
23	31.60	46	65.48	69	83.44	92	119.95

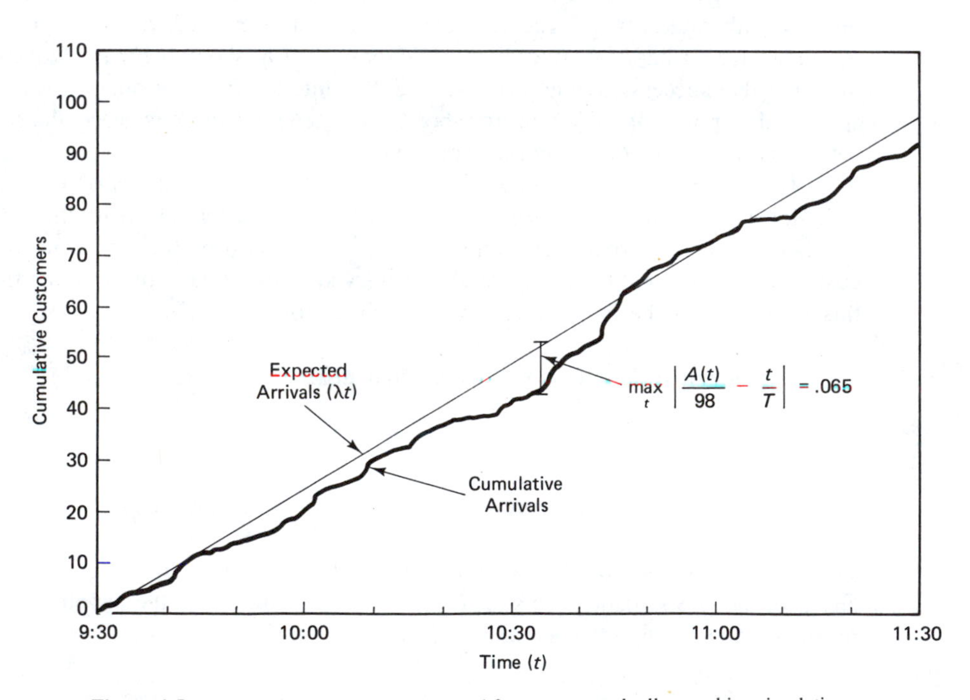

Figure 4.5 Cumulative arrival curve created from automated teller machine simulation.

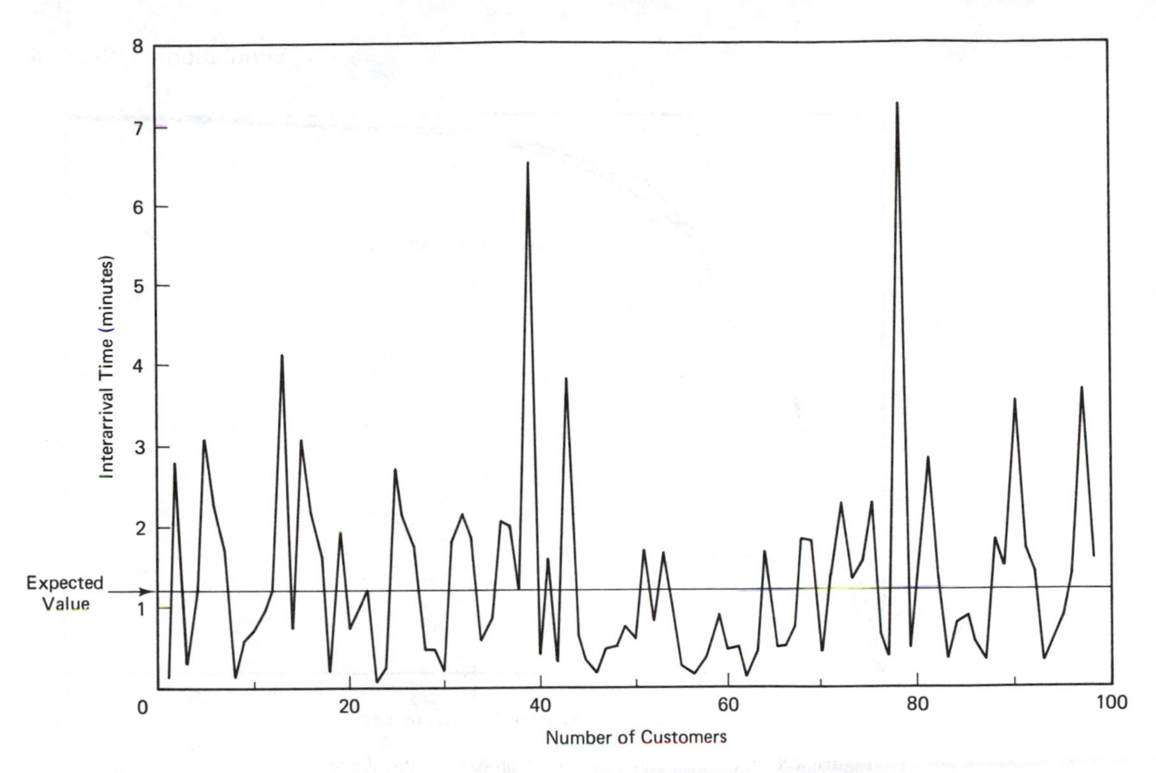

Figure 4.6 Interarrival times created from automated teller simulation.

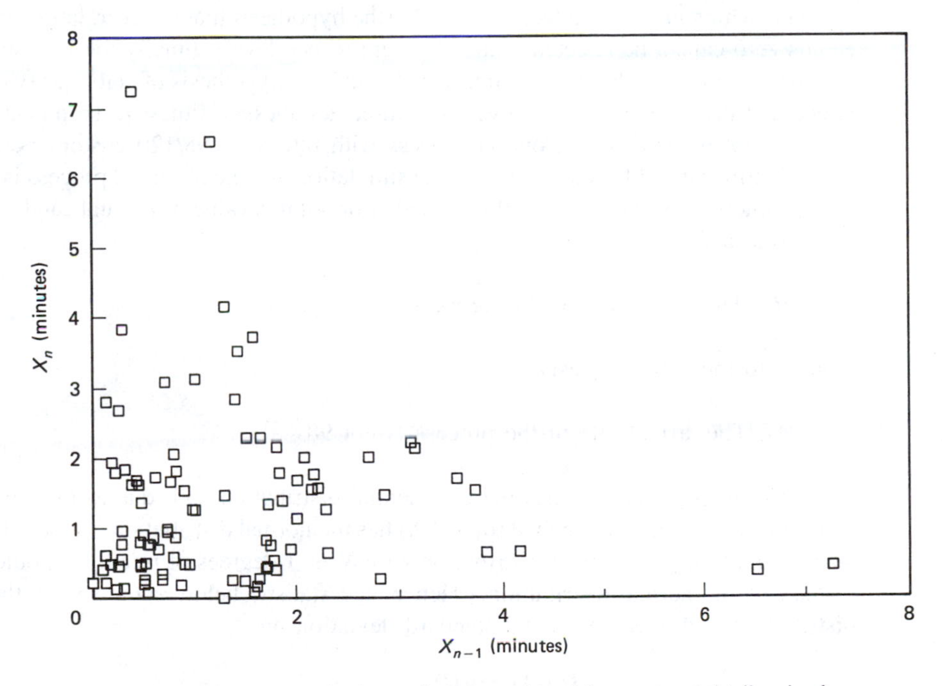

Figure 4.7 Paired successive interarrival times created from automated teller simulation.

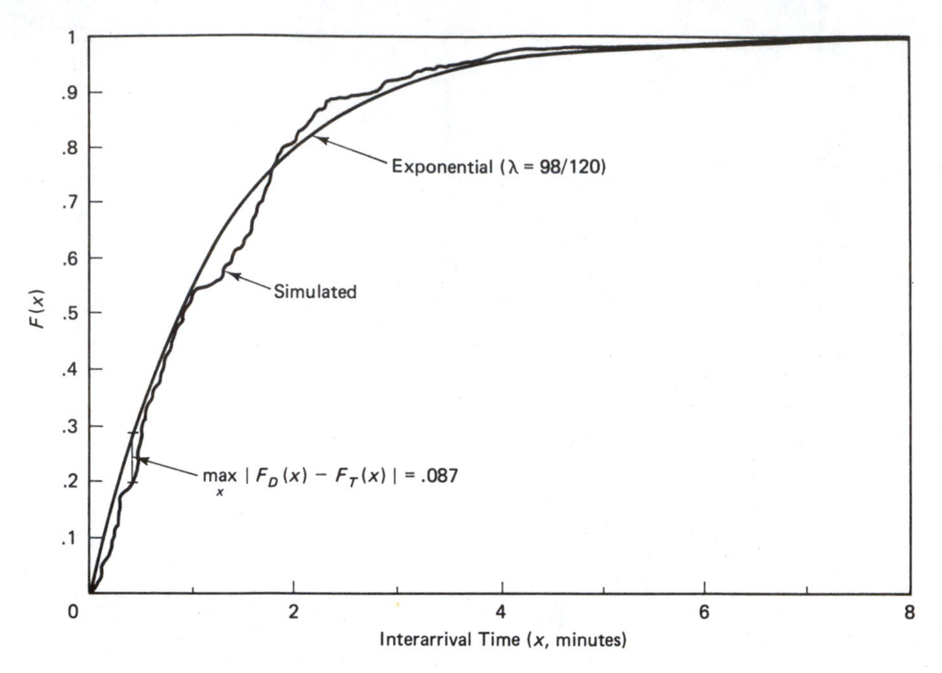

Figure 4.8 Simulated and exponential probability distributions for interarrival times.

$-.034$, which gives a t statistic of $-.321$ (the hypothesis that the correlation coefficient equals zero cannot be rejected at the 5% significance level). Finally the K-S statistic for the arrival times equals .065 (at time 10:34), and the hypothesis of stationarity cannot be rejected at the 5% significance level. The three goodness of fitness tests indicate that the data are not unusual for a Poisson process with rate $\lambda = 98/120$ customers/minute.

A noticeable difference between the simulation and the observed process is that only 92 customers arrived (instead of 98). Whether or not this value is unusual can be tested by evaluating a fourth hypothesis:

H_4: The arrival rate of the process is 98

relative to the antihypothesis:

A_4: The arrival rate of the process is not 98.

The interarrival time has the exponential distribution and, due to the central limit theorem, the average interarrival time $(1/\lambda)$ has the normal distribution. (When the sample size is small, under 50, the t distribution with $N - 1$ degrees of freedom should be used instead of the normal distribution.) Hence, the statistic below should have the normal distribution with mean zero and standard deviation one:

$$\frac{T/A(T) - 1/\lambda}{(1/\lambda)/\sqrt{A(T)}} \qquad \text{[normal } (0,1)] \tag{4.15}$$

For the simulated data

$$\frac{(120/92) - (120/98)}{(1/\lambda)/\sqrt{92}} = .626$$

Reading from Table A.2 for the normal distribution in the appendix to this book, we find the probability that $-.63 \leq Y \leq .63$ is .46. Thus, there was only a 46 percent chance that $A(T)/T$ would be within (120/92) − (120/98) of the value (120/98) customers/minute.

All in all, the tests show that the data are entirely ordinary for the Poisson process. But how would a data set appear that is abnormal? The simulation was repeated more than 50 times until a data set was created that was rejected at the 1% significance level by the K-S test for the interarrival distribution. The four graphs of cumulative arrivals, interarrivals, successive interarrivals, and the interarrival distribution are shown in Figs. 4.9 through 4.12.

The data are every bit a product of the Poisson process, just as Figs. 4.5 through 4.8 are. Yet they represent a more extreme case, something that would occur only once in 100 times. The K-S statistic for the interarrival distribution is .153, based on 124 observations. This means that the hypothesis that the interarrival distribution is exponential is rejected at the 1% significance level.

4.5 SIMULATION OF A QUEUEING PROCESS

The queueing process involves customer service as well as customer arrivals. Both are part of the queueing simulation. The easiest type of process to simulate, which is discussed in this section, is where service time does not depend on arrival times or queue lengths and where customers arrive by the Poisson process. More complicated processes can also be simulated. These are discussed in later chapters.

The following procedure is called an ***event-oriented*** simulation, because the system is only evaluated at event times (arrival or departure time). This contrasts with the ***activity scanning*** method, in which the system is evaluated after fixed time increments, independent of whether or not events occur at these times. The event-oriented simulation really amounts to two different simulations, one for arrival times and the other for service times. The arrival times determine the arrival curve, $A(t)$, and the arrival and service data are combined to determine the departure curves, $D_q(t)$ and $D_s(t)$. First, a few terms must be defined:

Definitions 4.1

$\mathrm{A}(t)$ = set of customers that have arrived by time t.

$\bar{\mathrm{A}}(t)$ = set of customers that have not arrived by time t.

$\mathrm{D}(t)$ = set of customers that have departed from the system by time t

$\bar{\mathrm{D}}(t)$ = set of customers that have not departed from the system by time t

$\mathrm{S}(n)$ = service time for customer n

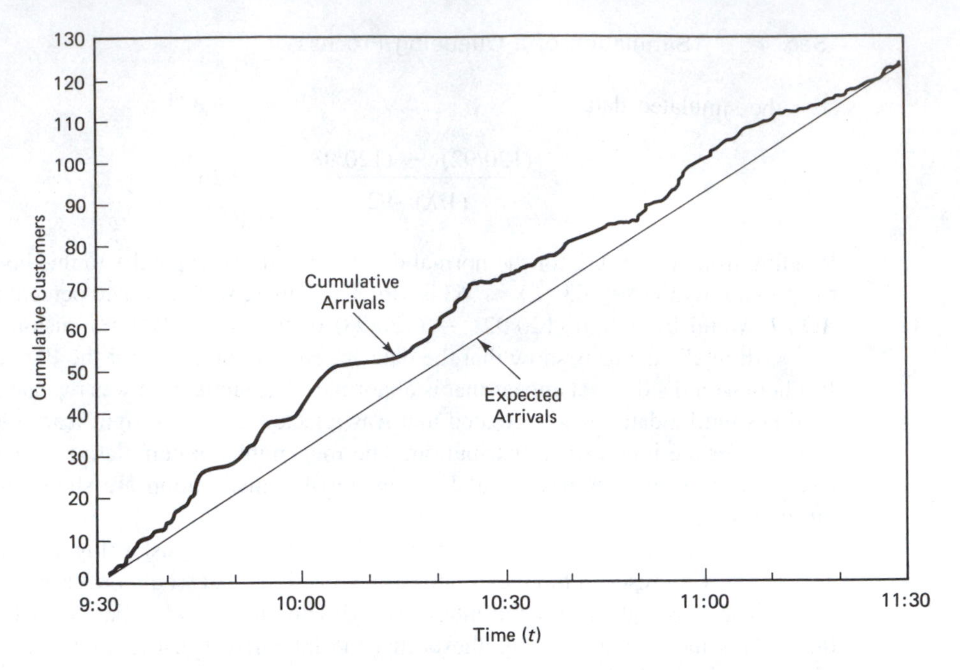

Figure 4.9 Cumulative arrival curve showing abnormal simulation results.

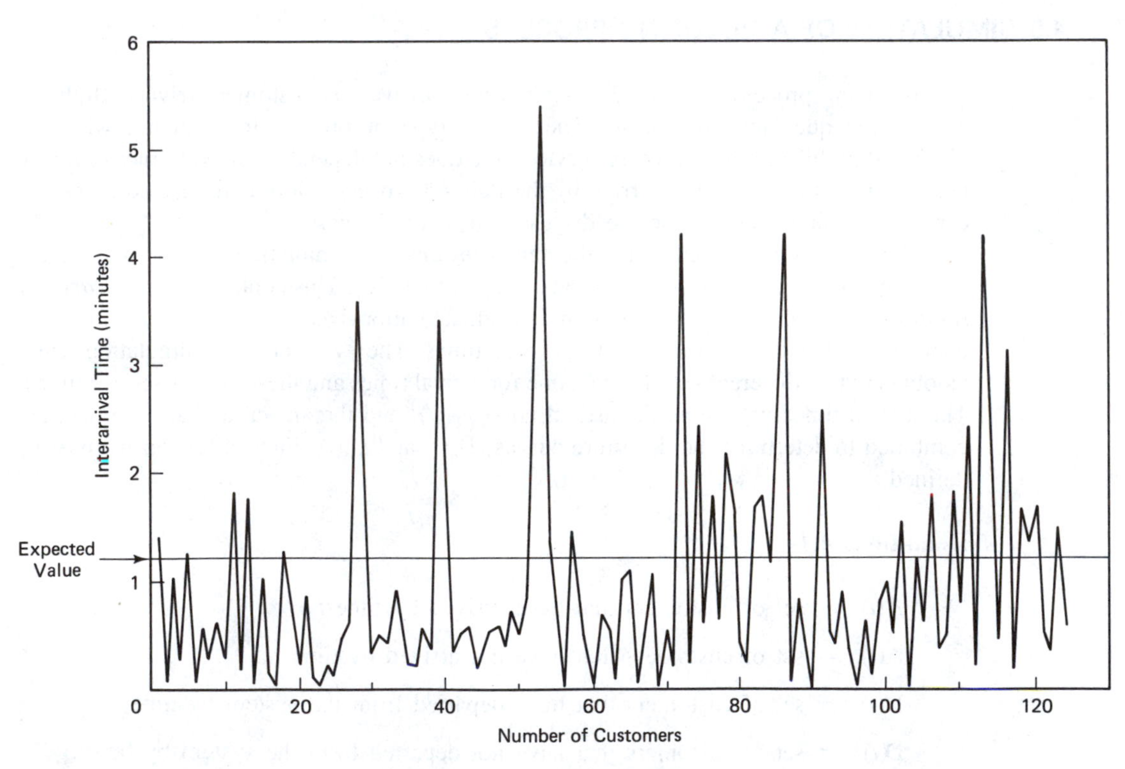

Figure 4.10 Interarrival times showing abnormal simulation results.

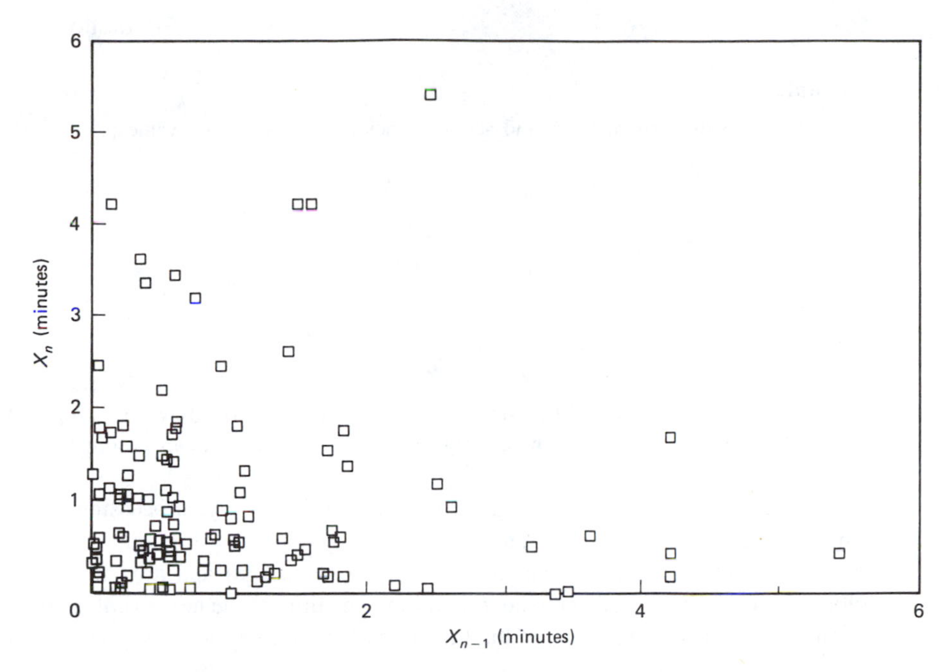

Figure 4.11 Paired successive interarrival times showing abnormal simulation results.

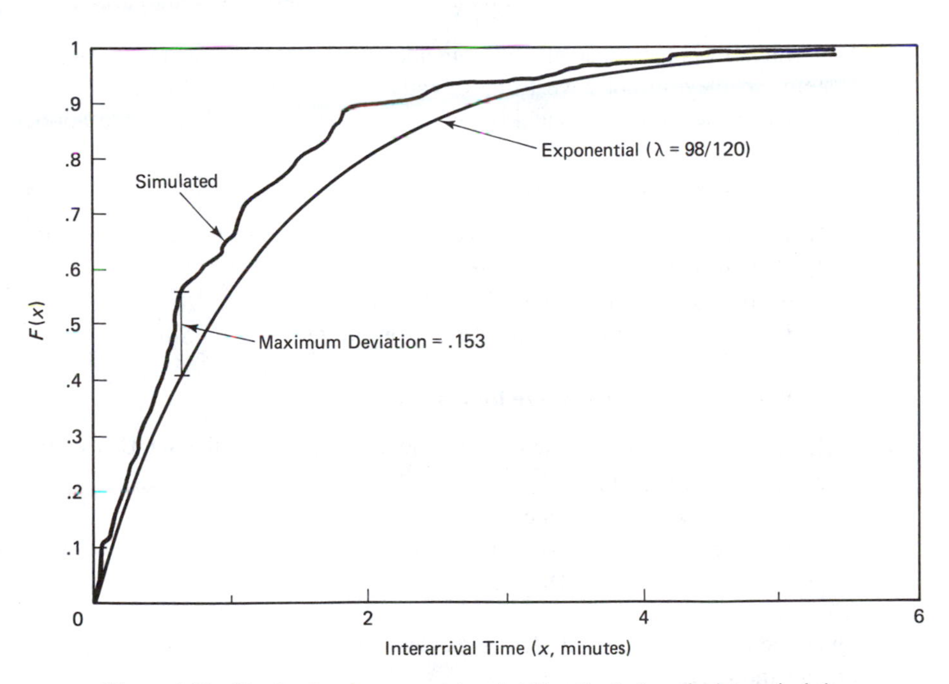

Figure 4.12 Simulated and exponential probability distributions for interarrival times showing abnormal simulation results. Exponential distribution is rejected by Kolmogorov-Smirnov test.

Example

Suppose that arrival times and service times are the following values:

n	$A^{-1}(n)$	$S(n)$
1	9:05	4 min
2	9:08	7
3	9:12	2
4	9:20	6
5	9:25	7

Then A(9:15) = {1,2,3} (the set of customers that have arrived by 9:15), $\bar{A}$(9:15) = {4,5}. D(9:15) and $\bar{D}$(9:15) will be determined through the calculations described below.

The algorithm for combining the data works iteratively, processing one event at a time. At the end of each iteration, and at the very start, the time until the next event is calculated. This determines which event is processed next and how far the simulation clock should be advanced at the next iteration. The time of the next event is either the time of the next arrival or the time of the next departure from system, whichever comes first:

$$\begin{aligned} T(t) &= \text{time of next event after time } t \\ &= \min \{ \min_{n \in \bar{A}(t)} A^{-1}(n) \qquad \min_{n \in \bar{D}(t)} D_s^{-1}(n) \} \end{aligned} \tag{4.16}$$

Departures from the queue always coincide with either an arrival or a departure from the system, so they do not have to be directly accounted for in Eq. (4.16).

Once the simulation clock is advanced, the following steps are applied:

If next event is an arrival

1. Increase $A(t)$ by one and move the customer from the set $\bar{A}(t)$ to the set $A(t)$.
2. If there is an available server
   a. Increase $D_q(t)$ by one.
   b. Set customer's departure time from system at $t + S(n)$.

If next event is a departure from system

1. Increase $D_s(t)$ by one and move the customer from the set $\bar{D}(t)$ to the set D(t).
2. If there is a customer in queue
   a. Increase $D_q(t)$ by one.
   b. Select highest priority customer from queue and set its departure time from system at $t + S(n)$.

Once the event is processed, $T(t)$ is recalculated and the procedure is repeated until there are no more events to process.

These steps are general, and apply to multiple servers and any queue discipline that does not interrupt customers already in service. The following example is more specific:

Example

Suppose that the arrival and service data in the previous example apply to a single server FCFS queueing system. The data are processed in the following steps, beginning at time 9:00:

Time	A(*t*)	D(*t*)	*T*(*t*)	*A*(*t*)	$D_q(t)$	$D_s(t)$	Event Type
9:00	{ø}	{ø}	9:05	0	0	0	—
9:05	{1}	{ø}	9:08	1	1	0	Arrival
9:08	{1,2}	{ø}	9:09	2	1	0	Arrival
9:09	{1,2}	{1}	9:12	2	2	1	Departure
9:12	{1,2,3}	{1}	9:16	3	2	1	Arrival
9:16	{1,2,3}	{1,2}	9:18	3	3	2	Departure
9:18	{1,2,3}	{1,2,3}	9:20	3	3	3	Departure
9:20	{1,2,3,4}	{1,2,3}	9:25	4	4	3	Arrival
9:25	{1,2,3,4,5}	{1,2,3}	9:26	5	4	3	Arrival
9:26	{1,2,3,4,5}	{1,2,3,4}	9:33	5	5	4	Departure
9:33	{1,2,3,4,5}	{1,2,3,4,5}	—	5	5	5	Departure

The table in the example contains all of the necessary information to plot $A(t)$, $D_q(t)$, and $D_s(t)$, as shown in Fig. 4.13. The simulation is complete.

For *FCFS single server queues*, there is actually a simpler way to process the data: Process the data one customer at a time, in order of arrival, rather than one event at a time, in order of events. This method only works when customers depart the queue and depart the system in the same order that they arrived, as is the case for FCFS single server queues. If the customer arrives when no other customer is in the system, the customer begins service immediately. Otherwise, the customer begins service when the previous customer completes service:

$$D_q^{-1}(n) = \max \{A^{-1}(n), D_s^{-1}(n - 1)\} \tag{4.17}$$

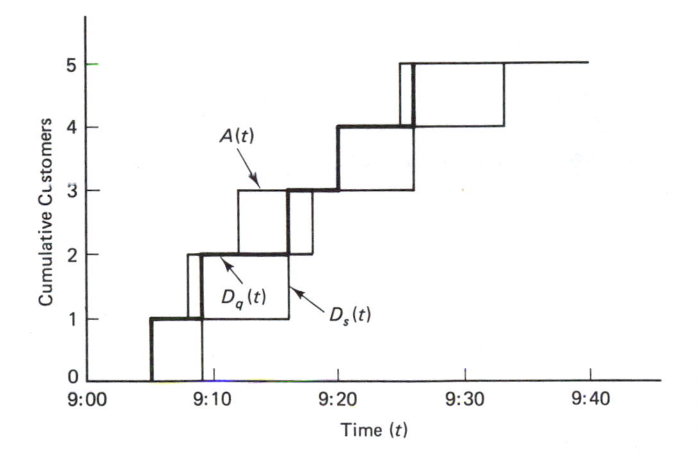

Figure 4.13 Cumulative arrival and departure diagram created from queue simulation.

The simulation proceeds in the following five steps, one for each customer:

1. $D_q^{-1}(1) = \max(9{:}05,0) = 9{:}05;\ D_s^{-1}(1) = 9{:}05 + 4 = 9{:}09$
2. $D_q^{-1}(2) = \max(9{:}08,9{:}09) = 9{:}09;\ D_s^{-1}(2) = 9{:}09 + 7 = 9{:}16$
3. $D_q^{-1}(3) = \max(9{:}12,9{:}16) = 9{:}16;\ D_s^{-1}(3) = 9{:}16 + 2 = 9{:}18$
4. $D_q^{-1}(4) = \max(9{:}20,9{:}18) = 9{:}20;\ D_s^{-1}(4) = 9{:}20 + 6 = 9{:}26$
5. $D_q^{-1}(5) = \max(9{:}25,9{:}26) = 9{:}26;\ D_s^{-1}(5) = 9{:}26 + 7 = 9{:}33$

As expected, the result is the same as before.

4.6 SERVICE TIME

The service times are simulated in the same manner that any other random variable is simulated, by first generating $U[0,1]$ random variables and then inverting the probability distribution function. Unlike interarrival times, however, there is seldom a theoretical basis for selecting any particular probability distribution. Studies of queueing systems have found many different distributions—gamma, normal, exponential, and so on—to be applicable under different circumstances. Further, the service time distribution does not necessarily conform to any of the familiar theoretical distributions. When used, theoretical distributions are primarily approximations to the empirical distribution. If the data appear to be exponential, then, for the sake of simplicity, it might be approximated by the exponential distribution. This does not necessarily mean that the exponential distribution is correct, only that the true distribution is something similar to the exponential distribution.

An important feature of the service time is that it can often be predicted in advance, should the server know something about the customer. As an example, the service time at a supermarket checkout counter is highly correlated with the number of items purchased. If the store's manager surveys the shopping carts in the checker queues, he or she can fairly accurately predict how long it will take to serve the customers. The same holds true for any queue where the amount of work can be ''seen'' in advance.

Awareness of the factors that cause service time variability changes the nature of the randomness. Rather than attribute the randomness directly to the service time, the randomness is attributed to the casual factor or factors. In the supermarket example, randomness is attributed to the variation in the number of items purchased. This, in turn, changes the nature of the simulation. Instead of simulating service times directly, the purchase quantities are first simulated and then transformed into service times.

In some situations, the causal factor might simply be a customer type.

Example

A firm has two categories of clients: commercial and general public. Commercial clients generally purchase goods in large quantities, while the general public purchases in smaller quantities. The probability distribution function for the service time is shown in Fig. 4.14.

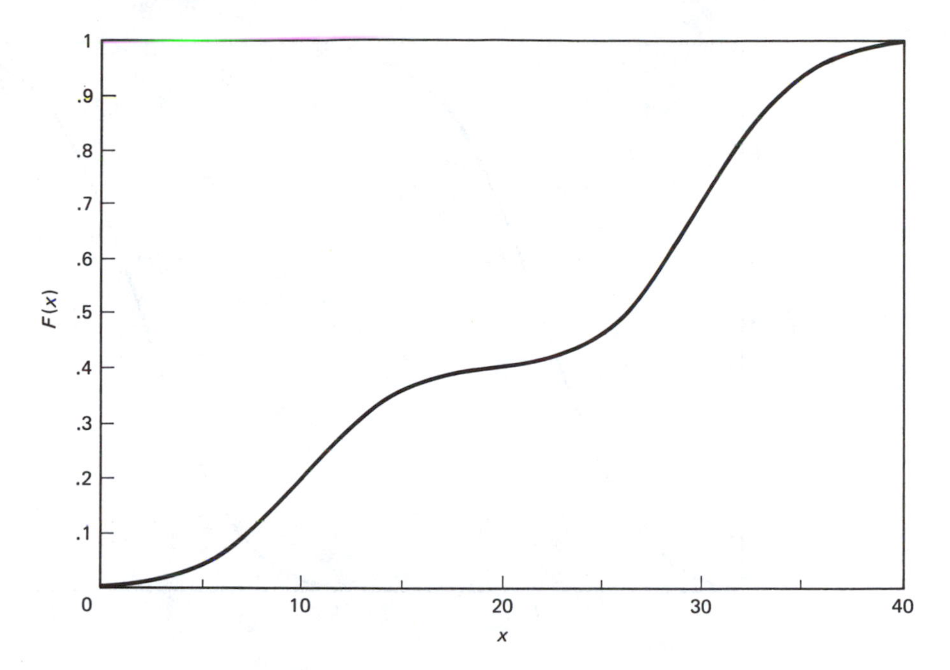

Figure 4.14 Bimodal probability distribution function.

> The bimodal (two-peak) distribution function reflects the aggregation of two distinct customer types into a single category. If the two types are examined separately, the distributions look like Fig. 4.15. These distributions are approximately normal. The simulation is then performed for each class of customer separately and the data are combined to obtain the complete simulation.

In the literature on queueing, the exponential distribution is used far more often than any other distribution to approximate the service time distribution. The reason for this is partly empirical. As early as Erlang's work (see Brockmeyer et al. 1948), it has been observed that the exponential distribution is a reasonable approximation for the length of telephone calls. But the widespread use of the exponential distribution is much more related to its inherent simplicity (and especially its memoryless property) than to any empirical evidence supporting its application. From the analytical perspective, the exponential distribution is by far the easiest service time distribution to evaluate, as is shown in Chap. 5. However, should one go to the trouble of simulating a queueing process, there is little to be gained by assuming the service time distribution is exponential or, for that matter, any other familiar distribution. The inherent strength of simulation is that it can deal with almost any distribution. It is a strength that should be exploited to the fullest.

If there is no reason to believe that the service time distribution conforms to any of the familiar theoretical distributions, the best approach may be to base the simulation on the empirical distribution or a smooth approximation to it. Unfortunately, organizations

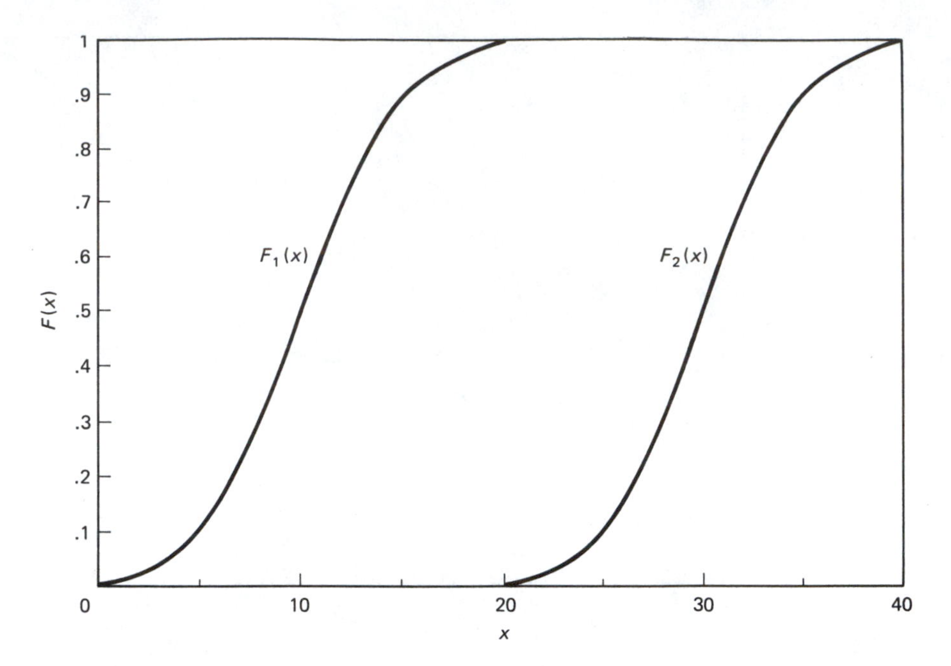

Figure 4.15 When the bimodal random variable is classified, two distinct normal distributions result.

do not always collect detailed data on service times. Perhaps they have a record of average service time and service time standard deviation, but they have no record of the service time distribution. If no further data can be collected, then one has no choice but to resort to one of the standard distributions. As a general rule, a good choice is the gamma distribution. Unlike the normal distribution, the gamma distribution is only defined for non-negative random variables, which must apply to service times. The two-parameter gamma distribution is also more general than the exponential distribution, which has a single parameter.

One should keep in mind that the comments made in Chap. 3 with regard to parameter estimation and confidence intervals also apply to service times. The precision of the estimate depends on the size of the data sample, and the more data the better. Unfortunately, service times are somewhat harder to measure than arrival times. (The estimate of the arrival rate only depends on a count of customers.) The mean service time can be estimated by dividing the total amount of busy server time by the number of customers served. Therefore, one must keep track of how much time the server is busy and how much time the server is idle, waiting for customers.

4.6.1 Simulation Comparison

As a point of contrast, Table 4.2 presents the results of three service time simulations for the automatic teller machine queue. The first uses the exponential distribution, the second the gamma distribution, and the third the empirical distribution. All three distributions

TABLE 4.2 SIMULATED SERVICE TIMES AT AUTOMATIC TELLER MACHINE (SECONDS)

U	Gamma	Empirical	Exponential	U	Gamma	Empirical	Exponential
0.090	27	34	5	0.694	57	52	59
0.614	53	49	47	0.334	40	41	20
0.899	73	75	114	0.428	45	43	28
0.587	51	48	44	0.899	73	75	114
0.926	78	82	130	0.101	28	34	5
0.715	59	52	62	0.797	64	57	79
0.932	80	87	134	0.449	46	59	30
0.387	43	43	24	0.077	26	33	4
0.900	73	75	115	0.173	33	30	9
0.222	35	38	12	0.276	37	42	16
0.924	77	82	128	0.909	75	78	119
0.569	51	46	42	0.606	52	49	46
0.674	56	51	56	0.560	50	46	41
0.395	43	43	25	0.458	46	46	31
0.151	31	36	8	0.352	41	41	22
0.990	105	141	229	0.033	38	29	2
0.827	66	60	87	0.570	51	47	42
0.388	43	43	24	0.368	42	42	23
0.312	39	40	19	0.284	38	40	17
0.964	88	105	166	0.048	24	31	2
0.148	31	35	8	0.279	38	39	16
0.317	39	40	19	0.486	47	44	33
0.033	38	29	2	0.621	53	50	48
0.396	43	43	25	0.800	65	57	80
0.769	62	54	73	0.329	40	41	20
0.907	75	78	118	0.243	36	39	14
0.386	43	43	24	0.055	24	32	3
0.425	45	43	28	0.316	39	40	19
0.537	49	46	38	0.219	34	38	12
0.394	43	43	25	0.794	64	57	79
0.269	37	39	16	0.671	56	51	55
0.379	43	42	24	0.485	47	44	33
0.387	43	43	24	0.523	49	46	37
0.608	52	49	47	0.803	65	57	81
0.772	62	55	74	0.298	39	40	18
0.087	27	33	5	0.097	27	34	5
0.845	68	60	93	0.325	40	41	20
0.397	43	43	25	0.687	57	51	58
0.874	71	67	103	0.003	13	27	0
0.839	67	60	91	0.742	60	54	68
0.638	54	50	51	0.667	56	51	55
0.829	67	60	88	0.225	43	38	13
0.086	27	33	4	0.518	46	46	36
0.557	50	46	41	0.683	57	51	57
0.980	95	111	195	0.691	57	52	59
0.674	56	51	56	0.253	37	39	15
0.760	61	54	71				

have the same mean (49.8 seconds). The gamma distribution has the same standard deviation as the empirical distribution (18.96 seconds), but the exponential distribution has a different standard deviation, because the parameter $1/\lambda$ is set to equal the mean. The three distributions are compared in Fig. 4.16. Note that the gamma distribution provides a much better fit than the exponential distribution to the empirical distribution. An even better fit could be obtained from a "shifted" gamma distribution—that is, a gamma distribution whose minimum value is shifted above the value zero (in Fig. 4.16, that value would be approximately 20). For the sake of simplicity, only the standard gamma distribution is used here.

The same set of $U[0,1]$ random variables was used to generate all three sets of service times in Table 4.2. So, to take an example, the first service time occurs at the 9th percentile for all three distributions (the gamma random variables are derived directly from the cumulative graph, not by the method in Sec. 4.3). Note that the simulated values are nearly the same for the gamma and the empirical distributions, but the exponential distribution shows greater variation. This is because the exponential distribution has a much larger standard deviation than the empirical distribution.

Cumulative arrivals and cumulative departures from the system are shown for the exponential distribution in Fig. 4.17. All three distributions use the same arrival data and differ only slightly in the departure curve. However, the larger service time variance of the exponential distribution leads to greater delay than by the other two distributions:

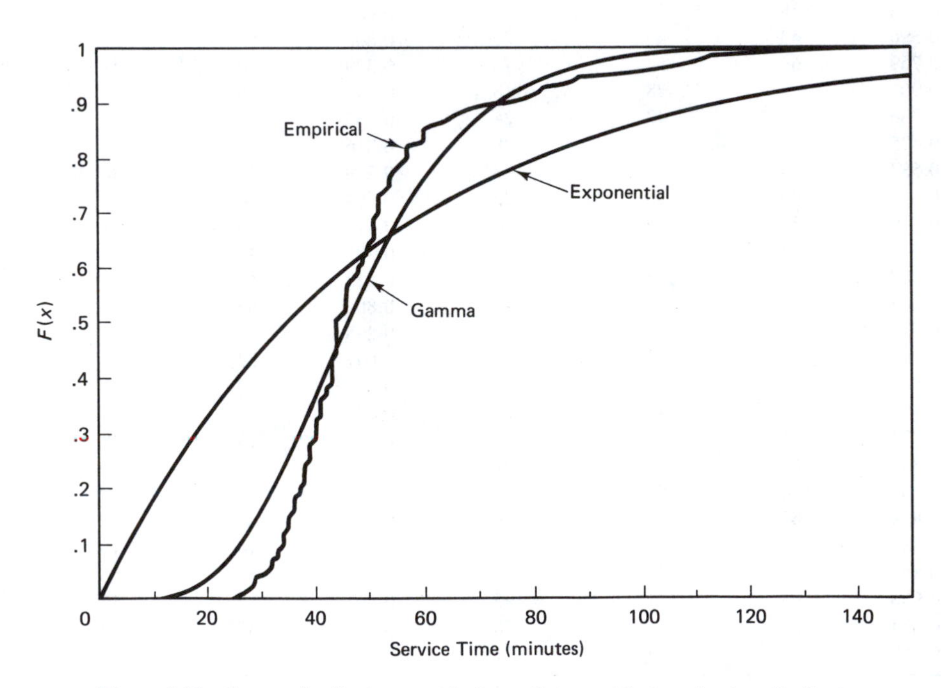

Figure 4.16 Gamma distribution provides better fit to empirical service time distribution than the exponential distribution.

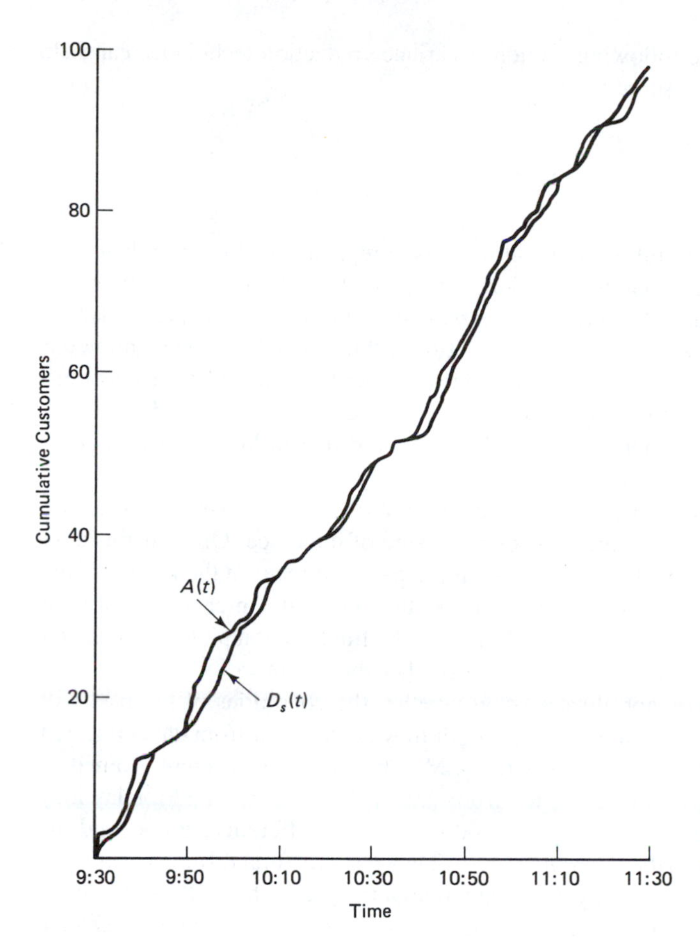

Figure 4.17 Simulation of arrival and departures with exponential service time distribution (diagrams for empirical and gamma distributions are similar).

Average Time in the Queue (W_q, minutes)

Exponential	Gamma	Empirical	Observed
1.302	1.075	1.105	.800

Not surprisingly, the gamma and empirical results are nearly the same, while the time in the queue is larger for the exponential distribution. Clearly, larger variation in service times leads to longer waits. It is somewhat of a surprise that the observed time in queue is smaller than all three simulations. This difference can be attributed to simple randomness. Because service times and arrival times vary, the performance measures will not be the same every time. This is especially true because waits and queue lengths exhibit a high degree of correlation—the wait for one customer is correlated with the wait for the previous customer. What all this means is that the simulation must be run for a long time in order to obtain a precise estimate for system performance. The accuracy of these estimates is discussed in the following section. (Variance reduction techniques can help here; see Wilson 1984 for a survey.)

4.7 PREDICTION

A model would not be very useful if it only allowed us to replicate the historical behavior of a system. A much more important function is to predict the future behavior of the system, especially the impact of alternative courses of action on system performance. Predictive models help choose the best course of action without actually implementing the actions or disturbing the system under study. Before a server is added to a queueing system, one would like to know how much waiting times will improve. Before the configuration of a queueing system is changed, one would like to know the change in service cost and the change in waiting time.

It is a far easier thing to model the past behavior of a system than to predict its future behavior. To predict the future requires a strong measure of inference. One must usually infer that the conditions that created the process in the past will recur in the future. If the arrival process last Monday was Poisson with rate λ, then one must infer that the arrival process next Monday will also be Poisson with rate λ. The fundamental philosophy of any predictive model is that the past can be used to predict the future.

Unfortunately, one does not always know whether the conditions of the past will recur. Data collected on the last Monday of a month may be different from data collected on the first Monday. Data collected on a rainy Monday may be different from data collected on a sunny Monday. Data collected from 9:00 to 10:00 A.M. on a Monday may be different from data collected from 2:00 to 3:00 P.M. These differences go beyond the simple randomness inherent to the Poisson process. They are structural differences that, at the very least, can cause a substantially different value of the parameter λ (λ itself may be treated as a random variable). One must also hope that the courses of action do not change the presumed conditions (for example, that adding servers does not change the arrival process). Applying a model designed for one set of conditions to an entirely different set of conditions is one of the dangers in prediction.

Simulation is an important method for predicting the performance of a system. But we must recognize that a single simulation of the process will not suffice. The outcome of a queueing process, as with the process it imitates, can differ from time to time, even though the process stays fundamentally the same. Prediction requires that we somehow combine the results of the simulations to estimate the performance of the system.

Suppose that a firm would like to predict the performance of a queueing system over a 2-hour period, from 9:00 to 11:00 A.M. This can be accomplished by repeatedly simulating arrivals and service times over 2-hour periods. Each simulation represents one day, and the more days that are simulated, the more precise will be the prediction of system performance. But there is usually a simpler alternative. Instead of simulating the same period again and again, a single simulation can be performed over a longer period of time. Instead of simulating the 2-hour period ten times, a 20-hour period can be simulated once. If a single simulation will have the correct properties of the random process, then the process is called ***ergodic***.

The accuracy of the second approach depends on the sensitivity of the simulation to starting conditions. Specifically, it depends on whether waiting time and queue length are sensitive to the number of customers in the system at the start of the 2-hour period. For example, if 9:00 is the opening time for a store, then the system will likely begin the period with no customers in the system. This means that waiting time will start out small and gradually increase. The single long simulation will not account for the lower waiting time at the start of the period and will overestimate waiting time (it also will not account for an arrival rate that varies from day to day).

Another concern is that there may still be customers in the system when the simulation ends. These customers will have received partial waits and partial services. Including these times in the estimates for average time in system or in queue will cause a downward bias. However, if the simulation is run over a long time period, the downward bias will be negligible.

Fortunately, many systems are not sensitive to the starting and ending conditions (particularly when the system operates below capacity), and the waiting time quickly approaches the long-run average, in which case either simulation approach works. Chapter 5 shows how to determine whether the system is sensitive to starting conditions. For the moment, we will be content with assuming that the system is insensitive to starting conditions.

4.7.1 Prediction of Performance Measure

Performance measures can be predicted directly or indirectly from the simulation. In the former case, predicting the performance measure is no different from estimating a performance measure from actual observations. The average waiting time from the simulation is the prediction for average waiting time in the future. The average queue length from the simulation is the prediction for average queue length in the future, and so on.

Performance measures can also be predicted indirectly. Average waiting time can be derived from average queue length with Little's formula. Dividing the average queue length by the *observed* arrival rate provides a prediction of the waiting time that is less sensitive to partial waits. Consider another example:

Example

A company would like to predict the percentage of customers who are served immediately (no wait in queue) at a single server queue, with Poisson arrivals. In a simulation, 57 percent of the arrivals were served immediately. However, in the same simulation, the system was empty 53 percent of the time. Fifty-three percent is the prediction.

Because customer arrivals are time and state independent, 53 percent of the customers should, on average, find the system empty upon arrival. The 57 percent figure is sensitive to the times that customers happened to arrive in the simulation. Fifty-three percent provides a more precise estimate of the performance measure.

4.7.2 Confidence Intervals

Conceptually and mathematically, this section is one of the most difficult in this book. It may take several readings to grasp its content fully.

We will consider only one technique for estimating the confidence interval for the mean value of a performance measure. The books cited at the end of this chapter expand this topic to other statistics and techniques. The following will distinguish the average value of a simulation from the long-run average through use of the bar symbol. For example

$$\bar{W}_s = \text{average time in system for simulation}$$

$$W_s = \text{expected long-run average time in system}$$

4.7.2.1. Confidence Interval for Average Time in System. The confidence interval for a prediction is similar to the confidence interval for a parameter estimate. However, one must be especially careful to account for dependence between random variables. In Table 4.3, two data sets are presented, one independent and the other dependent. The independent data are drawn from a uniform [0,1] distribution. The dependent data comprise a series of six mutually independent simulation runs. Each column is independent of the others, but the data within each column are dependent. Each data point is created by adding a uniform [0,1] random variable to the number above it. Hence, small values tend to be followed by small values and large values tend to be followed by large values.

The last number in each column is the average for the data points in the simulation run. Also shown are the standard deviations of the averages. The thing to note is that the averages show far greater variation in the dependent data than in the independent data. That is, a simulation of independent data is bound to produce more precise estimates than a simulation of dependent data (of identical size). Also note that ignoring dependencies in the calculation of a standard error grossly underestimates the true value.

Suppose that we would like to estimate the standard deviation of average time in system from a single simulation run, accounting for dependencies. Let $W_s(n)$ be the time in system for the nth customer. For large data sets, the average of $W_s(n)$, $\bar{W}_s$, should have a normal distribution. The variance of $\bar{W}_s$ is defined by the summation of all pairs of covariances:

$$V(\bar{W}_s) = \frac{1}{N^2} \sum_{n=1}^{N} \sum_{j=1}^{N} \text{COV}\{W_s(n),\ W_s(j)\} \tag{4.18}$$

$\text{COV}\{W_s(n), W_s(j)\}$ is the covariance between the nth and jth customers. The covariance between $W_s(n)$ and itself can be interpreted as the variance of $W_s(n)$.

TABLE 4.3 COMPARISON OF INDEPENDENT AND DEPENDENT DATA SETS

Independent Data

	Run number						
Observation	1	2	3	4	5	6	
1	.491	.895	.316	.410	.242	.221	
2	.869	.132	.962	.735	.800	.259	
3	.491	.210	.855	.625	.937	.530	
4	.639	.953	.102	.895	.946	.612	
5	.505	.101	.239	.488	.942	.819	
6	.092	.382	.663	.276	.869	.432	
7	.709	.341	.099	.417	.885	.453	
8	.308	.569	.737	.921	.175	.028	
9	.828	.285	.766	.938	.022	.648	
10	.224	.649	.398	.423	.638	.146	Standard deviation of averages
Average	.516	.452	.514	.613	.646	.415	.082

Standard Error Based on Equation 3.29 and First Run

Standard Deviation = .255

Standard Error = $.255/\sqrt{10} = .081 \approx .082$

Dependent Data

	Run number						
Observation	1	2	3	4	5	6	
1	.500	.500	.500	.500	.500	.500	
2	.417	.977	.593	.186	.755	.654	
3	.556	1.077	.215	.168	.716	.636	
4	.172	1.544	−.211	−.113	.225	.966	
5	−.114	1.100	−.154	−.021	.591	.788	
6	.111	.700	−.219	.300	.932	.652	
7	.429	.852	.180	.219	1.290	.832	
8	.600	.958	−.313	.280	1.352	.738	
9	.974	.737	−.230	.747	1.274	1.037	
10	.499	.538	−.544	1.122	.841	.584	Standard deviation of averages
Average	.414	.898	−.018	.339	.847	.739	.324

Incorrect standard error based on Eq. (3.29) and first run: Standard deviation = .300; standard error = $.300/\sqrt{10} = .095 < .324$.

Note: Failure to account for positive correlations will underestimate the standard error.

Should the random variables $W_s(n)$ be independent and identically distributed, all but the variance terms will equal zero, and Eq. (4.18) reduces to the form seen earlier in Chap. 3.

Independent Identically Distributed

$$V(\bar{X}) = \sigma_{\bar{X}}^2 = \frac{\sigma^2}{N} \tag{4.19}$$

where

$\bar{X}$ = the average of a set of N random variables

σ = the standard deviation of the random variable

$\sigma_{\bar{X}}$ = the standard error of $\bar{X}$ (the standard deviation of the average)

We have already seen that Eq. (4.19) applies to interarrival times from the Poisson process because interarrival times are independent. However, Eq. (4.19) does not apply to most queueing measures of performance because the measures are not independent. If one customer endures a long wait, the subsequent customer will also likely endure a long wait. If the queue is long at 9:00, the queue will also likely be long at 9:01, and so on. This type of dependence must be accounted for.

If a queueing simulation is allowed to run for a long period of time, it may enter a condition known as steady state. ***Steady state*** means that the probability distribution for $L_s(t)$ and $L_q(t)$ does not vary with time. Once this condition occurs, Eq. (4.18) might be simplified by focusing on dependencies between $W_s(n)$ and a limited set of values $\{W_s(j)\}$, $j \geq n - M$. The value M reflects a ***relaxation time***, the time required for the system to "forget" its initial state. Waiting times prior to $n - M$ only have a small impact on $W_s(n)$, and the dependency can be ignored. This leads to the following equation for the variance of the average of N identically distributed random variables.

Identically Distributed/Not Independent ($N>>M$)

$$V(\bar{X}) = \frac{\sigma^2}{N} \left[1 + 2 \sum_{m=1}^{M} \rho_m\right] \tag{4.20}$$

where

ρ_m = the correlation coefficient between $W_s(n)$ and $W_s(n - m)$

$$= \frac{\text{COV}\{W_s(n), W_s(n - m)\}}{\sigma^2}$$

M = maximum value of m for which $\rho_m \neq 0$

$V(\bar{W}_s)$ can be estimated conservatively by calculating the standard deviation of a limited set of waiting times that are sufficiently far apart to be nearly independent. That is, $V(\bar{W}_s)$ can be estimated by first estimating the variance among a data set comprising every Mth data point $[W_s(M), W_s(2M), W_s(3M), \ldots]$, then dividing the result by the number of observations in the reduced data set. This is done in the following complicated expression:

$$\hat{V}(\bar{W}_s) = \frac{1}{k(k - 1)} \sum_{i=1}^{k} \left[W_s(iM) - \frac{1}{k} \sum_{j=1}^{k} W_s(jM)\right]^2 \tag{4.21}$$

where k is the largest integer less than or equal to N/M (which is the size of the reduced data set).

Equation (4.21) is actually based on the simpler Eq. (4.19). The outside summation, divided by $k - 1$, is the estimator for the variance, and k is equivalent to N, the number of random variables. Equation (4.21) applies because the data points are spread sufficiently far apart to be effectively independent.

The value of M should represent the length of time that it takes a queue of average length to reach zero and re-form to a length that is independent of its initial state. Newell (1982) created the following approximation:

$$M \geqslant \frac{\lambda/c}{[1 - (\lambda/c)]^2} \tag{4.22}$$

where c is the combined service rate among all servers (that is, the system capacity, in customers per unit time).

If the correlation coefficients are positive (as should be the case for all queueing systems), Eq. (4.20) provides a larger estimate of the variance than does Eq. (4.19). This means that the confidence interval is larger than it would be if the random variables were independent. Put another way, *positive correlations reduce the effective size of the data set*. If $M = 10$, the number of "good" data points is one-tenth of the total number of data points. This, again, means that the simulation must be run for a long time to obtain precise estimates of system performance.

The following example uses the same method for W_q as was used for W_s. $W_q(n)$ represents the time in queue for the nth customer.

Example

The average time in queue from the simulation of the teller queue was 1.105 minutes (empirical service time distribution). The arrival rate, λ, is $98/120 = .817$ customers per minute. The average service time, $1/\mu$, is 49.8 seconds, or .83 minutes. Thus, M is approximately

$$M \approx \frac{(.817)(.83)}{[1 - (.817)(.83)]^2} = 6.5 \text{ observations}$$

If M is rounded up to 7, $V(\bar{W}_q)$ is estimated by calculating the variance along the seventh observation, the fourteenth observation, twenty-first observation, and so on, with Eq. (4.21). The data are provided below:

n	7	14	21	28	35	42	49
$W_q(n)$	0	.029	.642	.257	.416	.722	3.015

n	56	63	70	77	84	91
$W_q(n)$	2.562	4.926	2.343	.585	.188	0

First, the sample variance is computed among the 13 data points (1.53^2), then divided by the size of the data set:

$$\hat{V}(\bar{W}_q) = 1.53^2/13 = .18 \text{ minute}^2$$

Due to the small sample size (13 data points, 12 degrees of freedom), a 95% confidence interval should be constructed from the t distribution (Table A.3 in the appendix) rather than the normal distribution (these distributions are nearly identical when the sample size is larger than 50). Applying the methodology of Sec. 3.6.1, we get

$$P(1.105 - 2.18\sqrt{.18} \leq W_q \leq 1.105 + 2.18\sqrt{.18}) = .95$$

$$P(.180 \leq W_q \leq 2.03) = .95$$

Note that the confidence interval is quite large relative to W_q, so a much longer simulation must be carried out to obtain a precise estimate of time in queue. Had the correlations not been accounted for—that is, had all the data been used in estimating $V(\bar{W}_q)$—the confidence interval would have been [.811,1.40], 67% smaller.

Whether or not the data are independent, it is important to keep in mind that the standard error of an average is less than the standard deviation of the individual data points. That is, an average among N data points will show less variation than the points constituting the average.

It is also important to keep in mind that treating dependent data as independent data will underestimate the standard error of an average. Table 4.4 gives the average time in queue from 32 one-hour simulations of a single server queue. The system has Poisson arrivals (40/hour) and a shifted exponential service time distribution (mean 60 seconds and standard deviation 30 seconds). The standard deviation among the 32 averages is .011 hour, and the overall average time in queue is .020 hour.

In Table 4.4, Eq. (4.21) is applied to the first of the 32 runs. The standard deviation among every sixth data point is .027 hour. Dividing this result by the square root of the sample size (7) yields the standard error: 0.10 hour. This number is similar to the standard deviation among the averages provided earlier.

TABLE 4.4 AVERAGE WAITING TIMES FROM INDEPENDENT QUEUE RUNS

Average Time in Queue for One-Hour Simulation*

.0328	.0057	.0023	.0104	.0084	.0213	.0398	.0282	.0327	.0100
.0080	.0180	.0131	.0127	.0057	.0190	.0321	.0146	.0125	.0065
.0090	.0120	.0050	.0100	.0110	.0020	.0140	.0065	.0195	.1572
.0327	.0225								

Standard deviation among averages = .011 hour
Overall average = .020 hour
Standard error for overall average = $.011/\sqrt{32}$ = .002 hour

Estimation of Standard Deviation Among Averages from Single Run

Wait in Queue for Every Sixth Customer

.0185	.0372	.0704	.0707	.0748	.0630	.0090

Standard deviation among waits = .0272 hour
Standard error = $.0272/\sqrt{7}$ = .010 ≈ .011 hour

*Poisson arrivals (40/hour).
Shifted exponential service times (mean 60 seconds, s.d. 30 seconds).

4.7.2.2 Queue Lengths. Extra care must be taken in deriving the confidence interval for average queue length (customers in queue or customers in system). Whereas the number of data points for the waiting time is limited by the number of customers served, the number of data points for queue length is unlimited. Queue length can be measured at 9:00, 9:01, 9:01 and 10 seconds. Fortunately, all of these data points do not have to be considered.

L_q and L_s are most easily estimated by first calculating average time in system and average time in queue and then multiplying the results by $\lambda = A(T)/T$, where $A(T)$ is the number of customers served during the simulation. This is a natural application of Little's formula. Determining $\hat{V}(\bar{L}_q)$ and $\hat{V}(\bar{L}_s)$ requires separate calculations, based directly on queue lengths. This is accomplished by dividing the time period into equal sized time increments, of sufficient length that the queue size at the end of any increment is nearly independent of the queue size at the end of the next increment. The minimum size of these increments, T_0, is the relaxation time, which is defined as follows:

$$T_0 = M/\lambda = \frac{1/c}{[1 - (\lambda/c)]^2} \tag{4.23}$$

Thus, the estimate for $V(\bar{L}_q)$ is based on the variance of a data set comprising $L_q(T_0)$, $L_q(2T_0)$, $L_q(3T_0)$, . . . , and is calculated in the same manner as Eq. (4.21). The same approach is used for $V(\bar{L}_s)$.

Example

From the simulation of the teller queue (with empirical distribution), the average time in queue was estimated to be 1.105 minutes. Applying Little's formula, we find the average number of customers in queue is $1.105 \cdot (92/120) = .847$ customer (the number of arrivals from the simulation is used here). To calculate $\hat{V}(\bar{L}_q)$, T_0 must be calculated first:

$$T_0 = M/\lambda = 6.5/(98/120) \approx 8 \text{ minutes}$$

The number of customers in the queue was determined for the following times:

Time	9:38	9:46	9:54	10:02	10:10	10:18	10:26	
$L_q(t)$	0	0	0	2	1	0	0	
Time	10:34	10:42	10:50	10:58	11:06	11:14	11:22	11:30
$L_q(t)$	0	2	7	0	0	0	0	0

Calculating the variance of the above data set provides the following: $\hat{V}(\bar{L}_q) = .246$. A 95% confidence interval can be constructed from the t distribution with 14 degrees of freedom:

$$P(.847 - 2.145\sqrt{.246} \leq L_q \leq .847 + 2.145\sqrt{.246}) = .95$$

$$P(-.218 \leq L_q \leq 1.91) = .95$$

The fact that the lower end of the confidence interval is negative indicates that the simulation would have to be run over a longer period of time to obtain a precise estimate of L_q. It also indicates that the normal distribution may not be applicable to such a small sample.

4.7.3 Confidence Intervals from Many Short Simulations

The approach of running many short simulations rather than one long one has the advantage of eliminating some of the dependency issues. The average performance for any run is independent of the average performance for any other run. These averages constitute a set of independent data. Therefore, the confidence interval for the overall average performance (the average of the averages) can be computed directly from Eq. 4.19).

Example

Returning to Table 4.4, the overall average time in queue is .020 hour. The standard deviation among the averages is .011 hour. Dividing this value by the square root of the number of observations (32) yields .002 hour. This is the standard error for the prediction of average time in queue, based on all 32 runs.

4.7.4 Parameter Uncertainty

A prediction is only meaningful if the model is structurally correct. If the assumed conditions do not materialize in the future, then the prediction will be wrong—wrong no matter how much care is invested in the simulation. One source of structural error is parameter misestimation. As seen in Chap. 3, the arrival rate has a confidence interval just as the prediction has a confidence interval. The 95% confidence range for $1/\lambda$ from the teller data presented in Sec. 3.6.1 was from .98 to 1.46. Without knowing the true value of λ, it is impossible to predict exactly the performance of the system.

The confidence intervals presented in the previous section have little meaning if the parameter is not known with certainty. For example, Eq. (4.20) implies that the confidence interval can be made as small as desired, simply by repeating the simulation. One might get the impression that simulating a system over a long enough time period can compensate for imprecise parameter estimates. This surely is not the case. *If the parameter is uncertain, then the prediction must also be uncertain, no matter how much care is invested in the simulation*. Furthermore, uncertainty in the value of a parameter is magnified when the performance measure is calculated. If λ is estimated to be some value, plus of minus 10 percent, then time in queue might only be known with a confidence interval of plus or minus 20 percent.

The simulation should account for two sources of uncertainty. The first, structural uncertainty associated with model parameters, cannot be eliminated through extended simulation. The second, uncertainty inherent to the model (for example, uncertainty as to when customers arrive), *can* be eliminated through extended simulation. The confidence interval for the prediction should account for both sources of uncertainty.

The task is accomplished through a series of simulations, each for a different value of the parameter (or parameters) that is not known with certainty. For the purpose of demonstration, suppose that the mean service time is known to be 49.8 seconds, but the interarrival time, $1/\lambda$, is not known with certainty. Each simulation is based on a different value of $1/\lambda$, denoted by $1/\lambda_i$. For each value of $1/\lambda_i$, the simulation is performed in the exact manner described in Secs. 4.5 and 4.6. The performance measures are estimated in the manner mentioned in Sec. 4.7.1, and the confidence interval is estimated in the exact manner mentioned in Sec. 4.7.2. Where the simulations diverge is in the interpretation of the confidence interval. Rather than apply to the prediction of the performance measure, the confidence interval applies to the prediction of the performance measure *given that the assumed value of $1/\lambda$ is correct*. Should $1/\lambda$ not be the true value of $1/\lambda$, then the confidence interval is moot.

It has already been mentioned that the simulation is repeated for different values of $1/\lambda$. These values are not arbitrarily selected, but systematically selected from the probability distribution for $\widehat{1/\lambda}$. Because $\widehat{1/\lambda}$ is the average of a large sample, it must have the normal distribution, with mean and standard deviation estimated from the sample average and sample standard deviation. The values of $1/\lambda$ used in the simulation should come from equally spaced percentiles of this probability distribution. If ten simulations are performed, $1/\lambda$ should come from 5%, 15%, . . . , 95%.

Each simulation defines one point on a probability distribution for the performance measure. When drawn properly, the distribution will also indicate both types of uncertainty—uncertainty associated with misestimation of model parameters and uncertainty inherent to the model. This is best appreciated through an example.

Example

The simulation for the teller queue was repeated 19 times for 19 different values of $1/\lambda$ (with exponential service time distribution). The values of $1/\lambda$ are drawn from equally spaced percentiles of the normal (1.22,.123) distribution. For each value of $1/\lambda$, $\bar{W}_q$ and $\hat{V}(\bar{W}_q)$ were calculated in the manner described in Sec. 4.7.2.1. The results follow:

Percentile	$1/\lambda$	$\bar{W}_q$	$\sqrt{\hat{V}(\bar{W}_q)}$
5%	1.02	4.20	.91
10	1.06	2.70	.60
15	1.09	1.56	.44
20	1.12	1.87	.60
25	1.14	1.21	.45
30	1.16	1.88	.38
35	1.17	1.03	.24
40	1.19	1.34	.24
45	1.20	1.23	.28
50	1.22	1.96	.58
55	1.24	.79	.19
60	1.25	2.11	.53
65	1.27	.98	.37
70	1.28	1.07	.15
75	1.30	.89	.35
80	1.32	1.39	.37
85	1.35	.70	.13
90	1.38	.89	.17
95	1.42	.85	.23

The data are plotted as a cumulative probability distribution function in Fig. 4.18. Percentiles are shown on the vertical axis and waiting times are shown on the horizontal axis. The 19 points on the distribution function correspond to the 19 estimates for W_q provided in the table. As a measure of confidence in the estimates, $\bar{W}_q \pm \sqrt{\hat{V}(\bar{W}_q)}$ are also shown (this is a 70% confidence interval). The graph indicates that there is about a 95% chance that W_q is less than 4.5 minutes and about a 50% chance that W_q is less than 1.5 minutes.

The shape of the distribution function represents uncertainty associated with the estimation of the model parameters; the confidence intervals represent uncertainty inherent to the simulation. Extending the simulations can only reduce the latter source of uncertainty.

Figure 4.18 illustrates how a small amount of parameter uncertainty can lead to large uncertainty in W_q (as well as in L_s, L_q, and W_s). The 90% confidence range for $1/\lambda$ was approximately 1.22 ± 17%. The 90% confidence range for W_q runs from about .8 to 4.5, up to 200% more than the "best estimate" (the estimate based on $\lambda = 98/120$). Clearly, small changes in the arrival rate can lead to large changes in system performance. This subject is explored in greater detail in Chap. 5.

A slightly different approach may be called for if more than one parameter is not known with certainty. Instead of systematically selecting parameters from equally spaced percentiles, the parameters themselves may be randomly simulated. Repeatedly simulating the system for different combinations of parameters can then provide the desired distribution for the performance measure.

Keep in mind that the way to resolve parameter uncertainty is to collect more and/or better data. A more precise estimate of the parameter provides a more precise estimate of the performance measure. Repeating the simulation cannot overcome this uncertainty.

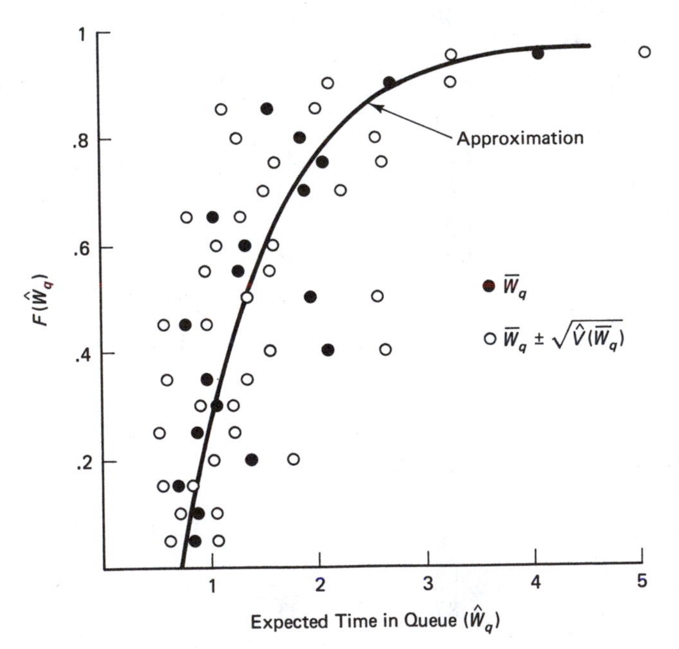

Figure 4.18 Probability distribution function for time in queue depends on two sources of randomness, parameter misestimation and simulation error. No matter how many replications of a simulation, parameter errors cannot be eliminated.

4.8 CHAPTER SUMMARY

Monte Carlo simulation is a powerful technique for evaluating stochastic models. It allows us to learn about the behavior of the system by experimenting with a copy of the system rather than the system itself. It does this by evaluating randomly created scenarios, which represent plausible sets of circumstances that the system might encounter.

Computer simulations begin by creating pseudo-random variables with the uniform [0,1] distribution. Though these variables possess the characteristics of the uniform distribution, they are not truly random, for the process that creates them is predictable. The desired probability distribution is obtained by substituting the uniform [0,1] variables in the inverse of the distribution function.

The Poisson process can be simulated by summing exponential random variables, each of which represents an interarrival time. Simple queueing processes can be simulated by simulating two separate data sets, one representing arrivals and the other representing service times. These data are iteratively combined to derive the departure curves and complete the simulation. The service time distribution can take on many different forms, depending on the process observed. Unlike the arrival process, there is usually no theoretical justification for choosing one distribution over another.

Simulation can be used to predict the future performance of a queueing system. However, one must usually infer that the conditions that created the process in the past will recur in the future. The fundamental philosophy of any predictive model is that the past can be used to predict the future. The average performance of the system is predicted by repeatedly simulating arrivals and departures. Confidence intervals for the simulation are created from the variance of a set of nearly independent data points.

Just as in a real process (the process being imitated), Monte Carlo simulation does not produce the same results every time. From time to time, a process may create unusual data that appear to have been produced by an entirely different process. This does not mean that the model is incorrect, only that something unusual has happened. By the same token, consistency in simulation results does not imply that the model is correct. As indicated in Chap. 3, prediction errors can result from the following:

1. The model is structurally incorrect (for example, customers do not arrive by a Poisson process).
2. Parameters are incorrectly estimated.

Hence, a simulation model can be incorrect for any of three reasons: structural error, parameter error, or simulation error. Repeatedly simulating a process only addresses the third source of error.

FURTHER READING

BANKS, J. and J. S. CARSON II. 1984. *Discrete-Event System Simulation*. Englewood Cliffs, N.J.: Prentice Hall.

BOX, G. E. P., and M. E. MULLER, 1958. ''A Note on the Generation of Random Normal Deviates,'' 29, 610–611.

BROCKMEYER, E., H. A. HALSTROM and A. JENSEN. 1948. *The Life and Works of A. K. Erlang*, Copenhagen: Danish Academy of Technical Science.

FISHMAN, GEORGE S. 1978. *Principles of Discrete Event Simulation*, New York: John Wiley.

IGNALL, E. J., P. KOLESAR, and W. E. WALKER. 1978. "Using Simulation to Develop and Validate Analytic Models: Some Case Studies," *Operations Research*, 26, 237–253.

LAW, A. M., AND W. D. KELTON. 1982. *Simulation Modeling and Analysis*. New York: McGraw-Hill.

NEWELL, G. F. 1982. *Applications of Queueing Theory*. London: Chapman and Hall.

PANKRATZ, A. 1983. *Forecasting with Univariate Box-Jenkins Models, Concepts and Cases*. New York: John Wiley.

WILSON, J. R. 1984. "Variance Reduction Techniques for Digital Simulation," *American Journal of Mathematical and Management Sciences*, 4, 277–312.

ZEIGLER, B. P. 1984. *Multifaceted Modeling and Discrete Event Simulation*. New York: Academic Press.

PROBLEMS

1. Simulate three uniform [0,1] random variables using Eq. (4.1) with $m = 10{,}000$, $a = 101$, and $b = 7$.
   (a) Use 9500 as your random number seed.
   (b) Use 240 as your random number seed.
2. Simulate two values of the continuous probability distribution defined by Fig. 4.1 by transforming the uniform random variables: (a) .4, (b) .95.
3. Repeat Prob. 2 for the discrete probability distribution in Fig. 4.2 and the empirical probability distribution in Fig. 4.3.
4. Simulate a binomial random variable, with $p = 1$ and $n = 4$, from its distribution function. Transform the $U[0,1]$ random variables: (a) .2, (b) .348, and (c) .74.
5. Transform the $U[0,1]$ random variables .01 and .5 into the following distributions:
   (a) Uniform $[-5,100]$
   (b) Exponential, mean 100
   (c) Normal, mean 100, variance 100
6. Generate a gamma (3,50) random variable. Use the $U[0,1]$ random variables .4, .9, and .15 to generate the exponential random variables.
7. Simulate a Poisson process with rate 5 per hour, over a 1-hour period. Base your simulation on the generation of exponential random variables, using the following $U[0,1]$ random variables: .742, .683, .112, .420, .903, .163, .436, .984.
8. Again simulate a Poisson process with rate 5 per hour, over a 1-hour period. However, base your simulation on the generation of Poisson and uniform random variables. Use the $U[0,1]$ random variables from Prob. 7.
9. The interarrival times between airplanes at a major airport were found to be independent and identically distributed with a U[1 min,2 min] distribution. The last plane arrived at 1:43. Simulate the arrival times of the next four planes ($U[0,1]$: .742, .683, .112, .420).
10. At the airport in Prob. 9, the distribution for the number of passengers per plane is normal, with mean 150 and standard deviation 50. Using your answer to Prob. 9 as a starting point,

simulate the arrivals of passengers for the first four planes ($U[0,1]$: .903, .163, .436, .984). Give cumulative arrivals of passengers as a function of time.

***11.** Buses are scheduled to arrive on a busy line every 30 minutes, but half fall behind schedule. For the half that arrive late, the lateness has an exponential distribution, with mean 5 minutes. Simulate the arrival times of five buses, beginning with the bus scheduled to arrive at 9:00 A.M. Base your simulation on the transformation of the third through fifth digits of phone numbers that you select from the white pages of your local directory. (Show the phone numbers along with your answer.)

12. The arrivals described in Prob. 7 enter a queueing system in which the service time has the following probability distribution function:

$$P(S \leq s) = \begin{cases} (5s)^2 & 0 \leq s \leq .2 \text{ hour} \\ 1 & s > .2 \text{ hour} \end{cases}$$

(a) For each of your arrivals in Prob. 7, simulate the service time ($U[0,1]$ = .940, .154, .687, .432, .444, .775, .482, .982, .145).

(b) Assuming an FCFS discipline with a single server, plot cumulative arrivals, cumulative departures from queue, and cumulative departures from the system. (Assume system begins with no customers in the system.)

13. Repeat Prob. 12, using the arrival times from Prob. 8.

***14.** Upon arriving, suppose that customers examine a queue's length and decide whether or not it is worthwhile waiting. If one customer is in the system, the arriving customer will renege with probability .2; with two customers, probability .5; with three customers, probability .8; and with four customers, probability .9. Repeat Prob. 12 using this reneging process ($U[0,1]$ = .940, .154, .687, .432, .444, .775, .482, .982, .145). Do not use the same $U[0,1]$ variable to generate more than one random variable.

***15.** The following waiting times (in system) were found in a simulation of 100 arrivals over a 1.08-hour period at a queueing system.

0.46	1.15	0.82	0.64	1.44	3.02	1.56	1.23	0.03	1.24
1.35	1.56	1.23	1.82	1.26	2.80	2.27	2.19	0.06	1.50
1.38	0.37	1.62	2.00	1.24	4.01	2.12	2.21	0.10	1.53
1.23	0.42	1.65	2.10	0.71	3.32	2.36	2.06	0.29	2.27
1.02	0.32	2.04	2.69	1.86	2.09	2.29	1.96	0.27	2.48
0.75	0.04	2.14	2.62	2.00	0.57	3.85	2.34	0.07	1.70
0.97	0.90	1.68	2.38	2.12	0.06	3.59	2.09	0.14	2.27
0.54	0.55	2.70	1.73	3.01	0.75	2.23	1.72	0.67	2.34
0.68	0.09	2.37	2.72	3.27	1.55	1.46	0.05	1.13	1.90
0.75	1.28	1.37	3.30	3.05	1.96	1.51	.00	1.58	1.45

Average waiting = 1.58 minutes.

(a) Derive an estimate for mean time in system.

(b) If the service capacity is 140 customers/hour, what is the relaxation time?

(c) Derive a 95% confidence interval for your estimate of mean time in system.

16. For the data provided in Prob. 15, describe how you would develop a 90% confidence interval for customers in system.

*Difficult problem

EXERCISE: SIMULATION

The purpose of this exercise is to learn how to simulate a queueing process. Your simulation should be based on the observations recorded in Chap. 2. Parts 1–8 can be performed either by hand or with a computer. If performed by hand, use a random number table to generate the uniform random variables.

1. Create two tables of the form shown below:

Customer	$A^{-1}(n)$	$S(n)$	$D_q^{-1}(n)$	$D_s^{-1}(n)$

2. Assuming that customers arrive by a Poisson process, simulate customer arrivals for a 2-hour period.

3. Using the empirical probability distribution, simulate customer service times.

4. Combine the data sets to obtain $D_q^{-1}(n)$ and $D_s^{-1}(n)$.

5. Based on your simulation, calculate the proportion of time that there were no persons in the system, one person in the system,. . . . Repeat this calculation based on your observations in Chap. 2. List your results in a single table, and plot them on a piece of graph paper. Describe why or why not the lists are different.

6. Calculate $\bar{L}_q$, $\bar{L}_s$, $\bar{W}_q$, and $\bar{W}_s$. Assuming that your parameter estimates are correct, calculate 95% and 99% confidence intervals for W_q.

7. Repeat parts 2–6 with an exponential service time (with mean equal to the observed mean).

8. Compare your two estimates for L_q, L_s, W_q, and W_s to the values observed in Chap. 2 and to each other. Discuss why they are the same or different.

chapter 5

Steady-State Analysis

Each Monte Carlo simulation reflects a plausible outcome of a stochastic process. It represents something that might actually occur, given that the conditions underlying the model are correct. To predict the average performance, the stochastic process must be repeatedly simulated. The average performance among the simulations is the prediction for the expected performance of the system.

The strength of Monte Carlo simulation is that it can deal with virtually any probability distribution. It is a very robust technique. This strength comes about because simulation is, in essence, a brute force approach to evaluation. Simulations rely on computers to crunch through calculations, calculations that could never be performed by hand. While robust, the brute force approach lacks a certain elegance, an elegance often needed to derive meaning from the model. Should one wish to understand the relationship between time in system and the arrival rate, a separate simulation would be required for each arrival rate. Only after these points are plotted would the relationship be revealed. This contrasts with the analytical approach used to evaluate the Poisson process in Chap. 3, in which the key properties of the process were found through algebra and calculus. The facts that the number of events within a time interval has a Poisson distribution, that the interarrival time has an exponential distribution, and so on, were all found without resorting to simulation.

This raises the question of how a queueing process might be analyzed directly, without simulation. In this chapter, one such approach is presented. It applies to a limited class of queueing systems that are said to operate in ***steady state***. That is, it applies to queueing systems for which the probability distribution for the queue length is time independent. Of course, no system can enter steady state unless the arrival process is stationary. The service process must also be time independent (for example, service time does not vary with time). However, the arrival process does not have to possess independent increments, nor does it have to be Poisson, though the steady-state properties are much easier to analyze when it is Poisson.

Stationarity in the arrival and service processes still does not guarantee that the queue will be in steady state at any point in time. The probability distribution for the queue length depends on how many customers were in the system at the time the process began. Over time, the distribution of queue length should approach the steady-state distribution. But in the meantime, the queue length will display ***transient*** (passing away with time) behavior, which may be very different from the steady-state behavior. It is also possible that the stationary conditions end before the system leaves the transient phase, meaning that the system never enters steady state.

Despite the shortcomings of steady-state analysis, it is one of the few situations in which the properties of a queueing system can be analyzed in detail. With this in mind, steady-state models serve a pedagogical function more than a prescriptive function. Steady-state analysis is a way to understand how queues operate, but it cannot always be relied on for predicting queueing behavior.

This chapter begins by introducing the modeling concept of state probabilities. From these probabilities, general techniques are presented for evaluating the system measures of performance. The chapter next shows how to calculate the steady-state probabilities for a single server queue with Poisson arrivals and exponential service times. Then a more general model is presented for the state probabilities and applied to several difficult types of queueing systems. Next, the transient behavior of a queueing system is discussed. The chapter concludes with a model for alternating servers between direct customer service and ancillary activities. Chapter 5 is largely mathematical and may require more than one reading to grasp its content fully.

5.1 STATE PROBABILITIES

Before one speaks of state probabilities, one must first speak of the state. State is quite a general word. It can be used as a verb, as in "state your intentions." And it can be used as a noun, as in the "United States of America." In queueing, *state* is used in the context of "a condition or mode of being." It is a way to describe completely the condition of a queueing system. Certainly, the state must encompass one key variable: the number of customers in the system. The state may also encompass other pertinent factors, such as the number of servers working, or even the specific servers working.

The choice of state variables must always be guided by the application. And, considering that the objective here is to present pedagogical (rather than prescriptive) models, simplicity must sometimes outweigh realism. For the moment, this means that the state will be defined solely by the number of customers in the system. The number of servers working will be assumed to be constant and each server is assumed to work individually, one customer at a time, which means that the number of customers in the system also determines the number of customers in the queue.

The state probability is the probability that the system is in a given state at any point in time:

Definition 5.1

P_n = ***steady-state probability*** that there are n customers in the system

The state probability represents an *expectation* for the proportion of time that the system will be in state n. If the queueing process is ***ergodic***, then we know that if the system is operated over a long time interval, the proportion of time that the system is in any state n will approach P_n. Most queueing systems with stationary arrival and service processes are ergodic. However, one can create examples of stochastic processes that enter steady state but are still not ergodic.

Example

The first random variable of a sequence, X_1, is randomly selected from the set $\{1,2,3\}$ with equal probability. Each subsequent variable is selected from the set $\{X_1 - 1, X_1, X_1 + 1\}$ with equal probability.

The system enters steady state, but it is not ergodic. The state probabilities are $P_0 = 1/9$, $P_1 = 2/9$, $P_2 = 3/9$, $P_3 = 2/9$, $P_4 = 1/9$. However, no matter which value of X_1 is selected, the proportion of time that the system is in state n will not approach P_n (for example, if $X_1 = 1$, the system will be in state 0, 1, and 2 one-third of the time each). Again, the nonergodic steady-state process is more the exception than the rule.

5.2 GENERAL TECHNIQUES FOR CALCULATING MEASURES OF PEFORMANCE

This section shows how to use the state probabilities to derive important measures of system performance, such as expected time in system or expected customers in queue. The techniques provided here will be general, so it is important to understand them before proceeding to subsequent sections. In those subsequent sections, techniques are provided for calculating the state probabilities for different types of queueing systems.

The expected number of customers in the system, L_s, is the easiest performance measure to derive. It is simply the mean value of the state:

$$L_s = \sum_{n=0}^{\infty} nP_n \tag{5.1}$$

The expected number of customers in queue is also easily derived. Suppose that there are m servers. If m or fewer customers are in the system, then no one is in queue. If more than m customers are in the system, then $n - m$ customers are in queue. Thus, the expected number of customers in the queue is

$$L_q = \sum_{n=m}^{\infty} (n - m)P_n \tag{5.2}$$

The difference between L_s and L_q equals the expected number of customers in service, which also equals the expected number of busy servers.

The standard deviation of the number of customers in the queue and the number of customers in the system are also easily calculated:

$$\sigma_{L_s} = \sqrt{\sum_{n=0}^{\infty} (n - L_s)^2 P_n} \tag{5.3}$$

$$\sigma_{L_q} = \sqrt{\sum_{n=0}^{m-1} (0 - L_q)^2 P_n + \sum_{n=m}^{\infty} (n - m - L_q)^2 P_n} \tag{5.4}$$

Little's formula provides a simple way to convert expected queue lengths into expected waiting times:

$$W_s = L_s/\lambda = \frac{\sum_{n=0}^{\infty} nP_n}{\lambda} \tag{5.5}$$

$$W_q = L_q/\lambda = \frac{\sum_{n=m}^{\infty} (n - m)P_n}{\lambda} \tag{5.6}$$

In steady-state analysis, it is often convenient to think of the service process in terms of a rate, where the rate is how many customers that one server can serve per unit time. Assume for the moment that the service rate is the same for all servers, and is defined as follows:

Definitions 5.2

μ = service rate per server when busy (customers/unit time)
$1/\mu$ = mean service time

The expected time in queue can now be expressed as the expected time in system minus the mean service time:

$$\begin{aligned} W_q &= W_s - 1/\mu \\ &= L_s/\lambda - 1/\mu \end{aligned} \tag{5.7}$$

Using Little's formula again, we can express the expected number of customers in queue in terms of the expected number of customers in the system:

$$L_q = W_q\lambda = L_s - \lambda/\mu \tag{5.8}$$

The ratio of λ to μ has a special interpretation. It is the ratio of the rate at which customers arrive to the rate at which customers are served, and represents how close the system is operating to capacity:

Definition 5.3

$$\rho = \lambda/\mu$$

The parameter ρ is referred to as the absolute ***utilization***, because it also equals the expected number of busy servers:

$$E(\text{busy servers}) = L_s - L_q = \rho \tag{5.9}$$

ρ divided by m is the proportional utilization, the average proportion of servers that are busy.

Two performance measures that cannot always be derived from the state probabilities are the standard deviation of the time in system and standard deviation of the time in queue. These measures depend on the passage times between states, which are more difficult to calculate than the state probabilities alone. The passage times also depend on the queue discipline, whereas the other performance measures do not.

Example

The state probabilities for a two-server queue, with arrival rate 8 customers/hour, were found to be $P_0 = .4$, $P_1 = .3$, $P_2 = .2$, $P_3 = .1$. All other state probabilities equal zero. The system performance measures are calculated as follows:

$$
\begin{aligned}
L_s &= 0(.4) + 1(.3) + 2(.2) + 3(.1) &&= 1.0 \text{ customer}\\
L_q &= 0(.4 + .3 + .2) + 1(.1) &&= .1 \text{ customer}\\
\rho &= L_s - L_q = 1.0 - .1 &&= .9 \text{ customer}\\
\sigma_{L_s} &= \sqrt{1(.4) + 0(.3) + 1(.2) + 4(.1)} &&= 1.0 \text{ customer}\\
\sigma_{L_q} &= \sqrt{.01(.9) + .81(.1)} &&= .3 \text{ customer}\\
W_s &= L_s/\lambda = 1.0/8 &&= .125 \text{ hour}\\
W_q &= L_q/\lambda = .1/8 &&= .0125 \text{ hour}\\
1/\mu &= W_s - W_q = .125 - .0125 &&= .1125 \text{ hour}
\end{aligned}
$$

Note in the example that the service rate is not needed to calculate the measures of performance. In fact, the service rate is derived from the state probabilities and the arrival rate.

The equations presented in this section apply to any queueing system in steady state, where servers work individually, one customer at a time. They do not depend on the queue discipline. In the following sections, models are presented for specific types of queueing systems. In all cases, the performance measures are calculated with the equations presented here. But, before proceeding, some nomenclature must be introduced.

There is a long-standing tradition to classify queueing systems according to four factors: arrival process, service process, number of servers, and maximum queue size. Each classification is represented by a series of four letters separated by slashes, as shown below:

$$X_1/X_2/m/b$$

where

X_1 represents the probability distribution of interarrival times
X_2 represents the probability distribution of service times
m represents the number of servers working in parallel
b represents the maximum number of customers in the queue

Some of the standard probability distributions are denoted by the following:

M = exponential (Markovian) distribution
D = constant service times (deterministic)
E_n = gamma distribution with shape parameter n (also called an Erlangian distribution)
G = general distribution

For example, an $M/M/1/\infty$ queue has exponential interarrival times (suggesting a Poisson process), exponential service times, one server, and infinite queue capacity. Sometimes, additional terms are provided to represent the number of customers in a "calling population" and the queue discipline.

5.3 THE *M/M/1/∞* QUEUE

In this section, the steady-state performance of the $M/M/1/\infty$ queue is determined. To be more specific, a queueing system with Poisson arrivals, exponential service times, a single server, and infinite queue capacity is evaluated. The analysis is independent of the queue discipline, provided that the discipline does not account for the service time (put another way, service time cannot be predicted in advance).

Here, as well as in the remainder of the chapter, performance is determined by calculating the state probabilities and applying the equations provided in Sec. 5.2. The state probabilities are found by formulating and solving ***balance equations*** that equate the rate at which transitions occur into a state (or group of states) to the rate at which transitions occur out of the state (or group of states). This method relies on the memoryless property of the exponential distribution and does not apply to systems that do not have exponential service times and Poisson arrivals.

To take an example, consider the rate at which transitions occur into state 2, that is, into the state 2 customers in the system. Because customers are served and arrive one at a time, the transition can occur in either of two ways. Either the prior state was 1 and a customer arrived or the prior state was 3 and a customer departed. For the $M/M/1/\infty$ queue, the first type of transition occurs with the rate λP_1, the arrival rate multiplied by the probability that the system is in state 1. The second type of transition occurs with the rate μP_3, the service rate multiplied by the probability that the system is in state 3. Thus

$$\text{Transition rate into state } 2 = \lambda P_1 + \mu P_3 \tag{5.10}$$

Now consider transitions *out* of state 2. Again, there are two possibilities, arrival or departure. The combined transition rate out of state 2 must therefore be

$$\text{Transition rate out of state 2} = \lambda P_2 + \mu P_2 \tag{5.11}$$

Thus, to be in steady state

$$\text{Transition rate out of state 2} = \text{Transition rate into state 2}$$

$$\lambda P_2 + \mu P_2 = \lambda P_1 + \mu P_3 \tag{5.12}$$

The transition rates can be summarized in a diagram, as in Fig. 5.1. To be in steady state, the transition rate out of each state must be equal to the transition rate into each state. This condition defines a set of balance equations (one for each state) and provides a method for determining the state probabilities:

Balance Equations for $M/M/1/\infty$ Queue

State	Rate In = Rate Out
0	$\mu P_1 = \lambda P_0$
1	$\lambda P_0 + \mu P_2 = \lambda P_1 + \mu P_1$
2	$\lambda P_1 + \mu P_3 = \lambda P_2 + \mu P_2$
⋮	
n	$\lambda P_{n-1} + \mu P_{n+1} = \lambda P_n + \mu P_n$ (5.13)

Note that the balance equations have the same form for all states, except state 0. State 0 is an exception because it is impossible to have a transition from state -1 to state 0 (and vice versa). State -1 does not exist.

There is actually an easier way to write the balance equations. For any "cordon" line drawn across the transition diagram, the transition rate must be the same in both directions across the line. This leads to the following set of balance equations:

Balance Equations for $M/M/1/\infty$ Queue

Line	
0,1	$\mu P_1 = \lambda P_0$
1,2	$\mu P_2 = \lambda P_1$
2,3	$\mu P_3 = \lambda P_2$
⋮	
$n,n+1$	$\mu P_{n+1} = \lambda P_n$ (5.14)

At this point, the general relationship between state probabilities should be apparent:

$$P_n = \frac{\lambda}{\mu} P_{n-1} = \rho P_{n-1} \tag{5.15}$$

Furtherfore, it should also be apparent that all of the state probabilities can be expressed in terms of a single state probability:

$$P_n = \rho^n P_0 \tag{5.16}$$

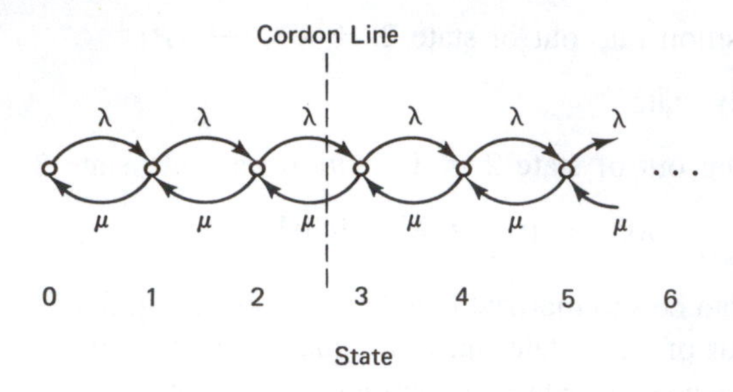

Figure 5.1 Transition rate diagram for the $M/M/1/\infty$ queue. Transition rates must be equal across any cordon line.

Readers may wish to verify that Eq. (5.16) also satisfies the first set of balance equations (5.13).

All that is needed to complete the calculations for P_n is to determine P_0. This is easily accomplished by noting one additional characteristic of P_n. To be a proper probability distribution, the sum of the state probabilities must equal 1. That is

$$\sum_{n=0}^{\infty} P_n = 1 \tag{5.17}$$

Therefore, substitution of Eq. (5.16) into Eq. (5.17) gives

$$\sum_{n=0}^{\infty} \rho^n P_0 = 1 \tag{5.18}$$

Equation (5.18) is the sum of a geometric series. Using the standard equation for the sum of a geometric series, we can write Eq. (5.18) as

$$P_0 \frac{1}{1 - \rho} = 1 \rightarrow P_0 = 1 - \rho \qquad \rho < 1 \tag{5.19}$$

P_0 is just 1 minus the utilization, ρ. Substitution of P_0 in Eq. (5.16) provides all of the state probabilities:

$$P_n = (1 - \rho)\rho^n \qquad \rho < 1 \tag{5.20}$$

The substitution reduces the state probabilities to a simple function of one parameter, ρ. The state probabilities do not depend on the exact values of λ and μ, only on the ratio of λ to μ. This greatly simplifies the analysis of the queueing system.

Equation (5.20) only applies to values of ρ less than 1. What if ρ is greater than 1? Then the arrival rate is larger than the service rate and the queue will continuously grow larger and larger. Hence, the system will never enter steady state. In reality, such a system could never exist, for customers would begin to renege once the queue becomes sufficiently large. Queues never grow infinitely large.

To summarize, the state probabilities for the $M/M/1/\infty$ system are determined by (1) writing balance equations for the transition rates; (2) expressing P_n as a function of P_{n-1}; (3) combining the expressions to obtain P_n as a function of P_0; and (4) deriving P_0, based on the requirement that the P_n must sum to 1.

5.3.1 Performance Measures

Consider how well the $M/M/1/\infty$ queue performs. The expected number of customers in the system is found by substituting the state probabilities in Eq. (5.1), and the expected number of customers in the queue is found from Eq. (5.8):

$$L_s = \sum_{n=0}^{\infty} n(1 - \rho)\rho^n = (1 - \rho) \sum_{n=0}^{\infty} n\rho^n = \frac{\rho}{1 - \rho} \qquad \rho < 1 \quad (5.21)$$

$$L_q = L_s - \rho = \frac{\rho}{1 - \rho} - \frac{\rho(1 - \rho)}{1 - \rho} = \frac{\rho^2}{1 - \rho} \qquad \rho < 1 \quad (5.22)$$

The calculation for L_s exploits the fact that the summation of $n\rho^n$ equals $\rho/(1 - \rho)^2$.

Note that just because customers are served faster than they arrive does not guarantee that queues never exist. The randomness associated with service times and arrival times always creates transient queues from time to time. Even when $\rho = .5$, there is, on average, .5 customer in the queue. As ρ approaches 1, queue length rapidly grows. With $\rho = .9$, the average number of customers in the queue is 8.1, a considerable number. The important message is that the capacity of the system must be larger than the rate at which customers arrive to ensure that queues are small, if the arrival and service processes are random.

The expected time in queue and time in system are not uniquely defined by ρ; they also depend on the arrival rate. From Little's formula

$$W_s = \frac{1}{\lambda} \frac{\rho}{1 - \rho} \qquad \rho < 1 \qquad (5.23)$$

$$W_q = \frac{1}{\lambda} \frac{\rho^2}{1 - \rho} \qquad \rho < 1 \qquad (5.24)$$

For example, if ρ equals .6, then L_s equals 1.5 customers and L_q equals .9 customer. The expected time in system equals L_s multiplied by $1/\lambda$. If $1/\lambda$ equals 1 minute, W_s would equal 1.5 minutes and W_s .9 minute, but if $1/\lambda$ equals 30 minutes, W_s would equal 45 minutes and W_q equal 27 minutes. The arrival rate determines how often the queue turns over, that is, whether the queue is created by many customers waiting a short time each or a few customers waiting a long time each.

5.3.2 FCFS Discipline

If the queue discipline is FCFS (first-come, first-served), then the probability distribution for the waiting times can also be derived. Suppose that there are n customers in the queue after customer X arrives. Then customer X will not begin service until the previous $n - 1$ customers complete service, and will not complete service until n customers (including itself) complete service. If the service times have an exponential distribution (as already assumed), the time in system for the customer is the sum of n exponential random variables, which must have a gamma distribution with parameters n and λ. Considering all of the possible states in which the customer might arrive, the probability distribution for the time in system must therefore be

$$P(W_s(n) \leq t) = \sum_{n=0}^{\infty} P_n \int_0^t \frac{\lambda^n}{\Gamma(n)} x^{n-1} e^{-\lambda x} dx \qquad t \geq 0 \tag{5.25}$$

Fortunately, Eq. (5.25) can be reduced to a much simpler form:

$$P(W_s(n) \leq t) = 1 - e^{-(1/W_s)t} \qquad t \geq 0 \tag{5.26}$$

That is, $W_s(n)$ has an *exponential distribution* with mean defined by Eq. (5.23). The distribution for the time in queue is also easily defined, though more complicated because it is partially continuous and partially discrete:

$$P(W_q(n) \leq t) = (1 - \rho) + \rho[1 - e^{-(1/W_s)t}] \qquad t \geq 0 \tag{5.27}$$

It is much more difficult to derive the probability distributions for $W_q(n)$ and $W_s(n)$ for queue disciplines other than FCFS and for queueing systems other than $M/M/1$. They are not covered here.

Example

Sports fans arrive at a ticket counter at the Wayout Arena by a Poisson process with rate 105 customers per hour. Customers are served by a single cashier. The service time has an exponential distribution with mean 30 seconds.

The service rate is 1/(30 seconds), or 120 customers per hour. The utilization is λ/μ, or $105/120 = .875$. The performance measures are calculated below.

$$L_s = \frac{.875}{(1 - .875)} = 7 \text{ customers}$$

$$L_q = \frac{.875^2}{(1 - .875)} = 6.125 \text{ customers}$$

$$W_s = (1/105)(7) = .0667 \text{ hours} = 4 \text{ minutes}$$

$$W_q = (1/105)(6.125) = .0583 \text{ hours} = 3.5 \text{ minutes}$$

5.4 GENERAL EVALUATION APPROACH FOR POISSON ARRIVALS AND EXPONENTIAL SERVICE TIMES

The $M/M/1/\infty$ queue is unique in that the service rate and arrival rate are independent of the state of the system (with the exception of state 0). Customers arrive at the rate λ and customers are served at the rate μ, whether there is 1 or 20 customers in the system. A more general model would still have exponential service times and Poisson arrivals, but allow λ and μ to be different for each state. This may account for such factors as reneging when queues become large, increased service rate when more servers operate, and the like. Again, the state probabilities are independent of the queue discipline, provided that service times cannot be predicted in advance.

Definitions 5.4

λ_n = arrival rate when the state equals n

μ_n = service rate when the state equals n

If we assume that arrivals occur by a Poisson process, and service times are exponential and independent, then the balance equations across the cordon lines (see Fig. 5.2) can be written as below:

Balance Equations: General

Line	
0,1	$\mu_1 P_1 = \lambda_0 P_0$
1,2	$\mu_2 P_2 = \lambda_1 P_1$
2,3	$\mu_3 P_3 = \lambda_2 P_2$
⋮	
$n, n+1$	$\mu_{n+1} P_{n+1} = \lambda_n P_n$ (5.28)

The balance equations are much the same as they were for the $M/M/1/\infty$ queue, only that λ_n and μ_n substitute for λ and μ. The equations are also solved in much the same way. However, the result is more complicated:

$$P_1 = \frac{\lambda_0}{\mu_1} P_0 \tag{5.29a}$$

$$P_2 = \frac{\lambda_1}{\mu_2} P_1 = \left[\frac{\lambda_1 \lambda_0}{\mu_2 \mu_1}\right] P_0 \tag{5.29b}$$

$$P_3 = \frac{\lambda_2}{\mu_3} P_2 = \left[\frac{\lambda_2 \lambda_1 \lambda_0}{\mu_3 \mu_2 \mu_1}\right] P_0 \tag{5.29c}$$

$$\vdots$$

$$P_{n+1} = \frac{\lambda_n}{\mu_{n+1}} P_n = \left[\frac{\lambda_n \lambda_{n-1} \cdots \lambda_0}{\mu_{n+1} \mu_n \cdots \mu_1}\right] P_0 \tag{5.29d}$$

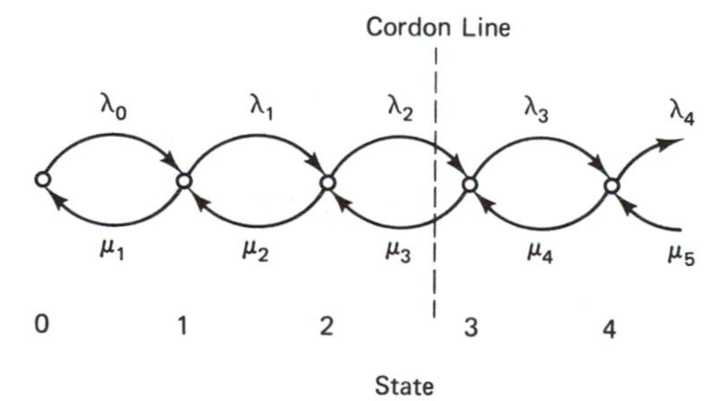

Figure 5.2 Transition rate diagram for general exponential queue. Transition rates must still be equal across any cordon line.

As before, the summation of the state probabilities must equal 1. To keep notation in this calculation simple, the following symbol is introduced:

$$B_n = \begin{cases} \dfrac{\lambda_{n-1}\lambda_{n-2}\cdots\lambda_0}{\mu_n\mu_{n-1}\cdots\mu_1} & n \geq 1 \\ 1 & n = 0 \end{cases} \tag{5.30}$$

Then

$$\sum_{n=0}^{\infty} B_n P_0 = 1 \rightarrow P_0 = \frac{1}{\sum_{n=0}^{\infty} B_n} \tag{5.31}$$

Unfortunately, the calculations are not as straightforward as they may appear, for the summation of B_n does not necessarily result in a simple equation, as was the case for the $M/M/1/\infty$ queue. Nevertheless, Eq. (5.31) is not difficult to calculate with a computer.

For the general Poisson/exponential model, the average arrival rate and average service rate per server may also depend on the state probabilities. They are calculated as follows:

$$\bar{\lambda} = \sum_{n=0}^{\infty} \lambda_n P_n \tag{5.32}$$

$$\bar{\mu} = \left[\sum_{n=1}^{\infty} [\mu_n/m(n)]P_n \right] / (1 - P_0) \tag{5.33}$$

where $m(n)$ is the number of servers operating when there are n customers in the system. It is the average arrival rate that should be used in Little's formula, and it is the average service rate per server that should be used in Eq. (5.7) to derive the time in queue from the time in system. Note that the average arrival rate is a measure of performance in itself, for it measures how many customers are actually served per unit time—a factor that may depend on reneging.

5.4.1 *M/M/m/∞* Queue

Suppose that a queueing system has m servers working in parallel, that servers operate at the same rate, μ, and servers work on one customer at a time. And suppose that the arrival rate is not influenced by the state of the system. Then λ_n and μ_n equal the following quantities:

$$\lambda_n = \lambda \qquad n = 0, 1, 2, \ldots \tag{5.34}$$

$$\mu_n = \begin{cases} n\mu & n = 0, 1, \ldots, m \\ m\mu & n = \mathrm{m} + 1, \mathrm{m} + 2, \ldots \end{cases} \tag{5.35}$$

Substitution of λ_n and μ_n in Eq. (5.29) provides the state probabilities as a function of P_0, and using ρ to represent the ratio λ/μ, the state probabilities can be written as follows:

$$P_n = \begin{cases} \dfrac{\rho^n}{n!} P_0 & n = 0, 1, \ldots, m-1 \\[2ex] \dfrac{\rho^n}{m!m^{n-m}} P_0 & n = m, m+1, \ldots \end{cases} \tag{5.36}$$

Hence, P_0 must equal

$$P_0 = \frac{1}{\sum_{n=0}^{m-1} \rho^n/n! + \sum_{n=m}^{\infty} \rho^n/m!m^{n-m}}$$

$$= \frac{1}{\sum_{n=0}^{m-1} \rho^n/n! + \dfrac{\rho^m}{m!} \dfrac{1}{1 - \rho/m}} \qquad (\rho/m < 1) \tag{5.37}$$

The equation for P_0 cannot be written as a simple equation without a summation. (However, Eq. (5.37) is not difficult to calculate.) For this reason, the performance measures will be written as a function of P_0. The measures are derived from the equations presented in Sec. 5.2. The expected number of customers in queue and the expected number of customers in the system are the following:

$$L_q = \frac{\rho^{m+1}/m}{m!\,(1 - \rho/m)^2} P_0 \qquad \rho/m < 1 \tag{5.38}$$

$$L_s = L_q + \rho \qquad \rho/m < 1 \tag{5.39}$$

The waiting times are calculated from Little's formula:

$$W_q = L_q/\lambda = \frac{1}{\lambda} \frac{\rho^{m+1}/m}{m!(1 - \rho/m)^2} P_0 \qquad \rho/m < 1 \tag{5.40}$$

$$W_s = L_s/\lambda = \frac{1}{\lambda} \frac{\rho^{m+1}/m}{m!(1 - \rho/m)^2} P_0 + \frac{1}{\mu} \qquad \rho/m < 1 \tag{5.41}$$

Figure 5.3 shows the expected number of customers in the queue as a function of ρ and the number of servers. Note that the general shape is the same for all values of m. L_q becomes very large as ρ approaches m, and approaches zero as ρ becomes small. In fact, *the equation for L_q for the single server queue can provide a reasonable approximation for the multiserver queue*, provided that ρ/m is substituted for ρ in Eq. (5.22). Then L_s can

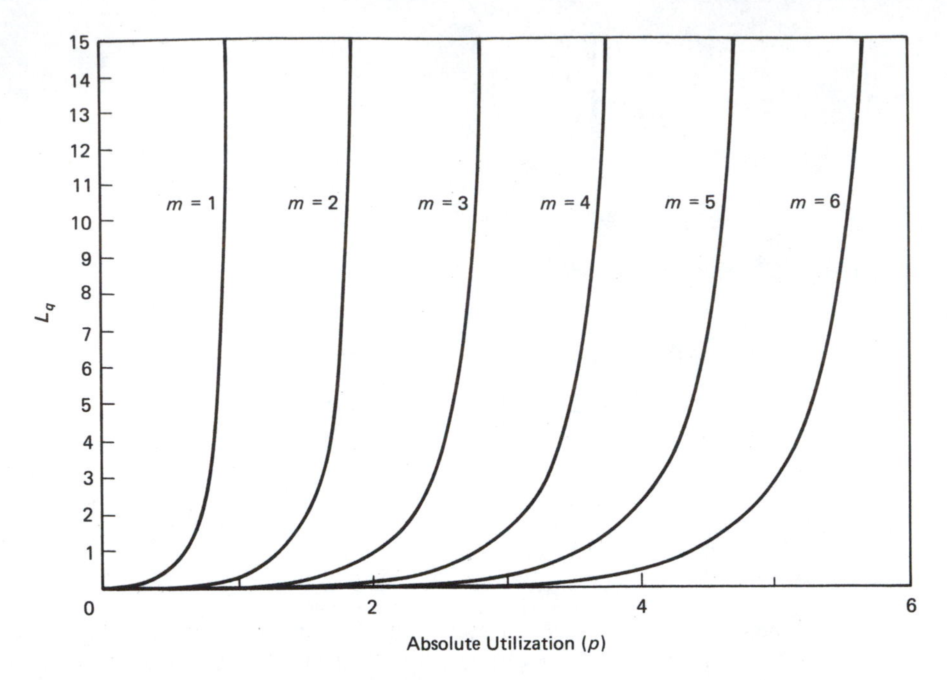

Figure 5.3 Expected queue length for the $M/M/m/\infty$ queue. L_q will be small when the number of servers equals or exceeds $p + \sqrt{p}$.

be estimated by adding ρ (not ρ/*m*) to L_q, and W_s and W_q can be estimated in the usual manner from Little's formula. This approximation is most accurate for large values of ρ/*m*, close to 1 (which is to say it is a ***heavy traffic approximation***. See Köllerström 1974.)

Figure 5.3 also illustrates the fundamental property that adding servers reduces waiting time. However, note that adding servers does not always provide an appreciable benefit. For ρ less than 1, L_q is nearly the same for any number of servers greater than or equal to 2, implying that one or two servers is all that is ever needed.

Example

The Wayout Arena (see example in Sec. 5.3.2) would like to evaluate the benefits of adding a second server. If $m = 2$, $\mu = 120$ customers/hour per server, $\lambda = 105$ customers/hour, and $\rho = .875$:

$$P_0 = \frac{1}{1 + \rho + \dfrac{\rho^2/2}{1 - \rho/2}} = \frac{1 - \rho/2}{1 + \rho/2} = .391$$

$$L_q = \frac{.875^3/2}{2!(1 - .875/2)^2}\,.391 = .529 \cdot .391 = .206 \text{ customer}$$

$$L_s = .206 + .875 = 1.08 \text{ customers}$$

$$W_q = .206/105 = .00196 \text{ hour} = .12 \text{ minute}$$

$$W_s = 1.08/105 = .0103 \text{ hour} = .62 \text{ minute}$$

Adding a server reduces L_q from 6.125 (the one server estimate) to .206.

As a rule of thumb, if the value of a server's time is comparable to the value of the customer's time, then adding a server is justified if L_q declines by at least 1. That is, one more server should be added if it reduces the number of people waiting by at least 1. For the example, the added server seems highly justified, for it reduces L_q by nearly 6.

5.4.2 *M/M/∞* Queue and the *M/G/∞* Queue

If the number of servers is very large, queues will rarely materialize. Then the system will perform as though the number of servers is infinite. The performance of a system with a large number of servers can be approximated by the infinite server *M/M/∞* queue. It happens that the *M/M/∞* queue is more easily evaluated than the *M/M/m/∞* queue.

The arrival rate is the same for all states; the service rate is proportionate to the state

$$\lambda_n = \lambda \qquad n = 0, 1, 2, \ldots \tag{5.42}$$

$$\mu_n = n\mu \qquad n = 0, 1, 2, \ldots \tag{5.43}$$

This leads to the following equation for the state probabilities:

$$P_n = \rho^n/n! \, P_0 \qquad n = 0, 1, 2, \ldots \tag{5.44}$$

where

$$P_0 = \frac{1}{\sum_{n=0}^{\infty} \rho^n/n!} = e^{-\rho} \tag{5.45}$$

Substitution of P_0 in Eq. (5.44) provides the state probabilities as a function of ρ alone:

$$P_n = \frac{\rho^n}{n!} e^{-\rho} \tag{5.46}$$

Unlike the *M/M/m/∞* queue, P_n does not have to be expressed as a function of P_0. It should also be apparent that the state, n, has a familiar probability distribution, Poisson, with mean ρ. Therefore, the expected number of customers in the system is simply

$$L_s = \rho \tag{5.47}$$

And, from Little's formula

$$W_s = 1/\mu \tag{5.48}$$

Of course, L_q and W_q both equal zero because queues never appear when the number of servers is infinite.

The $M/M/\infty/\infty$ queue defines the limiting performance of the $M/M/m/\infty$ queue as the number of servers, m, becomes large. The $M/M/\infty/\infty$ queue also defines the limiting performance of the $M/G/m/\infty$ queue as the number of servers become large (see Takacs 1969). Thus, Eqs. (5.46)–(5.48) also apply to the $M/G/\infty/\infty$ queue.

The implication of Eq. (5.46) is not only that the state has a Poisson distribution but that the variance and mean of the number of customers in the system both equal ρ, independent of the service time distribution. Further, the number of customers in the system should only occasionally exceed the sum of the mean, ρ, and the standard deviation $\sqrt{\rho}$. This result is the basis of the following rule of thumb for deciding how many servers to use:

Queues should be small if

$$m \geqslant \max\{1, \rho + \sqrt{\rho}\} \tag{5.49}$$

For example, if $\rho = 5$, eight servers would provide negligible waits, and when $\rho = 20$, 25 servers would provide negligible waits.

5.4.3 *M/M/m/b* Queue

Some queues are restricted by a capacity, referred to as the ***buffer size***, b. Suppose that you phone an airline to make a reservation. If all of the agents are busy, you will be put on hold. The maximum number of callers that can be put on hold at any one time is the buffer size. Once the buffer is full, subsequent callers will receive a busy signal and will not be allowed to join the queue. For such a system, the arrival rates and service rates might be defined as follows:

$$\lambda_n = \begin{cases} \lambda & n = 0, 1, \ldots, m + b \\ 0 & n = m + b + 1, m + b + 2, \ldots \end{cases} \tag{5.50}$$

$$\mu_n = \begin{cases} n\mu & n = 0, 1, \ldots, m \\ m\mu & n = m + 1, m + 2, \ldots \end{cases} \tag{5.51}$$

where m is the number of servers and b is the buffer size.

The equation for λ_n, perhaps unrealistically, assumes that a customer that finds the system busy upon arrival leaves the system entirely, never to return. One might debate the realism of this assumption. Unfortunately, statistical analysis of more complicated models, which account for customers returning later, would be too difficult to undertake here.

Because of lost customers, the average arrival rate $\bar{\lambda}$, does not equal the rate λ; it is some value less than λ. Unlike all of the previous examples, this means that the utilization, ρ, does not equal λ/μ, but instead equals $\bar{\lambda}/\mu$. The ratio of λ to μ will still be important in the following calculations, so it will be denoted by a new symbol.

Definition 5.5

$$r = \lambda/\mu = \textbf{intensity}$$

The intensity can be thought of as the ratio of the maximum arrival rate to the maximum service rate. In the previous examples ($M/M/1/\infty$ and $M/M/m/\infty$), the utilization and the intensity were the same ($r = \rho$) because $\bar{\lambda} = \lambda$.

The state probabilities are found, as usual, from Eq. (5.29), which results in the following:

$$P_n = \begin{cases} \dfrac{r^n}{n!} P_0 & n = 0, 1, \ldots, m \\ \dfrac{r^n}{m!m^{n-m}} P_0 & n = m + 1, \ldots, m + b \end{cases} \tag{5.52}$$

Perhaps it is surprising that P_n is nearly the same as Eq. (5.36) for the $M/M/m/\infty$ queue. The only difference is that the second line is limited, by $m + b$. The reason why the results are so similar can be deduced from the way that P_n is calculated, successively beginning from P_1. The buffer size has no impact on the calculation until the capacity is reached, after which P_n equals zero (the number of customers in the system cannot exceed $m + b$). This does not, by the way, prevent P_0 from changing, which in turn affects all of the P_n.

P_0 is found in the usual manner, by ensuring that the probabilities sum to 1:

$$P_0 = \frac{1}{\displaystyle\sum_{n=0}^{m} \frac{r^n}{n!} + \sum_{n=m+1}^{m+b} \frac{r^n}{m!m^{n-m}}}$$

$$= \frac{1}{\displaystyle\sum_{n=0}^{m-1} \frac{r^n}{n!} + \frac{r^m}{m!} \frac{1 - (r/m)^{b+1}}{1 - r/m}} \qquad (r/m \neq 1) \tag{5.53}$$

As with the $M/M/m/\infty$ queue, it is impossible to express P_0 as a simple expression without summations. Therefore, P_0 will be included in the equations for the performance measures:

$$L_q = \left[\frac{r^{m+1}/m}{m!(1 - r/m)^2} P_0 \right] \left[1 - (r/m)^b - b(1 - r/m)(r/m)^b \right] \qquad r/m \neq 1 \tag{5.54}$$

The first part of Eq. (5.54) is identical to Eq. (5.38), which gives the expected number of customers in queue for the $M/M/m/\infty$ queue. The second part of Eq. (5.54) can be viewed as a correction factor, to account for the limited buffer size. This factor must always be less than 1, meaning that the buffer limitation reduces the number of customers in the queue.

The other performance measures are somewhat more difficult to derive than in the other examples, due to a more complicated calculation for the average arrival rate and the average service time. Without showing the derivation, the equation for the expected number of customers in the system is

$$L_s = L_q + \left[m - P_0 \sum_{n=0}^{m-1} \frac{(b - m)r^n}{n!} \right] \tag{5.55}$$

As usual, the term in brackets equals ρ ($L_s - L_q = \rho$), which, as a point of contrast, is not the same as r. The calculations for the waiting times depend on the average arrival rate, denoted by $\bar{\lambda}$, which equals

$$\bar{\lambda} = \lambda \sum_{n=0}^{m+b-1} P_n = (1 - P_{m+b})\lambda \tag{5.56}$$

That is, the arrival rate equals 0 when the state is $b + m$ and λ otherwise. This provides an alternative way to express ρ:

$$\rho = r\,(1 - P_{b+m}) \tag{5.57}$$

Furthermore

$$W_q = \frac{L_q}{(1 - P_{b+m})\lambda} = L_q/\bar{\lambda} \tag{5.58}$$

$$W_s = \frac{L_s}{(1 - P_{b+m})\lambda} = L_s/\bar{\lambda} \tag{5.59}$$

Note that $\bar{\lambda}$ (average arrival rate) is used instead of λ (maximum arrival rate) in Little's formula.

It may seem somewhat surprising, but the finite buffer size has the impact of reducing the average waiting time. However, this reduction does not come without a cost, for customers are lost to the system when the buffer becomes full. The arrival rate is not λ, but some value less than λ, as indicated by Eq. (5.56). These facts are illustrated in Figs. 5.4 and 5.5, which show L_q and ρ as a function of r for the $M/M/1/\,b$ queue. Note that the impact of the buffer is fairly small for $b \geqslant 10$, provided that r is less than .7. When the system operates close to capacity, the buffer size has a large impact on performance: Large numbers of customers are lost to the system, and L_q differs substantially from the $M/M/1/\infty$ queue. The same basic pattern applies to queues with more than one server and for other service time distributions.

5.4.4 Finite Calling Population: *M/M/m/∞*

We will evaluate one final application of the general M/M model, this time to a finite calling population. The ***calling population*** represents the group of potential customers. Consider an example. Within a factory, machines malfunction according to a random process. Once a machine malfunctions, it enters a queue to wait for repair. During this time, the machine is not running and cannot malfunction a second time.

Suppose that there are N machines in the factory and m repair crews, and that λ is the maximum arrival rate (the arrival rate when all machines are functioning). The times

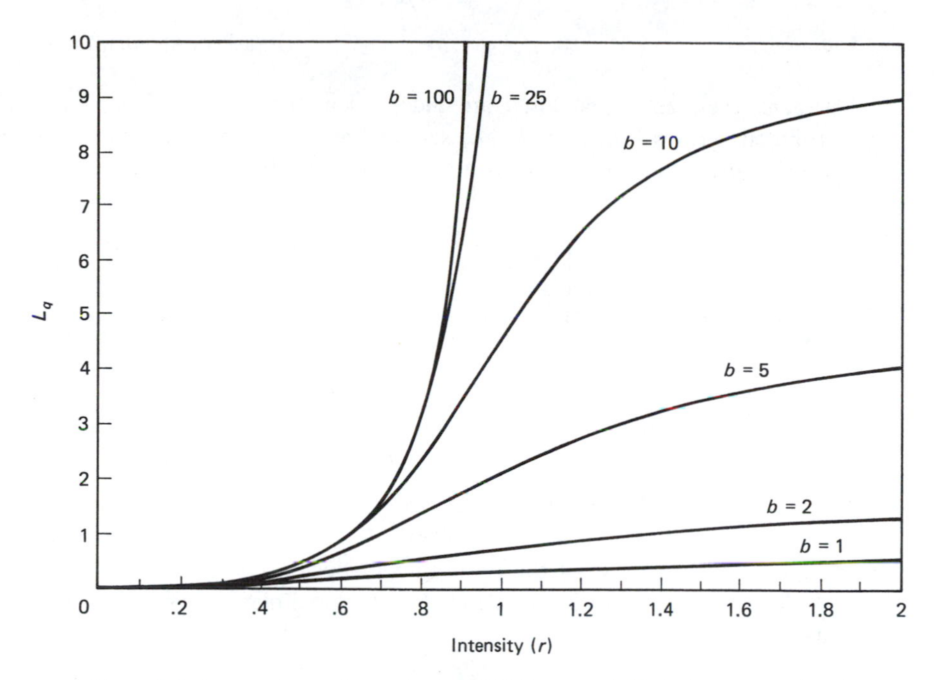

Figure 5.4 Expected queue length for the $M/M/1/b$ queue. Buffer limits the queue length to b, so L_q does not grow toward infinity as the intensity becomes large.

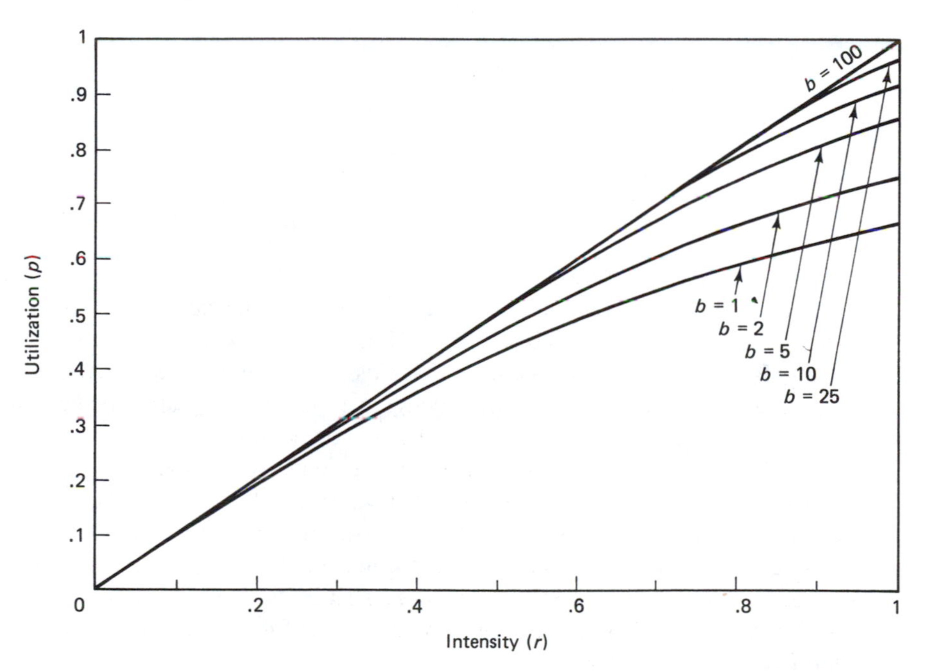

Figure 5.5 Utilization for the M/M/1/b queue. Limited buffer size results in lost customers and reduced server utilization.

between repair and breakdown are independent among machines and exponentially distributed, each will mean N/λ, and service times are also exponential, with mean $1/\mu$. Then the arrival and service rates are the following functions of n:

$$\lambda_n = \begin{cases} \dfrac{N-n}{N}\lambda & n = 0, 1, \ldots, N \\ 0 & n = N+1, N+2, \ldots \end{cases} \tag{5.60}$$

$$\mu_n = \begin{cases} n\mu & n = 0, 1, \ldots, m \\ m\mu & n = m+1, m+2, \ldots \end{cases} \tag{5.61}$$

If N is a very large number, the arrival process is approximately Poisson and the queueing model is no different from the standard $M/M/m/\infty$ model. The finite model is only important when the size of the calling population, N, is small.

The state probabilities are derived in the usual manner, and result in the following equations:

$$P_n = \begin{cases} P_0 \dfrac{N!}{(N-n)!n!} (r/N)^n & n = 1, 2, \ldots, m \\ P_0 \dfrac{N!}{(N-n)!m!m^{n-m}} (r/N)^n & n = m+1, \ldots, N \end{cases} \tag{5.62}$$

And consequently

$$P_0 = \frac{1}{\displaystyle\sum_{n=0}^{m} \frac{N!}{(N-n)!n!} \left[\frac{r}{N}\right]^n + \sum_{n=m+1}^{N} \frac{N!}{(N-n)!m!m^{n-m}} \left[\frac{r}{N}\right]^n} \tag{5.63}$$

The performance measures are calculated in the usual manner. Because the results are not simple, they are not presented here. Keep in mind that ρ again does not equal r, because the arrival rate is influenced by the state. Hence, ρ must be calculated from Eq. (5.32). In 1980, Bunday and Scraton showed that Eqs. (5.62) and (5.63) do not depend on the probability distribution for the time between machine repair and breakdown. They do, however, depend on the service time being exponential.

Figure 5.6 illustrates the impact of the finite population on L_q for the $M/M/1/\infty$ queue. The upper curve is the standard value for L_q, Eq. (5.22). As expected, L_q approaches the upper curve as the population size increases. Overall, the impact of a finite calling population is much the same as the impact of a finite buffer. It tends to moderate the arrival rate when queues become long. That is, for a given value of r, queue lengths tend to be smaller. This general property holds true for any number of servers and for any service time distribution.

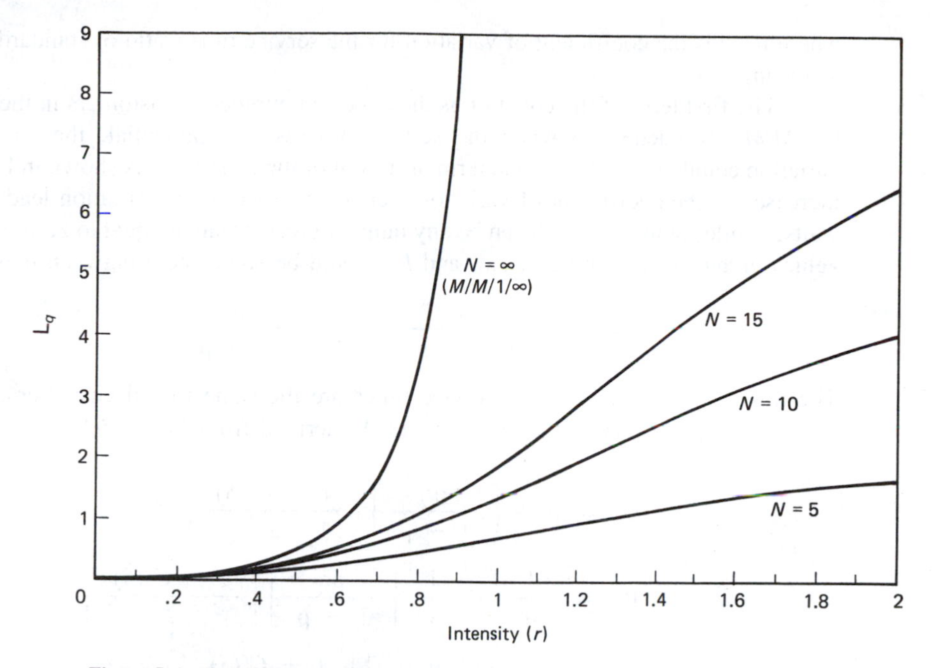

Figure 5.6 Expected queue length for the $M/M/1/\infty$ queue with finite calling population. As queue length grows, customer arrival rate declines. The maximum number of customers in the system equals the population size. If the population is large, results are the same as standard $M/M/1/\infty$ queue.

5.5 GENERAL SERVICE TIME DISTRIBUTIONS

It is ordinarily quite difficult to evaluate systems that do not have Poisson arrivals and exponential service times. However, there are a few exceptions. Three examples are considered in this section.

5.5.1 Pollaczek-Khintchine Formula for the $M/G/1/\infty$ Queue

If the arrival process is Poisson, and there is just one server, the expected performance can be calculated for *any service time distribution* that is independent of the state and the interarrival times. All that one needs to know to calculate the expected number of customers in the queue are the utilization and the coefficient of variation for the service times. The derivation of the following is provided in the appendix to this chapter:

Pollaczek-Khintchine Formula

$$L_q = \left[\frac{\rho^2}{1-\rho}\right]\left[\frac{1 + C^2(S)}{2}\right] \qquad \rho<1 \tag{5.64}$$

where $C(S)$ is the coefficient of variation for the service time (ratio of standard deviation to mean).

The first term of the equation is the expected number of customers in the queue for the $M/M/1/\infty$ queue. So when the service process is exponential, the coefficient of variation equals 1 and the second term drops out of the equation. As shown in Fig. 5.7, L_q increases as the coefficient of variation increases; hence, more variation leads to longer waits. While, in theory, $C(S)$ can be any number greater than or equal to zero, in practice, values greater than 1 are unusual, and L_q should be somewhere between these bounds:

$$.5 \frac{\rho^2}{1-\rho} \leqslant L_q \leqslant \frac{\rho^2}{1-\rho} \tag{5.65}$$

The lower bound occurs when service times are the same for all customers.

Other performance measures are easily derived from Eq. (5.64):

$$L_s = \rho + \left[\frac{\rho^2}{1-\rho}\right]\left[\frac{1+C^2(S)}{2}\right] \tag{5.66}$$

$$W_s = \frac{1}{\mu} + \frac{1}{\lambda}\left[\frac{\rho^2}{1-\rho}\right]\left[\frac{1+C^2(S)}{2}\right] \tag{5.67}$$

$$W_q = \frac{1}{\lambda}\left[\frac{\rho^2}{1-\rho}\right]\left[\frac{1+C^2(S)}{2}\right] \tag{5.68}$$

Example

From the automatic teller machine data (Sec. 3.5), customers arrive at a rate of 49 per hour. The average service time was estimated to be 49.8 seconds ($\mu = 72$ customers per hour) with a standard deviation of 19.0 seconds.

The coefficient of variation is the standard deviation divided by the mean service time, which, for the teller data, equals .38. The utilization is the arrival rate divided by the service rate, or 49/72 = .68. The performance measures are calculated below:

$$L_q = \left[\frac{.68^2}{1-.68}\right]\left[\frac{1+.38^2}{2}\right] = .83 \text{ customer}$$

$$L_s = \rho + L_q = .68 + .83 = 1.51 \text{ customer}$$

$$W_q = L_q/\lambda = .83/49 = .017 \text{ hour} = 1.02 \text{ minutes}$$

$$W_s = L_s/\lambda = 1.51/49 = .031 \text{ hour} = 1.85 \text{ minute}$$

Note that these performance measures do not depend on the distribution of service times, only the mean and coefficient of variation. Therefore, for the simulation performed in Sec. 4.6, the expected result is the same for either the empirical service time distribution or the gamma service time distribution (both had the same mean and standard deviation). In fact, this was borne out in the simulation—in both cases W_q was about 1.1 minutes.

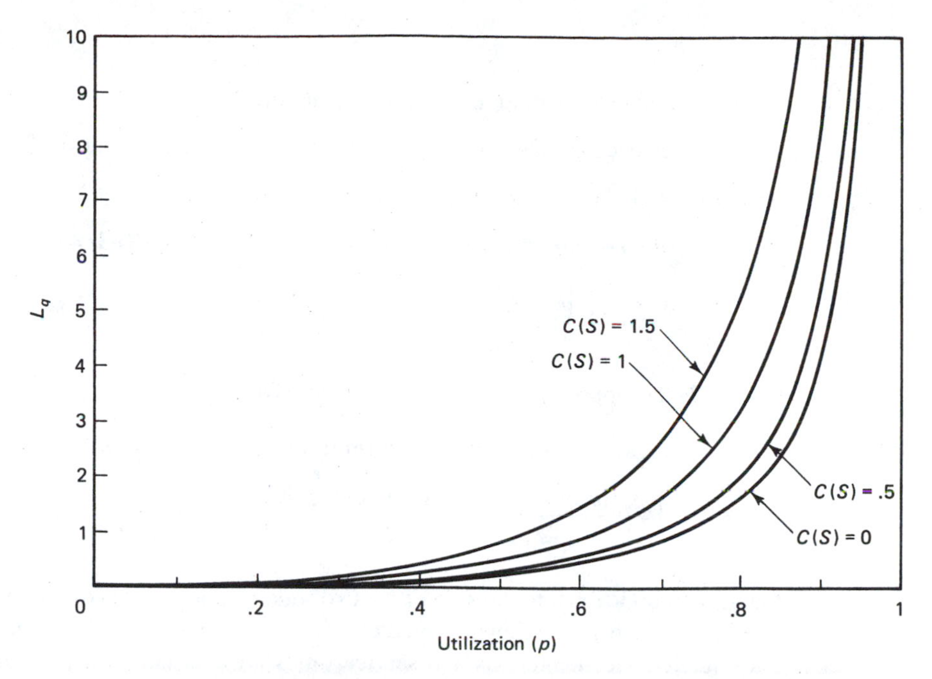

Figure 5.7 Pollaczek-Khintchine formula showing expected queue length for the $M/G/1/\infty$ queue. Increases in the coefficient of variation of the service times increase queue lengths.

Compared to the simulated values of W_q—even compared to the observed value of W_q—the P-K formula provides the best estimate of $M/G/1/\infty$ performance. It is best because it is least susceptible to random fluctuations. That is, it is equivalent to running the simulation for an infinite length of time. Nevertheless, as stated before, this does not guarantee accuracy. The P-K formula cannot correct for errors in model structure or errors in parameter estimation.

5.5.2 Approximation for $G/G/m/\infty$ Queue

No simple exact formula exists for multiple server queues. However, there are a few useful approximations. The one presented here was developed by Allen and Cunneen (Allen 1978). The approximation takes the value for L_q from the $M/M/m/\infty$ queue, Eq. (5.38), and multiplies the result by an adjustment factor accounting for variations in the interarrival time and service time. The approximation is exact for the $M/G/1/\infty$ queue.

Allen-Cunneen Approximation ($G/G/m/\infty$)

$$L_q = L_{q,M/M/m} \cdot \left[\frac{C^2(A) + C^2(S)}{2} \right] \tag{5.69}$$

where:

$C(A)$ = coefficient of variation for interarrival time

$L_{q,M/M/m}$ = expected customers in queue for an $M/M/m/\infty$ system, Eq. (5.38)

When $m = 1$, the equation simplifies to the following:

Allen-Cunneen Approximation: One Server ($G/G/1/\infty$)

$$L_q = \left[\frac{\rho^2}{1 - \rho}\right]\left[\frac{C^2(A) + C^2(S)}{2}\right] \qquad \rho < 1 \tag{5.70}$$

When $m = 2$, the approximation can be written as follows:

Allen-Cunneen Approximation: Two Servers ($G/G/2/\infty$)

$$L_q = \left[\frac{(\rho/2)^2}{1 - \rho/2}\right]\left[\frac{C^2(A) + C^2(S)}{2}\right]\left[\frac{\rho}{1 + \rho/2}\right] \qquad \rho < 2 \tag{5.71}$$

Note that L_q is proportional to the sum of the two squared coefficients of variation, $C^2(A)$ and $C^2(S)$. Thus, the approximation predicts that queues will never exist if customers arrive at equally spaced intervals and service times are constant (for $\rho < m$). Queues materialize when service times and arrival intervals vary, even when ρ/m is less than 1 and the system has sufficient capacity to handle the arrivals.

Example

The Allen-Cunneen approximation is to be used to predict L_q and W_q for the teller system with two servers. Because the arrival process appears to be Poisson, $C(A) = 1$. The coefficient of variation in the service times was found to be .382, and utilization was found to be .68. Therefore, the predictions are as follows:

$$L_q = \left[\frac{(.68/2)^2}{(1 - .68/2)}\right]\left[\frac{1 + .382^2}{2}\right]\left[\frac{.68}{1 + .68/2}\right] = .050 \text{ customer}$$

$$L_s = L_q + \rho = .050 + .68 = .73 \text{ customer}$$

$$W_q = L_q/\lambda = .050/(98/120) = .061 \text{ minute}$$

$$W_s = L_s/\lambda = .73/(98/120) = .89 \text{ minute}$$

Adding a server reduces the number of customers in the system from .83 (see the example in Sec. 5.5.1) to .05, a difference of less than one customer. Thus, the added server would only be justified if the customer's time is more valuable than the server's time.

Other, more specialized, approximations can sometimes provide better estimates. In particular, better approximations exist for systems for which ρ/m is small (Newell 1982 discusses ***light traffic approximations***).

5.5.3 Erlang Loss Formula: *M/G/m/0*

Another simple exact equation that applies to a variety of queueing systems is the ***Erlang loss formula*** (see Brockmeyer et al. 1948). Consider an *M/G/m/0* queue, that is, a queueing system with *m* servers and a buffer size of zero, with general service times and Poisson arrivals. Without showing the derivation, the state probabilities are defined by

$$P_n = \frac{r^n/n!}{\sum_{i=0}^{m} r^i/i!} \qquad n = 0, 1, \ldots, m \tag{5.72}$$

Equation (5.72) was originally derived by Erlang for an exponential service time distribution, so it was unexpected when it was later shown that Eq. (5.72) applies to any service time distribution (see Takacs, 1969, for example). Erlang loss formula specifically refers to an equation for the proportion of customers that are "lost" because all servers are busy. With Poisson arrivals, this is simply P_n evaluated at $n = m$ (the probability that all servers are busy):

Erlang Loss Formula

$$P(\text{lost customer}) = P_m = \frac{r^m/m!}{\sum_{i=0}^{m} r^i/i!} \tag{5.73}$$

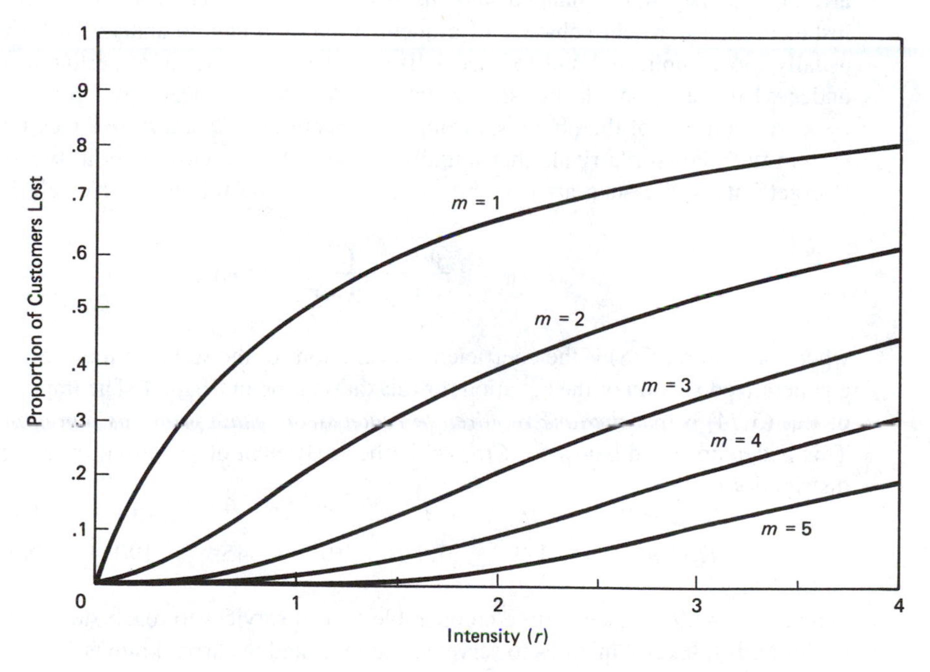

Figure 5.8 Erlang loss formula for the *M/G/m/0* queue.

One must remember that the loss formula assumes that a customer who arrives when all servers are busy is indeed lost and does not return later. In reality, the number of lost customers should be somewhat less than the number predicted by Eq. (5.73). The loss calculations are shown in Fig. 5.8 for one to five servers as a function of the intensity, r. Further note that with a buffer size of zero, queues can never exist. Thus, the number of lost calls is the pertinent performance measure.

Example

Telephone calls arrive at the rate of 55 per hour, and are served at the rate of 10 per hour per telephone line. There are seven lines and there is no buffer. With $r = 5.5$, the proportion of calls lost is

$$P(\text{loss}) = P_7 = \frac{5.5^7/7!}{\sum_{i=0}^{7} (5.5)^i/i!} = \frac{30.2}{198} = .153$$

5.6 TRANSIENT QUEUE BEHAVIOR

The equations derived in this chapter can only be applied if the system enters into steady state. But this may never occur. The arrival rate may stay fairly constant over short time intervals, but not over long intervals. By the time the system approaches steady state, the arrival rate may have changed and the result no longer applies. Although, in many instances, the transient behavior of a queueing system can be analyzed, the analysis is usually too complicated and too specialized to be useful. A more pertinent question is under what conditions do the steady-state results provide meaningful predictions?

A good rule of thumb comes from the concept of ***relaxation time*** (see Chap. 4). A system with Poisson arrivals that initially has less than L_s customers in the system will "forget" its initial state after a time comparable to the following (Newell 1982):

$$T_0 = \frac{\rho/m + C^2(S)}{(1 - \rho/m)^2}\ (1/m\mu) \tag{5.74}$$

where, as before, $C(S)$ is the coefficient of variation for the service time. This equation is a generalized version of the equation for relaxation time in Chap. 4. The important feature of Eq. (5.74) is that *the time required to reach steady state grows as ρ/m approaches 1.* This is demonstrated below for $C(S) = 1$ (the coefficient of variation for the exponential distribution):

ρ/m	.2	.4	.6	.8	.9	.95
$T_0/(1/m\mu)$	1.9	3.9	10	45	190	780

When $\rho/m = .9$, it takes a time comparable to 190 services to reach steady state. As an example, if it takes 2 minutes to serve a customer, and the arrival rate is 27 customers per hour, then it will be more than 6 hours before steady state is reached. Because few systems maintain the same arrival rate over such a long period of time, one can conclude that steady-state models are not generally applicable when ρ/m is at all close to 1. On the other hand, if $\rho/m = .4$, it takes only four service times to reach steady state, a small number indeed. So the results might well be meaningful if ρ/m is a small number (less than .5).

Should the initial state of the system not be comparable with L_s—say, if the system begins with a very large number of customers—then the time required to reach steady state would be somewhat larger than T_0. As a rule of thumb, if the initial state is larger than L_s, then T_0 should be incremented by the expected length of time required to make the transition from the initial customers in system, $L_s(0)$, to the expected customers in system, L_s:

$$T_0' = \frac{L_s(0) - L_s}{m\mu - \lambda} + T_0 \qquad L_s(0) > L_s \tag{5.75}$$

Of course, the impact of the correction factor is negligible unless the difference between $L_s(0)$ and L_s is large.

5.7 A SYSTEM WITH ANCILLARY ACTIVITIES

So far, a variety of models have been presented for systems operating in steady state with a fixed number of servers. In contrast, many queueing systems follow a somewhat different policy, whereby servers alternate between direct customer service, and, perhaps, some ancillary activity. To take one common example, employees in a supermarket may alternate between customer checkout and backroom activities, such as stocking shelves. When queues become long, checkers are diverted to direct customer service; when queues dissipate, checkers return to the ancillary activity. The efficiency gains from this policy should be obvious. What would otherwise be idle time becomes productive time. Why leave a server idle when there is other work to do?

An effective operating policy is to add a server whenever the number of customers in the queue divided by the number of servers exceeds a set value (usually 2 or 3), and remove servers when the queue dissipates.

Definitions 5.6

$$K = \text{maximum number of customers in the queue per server}$$

$$m(t) = \text{number of servers operating at time } t$$

Example

A grocery store alternates employees between stocking shelves and customer checkout, with $K = 3$. At 10:30, 3 employees are working at the checkout counters and 12 customers are in the queueing system. At 10:31, one more customer arrives. Because $L_q(t)/m(t) = 10/3 > 3$, an employee is diverted from stocking to checkout [$m(10{:}31) = 4$]. Later on, at 10:35, the number of customers in the system has declined to 4 (one per server). At 10:36, one of the customers completes service and one of the employees is returned to stocking.

Note that the system manages to keep employees busy all of the time, either in checkout or in stocking. However, this does not come without a cost, for each time an employee is diverted between activities, some loss in productivity results. The more frequent these interruptions occur, the less productive will be the employee. Therefore, it is important to set K at a value that does not disrupt the employee too often, yet still keeps queue lengths reasonable.

The trade-off between interruptions and waiting time can be analyzed with a steady-state model, assuming as before that the arrival process is Poisson and that service times are independent exponential random variables. However, the analysis is more complicated because the service rate is not uniquely determined by the number of customers in the system. For example, if $K = 1$ and $L_s(t) = 2$, one server might be operating or two servers might be operating. Two servers would be operating if the number of customers in the system previously exceeded two and the second server had not yet been removed.

The state of system can be uniquely defined by two variables: the number of customers in the queue and the number of customers in service. Figure 5.9 shows the two dimensional state-transition diagram for $K = 2$. The figure assumes that the pool of available servers is unlimited, an assumption that will be discussed later.

Earlier, state probabilities were found by solving balance equations, which equated the transition rates across cordon lines. The same approach applies here, except three types of cordon lines are needed, as illustrated in Fig. 5.9.

Definition 5.7

$\tilde{P}_{n,s}$ = steady-state probability that there are n customers in the queue and s customers in service

The calculations for the state probabilities are messy, but require no more than basic algebra to solve balance equations. Details of the calculations are provided in Moder and Phillips (1962), Yadin and Naor (1967), and Hall (1989). Just keep in mind that the performance measures are easily calculated from the state probabilities, though the

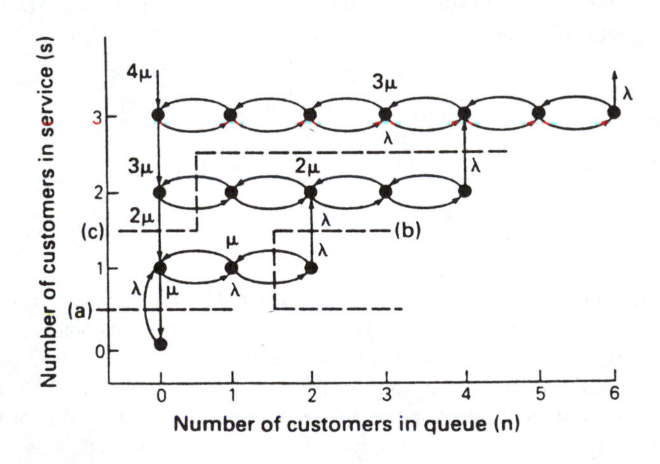

Figure 5.9 Transition rate diagram for queue with ancillary activities. Transition rates must be equal across any cordon line.

calculations are a bit different from before. First, note that the number of customers in the system is $n + s$, and the number of customers in the queue is n. Therefore

$$L_s = \sum_{n=0}^{\infty} \sum_{s=0}^{\infty} (n + s)\tilde{P}_{n,s} \tag{5.76}$$

$$L_q = \sum_{n=0}^{\infty} \sum_{s=0}^{\infty} n\tilde{P}_{n,s} \tag{5.77}$$

The expected waiting times can be calculated, as usual, from Little's formula:

$$W_s = L_s/\lambda \qquad W_q = L_q/\lambda \tag{5.78}$$

The difference between W_s and W_q is also, as usual, the average service time.

In addition to these standard measures of performance, two more should be calculated. The first is the interruption probability—the probability that an arriving customer causes a server to interrupt an ancillary activity. Note that interruptions occur only when the number of customers in the queue is an integer multiple of K. Because arrivals are independent of the state, this translates into the following.

Definition 5.8

$$\begin{aligned} \mathrm{I} &= \text{probability of interruption} \\ &= \sum_{i=1}^{\infty} \tilde{P}_{Ki,i} \end{aligned} \tag{5.79}$$

Thus, the interruption probability is the probability of being in state K, 1 or state $2K$, 2 or $3K$, 3, ... The calculation presumes that one server must always be on duty, so interruptions do not occur in state 0. With this in mind, the expected number of employees serving customers can also be calculated. At most, only one server can be idle, which only occurs when there are no customers in the system. This idle time applies to a "full-time" server who must always be on duty to greet customers. The expected number of employees serving customers is the idle time plus the utilization.

Definition 5.9

$$\begin{aligned} B &= \text{expected number of employees serving customers} \\ &= \rho + \tilde{P}_{0,0} \end{aligned} \tag{5.80}$$

The performance measures are summarized in Figs. 5.10 through 5.12, which show the expected number of customers in the system, the probability of interruption, and the expected number of employees serving customers, as a function of K and ρ. As might be expected, increasing K increases the waiting time (and customers in the system) and reduces the interruption rate. However, once K reaches 3 or 4, the interruption rate becomes small (less than once per ten arrivals), and further increases in K do not appear warranted, even for large values of ρ. Therefore, *a policy whereby servers are added*

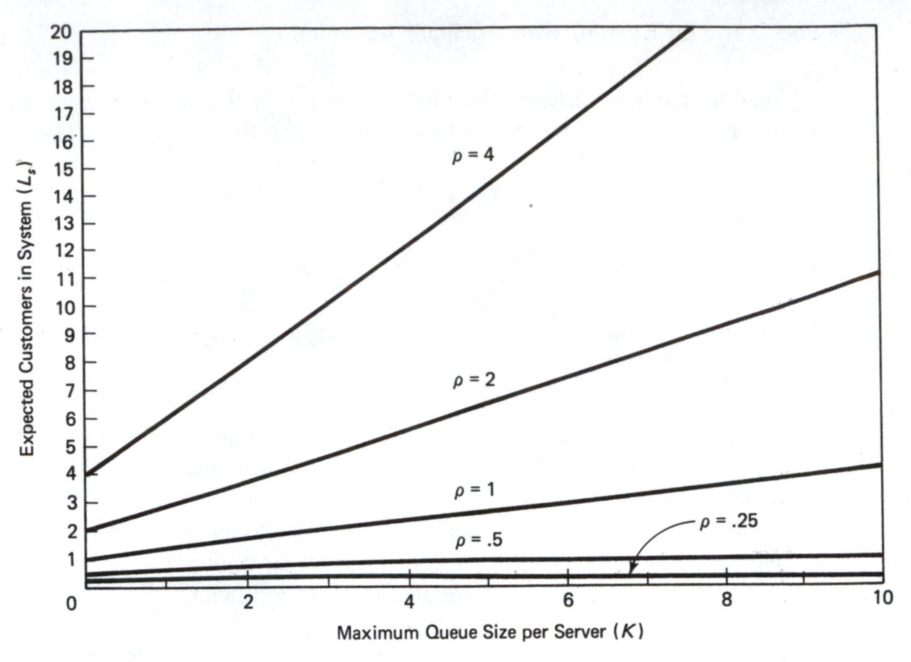

Figure 5.10 Increasing the maximum queue size per server increases L_s at a nearly linear rate when p is greater than 1. For $p < 1$, L_s approaches $p/(1 - p)$ as K becomes large.

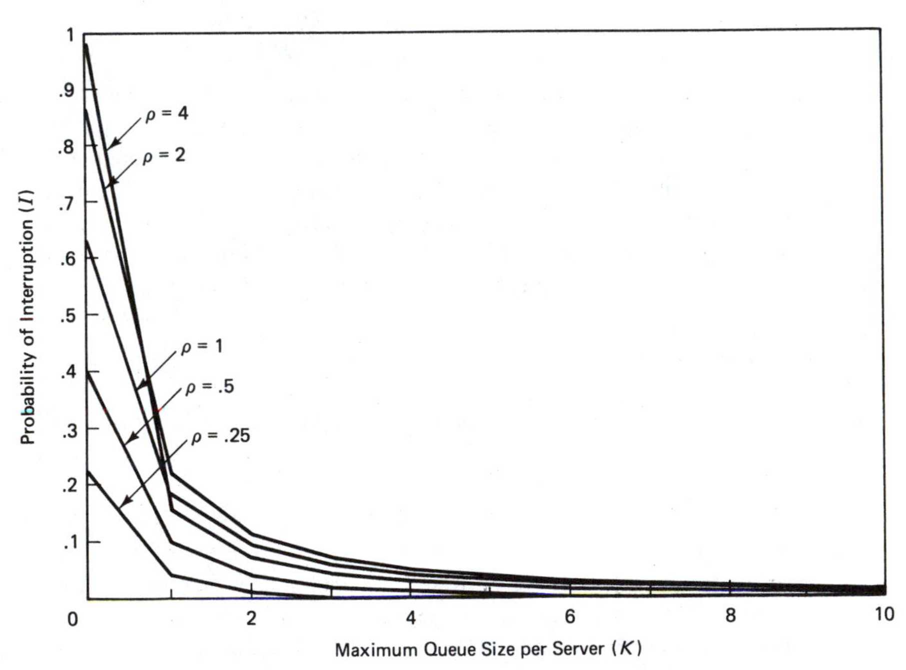

Figure 5.11 The probability of an interruption declines toward zero as the maximum queue size per server increases. For $K \geq 3$, the probability of interruption does not exceed .1.

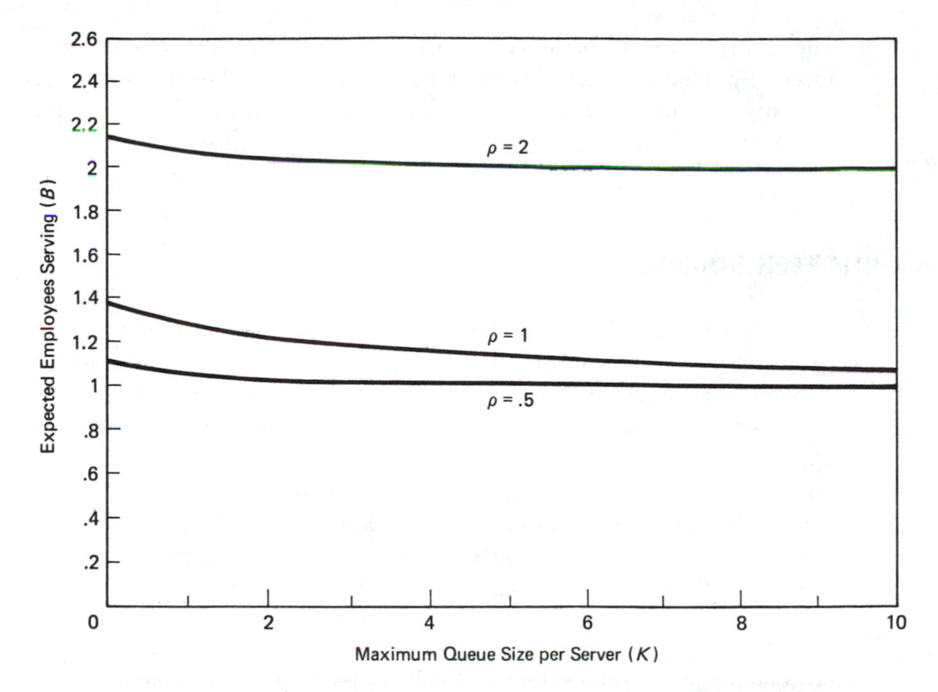

Figure 5.12 The expected number of employees serving declines as the maximum queue size per server increases. For $p > 1$, the limit is p. For $p \leq 1$, the limit is one.

when the number of customers in queue per server exceeds three should provide good results in most instances.

By comparison to the $M/M/m/\infty$ queue, the same number of "busy" servers provides far less delay. But, perhaps more importantly, the ancillary policy is far more robust with respect to changes in the arrival rate. With fixed servers, even small changes in ρ can produce enormous increases in delay. This is not true with ancillary servers, for the time in queue never exceeds K/μ. In fact, for large values of ρ, the expected time in queue is approximately:

$$W_q \approx \frac{K}{2\mu} \qquad (\rho \geqslant 2) \tag{5.81}$$

or just one-half K multiplied by the average service time. This is predicated on having an "infinite" reserve of ancillary employees from which to draw upon. The realism of this assumption depends on the ability to find things for employees to do when they are not serving customers and the ease at which employees can alternate between tasks. So long as there are plenty of employees in the "back room," the ancillary policy is far superior to the $M/M/m/\infty$ system. In fact, if ancillary activities are available, there is little reason why a queue should ever become large.

The main obstacle to eliminating queues is the difficulty associated with diverting servers back and forth between activities. This is especially problematic (though not impossible) when the server is an object rather than a person. For example, a specialized piece of production machinery cannot be readily diverted to an ancillary activity. Hence, the only way to eliminate the queue may be to incur the expense of purchasing more equipment and allowing the equipment to remain idle for some portion of the time.

5.8 CHAPTER SUMMARY

This chapter has undertaken the task of determining the expected performance of a queueing system operating in steady state. To a great extent, the discussion has only scratched the surface of a great body of literature, which can be found in the many books on queueing theory (a few are listed at the end of the chapter). In all cases, the basic approach is to calculate state probabilities and convert these probabilities into performance measures. The calculations for expected performance apply to any queue discipline, *provided that the service time cannot be predicted in advance*. When service times can be predicted, waiting times and queue lengths can be reduced below the values provided in this chapter. This subject is taken up in Chap. 9.

On a more conceptual level, it is worthwhile to contrast the models developed in this chapter to the simulations in Chap. 4 and the observations in Chap. 2. The steady-state predictions represent the expected performance of the system, should the simulation be run for a very long time. But they do not exactly match simulation results because simulations are susceptible to random fluctuations. The steady-state predictions also do not exactly match observations—but for different reasons. The steady-state predictions are based on parameter estimates from a limited set of observations. These estimates may be different from the true parameter values. While differences between simulation and steady-state analysis are due to randomness *inherent* to the simulation, differences between steady-state analysis and reality are due to structural or parameter errors in the model. Thus, compared to simulation, steady-state analysis removes one source of uncertainty (simulation error) but it does not remove the second or third.

Steady-state analysis has another advantage: it highlights important relationships between the queue's performance and the queue's characteristics. These relationships are nearly impossible to identify from simulation. Despite these comments, simulation does not lack appeal, for it is a much more robust technique, capable of dealing with virtually any probability distribution. Hence, with simulation, one is less likely to compromise the accuracy of the model structure, as in adopting the exponential service time distribution when it is not correct.

When a system operates in steady state, queues do not materialize because the system is oversaturated, but because service times and interarrival times randomly vary. If the randomness is eliminated, then the queue will also be eliminated (this conclusion does not apply to systems that do not operate in steady state). Queue lengths can also be reduced by increasing the number of servers. Unfortunately, this means that servers must be idle much of the time as they wait for customers to arrive. Queues tend to be small whenever the number of servers is larger than $\rho + \sqrt{\rho}$. There is seldom justification for providing more servers than this. When queue length is restricted by a buffer capacity, some of the arrivals are lost to the system. However, a restricted buffer capacity has the positive impact of reducing the expected waiting time (excessively large queues are limited). When customers arrive from a finite population (as in the machine breakdown example), queue length also tends to be moderated because the arrival rate declines as the queue length increases. These results are summarized in Table 5.1.

Before a system reaches steady state, it will pass through a transient phase that may be very different from the steady state. The length of time required to reach steady state is inversely related to $(1 - \rho/m)^2$. When ρ/m is close to 1 (larger than .8), the system may

TABLE 5.1 SUMMARY OF STEADY-STATE RESULTS

Characteristic	Features
Exponential Service, Poisson Arrivals, Single Server	$L_s = \rho/(1 - \rho)$, $L_q = \rho^2/(1 - \rho)$ Average queue size becomes infinite as ρ approaches 1
Infinite Servers	Poisson state distribution, mean ρ, for $M/G/\infty/\infty$ Queue size equals zero State distribution for $M/G/m/\infty$ for large m ($m >> \rho$)
Finite Buffer Size	Lost customers reduce arrival rate Average queue size declines Average queue size approaches b as ρ becomes large
Finite Calling Population	Arrival rate declines as state increases Average queue size declines Average customers in system approaches population size as ρ becomes large
Nonexponential Service Times, Poisson Arrivals, Single Server	L_q defined by Pollaczek-Khintchine formula Average queue size becomes infinite as ρ approaches 1 Average queue size grows as $C(S)$ increases
General Service and Interarrival Distributions	L_q approximated by Allen-Cunneen Formula Average queue size becomes infinite as ρ approaches m Average queue size grows as $C(S)$ and $C(A)$ increase
Zero Buffer Size, Multiple Servers, Poisson Arrivals	State probabilities are independent of service time distribution Probability that customer is lost given by Erlang loss formula
Ancillary Activities	Average queue size declines substantially Interruption may reduce efficiency of activities

never achieve steady state. The steady-state models are much more accurate for systems that are not overly strained and, consequently, tend to have much smaller queue lengths.

As a final comment, delays can be eliminated (or at least made very small) if servers alternate between direct customer service and ancillary activities. In most instances, an effective policy is to add a server whenever the number of customers in the queue divided by the number of servers exceeds three and to remove servers when the queue dissipates. This policy achieves a small waiting time without incurring a large server idle time. It is also very robust; it works well for virtually any arrival rate. In fact, so long as the cost of alternating between the activities is not large, there is no reason why queues should ever become excessive. Queueing problems are difficult to solve only when the cost of alternating servers between activities is large. This subject is addressed in later chapters.

FURTHER READING

ALLEN, A. O. 1978. *Probability, Statistics, and Queueing Theory, with Computer Science Applications*. New York: Academic Press.

BROCKMEYER, E., H. A. HALSTROM, and A. JENSON. 1948. *The Life and Works of A. K. Erlang*. Copenhagen: Danish Academy of Technical Science.

BUNDAY, B. D., and R. E. SCRATON. 1980. "The *G/M/r* Machine Interference Model," *European Journal of Operational Research*, 4, 399–402.

GROSS, D., and C. M. HARRIS. 1985. *Fundamentals of Queueing Theory*. New York: John Wiley.

HALL, R. W. 1989. "Expected Performance of a Queueing System with Ancillary Activities," *Journal of the Operational Research Society*, v. 40, pp 741–750.

KLEINROCK, L. 1975, 1976. *Queueing Systems*, vols. 1 and 2. New York: John Wiley.

KÖLLERSTRÖM, J. 1974. "Heavy Traffic Theory for Queues with Several Servers. I," *Journal of Applied Probability*, 11, 544–552.

MODER, J. J., and C. R. PHILLIPS. 1962. "Queueing with Fixed and Variable Channels," *Operations Research*, 10, 218–231.

NEWELL, G. F. 1982. *Applications of Queueing Theory*. London: Chapman and Hall.

TAKACS, L. 1969. "On Erlang's Formula," *Annals of Mathematical Statistics*, 40, 71–78.

WOLFF, R. W. 1989. *Stochastic Modeling and the Theory of Queues*. Englewood Cliffs, N.J.: Prentice Hall.

YADIN, M., and P. NAOR. 1967. "On Queueing System with Variable Service Capacities," *Naval Logistics Research Quarterly*, 14, 43–54.

PROBLEMS

1. Give two examples of systems that ordinarily have a finite buffer size. Give two examples of systems that have a finite calling population.

2. After observing a single server queue for several days, the following steady-state probabilities have been determined: $P_0 = .4, P_1 = .3, P_2 = .2, P_3 = .05, P_4 = .05$. The arrival rate is 10 customers/hour.

(a) Determine L_s and L_q.

(b) Using Little's formula, determine W_s and W_q.
(c) Determine σ_{L_s} and σ_{L_q}.
(d) Determine the service time and utilization.

3. Repeat Prob. 2, now assuming there are two servers.

4. Repeat Prob. 3, with state probabilities $P_0 = .2$, $P_1 = .4$, $P_2 = .2$, $P_3 = .2$ ($\lambda = 10$/hour).

5. An $M/M/1/\infty$, FCFS, queue has an arrival rate of 100/hour and a service rate of 140/hour.
(a) Write the equation for P_n, the probability of being in state n.
(b) Calculate L_q,L_s,W_q, and W_s.
(c) What is the probability that a customer's waiting time in queue is less than 3 minutes?
(d) What is the probability that a customer's waiting time in system is less than 3 minutes?
(e) Compare your answer to part b to the solution to Prob. 15 of Chap. 4. Does your answer make sense, if both queues are $M/M/1/\infty$?

6. An $M/M/1/\infty$ queue has been found to have an average waiting time in queue of 1 minute. The arrival rate is known to be 5 customers/minute.
(a) What are the service rate and utilization?
(b) Calculate L_q, L_s, and W_s.
(c) The queue operator would like to provide chairs for waiting customers. He would like to have a sufficient number so that all customers can sit down at least 90 percent of the time. How many chairs should he provide?

7. A single server queueing system is known to have Poisson arrivals and exponential service times. However, the arrival rate and service time are state dependent. As the queue becomes longer, servers work faster, and the arrival rate declines, yielding the following functions:
$\lambda_0 = 5$/hour $\lambda_1 = 3$/hour $\lambda_2 = 2$/hour $\lambda_n = 0$/hour $n \geqslant 3$
$\mu_0 = 0$/hour $\mu_1 = 2$/hour $\mu_2 = 3$/hour $\mu n = 4$/hour $n \geqslant 3$
(a) Calculate the state probabilities.
(b) What are the average arrival and service rates?
(c) Calculate L_q,L_s,W_q,W_s.

8. The queueing system at a fast-food stand behaves in a peculiar fashion. When there is no one in the queue, people are reluctant to use the stand, fearing that the food is unsavory. People are also reluctant to use the stand when the queue is long. This yields the following arrival rates:

$$\lambda_0 = 10\text{/hour} \quad \lambda_1 = 15\text{/hour} \quad \lambda_2 = 15\text{/hour} \quad \lambda_3 = 10\text{/hour} \quad \lambda_4 = 5\text{/hour}$$
$$\lambda_n = 0\text{/hour} \quad n \geqslant 5$$

The stand has two servers, each of which can operate at 5/hour. Service times are exponential, and the arrival process is Poisson.
(a) Calculate the state probabilities.
(b) What are the average arrival and service rates?
(c) Determine L_q, L_s, W_q, and W_s.

9. The system in Prob. 6 has decided to add a second server. Calculate the new values of L_q and W_q.

***10.** A repair/maintenance facility would like to determine how many employees should be working in its tool crib. The service time is exponential, with mean 4 minutes, and customers arrive by a Poisson process, with rate 28/hour. The customers are actually maintenance workers at the facility, and are compensated at the same rate as the tool crib employees.
(a) Using Fig. 5.3 as a guide, plot W_s versus number of servers, for one to four servers.
(b) How many employees should work in the tool crib?

*Difficult problem

11. An $M/M/2/\infty$ queueing system is very heavily utilized, with an arrival rate of 11 customers/hour, and a service rate of 6 customers/hour per server.

(a) Determine L_q and W_q.

(b) Compare your solution to part a to an $M/M/1/\infty$ queue with $\rho = 11/12$. Explain why your answers are similar or different.

12. A telephone system has a very large number of lines, far exceeding the maximum number ever required. The length of telephone calls has a gamma distribution, with mean of 4 minutes and standard deviation of 3 minutes. The arrival rate of calls is 50 per hour.

(a) What is the expected number of lines in service at any given time?

(b) What is the probability that five or fewer lines are in service at any given time?

(c) Suppose that the telephone operator would like to eliminate some of the lines, but still have a sufficient number that 98 percent of the time, at least one line is free. Using the $M/G/\infty$ model as an approximation, how many lines are needed?

13. The telephone system manager has decided to repeat the calculation from Prob. 12, part b, using the Erlang loss formula. Based on this exact calculation, how many lines are needed? (No buffer is possible.)

***14.** An airline phone reservation line has one server, and a buffer for two customers. The arrival rate is 6 customers/hour, and the service rate is just 5 customers/hour. Arrivals are Poisson and service times are exponential.

(a) Estimate L_q and the average number of customers served per hour.

(b) From Fig. 5.4, estimate L_q for a buffer of size 5. In words, what is the impact of the increased buffer size on the number of customers served per hour?

(c) Repeat part a for a buffer of size 2, with two servers.

(d) If employees are paid $15/hour, how large would the profit per customer have to be to justify adding a second server (with a buffer of size 2)?

***15.** An operator services two machines. At the end of a run, she has to set up the machine for the next run. Run lengths have an exponential distribution, as do the service times. The mean run length is 30 minutes, and the mean setup time is 5 minutes.

(a) What proportion of the time will each machine be running? What proportion of the time will the operator be busy with setups?

(b) Suppose that the cost of owning a machine amounts to $100/hour, and labor cost amounts to $25/hour. On the basis of cost per unit of capacity, would it be more economical for the operator to service one machine rather than two?

16. The manager of a small firm would like to determine which of two people to hire. One employee is fast, on average, but is somewhat inconsistent. The other is a bit slower, but very consistent. The first has a mean service time of 2 minutes, with a standard deviation of 1 minute. The second has a mean service time of 2.1 minutes, with a standard deviation of .1 minute. If the arrival process is Poisson with rate 20/hour, which employee would minimize L_q? Which would minimize L_s?

17. A queueing system receives Poisson arrivals at the rate of 5/hour. The single server has a uniform service time distribution, with a range from 4 minutes to 6 minutes. Determine L_q, L_s, W_q, and W_s. (Hint: Properties of the uniform distribution are given in Chap. 3.)

***18.** Figure 4.18 (Chap. 4) used simulation to develop a probability distribution for the estimation of W_q for an uncertain arrival rate. Using the $M/M/1$ formula ($\mu = 1.20$/min), redraw the graph, at the 5%, 15%, . . . , 95% points. (Note that simulation error is now eliminated.) How does your graph compare to the simulation?

*Difficult problem

19. When a bus reaches the end of its line, it undergoes a series of inspections. The entire inspection takes 5 minutes on average, with a standard deviation of 2 minutes. Buses arrive at uniform intervals of 6 minutes.

(a) Assuming a single server, use the Allen-Cunneen approximation to estimate W_q.

(b) Compare this result to an $M/G/1$ system, with arrival rate of 10/hour and the same service time distribution as above. Explain why your answer to part a is smaller.

(c) Returning to the original arrival process, estimate W_q for two servers.

20. Estimate the relaxation time for the systems defined by the following problems:

(a) i. Prob. 5 ii. Prob. 11 iii. Prob. 17

(b) Given that most businesses have a 9-hour workday, for which of these systems would a steady-state model provide meaningful results? Discuss.

21. The queueing operator in Prob. 11 has decided to institute a policy whereby employees are removed from service, to perform cleanup work, whenever they complete service and no one is in queue. Employees are brought back into service when the number of customers in queue, per server, exceeds two. Using Figs. 5.10 to 5.12, estimate W_q, the interruption rate, the expected time between interruptions, and the expected number of busy servers. Do you believe this policy is a good alternative?

***22.** At a large hotel, taxicabs arrive at the rate of 15/hr, and parties of riders arrive at the rate of 12/hr. Whenever taxicabs are waiting, riders are served immediately upon arrival. Whenever riders are waiting, taxicabs are loaded immediately upon arrival. A maximum of three cabs can wait at a time (other cabs must go elsewhere).

(a) Let $P_{i,j}$ be the steady-state probability of there being i parties of riders and j taxicabs waiting at the hotel. Write the state transition equations for the system.

(b) Calculate the expected number of cabs waiting and the expected number of parties waiting.

(c) Calculate the expected waiting time for cabs and the expected waiting time for parties. (For cabs, compute average among those that do not go elsewhere.)

(d) In words, what would be the impact of allowing four cabs to wait at a time?

EXERCISE: APPLICATION OF STEADY-STATE MODELS

The purpose of this exercise is to learn how to use steady-state analysis to evaluate changes in a queueing system. Your answers should be based on your observations from Chap. 2.

1. For the $M/M/1/\infty$ queue, calculate the probability of having zero customers in the system, one customer in the system, . . . , up to ten customers in the system. Also calculate L_q, L_s, W_q, and W_s.

2. Repeat your calculations for the $M/M/2/\infty$ queue.

3. List your answers to part 1 and part 2 in a single table. In the same table, list the proportion of time that you observed zero customers in the system, one customer, and so on. Also list the proportions obtained from the simulation in Chap. 4.

(a) Explain why, or why not, your observed proportions are different from your answer to part 1.

*Difficult problem

(b) Explain why, or why not, your simulated proportions are different from your answer to part 1.

4. Calculate the approximate time for the system to reach steady state for one server and for two servers. Is this an important factor in answering Prob. 3a or 3b?

5. Use the Pollaczek-Khintchine formula to estimate L_q, using the observed coefficient of variation for the service times.

(a) Compare this calculation to your estimate of L_q obtained from simulation. Why are the numbers different or the same? Does the P-K calculation fall within the confidence interval obtained from simulation?

(b) Compare this calculation to your estimate from the $M/M/1/\infty$ model. Why are the numbers different or the same?

(c) Make the same comparison to the observed value of L_q.

6. Use the Allen-Cunneen approximation to estimate L_q for two servers, using the observed coefficient of variation for the service times and the observed coefficient of variation for the interarrival times. Compare this calculation to your estimate from the $M/M/2/\infty$ model. Why are the numbers different or the same?

7. Using your answers to Probs. 5 and 6, calculate the difference between L_q with two servers and L_q with one server. If a server's time is worth \$*a* per hour and a customer's time is worth \$*b* per hour, for what values of *a* and *b* would a second server be justified? (Difficult)

APPENDIX: POLLACZEK-KHINTCHINE FORMULA

The Pollaczek-Khintchine formula is relatively easy to derive, using expected value calculations and algebra. The following will show how to derive W_q for the $M/G/1/\infty$ model and then apply Little's formula to obtain L_q. The discussion is based on an FCFS discipline, but the result is the same for other disciplines that are independent of service time.

The probability that the system is empty when a customer arrives is $1 - \rho$, and the probability that the system is not empty equals ρ. If the system is empty, the wait in queue is zero. If the system is not empty, the wait can be divided into two parts: wait until customer currently in service completes service, and wait until customers currently in queue complete service. Let

W_r = expected remaining service time for the customer currently in service

N_q = expected customers in queue, given system is not empty

Therefore

$$W_q = E[\text{wait} \mid \text{nonempty}]\cdot P(\text{nonempty}) = [W_r + N_q(1/\mu)](\rho) \qquad (5.\text{A}1)$$

For Poisson arrivals, the expected number of customers in queue at an arrival time, $N_q\rho$, is the same as L_q (this is called the PASTA, or Poisson arrivals see time averages, property; Wolff 1989). W_r is defined by the following equation, which accounts for variations in service times:

$$W_r = E(S)\left[\frac{1 + C^2(S)}{2}\right] \tag{5.A2}$$

Equation (5.A1) can now be expressed as

$$W_q = E(S)\left[\frac{1 + C^2(S)}{2}\right]\rho + L_q(1/\mu) \tag{5.A3}$$

From Little's formula, L_q can be replaced by λW_q. Making this substitution, and collecting terms yield the following:

$$W_q(1 - \rho) = E(S)\left[\frac{1 + C^2(S)}{2}\right]\rho \rightarrow W_q = \frac{\rho E(S)}{1 - \rho}\left[\frac{1 + C^2(S)}{2}\right] \tag{5.A4}$$

Equation (5.64), the Pollaczek-Khintchine formula, results from applying Little's formula to Eq. (5A.4).

The above calculation does not provide the state probability distribution. Determining this, and other complicated queue performance measures, requires advanced techniques in mathematical statistics, such as the Laplace transform, the z-transform, and generating functions. However, even the advanced techniques cannot be used on many realistic queueing systems. In the end, one often resorts to simulation.

chapter 6

Nonstationary Arrivals

If you were to think of the most frustrating, the most aggravating, and the most time-consuming sort of queue, there is a good chance that the evening rush-hour commute home from work would come to mind. The rush hour, both morning and evening, is the product of large numbers of people desiring to use the roads and highways at the same time. It is also an example of a nonstationary arrival pattern.

A ***nonstationary*** (also called *nonhomogeneous*) arrival pattern occurs when the customer arrival rate varies over time. It is a phenomenon that virtually all queueing systems experience to one extent or another. Restaurants have rush periods at lunchtime and dinnertime. Retail stores experience a rush period in the month before Christmas. Accountants experience rush periods prior to tax due dates. Rush periods are the consequence of the natural cycles in our lives. We orient ourselves toward daily, weekly, monthly, and yearly patterns. We tend to work at the same time, eat at the same time, sleep at the same time, shop at the same time, . . . And this puts a strain on queueing systems. It is far easier to serve customers when they arrive at an even rate than an uneven rate.

We have seen that when a queueing system operates in steady state, queues are produced by ***random variability*** in service times and arrival times. In steady state, there is no way to know when these queues will occur because they are totally random. A nonstationary arrival pattern presents an additional source of variability: ***predictable variability***. Predictable variability in the arrival pattern means that queues occur in a predictable fashion, at the same time every day, or every week, and so on. Predictable queues (such as the evening rush hour) tend to be much larger and costlier than random queues.

This chapter provides several approaches for modeling a nonstationary arrival process and queueing system. It begins by defining the nonstationary version of the Poisson process. Next, a procedure is presented for using steady-state equations to model certain, lightly used queueing systems. This is followed by a description of how to simulate a nonstationary queueing process. Then a much simpler model, known as a fluid approximation, is provided. The chapter concludes with an alternative way to define the arrival process, in terms of desired departure times from the system.

6.1 THE NONSTATIONARY POISSON PROCESS

The nonstationary Poisson process is a Poisson process for which the arrival rate varies with time. More specifically, it can be defined as follows:

Definitions 6.1

The counting process $N(t)$ is a ***non-stationary Poisson process*** if:

A. The process has independent increments

B. $$Pr\,[N(t+dt)-N(t) \begin{cases} =0] = 1-\lambda(t)dt \\ =1] = \lambda(t)dt \\ >1] = 0 \end{cases}$$

where

$$\lambda(t) = \text{the arrival rate at time } t$$

$$dt = \text{a differential sized time interval}$$

The definition is identical to the stationary Poisson process (Chap. 3), with the exception that the arrival rate, $\lambda(t)$, is now a function of time. As before, the arrival rate represents the *expected* number of customers to arrive per unit time. If λ (9:00) equals 10 per minute, then we would expect to see ten customers arrive in the 1-minute interval between 9:00 and 9:01 on average. The actual number of customers to arrive can be either smaller or larger than ten.

The arrival rate, having the dimensions *customers/time*, when integrated with respect to time, yields the expected number of customers to arrive over a *time interval*:

$$E(\text{arrivals between time } a \text{ and time } b) = \int_a^b \lambda(t)dt \qquad (6.1)$$

One can think of the stationary Poisson process as a special version of the nonstationary Poisson process. So, to take an example, if $\lambda(t) = \lambda = 10$ customers/hour, then the expected number of arrivals over a 1-hour period is

$$E(\text{arrivals over 1 hour}) = \int_0^1 10dt = 10t \,\Big|_0^1 = 10 \text{ customers}$$

And the expected number of arrivals over a half-hour period is

$$E(\text{arrivals over } \tfrac{1}{2} \text{ hour}) = \int_0^{.5} 10dt = 10t \,\Big|_0^{.5} = 5 \text{ customers}$$

As might be expected, these results are identical to what was found in Chap. 3 when the stationary Poisson process was presented. Of course, $\lambda(t)$ does not have to be constant. Suppose that customers arrive at a restaurant at the following rate:

$$\lambda(t) = 100\sin(t\pi/2) \qquad 0 \leq t \leq 2 \text{ (customers/hour)} \tag{6.2}$$

where time $t = 0$ is 11:30 A.M. and time $t = 2$ is 1:30 P.M., and the sine angle is measured in radians. With this function, customers arrive at the fastest rate at 12:30 (100 customers/hour) and the slowest rate at 11:30 and 1:30 (0 customers/hour).

Definition 6.2

$\Lambda(t)$ is the expected number of arrivals from time 0 to time t. $\Lambda(t)$ is calculated by integrating $\lambda(t)$ from 0 to time t:

$$\Lambda(t) = \int_0^t 100\sin(\pi\tau/2)d\tau = (200/\pi)[1 - \cos(\pi t/2)] \text{ customers} \tag{6.3}$$

Note that the variable used in the integrand (τ) must be distinguished from the variable used to bound the integral (t). Also note that $\Lambda(t)$ is measured in terms of customers, whereas $\lambda(t)$ is measured as a rate: customers per hour. $\Lambda(t)$ is plotted for the example in Fig. 6.1. The slope of $\Lambda(t)$ (that is, the derivative) is $\lambda(t)$. The figure shows that the arrival rate is largest at the center of the time interval and smallest at the ends, as already predicted. Remember that $\Lambda(t)$ represents the expected number of arrivals to occur by time t. The actual number of arrivals can be either larger or smaller than $\Lambda(t)$.

It is more common to base $\lambda(t)$ on ***interval counts*** than on an equation, as shown above. In the restaurant example, records might indicate that the average numbers of arrivals in each of four time periods are the following:

Time	Average arrivals
11:30–12:00	19
12:00–12:30	45
12:30–1:00	45
1:00–1:30	19

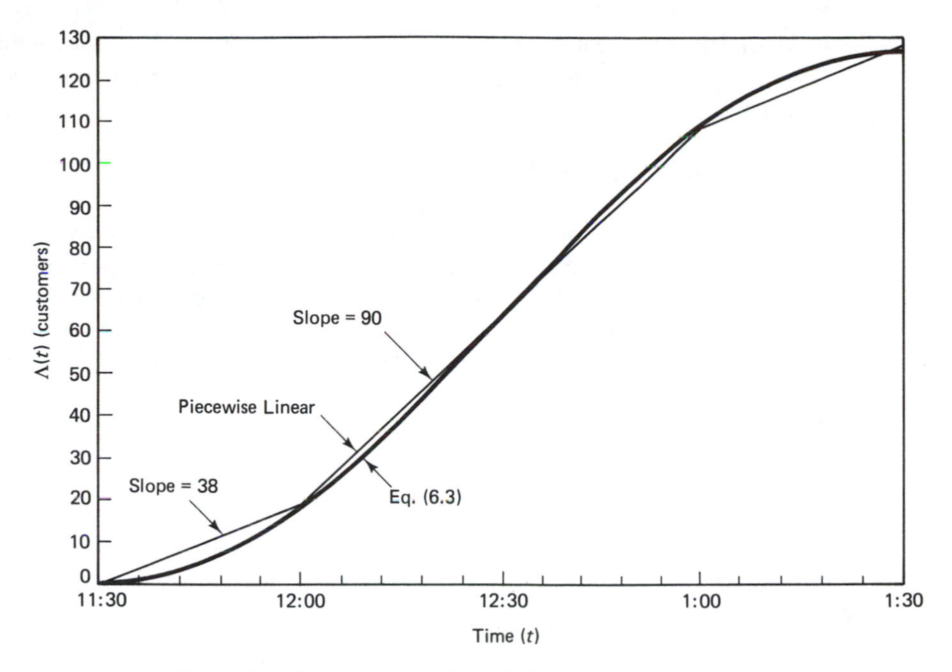

Figure 6.1 Expected cumulative arrivals at restaurant versus time.

These numbers can be translated into an arrival rate by dividing the number of arrivals by the size of the time interval. In all four cases, the time interval is one-half hour, so the arrival rate is

$$\lambda(t) = \begin{cases} 38 & 0 \leq t < .5 \\ 90 & .5 \leq t < 1.5 \\ 38 & 1.5 \leq t \leq 2 \end{cases} \qquad \text{(customers/hour)} \tag{6.4}$$

The units here are important. If the interval counts are divided by a time unit measured in hours, then $\lambda(t)$ is measured in terms of arrivals per hour. $\Lambda(t)$ can, as usual, be found by integrating $\lambda(t)$:

$$\Lambda(t) = \begin{cases} 38t & 0 \leq t < .5 \\ 19 + 90(t - .5) & .5 \leq t < 1.5 \\ 109 + 38(t - 1.5) & 1.5 \leq t \leq 2 \end{cases} \tag{6.5}$$

$\Lambda(.5)$ is the average number of arrivals recorded for the first half-hour, $\Lambda(1)$ is $\Lambda(.5)$ plus the average number of arrivals recorded in the second half-hour, and so on. So $\Lambda(t)$ can actually be calculated by summing the interval counts and interpolating between the points. This new version of $\Lambda(t)$ is shown in Fig. 6.1 next to the plot of the equation for $\Lambda(t)$. For the interval counts, $\Lambda(t)$ is a piecewise linear curve, and $\lambda(t)$ (the slope of $\Lambda(t)$) is a step curve, with discontinuities at the ends of the time intervals. In reality, the true arrival rate (the $\lambda(t)$ that generates the arrivals) would not have these discontinuities. The discontinuities are the unavoidable by-product of averaging the number of arrivals over time intervals.

6.1.1 Properties of the Non-Stationary Poisson Process

The nonstationary Poisson process does not possess the property that interarrival times are exponential random variables. Hence, it also does not possess the property that the time until the *n*th arrival is a gamma random variable. Yet it does have several properties in common with the stationary Poisson process. Most important of these is that the number of arrivals over any time interval is a Poisson random variable:

Property 1

The number of arrivals over the interval $[a,b]$ is Poisson with mean

$$E[A(b) - A(a)] = \int_a^b \lambda(t)dt = \Lambda(b) - \Lambda(a)$$

Example

The restaurant owner would like to determine the probability that three or fewer customers will arrive between 11:30 and 11:45, using the equation for $\lambda(t)$. The expected number of arrivals, from Eq. (6.3), is $\Lambda(.25) - \Lambda(0) = 4.85$ customers. The probability of n customers arriving is

$$P(n \text{ arrivals between 11:30 and 11:45}) = \frac{4.85^n}{n!} e^{-4.85} \quad n = 0, 1, \ldots$$

The probability of three or fewer arrivals is found by evaluating the above equation for $n = 0$, $n = 1, \ldots, n = 3$, which equals $.008 + .038 + .092 + .149 = .287$.

The event times within a time interval also have properties similar to the stationary Poisson process. As with the stationary Poisson process, the time of any event is independent of the time of any other event. The nonstationary process is also similar to the stationary process in that the probability distribution for the unordered event times is defined by $\Lambda(t)$:

Property 2

If $A(t)$ is the number of events in the interval $[0,\tau]$, the unordered event times are defined by $A(t)$ independent random variables with the probability distribution:

$$P(T \leq t) = \frac{\Lambda(t)}{\Lambda(\tau)} \tag{6.6}$$

where T is the random variable representing the event time.

With the stationary Poisson process, $\Lambda(t) = \lambda t$, so $P(T \leq t)$ is simply t/τ. This defines the uniform probability distribution over $[0,\tau]$. Again, the stationary Poisson process is a special case of the nonstationary Poisson process. More generally, the event times can have any distribution, as defined by the function $\Lambda(t)$. There is no reason to expect that it has any particular shape. The shape is determined from historical records of customer arrivals.

Example

The restaurant owner knows that five customers arrived between 11:30 and 11:45. He would now like to determine the likelihood that no one arrived before 11:35, using Eq. (6.3). The probability that any one of the five customers arrived before 11:35 is

$$P(\text{arrived before 11:35}) = \frac{\Lambda(1/12)}{\Lambda(.25)} \frac{.545}{4.85} = .112$$

The probability that no one arrived before 11:35 is $(1 - .112)^5 = .551$.

Keep in mind that because the interarrival times are not exponential, the nonstationary Poisson process does not possess the memoryless property.

6.1.2 Goodness of Fit

The basic concept of checking for goodness of fit is the same for a nonstationary Poisson process as for a stationary Poisson process. As always, this begins with a check of plausibility. Does the probability that a customer arrives at any time depend on the times when other customers arrived? Do customers arrive one at a time? The answers to these questions must be affirmative for both the nonstationary and the stationary processes. The major difference in the plausibility check is that the arrival rate does not have to be constant. Thus, the conditions for the nonstationary process are not as strict. Many real arrival processes satisfy the conditions underlying the nonstationary Poisson process.

The quantitative goodness of fit tests are somewhat different, because the interarrival times do not have to be independent exponential random variables, and the arrival times within a time interval do not have to be uniform. The primary check is for the hypothesis:

H_1: The number of events in any time interval has a Poisson distribution.

This test requires large quantities of data, representing the numbers of arrivals over many recurring cycles. It is not enough to know how many customers arrived over each time interval of a single cycle (a day, for example). One must know the number of arrivals over each time interval of many cycles. Then the number of arrivals within each interval should be a Poisson random variable. This test can be carried out through a slight modification of the Kolmogorov-Smirnov test (see the statistics texts cited at the end of Chap. 3). Practically speaking, however, it is virtually impossible to obtain sufficient data to carry out the test, and, in the end, one must rely on the plausibility check. If you believe that arrivals are independent and you believe that customers arrive one at a time, then it should be safe to assume that the arrival process is nonstationary Poisson.

6.1.3 Parameter Estimation

The nonstationary Poisson process is not defined by the single parameter, λ, but by a function, $\lambda(t)$. This makes parameter estimation more complicated. The most straightforward approach is to base $\lambda(t)$ on interval counts, as was already illustrated in this chapter. Suppose that f_n is the average number of arrivals in interval n, from time a to time b, over I cycles. Then

$$\hat{\lambda}(t) = \frac{f_n}{b - a} \qquad a \leq t \leq b \tag{6.7}$$

A confidence interval for $\lambda(t)$ can be formed under the hypothesis that the number of arrivals in any interval is a Poisson random variable. Hence, the variance of the number of arrivals is the same as the expected number of arrivals. Because f_n is the average of a set of random variables, it must have a normal distribution if I is large (central limit theorem). This leads to the following confidence intervals:

95% Confidence ($I \geqslant 50$)

$$P[f_n/(b - a) - 1.96\sqrt{f_n/I}/(b - a) \leqslant \lambda(t) \leqslant f_n/(b - a) + 1.96\sqrt{f_n/I}/(b - a)] = .95 \tag{6.8}$$

99% Confidence ($I \geqslant 50$)

$$P[f_n/(b - a) - 2.58\sqrt{f_n/I}/(b - a) \leqslant \lambda(t) \leqslant f_n/(b - a) + 2.58\sqrt{f_n/I}/(b - a)] = .99 \tag{6.9}$$

Example

Suppose that the interval counts for the restaurant are based on 60 days of records. The 95% confidence interval for the 11:30 to 12:00 period, in which 19 arrivals were observed on average, is calculated as follows:

$$P(19/(.5) - 1.96\sqrt{19/60}/.5 \leqslant \lambda(t) \leqslant 19/.5 + 1.96\sqrt{19/60}/.5) = .95$$

$$P(38 - 2.2 \leqslant \lambda(t) \leqslant 38 + 2.2) = .95$$

In the example, the 95% confidence interval is fairly small, but this relies on 60 days of records. Clearly, obtaining precise estimates of $\lambda(t)$ requires, at a minimum, detailed data on dozens of cycles. Yet, even doing this may not suffice, for there may be no way of guaranteeing that the arrival rate will stay the same every day, every week, or every month. The pattern may never recur. This presents a problem with no clear resolution. No matter what approach is used, predictions based on estimates of $\lambda(t)$ will usually be imprecise.

Another decision to consider is how large the time intervals should be. It is certainly much easier to obtain a precise estimate for the number of arrivals over 1-hour intervals than 1-minute intervals, yet, if the arrival rate truly varies over the hour interval, the variation will not be detected. As a rule, the intervals should be sufficiently small to detect any major changes in the arrival rate, but no smaller than necessary (unless there is an easy way to record arrival data). If a typical queue lasts 1 to 2 hours, then intervals of width 10 to 20 minutes should be sufficient. If a typical queue lasts an entire day, intervals of 1 hour should be sufficient; and if a typical queue lasts a week or more, then intervals of one day should be sufficient.

The alternative to interval counts is to derive $\lambda(t)$ from an estimate of $\Lambda(t)$. Suppose that $A_n(t)$ represents the cumulative arrivals to time t for cycle n of I total cycles. Then an estimate for $\Lambda(t)$ can be obtained as follows:

$$\hat{\Lambda}(t) = \frac{\sum_{n=1}^{I} A_n(t)}{I} = \bar{A}(t) \tag{6.10}$$

The natural estimate for $\lambda(t)$ would be the derivative of $\hat{\Lambda}(t)$. However, because $A_n(t)$ is a step function, $\hat{\Lambda}(t)$ must be too, meaning that the derivative of $\hat{\Lambda}(t)$ is undefined. As an alternative, $\hat{\Lambda}(t)$ can be set equal to a smooth approximation to the average of the arrival curves (a similar approach is shown in Fig. 4.4). The confidence interval for $\Lambda(t)$ is formulated in much the same way as the confidence interval for $\lambda(t)$:

95% Confidence (I ≥ 50)

$$P[\hat{\Lambda}(t) - 1.96\sqrt{\hat{\Lambda}(t)/I} \leq \Lambda(t) \leq \hat{\Lambda}(t) + 1.96\sqrt{\hat{\Lambda}(t)/I}] = .95 \tag{6.11}$$

99% Confidence (I ≥ 50)

$$P[\hat{\Lambda}(t) - 2.58\sqrt{\hat{\Lambda}(t)/I} \leq \Lambda(t) \leq \hat{\Lambda}(t) + 2.58\sqrt{\hat{\Lambda}(t)/I}] = .99 \tag{6.12}$$

The confidence interval is itself a function of t. The absolute width of the interval expands as t increases because the standard error for the estimator $\hat{\Lambda}(t)$ grows with t (but the relative width declines).

A third approach to estimating $\lambda(t)$ is to approximate the average of the cumulative arrival curves with an equation. The advantage of this approach is that it is much simpler to analyze an equation than a large data set. However, some loss in accuracy may result. The methodology for estimating such an equation is beyond the scope of this book, but can be found in texts on econometrics and statistical regression (see the end of this chapter).

An obvious disadvantage of the second and third approaches is that data on specific arrival times are required, whereas the interval approach only requires interval counts. The added data collection effort may not be justified in terms of increased accuracy.

Future arrival rates might also be partially predicted through a forecasting technique, of which there are many. A common approach is to base the shape of the curve $\Lambda(t)$ on the average of the historical arrival curves, but to scale the curve according to the forecast for the number of arrivals for a given cycle. That is, $\hat{\Lambda}(T)$ would equal a forecast for the number of arrivals during the cycle, and $\Lambda(t)$ would be scaled up or down from

$A(t)$ by the ratio $\hat{\Lambda}(T)/\bar{A}(T)$. (References on forecasting are provided at the end of this chapter.) Keep in mind that one's own judgment sometimes provides a good forecast, particularly when the arrival pattern is influenced by many external factors.

Finally, a nonstationary arrival process does not have to be cyclic; $\Lambda(t)$ can be any nondecreasing function. However, unless $\Lambda(t)$ is cyclic, it may be impossible to estimate $\Lambda(t)$, in which case the arrival pattern is not truly predictable. If the arrival pattern is not predictable, then it should not be modeled as a nonstationary Poisson process. The essence of the nonstationary Poisson process is *predictable* variability in the arrival process.

6.2 STEADY-STATE APPROXIMATION FOR A SLOWLY VARYING ARRIVAL RATE

A queueing system with a nonstationary arrival process will never enter steady state. The varying arrival rate constantly changes the probability distribution for the number of customers in the system. Yet this does not prevent the use of the steady-state equations to *approximate* the behavior of the system, particularly when the arrival rate is slowly changing and the system operates below capacity. When valid, the system will be said to be in ***quasi-steady state***.

Consider the performance of a single server queue, with exponential service times and a nonstationary Poisson arrival process.

Definitions 6.3

$\mu(t)$ = the service rate per server at time t

$\rho(t)$ = the absolute utilization at time t
= $\lambda(t)/\mu(t)$

$P_n(t)$ = the probability that n customers are in the system at time t

Then the steady-state approximation for $P_n(t)$ follows directly from the $M/M/1$ queueing equations:

$$P_n(t) \approx [1 - \rho(t)][\rho(t)]^n \qquad n = 0, 1, 2 \tag{6.13}$$

The validity of this approximation clearly depends on at least one factor: *$\rho(t)$ must be less than 1 for all t*. This factor in itself is quite restrictive, for systems with nonstationary arrival rates also tend to be overloaded from time to time. $\rho(t)$ must also be a slowly varying function. If $\rho(t)$ changes too quickly, then the system could not respond as fast as the steady-state model predicts. More precisely, the steady-state approximation can only be accurate if *the change in $\rho(t)$ during one relaxation time* (see Chap. 4) *is small in comparison to the average queue length*. Based on this principle, Newell (1982) provides the following rule for assessing the validity of the steady-state approximation:

$$\Delta = \left[\frac{1}{\mu(t)}\right]\left[\frac{1}{[1 - \rho(t)]^3}\right]\left|\frac{d\rho(t)}{dt}\right| \qquad \rho(t) < 1 \tag{6.14}$$

Single server approximation valid when $\Delta << 1$

One way to interpret Eq. (6.14) is that the amount that $\rho(t)$ changes during one service time $(d_\rho(t)/dt \mid \cdot 1/\mu)$ should be small compared to the quantity $[1-\rho(t)]^3$.

Example

Consider the restaurant example again, with $\lambda(t) = 100\sin(t\pi/2)$. Suppose that the operator is concerned with queues of customers waiting to be seated by the maître d'hôtel. Hypothetically, the time to serve a customer has an exponential distribution with mean 20 seconds (.00556 hour). $\rho(t)$ is defined as follows:

$$\rho(t) = \frac{\lambda(t)}{\mu} = \frac{100\sin(\pi t/2)}{180} = .556\sin(\pi t/2)$$

To evaluate Eq. (6.14), the derivative of $\rho(t)$ is calculated first:

$$\frac{d\rho(t)}{dt} = .556\ \frac{\pi}{2}\ \cos(\pi t/2) = .873\ \cos(\pi t/2)$$

Δ can now be written as

$$\Delta = \frac{1}{180}\ \frac{|.873\ \cos(\pi t/2)|}{[1 - .556\sin(\pi t/2)]^3}$$

The above is calculated for various values of t

t	0	.25	.5	.75	.9	1	1.1	1.25	1.5	1.75	2
Δ	.0048	.0092	.0213	.0161	.0083	0	.0083	.0161	.0213	.0092	.0048

In this instance, the quasi-steady-state model seems appropriate (largely because the service time is very small) and the system quickly adapts to changes in the arrival rate.

Should the steady-state approximation be valid, then the performance measure equations from Chap. 5 can be used. For the example of an $M/M/1$ system

$$E[L_s(t)] \approx \frac{\rho(t)}{1 - \rho(t)} \tag{6.15}$$

Figure 6.2 shows $E[L_s(t)]$ for the restaurant example. The multiple server steady-state results (the $M/M/m$ and $M/G/1$ models, for example), can also be used if Eq. 6.14 is satisfied, *provided the following substitutions are made for $\mu(t)$ and $\rho(t)$ in Eq. 6.14, respectively.*

Definitions 6.4

$c(t)$ = the *combined* service capacity among all servers at time t

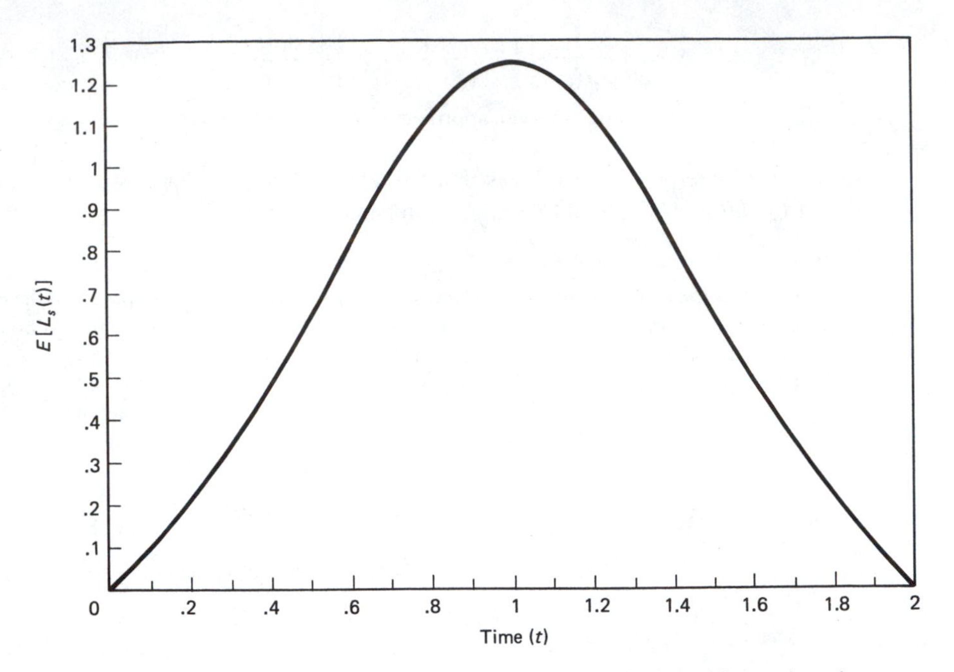

Figure 6.2 Expected customers in system for restaurant, determined by quasi-steady-state model.

$$\tilde{\rho}(t) = \text{proportional utilization at time } t$$
$$= \lambda(t)/c(t)$$

For example, if four servers work at the rate of ten customers per hour each, then $c(t) =$ 40 customers/hour. Unfortunately, the steady-state equations have limited validity because they are not accurate when $\tilde{\rho}(t)$ is close to or exceeds 1, which invariably occurs from time to time when the arrival rate is not stationary.

6.3 SIMULATION OF A NONSTATIONARY POISSON PROCESS

If quasi-steady-state analysis is not applicable, one alternative is to simulate the queueing system. As mentioned in Chap. 4, simulation is a very robust technique that can be applied to a variety of situations. However, simulation does not always provide as meaningful results as does direct analysis. In this section, three techniques are presented for simulating a queueing system with a nonstationary Poisson arrival process. The first two are similar to techniques used in Chap. 4 for the stationary Poisson process and the third is new. A fourth technique is presented for the special case where $\Lambda(t)$ is piecewise linear. In all but the third technique, the basic approach is first to simulate arrival times, second simulate service times, and third combine the data to form the queue simulation.

The second and third steps are no different from those in Sec. 4.5, so they will not be repeated. The emphasis here is on simulating the arrival times.

6.3.1 Simulation Method 1

Recall that there are two ways to simulate a stationary Poisson process. The first of these is to simulate exponential interarrival times and sum them to obtain arrival times. Clearly, this approach will not work for a nonstationary arrival process; the interarrival times are not exponential random variables. However, a modification will work. Let

$$\lambda_{max} = \max_t \lambda(t)$$

The nonstationary Poisson process is simulated as follows:

1. Simulate a stationary Poisson process with rate λ_{max} by summing exponential interarrival times.
2. For each arrival simulated, generate a Bernoulli {0,1} random variable, with $p = \lambda(t)/\lambda_{max}$; from a U[0,1] random variable, U:

$$U \leq p \text{ denotes a success:accept arrival}$$
$$U > p \text{ denotes a failure:reject arrival}$$

Example

The restaurant is to be simulated over the time interval from 11:30 to 11:45($t = 0$ to .25, with Eq. (6.3)). The maximum arrival rate over this period occurs at $t = .25$, and equals

$$\lambda_{max} = 100\sin(.25\pi/2) = 38 \text{ customers/hour} \quad \text{(Eq. (6.2))}$$

Taking $U[0,1]$ random variables (U_{n1} and U_{n2}) from a random number table, the simulation is summarized in the table below:

n	U_{n1}	X_n	Y_n	$p = \lambda(t)/\lambda_{max}$	U_{n2}	Accept
1	.2188	.040	.04	.165	.8479	No
2	.4846	.019	.059	.243	.6108	No
3	.9586	.001	.060	.248	.5703	No
4	.4061	.024	.084	.346	.3113	Yes
5	.1037	.060	.144	.590	.3349	Yes
6	.5104	.018	.162	.662	.4038	Yes
7	.6088	.013	.175	.714	.9031	No
8	.0707	.070	.245	.988	.7986	Yes
9	.7919	.006	—			

The series Y_n represents a stationary Poisson process with rate 38. The nonstationary simulation accepts four of these arrivals, yielding the arrival times .084, .144, .162, and .245 hours.

Note that the acceptance probability varies in proportion to the arrival rate, so the simulation is truly nonstationary.

6.3.2 Simulation Method 2

The second approach presented in Chap. 4 can also be modified for a nonstationary Poisson process. The arrival process is simulated in these three steps.

1. Simulate a Poisson random variable $A(T)$ representing the number of arrivals over the time interval $[0,T]$.
2. Simulate $A(T)$ random variables representing the arrival times.
3. Sort the arrival times in ascending order to obtain the function $A(t)$.

The mean of the Poisson random variable equals $\Lambda(T)$. The simulation in the second step follows from the probability distribution for the arrival times defined by $\Lambda(t)$. Suppose that Y represents an arrival time. Then Y is found by solving the following, where U represents a uniform [0,1] random variable:

$$U = \frac{\Lambda(Y)}{\Lambda(T)} \rightarrow Y = \Lambda^{-1}[U \cdot \Lambda(T)] \tag{6.16}$$

Example

The restaurant is to be simulated a second time over the interval from 11:30 to 11:45 ($t = 0$ to .25, with Eq. (6.3)). The expected number of arrivals over this interval equals 4.85. $A(.25)$ is found as follows:

a. $U = .435$ (from random number table)

b. From the Poisson distribution: $P[A(.25) \leq 3] = .287$
$P[A(.25) \leq 4] = .467 \rightarrow A(.25) = 4$

The arrival times are generated from solving the following:

$$U = \frac{\dfrac{200}{\pi}[1 - \cos(\pi Y/2)]}{4.85} \rightarrow Y = \frac{2}{\pi}\cos^{-1}\left[1 - U\frac{4.85\pi}{200}\right]$$

Taking $U[0,1]$ random variables from a random number table leads to the following values:

n	*U*[0,1]	*Y*
1	.8177	.226
2	.3677	.151
3	.2125	.115
4	.5474	.185

As in Chap. 4, the arrival times are sorted in ascending order to obtain $A(t)$.

The example can be visualized through Fig. 6.3, which shows how the arrival times are generated. The curve is $\Lambda(t)$, for t in the domain [0,.25].

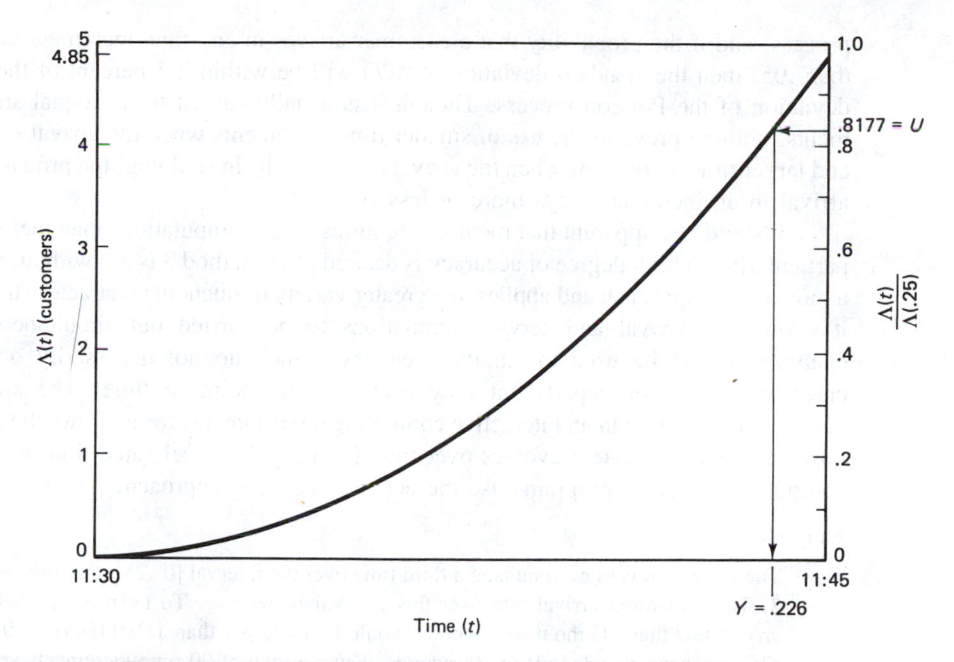

Figure 6.3 Simulation of a customer arrival time for a nonstationary process.

6.3.3 Simulation Method 3

An alternative simulation approach is to use an ***activity scanning*** procedure. This approach draws on the primary definition of the nonstationary Poisson process—that is, that the process has independent increments, and the probability of an arrival in a differential time interval dt is $\lambda(t)dt$. The simulation is an approximation to the Poisson process and is something like flipping coins whose probability of success $= \lambda(t)dt$. It is carried out in these steps:

1. Divide the time period $[0,T]$ into small time increments.
2. For each time increment, generate a $U[0,1]$ random variable.
3. For each increment simulate a Bernoulli random variable

$$\text{If } U \leq \lambda(t)dt \rightarrow \text{ then an arrival occurred in the increment.}$$
$$\text{If } U > \lambda(t)dt \rightarrow \text{ then no arrival occurred in the increment.}$$

The accuracy of the simulation depends on the size of the time increments. If these increments are very small, then the simulation will be indistinguishable from a nonstationary Poisson process; if the increments are large, then $A(T)$ will have a smaller variance than a nonstationary Poisson process (but usually the same mean). If the probability that a customer arrives in any time increment is no larger than .1, then the standard deviation of $A(T)$ will be within 5 percent of the standard deviation of the nonstationary Poisson process, and if the probability that a customer arrives in any time increment is no larger than .05, then the standard deviation of $A(T)$ will be within 2.5 percent of the standard deviation of the Poisson process. Though it is usually easiest to use equal sized increments, nothing prevents the use of smaller time increments when the arrival rate is large and larger time increments when the arrival rate is small. In so doing, the probability of an arrival in an increment stays more or less constant.

It should be apparent that method 3 requires more computations than method 1 or 2, particularly if a high degree of accuracy is desired. Yet method 3 is not without merit. It is a more robust approach and applies to a greater variety of queueing characteristics because it allows the arrival and service simulations to be carried out simultaneously. For example, it can be used to simulate reneges, which do not necessarily occur when customers arrive or depart, but may occur at any point in time. The approach is particularly effective in an interactive computing environment, for it allows the user to see how the queueing system evolves over time (at a rapidly accelerated time scale). Many computer simulation programs use the activity scanning approach.

Example

The restaurant is to be simulated a third time over the interval [0,.25], this time with method 3. The maximum arrival rate over this interval is 38/hour. To keep the probability of an arrival less than .1, the time intervals should be no larger than 1/380 hours = 9.5 seconds. This has been rounded off to 10 seconds. Thus, a total of 90 time increments are simulated over the 15 minute period, as shown in Table 6.1.

The queue can be simulated by generating arrivals, generating service times, and combining the data, just as before. However, should one go to the effort of using method 3, an alternative approach would likely be used. At each time increment, the following steps would be performed:

1. Determine whether an arrival occurs.
2. For each customer in service, determine whether service is completed in the time increment.
3. Determine whether any customer enters service in the time increment.

In addition, extra steps can be added to account for reneging or other factors. Method 3 amounts to a dynamic simulation, as opposed to the alternatives, which are more of a batch simulation.

6.3.3 Special Case: $\Lambda(t)$ Is Piecewise Linear

A nonstationary arrival process is easiest to simulate when $\Lambda(t)$ is a piecewise linear function, meaning that the arrival rate stays constant over each of several time intervals. This special case is not all that unusual, for if $\Lambda(t)$ is based on interval counts, the arrival rate must be assumed to be constant over each interval. In reality, the arrival rate is likely some smooth function of time—it is just that data are not available to determine its exact

TABLE 6.1 RESTAURANT ARRIVALS SIMULATION: METHOD 3

Time (sec)	U	$\lambda(t)dt$	1 = Arr	$A(t)$	Time (sec)	U	$\lambda(t)dt$	1 = Arr	$A(t)$
10	0.172	0.001	0	0	460	0.523	0.055	0	1
20	0.944	0.002	0	0	470	0.876	0.057	0	1
30	0.556	0.004	0	0	480	0.175	0.058	0	1
40	0.036	0.005	0	0	490	0.891	0.059	0	1
50	0.282	0.006	0	0	500	0.154	0.060	0	1
60	0.380	0.007	0	0	510	0.405	0.061	0	1
70	0.205	0.008	0	0	520	0.980	0.062	0	1
80	0.455	0.010	0	0	530	0.062	0.064	1	2
90	0.279	0.011	0	0	540	0.690	0.065	0	2
100	0.515	0.012	0	0	550	0.390	0.066	0	2
110	0.158	0.013	0	0	560	0.655	0.067	0	2
120	0.437	0.015	0	0	570	0.438	0.068	0	2
130	0.121	0.016	0	0	580	0.957	0.070	0	2
140	0.015	0.017	1	1	590	0.234	0.071	0	2
150	0.422	0.018	0	1	600	0.742	0.072	0	2
160	0.809	0.019	0	1	610	0.788	0.073	0	2
170	0.666	0.021	0	1	620	0.623	0.074	0	2
180	0.225	0.022	0	1	630	0.899	0.075	0	2
190	0.469	0.023	0	1	640	0.229	0.077	0	2
200	0.600	0.024	0	1	650	0.572	0.078	0	2
210	0.943	0.025	0	1	660	0.909	0.079	0	2
220	0.105	0.027	0	1	670	0.088	0.080	0	2
230	0.123	0.028	0	1	680	0.440	0.081	0	2
240	0.463	0.029	0	1	690	0.081	0.082	1	3
250	0.774	0.030	0	1	700	0.012	0.084	1	4
260	0.297	0.031	0	1	710	0.164	0.085	0	4
270	0.802	0.033	0	1	720	0.726	0.086	0	4
280	0.434	0.034	0	1	730	0.991	0.087	0	4
290	0.050	0.035	0	1	740	0.958	0.088	0	4
300	0.296	0.036	0	1	750	0.984	0.089	0	4
310	0.530	0.037	0	1	760	0.300	0.090	0	4
320	0.644	0.039	0	1	770	0.640	0.092	0	4
330	0.207	0.040	0	1	780	0.882	0.093	0	4
340	0.970	0.041	0	1	790	0.056	0.094	1	5
350	0.393	0.042	0	1	800	0.828	0.095	0	5
360	0.827	0.043	0	1	810	0.558	0.096	0	5
370	0.703	0.045	0	1	820	0.962	0.097	0	5
380	0.905	0.046	0	1	830	0.715	0.098	0	5
390	0.352	0.047	0	1	840	0.645	0.100	0	5
400	0.753	0.048	0	1	850	0.067	0.101	1	6
410	0.531	0.049	0	1	860	0.504	0.102	0	6
420	0.320	0.051	0	1	870	0.213	0.103	0	6
430	0.600	0.052	0	1	880	0.691	0.104	0	6
440	0.926	0.053	0	1	890	0.979	0.105	0	6
450	0.083	0.054	0	1	900	0.887	0.106	0	6

shape. Hence, a piecewise linear $\Lambda(t)$ can be viewed as an approximation to the true $\Lambda(t)$, which cannot be derived from the available data.

For this special case, the interarrival times within each piece of the piecewise linear curve must have an exponential distribution, with mean $1/\lambda_j$, where λ_j is the arrival rate for piece j. Let T_j represent the time that piece j begins. The simulation proceeds as follows.

0. Set $j = 1$ to simulate arrivals over the first period.

1. Simulate a series of exponential random variables, mean $1/\lambda_j$:$X_1, X_2, \ldots$ Sum the exponential random variables to obtain the arrival times for period j: $Y_{1j} = T_j + X_1$, $Y_{2j} = Y_{1j} + X_2$, $Y_{3j} = Y_{2j} + X_3, \ldots$

2. Stop the simulation of piece j when $Y_{nj} > T_{j+1}$.
Increment j to $j + 1$, and return to step 1.

The simulation is completed by catenating the arrival times within the periods.

A key difference between this simulation and the simulation of a stationary Poisson process is that the first arrival within a piece does not equal X_1 plus the time of the previous arrival. Instead, it equals X_1 plus the time the piece began. The validity of this approach relies on the memoryless property of the exponential distribution. For piecewise linear arrival rates, the distribution of the time until the first arrival in a time interval does not depend on the elapsed time since the last arrival.

Example

According to historical records, customers arrive by a nonstationary Poisson process at a small post office at the rate of 5/hour between 11:00 and 12:00 and the rate of 10/hour between 12:00 and 1:00. The simulation proceeds as follows:

Piece 1: 11:00–12:00

n	1	2	3	4	5	6
U_n	.5528	.6105	.9726	.8983	.1614	.0180
X_n (min)	7.1	5.9	.3	1.3	21.9	48.2
Y_n (min)	7.1	13.0	13.3	14.6	36.5	—

Piece 2: 12:00–1:00

n	6	7	8	9	10	11	12	13	14	15	16
U_n	.7521	.8912	.1927	.1579	.7268	.6218	.4875	.5608	.5403	.3921	.0480
X_n (min)	1.7	.7	9.9	11.1	1.9	2.9	4.3	3.5	3.7	5.6	18.2
Y_n (min)	61.7	62.4	72.3	83.4	85.3	88.2	92.5	96.0	99.7	105.3	—

The simulation produced 15 customers.

6.4 FLUID APPROXIMATIONS: SHORT SERVICE TIME

A nonstationary Poisson process encounters two types of variation: random variation and predictable variation. The *predictable* variation is reflected in the function $\Lambda(t)$, which gives the expected number of arrivals as a function of time. The *random* variation is reflected in the precise arrival times. Both are revealed in Fig. 6.4, which compares $\Lambda(t)$ to a simulation of $A(t)$ for the restaurant queue. While the simulation follows the same general pattern as $\Lambda(t)$, it is susceptible to minor perturbations reflecting random variations.

Because the number of arrivals in any time interval has a Poisson distribution, the mean, $\Lambda(t)$, must equal its variance. This means that the coefficient of variation (ratio of standard deviation to mean) is the following:

$$C[A(t)] = \frac{\sqrt{\Lambda(t)}}{\Lambda(t)} = \frac{1}{\sqrt{\Lambda(t)}} \tag{6.17}$$

Equation (6.17) states that the larger the value of $\Lambda(t)$, the smaller will be the random variations in the number of arrivals (in relation to the expected number of arrivals). For busy queueing systems, these random variations may be of minor importance relative to the predictable variations. To take an extreme example, a busy freeway toll plaza may have 8000 arrivals per hour, which would provide a coefficient of variation of just .011 for 1 hour. This means that a nonstationary Poisson arrival pattern can be accurately approximated with a ***deterministic*** model. The word "determinism" represents a belief that every event is the inevitable consequence of its antecedents. In the context of arrivals, determinism means that $A(t)$ is assumed to be known with certainty and equals $\Lambda(t)$. Though $\lambda(t)$ is not constant, the variations are entirely predictable.

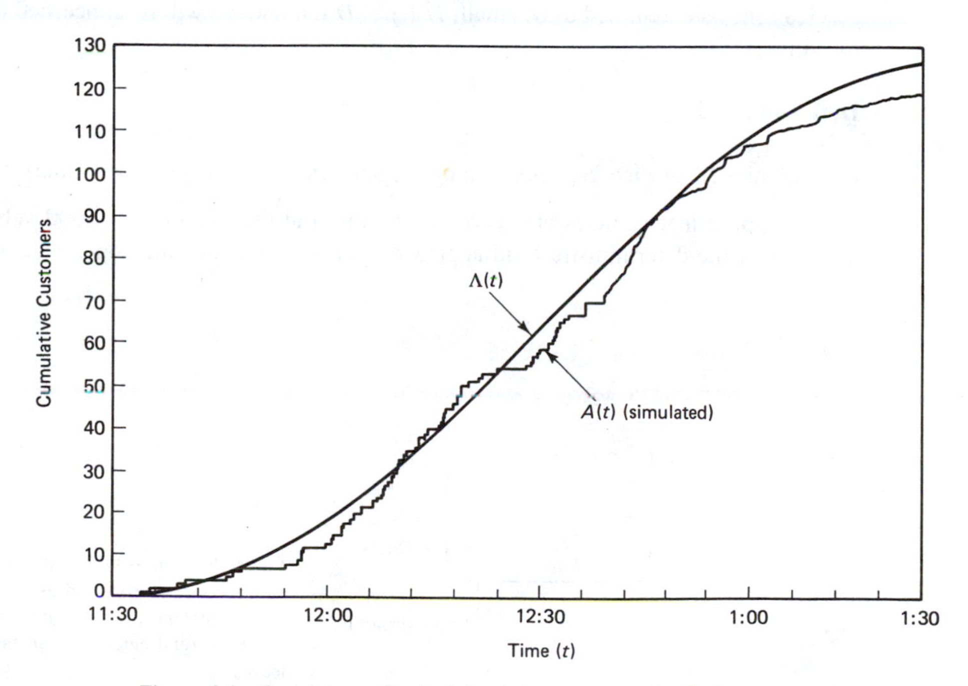

Figure 6.4 Comparison of arrival simulation to expected arrivals at restaurant.

Deterministic queueing models usually fall in the category of ***fluid approximations***. Whereas customers are discrete entities, fluids are not If a unit of fluid is divided into any proportion, the result will be a smaller quantity of the same entity—it is still a fluid. The same cannot be said of customers, for if a group of customers is divided into proportions, the result may no longer be a group of customers. Half a customer is not a customer at all. Customers are not infinitely divisible. Nevertheless, a quantity of customers can be approximated by a continuous variable (particularly if the quantity is large) and modeled as a fluid. If $\Lambda(t)$ equals 155.3, little harm is caused if $A(t)$ is also assumed to equal 155.3.

Deterministic fluid models are much simpler to use than simulation and also provide more meaningful results. They highlight the important relationships between system attributes and system performance. They should be used when random variation is small relative to predictable variation.

A useful way to think of the fluid queueing model is in terms of the illustration in Fig. 6.5. A faucet deposits water into a tub, and a drain empties water from the tub. The tub represents the queue, and the water represents customers. The arrival rate is the rate at which the water flows out of the faucet into the tub, and the service rate is the capacity of the drain for emptying water from the tub. If water enters the tub faster than it is drained, then the water level in the tub rises. Analogously, if customers arrive faster than they are served, the queue grows. And, if water is drained faster than it enters, the water level will drop, until no water is left in the tub.

Suppose that vehicles arrive at a freeway toll plaza according to the curve $A(t)$ in Fig. 6.6. Based on a fixed service capacity, we would like to determine $D_q(t)$ and $D_s(t)$, cumulative departures from queue and from system, respectively. However, because service times are assumed to be small, $D_q(t) \approx D_s(t)$, and we will be concerned only with the former.

Definition 6.5

c = combined service capacity among all servers (constant rate over time)

Suppose that vehicles are served at the plaza at the rate of $c = 3600$ vehicles per hour. Then the deterministic fluid approximation for $D_q(t)$ has the appearance shown in

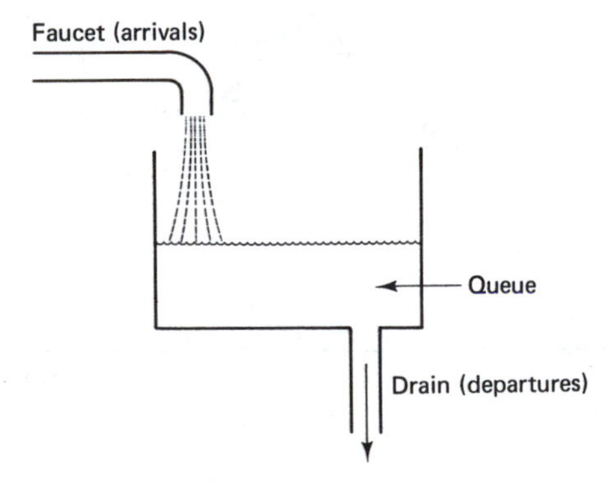

Figure 6.5 In a fluid model, the customers can be viewed as a liquid that accumulates in a tub. Queues increase when the fluid enters the tub faster than it leaves.

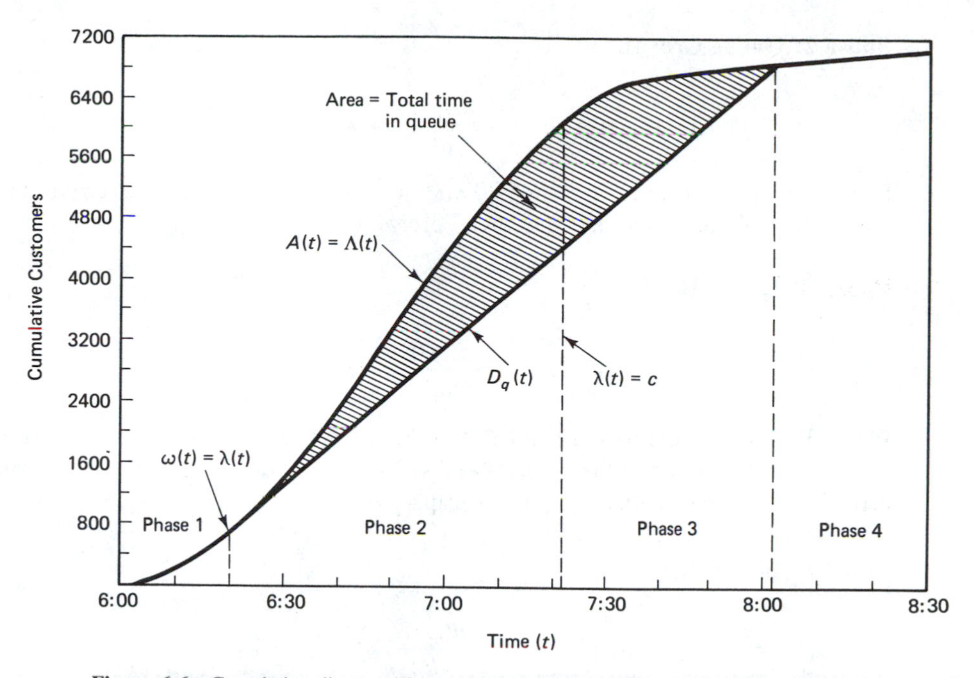

Figure 6.6 Cumulative diagram illustrating deterministic fluid model. When a queue exists, customers depart at a constant rate. Queues increase when the arrival rate exceeds the service capacity and decrease when the service capacity exceeds the arrival rate.

Fig. 6.6. Between the time 6:00 and 6:20 A.M., $D_q(t)$ is identical to $A(t)$ because vehicles can be served at a faster rate than they arrive. At 6:20 A.M., the arrival rate (slope of $A(t)$) has increased to the point where it exactly equals the service capacity. From this point on, the queue grows because vehicles arrive at a faster rate than they are served. The queue finally begins to decline when the arrival rate again equals the service capacity, which occurs at about 7:20. The queue eventually vanishes at about 8:00 A.M.

To draw $D_q(t)$, identify the point where $\lambda(t)$ first exceeds c (that is, the first point where the slope of $A(t) = c$, 6:20 in Fig. 6.6). From this point, draw a line tangent to $A(t)$, with slope c, forward until it again intersects $A(t)$ (8:00 in Fig. 6.6). From this second point onward, $A(t) = D_q(t)$ until such time that $\lambda(t)$ again exceeds c.

Over the period from 6:00 A.M. to 8:30 A.M., the system passes through four phases, which are identified as follows:

Phase 1: Stagnant

$$A(t) = D_q(t) \qquad \lambda(t) \leq c \qquad \frac{dL_q(t)}{dt} = 0 \tag{6.18}$$

Phase 1 represents the period from 6:00 to 6:20, when vehicles are served as fast as they arrive.

Phase 2: Queue Growth

$$A(t) > D_q(t) \qquad \lambda(t) > c \qquad \frac{dL_q(t)}{dt} = \lambda(t) - c > 0 \tag{6.19}$$

Phase 2 represents the period from 6:20 A.M. to 7:20 A.M., when vehicles arrive at a faster rate than they are served and the queue grows.

Phase 3: Queue Decline

$$A(t) > D_q(t) \qquad \lambda(t) < c \qquad \frac{dL_q(t)}{dt} = \lambda(t) - c < 0 \tag{6.20}$$

Phase 3 represents the period from 7:20 to 8:00 when customers arrive at a slower rate than they are served and the queue shrinks. Note that a queue can exist even when the arrival rate is smaller than the service capacity.

Phase 4: Stagnant

$$A(t) = D_q(t) \qquad \lambda(t) < c \qquad \frac{dL_q(t)}{dt} = 0 \tag{6.21}$$

The last phase begins at 8:00, when the queue finally vanishes.

In all four phases, the arrival rate and the service capacity determine the *rate* at which the queue grows or shrinks. When is the queue the largest? At the end of the phase 2, when the arrival rate equals the service capacity:

$$\text{Queue is largest when: } \lambda(t) = c \tag{6.22}$$

This is the time when the queue stops growing and begins shrinking. When does the queue vanish? When $A(t) = D_q(t)$. Note that the arrival rate can be much smaller than the service capacity when this happens.

Definition 6.6

$\omega(t)$ is the ***departure rate*** at time t

$$\omega(t) = \begin{cases} c & L_q(t) > 0 \\ \lambda(t) & L_q(t) = 0 \end{cases} = \frac{dD_q(t)}{dt}$$

The function $\omega(t)$ is the rate at which customers are departing at time t, whereas c is the *capacity* for serving customers. If there are zero customers in the queue, customers can depart no faster than the rate at which they arrive. $\omega(t)$ is also the slope (derivative) of $D_q(t)$. Figure 6.7 plots $\lambda(t)$ and $\omega(t)$ for the vehicle queue. In phases 1 and 4, $\omega(t) = \lambda(t)$, and in phases 2 and 3, $\omega(t) = c$. The division between phases 2 and 3 occurs when $\lambda(t) = \omega(t)$ (7:20), which signals the point where the queue begins to decline. By the time the queue vanishes (8:00), $\lambda(t)$ has dropped far below c. This creates a discontinuity in $\omega(t)$,

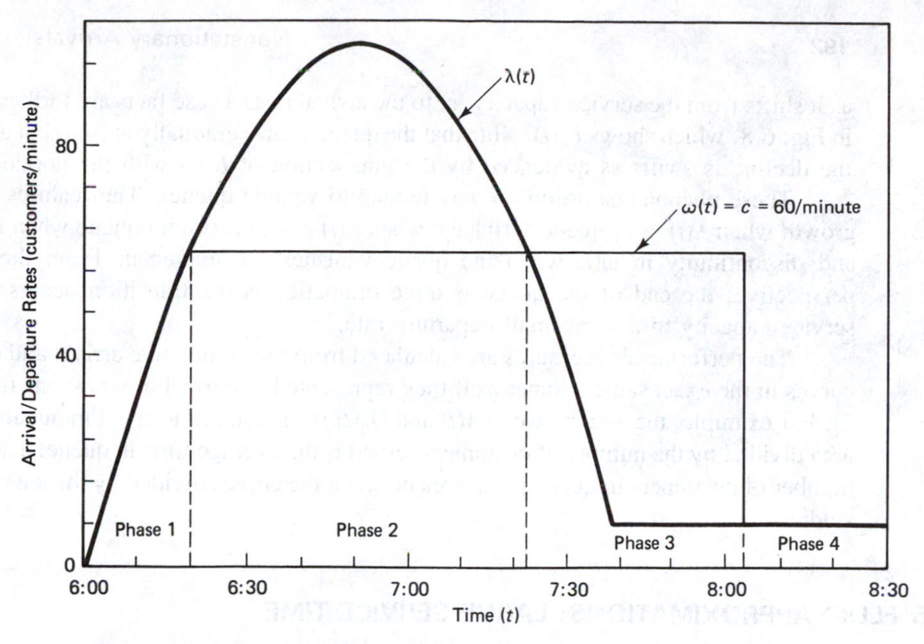

Figure 6.7 Arrival and departure rates versus time for a deterministic fluid model. When queue is at its maximum, the arrival and departure rates are equal. By the time the queue vanishes, the arrival rate is much less than the departure rate.

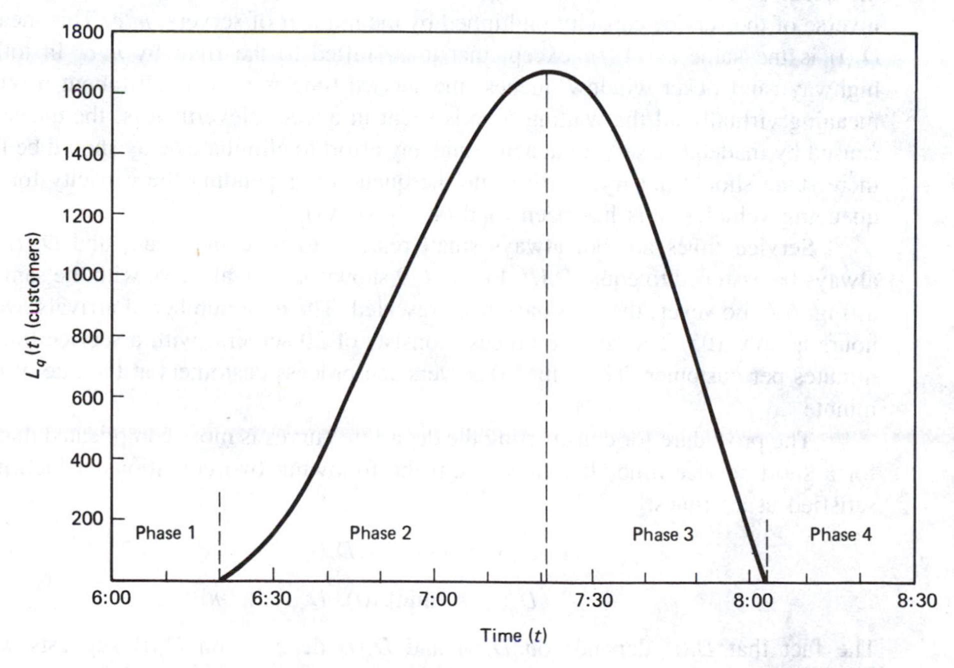

Figure 6.8 Customers in queue versus time for a deterministic fluid model. Queue size grows gradually at beginning, but declines rapidly at end.

as it shifts from the service capacity, c, to the arrival rate. These facts are further reflected in Fig. 6.8, which shows $L_q(t)$. Note that the queue grows gradually at first, but at the end, the decline is swift, as evidenced by the intersection of $L_q(t)$ with the horizontal axis.

These phenomena are in no way unique to vehicle queues. The features of queue growth when $\lambda(t) > c$, queue shrinkage when $\lambda(t) < c$, maximum queue when $\lambda(t) = c$, and discontinuity in $\omega(t)$ when the queue vanishes are universal. From the server's perspective, the end of the queue is quite dramatic, as the transition occurs from the service capacity to a very small departure rate.

The performance measures are calculated from the cumulative arrival and departure curves in the exact same manner as if they represented empirical observations (see Chap. 2). For example, the area between $A(t)$ and $D_q(t)$ is the total time spent in queue, and this area divided by the number of customers served is the average time in queue. The average number of customers in queue is the area between the curves divided by the length of time studied.

6.5 FLUID APPROXIMATIONS: LARGE SERVICE TIME

So far, no distinction has been made between departures from the system and departures from the queue. Earlier, we saw that $D_s^{-1}(n) = D_q^{-1}(n) + S(n)$, where $S(n)$ is the service time for customer n (that is, departure time from the system equals departure time from the queue plus the service time). With a deterministic approximation, $S(n)$ would be the inverse of the service capacity multiplied by the number of servers, m/c. This means that $D_s(t)$ is the same as $D_q(t)$, except that it is shifted to the right by m/c. In tollbooth, highway, and ticket window queues, the service time is very small (about 6 seconds), meaning virtually all the waiting time is spent in queue. Nevertheless, the queue is still caused by inadequate server capacity, and any effort to eliminate delay should be focused there. One should not try to eliminate the queue by expanding the capacity for storing queueing vehicles (this has been tried on highways).

Service times are not always small relative to time in queue, and $D_q(t)$ cannot always be assumed to equal $D_s(t)$. Figure 6.9 shows an arrival curve with the same shape as Fig. 6.6; however, the axes have been rescaled. The total number of arrivals over 2 1/2 hours is now 105. The service process consists of 20 servers, with a service time of 20 minutes per customer. Thus, the 20 servers can process customers at the rate of one per minute.

The procedure for constructing the departure curves is more complicated than it was for a short service time. It follows from the following two conditions, which must be satisfied at all times:

$$1.\ D_s(t + m/c) = D_q(t) \tag{6.23}$$

$$2.\ D_q(t) = \min\{A(t), D_s(t) + m\} \tag{6.24}$$

The fact that $D_s(t)$ depends on $D_q(t)$ and $D_q(t)$ depends on $D_s(t)$ suggests why the procedure is difficult. The procedure will be presented by way of an example. The segments below refer to portions of the curves $D_q(t)$ and $D_s(t)$, as indicated in Fig. 6.9.

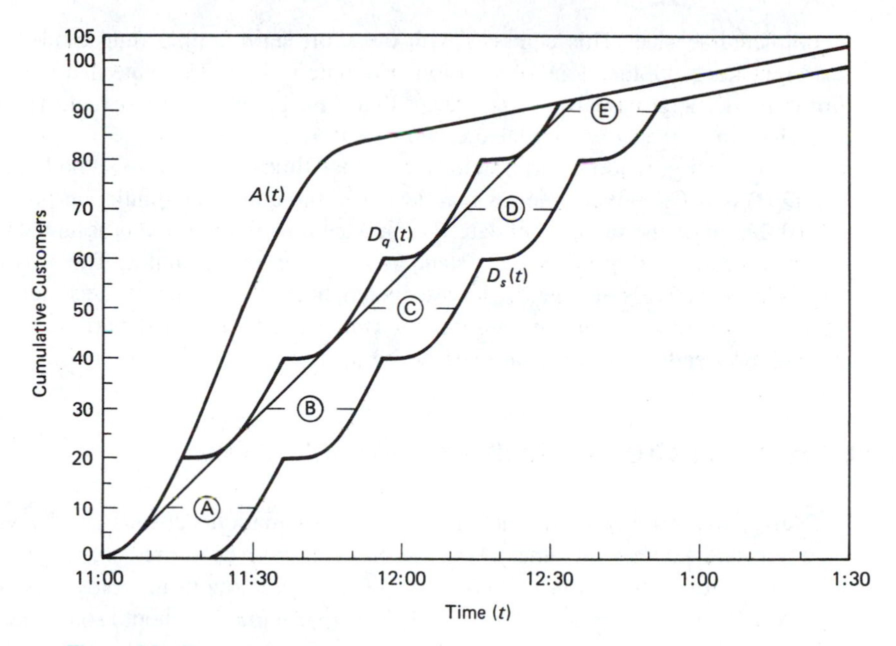

Figure 6.9 Deterministic fluid model with a long service time. Ripple pattern mirrors the arrival pattern immediately preceding the formation of a queue.

Segment A Until $A(t) = m$ ($t < 11{:}15$), customers enter service immediately and $D_q(t) = A(t)$. From condition 1, $D_s(t + m/c) = D_q(t) = A(t)$. That is, in segment A, $D_s(t)$ has the same shape as $A(t)$, but it is shifted right by 20 minutes.

Segment B At 11:15, 20 customers have arrived, but no customer has yet left the system, which means that a queue begins to form. $D_q(t)$ is now determined from the second part of condition 2: $D_q(t) = D_s(t) + m$. That is, segment B of $D_q(t)$ is segment A of $D_s(t)$ shifted upward by $m = 20$. From condition 1, segment *B* of $D_s(t)$ is found by shifting segment B of $D_q(t)$ to the right by 20 minutes.

The same procedure is repeated for segments C and D. Segment E is somewhat different.

Segment E From 12:15 until 12:30, $D_q(t) = D_s(t) + 20$. At 12:30, $D_q(t)$ intersects $A(t)$, meaning that the queue has vanished. From this point on, $D_q(t)$ is defined by the first part of condition 2: $D_q(t) = A(t)$. $D_s(t)$, as usual, is found by shifting $D_q(t)$ to the right by 20 minutes.

The ripple pattern in $D_q(t)$ and $D_s(t)$ is a distinct feature of the model. Customers are served in spurts, which parallel the arrival pattern in the first 15 minutes. These spurts alternate with 5-minute lulls, when $D_q(t)$ remains constant. The spurt/lull cycle occurs because a new batch of customers cannot begin service until the previous batch has completed service. This contrasts with the short service time fluid model, shown as a straight line with the slope of 1 customer/minute in Fig. 6.9. Note that the short service time model approximates $D_q(t)$, except that the ripples are eliminated. The smaller the service time, the more similar the two departure curves will be.

In reality, random perturbations in service times will tend to smooth out the ripples in $D_q(t)$ and $D_s(t)$ over time, as is reflected in the queueing simulation provided in Fig. 6.10, based on the same set of data. The service time distribution is assumed to be normal with a mean of 20 minutes and a standard deviation of 5 minutes. The arrival process is nonstationary Poisson. The ripples are evident at the beginning of the simulation, but are later smoothed. Despite the absence of ripples, the general departure patterns of the simulation and fluid model are nearly the same.

6.6 ADJUSTMENTS TO DETERMINISTIC APPROXIMATION

Clearly, the validity of the deterministic approximation depends on the variability in service and interarrival times. Whereas the approximation predicts that queues do not begin to form until $\tilde{\rho}(t) = \lambda(t)/c(t) > 1$, we already know from steady-state analysis that random queues will exist when $\tilde{\rho}(t) < 1$. The approximation should somehow account for these random queues.

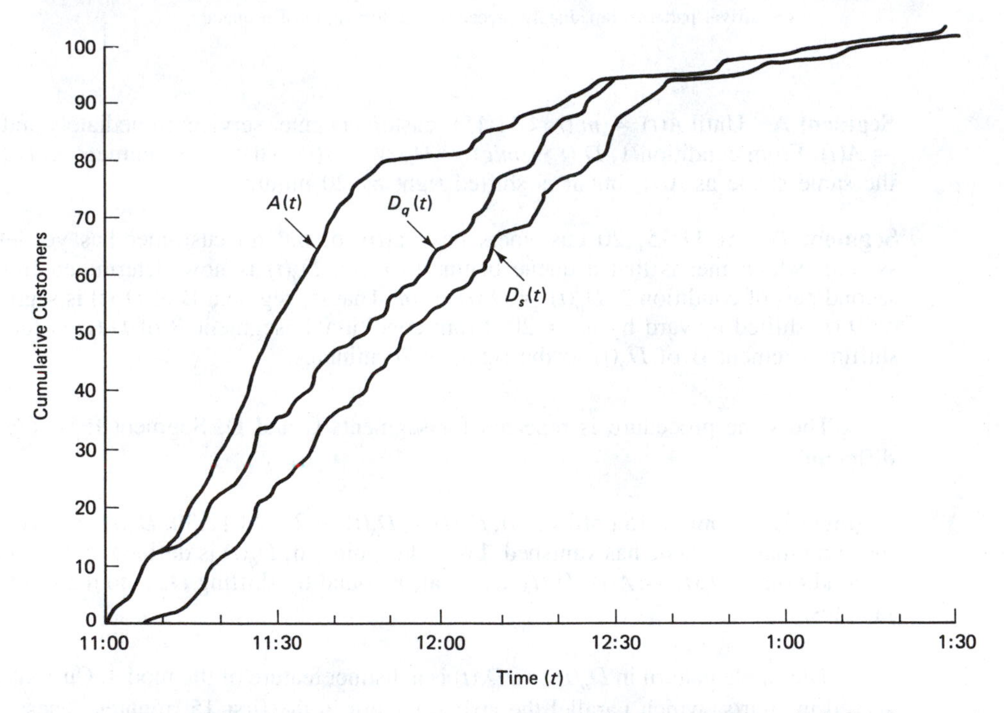

Figure 6.10 Simulation of a queueing system with long service time. Arrival and departure curves are similar to the deterministic fluid model (Fig. 6.9).

As $\tilde{\rho}(t)$ increases, from a value much less than 1 to a value much greater than 1, the expected queue length will pass through three stages:

Stage 1: $\tilde{\rho}(t) << 1$: In this stage, the quasi-steady-state model is valid, and provides a good prediction for queue length.

Stage 2: $\tilde{\rho}(t) \leq 1$, $(1 - \tilde{\rho}(t))$ is small: In this stage, queue lengths are difficult to predict. The quasi-steady-state model is not valid. Neither is the deterministic approximation, for it predicts zero queue lengths.

Stage 3: $\tilde{\rho}(t) > 1$: In this state, the *growth* in expected queue length is accurately predicted by the deterministic approximation.

In stage 2, two ways to predict queue length are ***diffusion models*** (see Newell, 1982, and the appendix to this chapter) and *simulation*. Neither approach is simple. In stage 3, queue length can be predicted from the *deterministic approximation*, provided that an estimate is made of the queue length when $\tilde{\rho}(t) = 1$. Newell provides the following approximation for nonstationary Poisson arrivals and independent service times:

$$E[L_q(t')] \approx \left[\left(\frac{1}{[1 + C^2(S)]^2}\right)\left(\frac{1}{c}\right)\left(\frac{d\tilde{\rho}(t')}{dt}\right)\right]^{-1/3} \tag{6.25}$$

where

$C(S)$ = the coefficient of variation in the service times
t' = the time when $\lambda(t) = c$

Example:

For the arrival curve in Fig. 6.7, $d\tilde{\rho}(t')/dt = (d\lambda(t)/dt)/c$, evaluated at 6:20, which is approximately 4/60 per minute ($c = 60$ customers/minute). If, for example, the coefficient of variation in the service times is .5, then

$$E[L_q(t')] \approx \left[\frac{1}{(1 + .25)^2} \; \frac{1}{60} \; \frac{4}{60}\right]^{-1/3} = 11.2$$

Thus, we would expect to have about 11 vehicles in the queue at time t', and $L_q(t)$, $t \geq t'$, would be shifted upward accordingly.

By comparison to the maximum queue length predicted by the deterministic model, $E[L_q(t')]$ appears to be small. The correction would have been much larger if the arrival rate changed slowly, which would have allowed the number of customers in the queue to approach the equilibrium steady-state distribution prior to t' (note that $E[L_q(t')]$ grows without bound as $d\tilde{\rho}(t')/dt$ approaches zero). Keep in mind that even though 11 vehicles is a relatively small number, its impact persists throughout the queueing period. It shifts the entire curve $L_q(t)$ upward by 11, not just a small portion. Thus, even a relatively small increase in queue length at or near t' can lead to large delays later on.

6.7 QUEUEING TO MEET A SCHEDULE

A good queueing model should not only replicate the behavior of a queueing system; it also should be capable of predicting the future behavior of the system should some change be made. If the service capacity is improved by adding servers, it should be able to predict the reduction in the waiting time. So far, we have assumed that the arrival pattern is a "given." This assumption deserves further examination.

For the sake of simplicity, it is worthwhile to consider two types of customers. The first customer will be labeled the *arrive when ready* customer. Take a post office queue as an example. To the postal customer, arrival time may not be influenced by queue lengths at various times of the day, because he or she arrives whenever he or she is ready to conduct business. The arrival pattern is fixed. Now consider a second type of customer. The *depart on schedule* customer can arrive at any time, but must make sure that he or she departs prior to a scheduled deadline. Take a commuter as an example. Commuters do not leave home "when ready"; commuters leave home at times that guarantee that they will arrive in time for work. Thus, the arrival time *is* influenced by the queue lengths encountered on the trip to work, for if queues become long, he or she will have to leave home earlier to guarantee that he or she "departs on schedule" (arrives at work on time).

By no means is this the only example of a customer that aims to depart on schedule. Perhaps the most pervasive example is in manufacturing, where companies place orders with their suppliers months in advance to ensure that components arrive when needed (that is, depart from the *supplier's* queue on schedule). Other examples include arrivals at sporting events and for airline flights.

A way to visualize the two customer types is that the "arrive when ready" customer is constrained to arrive at a certain time, and the "depart on schedule" customer is constrained to depart from the system at a certain time. The impact of a system improvement depends on which type (or types) of customer is using the system. In the former case, a change in service capacity will have no impact on customer arrival times. In the later case, a change in service capacity may encourage customers to change their arrival times.

Modeling customer behavior is a difficult thing. But, hypothetically, consider the consequences if the queue operator were to schedule "depart on schedule" customer arrivals with the objective of minimizing total waiting time. Then the cumulative arrival pattern would resemble Fig. 6.11. The operator would schedule arrivals at a constant rate, at times prior to the desired departure times. Although this pattern requires customers to depart from the system before desired, queues are eliminated (there is no way to serve all customers on time without having some depart from the system early).

Unfortunately, the optimal schedule is not how customers would actually arrive. Note in Fig. 6.11 that if any customer were to change unilaterally its arrival time, other customers would be forced to arrive late. Yet, one cannot deny that there is a strong incentive to do just that. A customer who arrives at 7:00 departs 30 minutes earlier than desired. Because the arrival pattern eliminates queues, this customer could easily change its arrival time to 7:30 and still depart on time. Therefore, if customers act individually in their own self-interest, then the operator's arrival pattern would be *unstable*. That is, the pattern could not be sustained because customers will change their departure times.

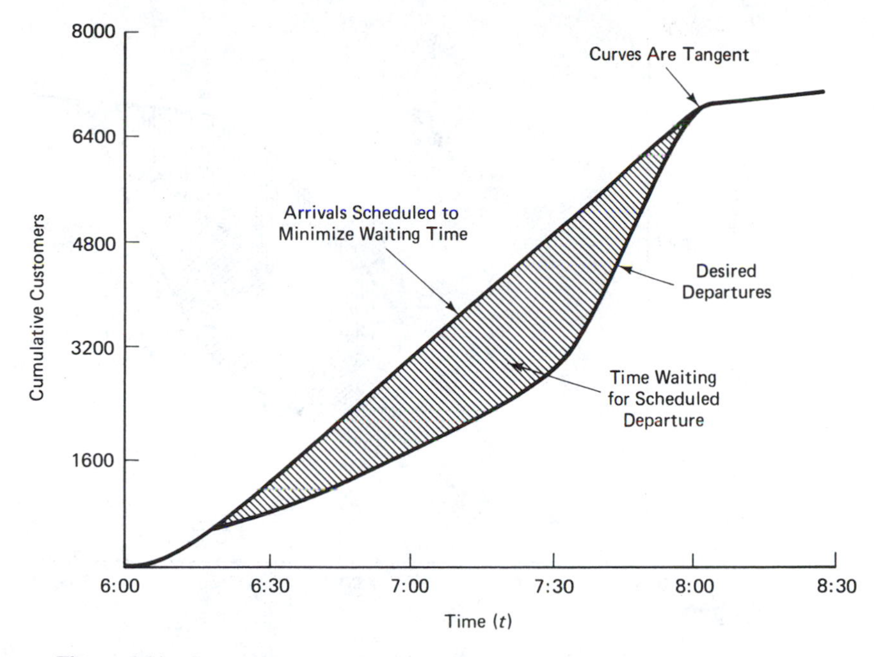

Figure 6.11 Customers must arrive early when they are required to depart from the queue on schedule. Ideally, customers should arrive at a constant rate and queues should not materialize.

A *stable* arrival pattern would be more like Fig. 6.12 (see Daganzo 1985; Hendrickson and Kocur 1981; and Newell 1987). Customers may at first choose to arrive later than the operator's schedule. But, with many customers changing their arrival times, queues will materialize and customers will be late. To compensate, the next time that customers arrive at the queue they will have to arrive earlier—even earlier than the original scheduled arrivals—and the end result: *Customers will not only depart from the queue earlier than desired, but they will also incur queueing delay.*

This is but one example of how queues are created by customers acting in their own self-interest, rather than working together for their common benefit. In Chap. 8, methods are presented for resolving such problems.

6.8 CHAPTER SUMMARY

The most severe queueing problems do not result from random variations in arrival times but, rather, from predictable variations in arrival rates. These predictable variations can be represented by the nonstationary Poisson process. Like the standard Poisson process, the nonstationary Poisson process possesses independent increments, and the probability of an arrival in a differential time increment equals $\lambda(t)dt$. Unlike the standard Poisson process, $\lambda(t)$ does not have to stay constant over time.

The arrival rate represents the expected number of arrivals to occur per unit time. The integral of $\lambda(\tau)$ from 0 to t, $\Lambda(t)$, equals the expected number of arrivals to occur by

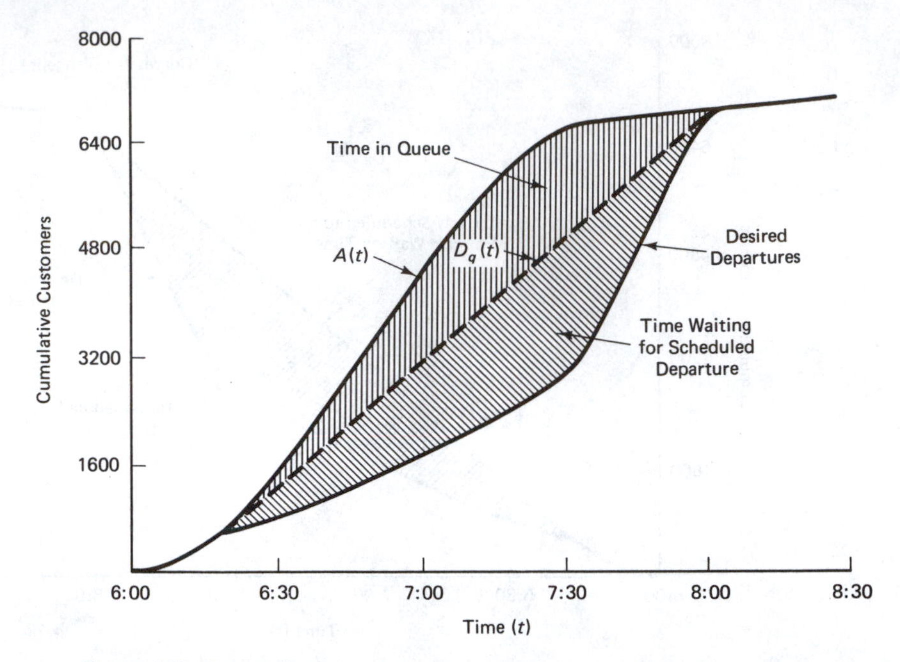

Figure 6.12 In actuality, when customers must "depart on schedule," customers will arrive earlier than necessary and queues will materialize.

time t. The actual number of arrivals to occur is a random variable, which can be more or less than $\Lambda(t)$. Like the standard Poisson process, this random variable has a Poisson distribution with mean $\Lambda(t)$. Nevertheless, the nonstationary Poisson process does not have exponential interarrival times and it is not memoryless. Among other things, this makes it difficult to determine whether a data set was or was not produced by a nonstationary Poisson process. Goodness of fit is primarily judged by plausibility.

A nonstationary Poisson process can be simulated in any of three ways: by simulating a stationary Poisson process and accepting arrivals with probability $\lambda(t)/\lambda_{max}$; by first generating a Poisson random variable representing the number of arrivals over a time interval and then simulating the arrival times; or by simulating Bernoulli random variables, representing whether or not arrivals occur in small time increments. If $\Lambda(t)$ is piecewise linear, arrivals can also be simulated by generating exponential random variables within each piece of the piecewise linear curve (see Sec. 6.3.3).

If the arrival rate is nonstationary, and much less than the service capacity, queueing behavior can be modeled with quasi-steady-state equations. When the function $\tilde{\rho}(t)$ is substituted for ρ/m, these equations provide an estimate for the queue's performance at any time t.

In most cases, the deterministic fluid model is the best way to visualize a nonstationary queueing system. The customer is represented by a fluid, which enters from a faucet (the arrival process) into a tub (the queue) and later empties from a drain (the service process). The queue grows whenever the arrival rate exceeds the service capacity, reaches a maximum when the arrival rate equals the service capacity, and declines when the arrival rate falls below the service capacity.

The fluid model does not account for random fluctuations in the arrival and service processes, just predictable fluctuations. These predictable fluctuations tend to overwhelm random fluctuations in arrival times. In Sec. 6.6, an adjustment factor was provided to estimate the size of the queue created by random fluctuations prior to the time $\tilde{\rho}(t)$ first exceeds 1. This adjustment factor is negligible when the rate of change in $\tilde{\rho}(t)$ is large.

Queues are most difficult to analyze when the arrival rate hovers at or near the service capacity, occasionally falling below or occasionally rising above. Neither steady-state analysis nor fluid approximations provide adequate predictions, the first overestimating queue lengths and the latter underestimating queue lengths. The most robust way to analyze such systems is through simulation.

FURTHER READING

BOX, G. E. P. and G. M. JENKINS. 1970. *Time Series Analysis: Forecasting and Control*. San Francisco: Holden-Day.

DAGANZO, C. F. 1985. "The Uniqueness of a Time-Dependent Equilibrium Distribution of Arrivals at a Single Bottleneck," *Transportation Science*, 19, 29–37.

FISHMAN, G. S. 1973. *Concepts and Methods in Discrete Event Digital Simulation*. New York: John Wiley.

HALL, R. W. 1987. "Passenger Delay in a Rapid Transit Station," *Transportation Science*, 21, 279–292.

HENDRICKSON, C., and G. KOCUR. 1981. "Schedule Delay and Departure Time Decisions in a Deterministic Model," *Transportation Science*, 15, 62–77.

HORONJEFF, R. 1969. "Analyses of Passenger and Baggage Flows in Airport Terminal Buildings," *Journal of Aircraft*, 6, 446–451.

KLEINROCK, L. K. 1976. *Queueing Systems, Vol. 2: Computer Applications*. New York: John Wiley.

NADDALA, G. S. 1977. *Econometrics*, New York: McGraw-Hill.

NEWELL, G. F. 1982. *Applications of Queueing Theory*. London: Chapman and Hall.

______. 1987. "The Morning Commute for Nonidentical Travelers," *Transportation Science*, 21, 74–88.

THEIL, H. 1971. *Principles of Econometrics*. New York: John Wiley.

PROBLEMS

1. Customers arrive at a cafeteria by a nonstationary Poisson process, with rates:

$$\lambda(t) = \begin{cases} 5/\text{hr} & 8{:}00\text{–}12{:}00 \\ 20/\text{hr} & 12{:}00\text{–}1{:}00 \\ 10/\text{hr} & 1{:}00\text{–}3{:}00 \end{cases}$$

(a) Write the function $\Lambda(t)$.
(b) What is the probability that exactly 30 customers arrived between 12:00 and 3:00?
(c) Suppose that 20 customers are known to have arrived between 8:00 and 1:00. What is the probability that no one arrived before 12:00? What is the probability that two customers arrived before 12:00?
(d) Suppose that the last customer to arrive before 12:00 arrived at 11:57. What is the probability that no customer arrived between 12:00 and 12:07?

2. The arrival rate in Prob. 1 is now defined by the function

$$\lambda(t) = 8 \sin(\pi t/7) \qquad 0 \leq t \leq 7$$

where $t = 0$ is equivalent to 8:00, time is measured in hours, and the angle is measured in radians. Repeat parts a–c from Prob. 1.

3. Figure 6.13 provides a cumulative arrival curve. From this curve, draw (approximately) $\lambda(t)$ on a piece of graph paper. At what time is the arrival rate the largest, and what is the largest arrival rate?

4. Figure 6.14 provides an arrival rate curve. From this curve, draw (approximately) $\Lambda(t)$ on a piece of graph paper.

5. The service time distribution at a single server queue is normal, with mean 5 minutes and variance 4 minutes^2. The arrival process is nonstationary Poisson, with the following arrival rate:

$$\lambda(t) = 50e^{.05t}/(100 + e^{.05t})^2 \qquad \lambda(t) \text{ in cust/min, } t \text{ in min.}$$

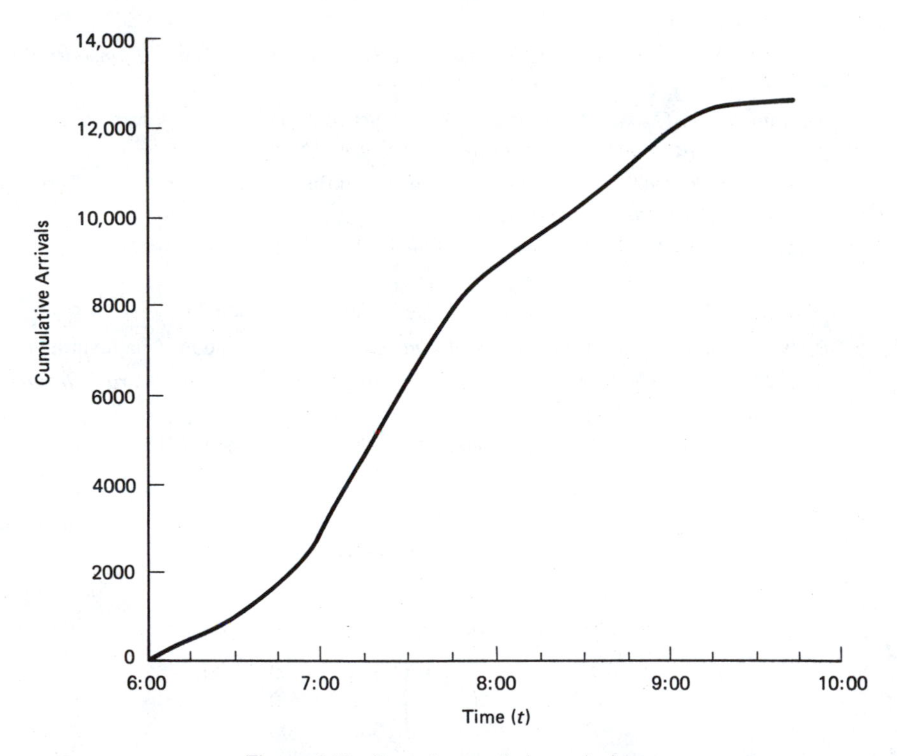

Figure 6.13 Example cumulative arrival curve.

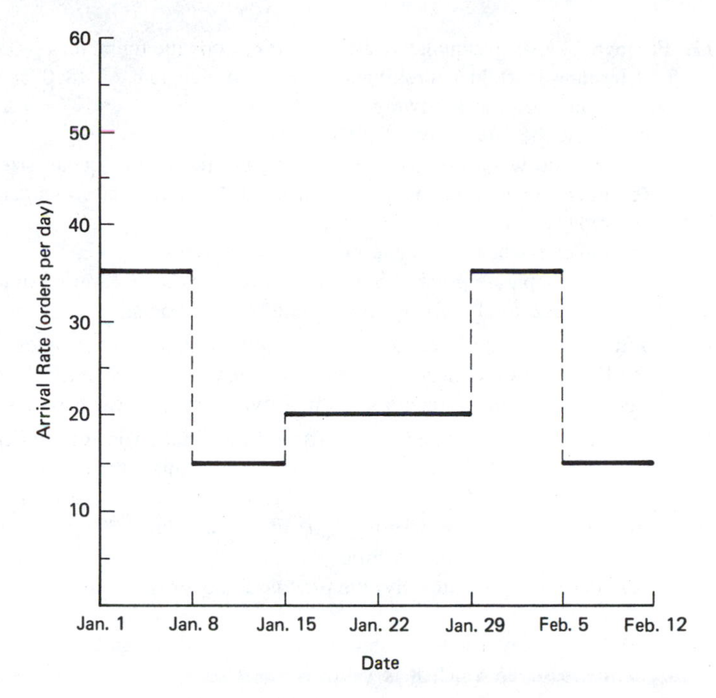

Figure 6.14 Example arrival rate curve.

(a) Use the quasi-steady-state approximation to write the function $L_q(t)$.

***(b)** Using a computer, plot the function $L_q(t)$ over the interval from $t = 0$ to 200 minutes. At what time is $L_q(t)$ the largest, and what is the largest value of $L_q(t)$?

***6.** For Prob. 5, determine whether the quasi-steady-state approximation is valid at all times (a computer may be helpful on this problem). If not, describe how the true performance would differ from the predicted performance.

7. Repeat Prob. 5, using the arrival rate function in Prob. 1 (over 8:00 to 3:00 interval) and two servers, each with exponential service time distribution and mean service time of 5 minutes. In words, discuss whether you believe your approximation is valid.

8. Suppose that the function given in Prob. 1 is based on the average number of arrivals over 20 days. Give 95% confidence intervals for the arrival rate in each interval.

9. Using a computer, simulate the arrival process in Prob. 1 by each of the following means:

(a) By generating a Poisson random variable, then generating the individual arrival times

(b) By generating exponential random variables

(c) By simulating a stationary Poisson process and randomly accepting or rejecting arrivals

***10.** Using a computer, simulate the queueing system described in Prob. 5, for one 200-minute period. Plot queue length as a function of time and compare your result to $L_q(t)$. Are your results reasonable?

*Difficult problem

11. Figure 6.14 gives a cumulative arrival curve. Copy the figure on a piece of graph paper. Using a deterministic fluid approximation, with service rate of 4800 customers/hour and short service time, do the following:
   (a) Draw the cumulative departure curve.
   (b) Indicate when the queue is largest, and the largest queue size.
   (c) Indicate when the queue is growing at the fastest rate. Estimate the maximum rate of growth.
   (d) Indicate when the queue begins and vanishes.
   (e) On a separate graph, show queue length as a function of time.
   (f) Estimate total waiting time among all customers.

***12.** A forms processing center has a constant service time of 15 minutes, and 12 servers operate. Copy Fig. 6.14 on a piece of graph paper, and rescale the vertical axis by dividing by 100. Based on the service process and the arrival curve, repeat parts a–f from Prob. 11.

***13.** For Prob. 11, suppose that the coefficient of variation in the service time is .5.
   (a) Using the adjustment factor in Sec. 6.6, estimate the queue size at the time $\lambda(t) = c$. Assume that $d\lambda(t)/dt = 20{,}000$ customers/hour^2 at the time $\lambda(t) = c$.
   (b) Draw a graph representing $L_q(t)$ for the period after $\lambda(t)$ first equals c. Compare and contrast your result to Prob. 11.
   (c) Identify (approximately) the time period over which a quasi-steady-state model would be valid.
   (d) Using the quasi-steady-state model, and assuming an $M/G/1/\infty$ queue, plot $L_q(t)$ over the period over which it is valid.
   (e) Discuss how your answers to parts b and d might be used to estimate $L_q(t)$ over the entire time range.

***14.** In Prob. 11, suppose that on one out of ten days the server malfunctions and can only process customers at the rate of 3600/hour. Estimate the total waiting time among all customers.

15. Figure 6.14 represents the cumulative number of jobs that a printer must deliver by time t. The company would like to begin processing jobs as late as possible, yet still deliver them on time. Jobs can be processed at the rate of 5500 per hour.
   (a) On a piece of graph paper, draw the curve representing cumulative job completions.
   (b) At what time is the number of jobs that have been completed, but not delivered, the largest?

16. Customers at a convenience market have a tendency to renege if their wait in line is sufficiently long.

$$P(\text{renege if wait is } t \text{ or less}) = \begin{cases} t/30 & 0 \leqslant t \leqslant 30 \text{ minutes} \\ 1 & t > 30 \text{ minutes} \end{cases}$$

Put another way, for each minute the customer waits (up to 30), there is a 1/30 chance of reneging. To take an example, if a customer has to wait 10 minutes, there is a one-third chance of reneging.

Describe in words how your simulation in Prob. 10 could be modified to account for this behavior.

QUEUE EXPERIMENT: FOR IN-CLASS DISCUSSION

The queue experiment can be completed in the privacy of your own home in about 20 minutes. You will need an ordinary bathroom sink (preferably with a screw-type water valve), a ruler, and a

*Difficult problem

timing device. Record your observations, but do not hand them in. Read the entire description before beginning.

Setup. Open the cold water valve to maximum flow, counting the number of revolutions of the handle. Record this number.

Push the drain control to its lowest height (open drain). Open the cold water valve to one-half the maximum flow. Now raise the drain control until you begin to see the water level rising in the sink. Leave the drain control at a level that maintains a constant water level. Place the ruler in the sink in a way that allows you to measure the depth of the water.

Experiment. Throughout the following, record the height of the water at 10-second intervals.

Time (sec)	Action
0	Open the cold-water valve to 3/4 maximum flow
10	Open the cold-water valve to maximum flow
20	Reduce the cold-water flow to 3/4 maximum flow
30	Reduce the cold-water flow to 1/2 maximum flow
40	Reduce the cold-water flow to 1/4 maximum flow
50	Turn the water off

Continue recording the water level until no water is left in the sink. (Absolute accuracy is not essential in this experiment.)

Questions

Based on the flow *only*, plot $A(t)$ and $D_q(t)$ on a piece of graph paper.
Based on the water height recordings *only*, plot $L_q(t)$ on a separate piece of graph paper.
When is $L_q(t)$ the largest?
When does the queue vanish?
Describe the function $\omega(t)$.
Are your data consistent? Explain why or why not.

EXERCISE: NONSTATIONARY MODELING

The purpose of this exercise is to learn how to simulate a nonstationary Poisson process and to use deterministic fluid approximations.

1. One of the local restaurants has been collecting data over the last year on the arrival pattern of customers during lunchtime. The data below provide the number of arrivals from the opening time (11:30) until 2:00.

Time period	Average number of arrivals
11:30–11:45	5
11:45–12:00	30
12:00–12:15	30
12:15–12:30	15
12:30–12:45	5
12:45–1:00	20
1:00–1:15	10
1:15–1:30	5
1:30–1:45	6
1:45–2:00	4
Total	130

Each arrival represents a "party" of customers. The restaurant has 35 tables. The service time (the time from when a party is seated at its table until the table is available to seat the next party) has an exponential distribution, with mean of 30 minutes.

1. Simulate the arrival pattern for one lunchtime (11:30–2:00), based on a nonstationary Poisson process. Plot $A(t)$ on a piece of graph paper.
2. Simulate the departure pattern for one lunchtime, based on the exponential distribution. Plot $D_q(t)$ and $D_s(t)$ on the same paper as $A(t)$.
3. Determine the average number of customers in the system and the average time in system. Answer the following:
   (a) At what time is the queue the longest?
   (b) At what time does queueing begin?
   (c) At what time does the queue vanish?
   (d) At what time is the wait (not counting service time) the longest?
   (e) What is the longest waiting time?
4. Now assume that the arrival pattern is deterministic, as defined by the average arrival rates. Also assume that all customers are served in exactly 30 minutes.
   (a) Draw $A(t)$, $D_q(t)$, and $D_s(t)$ on a single piece of graph paper. (Be careful. $D_q(t)$ and $D_s(t)$ are not easy to draw because of the long service time. Make sure that $D_q(t) - D_s(t) \leq 35$ for all t.)
   (b) Answer all the questions from part 3 for these new curves.
   (c) Explain why, or why not, your answers to part 4b are different from your answers to part 3.
   (d) Under what conditions would the deterministic approximation be more accurate? Explain.

APPENDIX: DIFFUSION MODELS

The term *diffuse* means "to cause to spread or disperse, as a gas or a liquid." The term *diffusion* means the process of diffusing. Diffusion models are used in physics to represent the molecular diffusion of fluids. But diffusion models are also applicable to queues, particularly in the analysis of the stochastic behavior of nonstationary queueing systems. Exact analysis of these systems is extremely difficult, and diffusion models provide both simple and robust results. As already seen in the chapter text, deterministic fluid models can be used to approximate queue behavior. Stochastic diffusion models can too: The rate at which a fluid diffuses across a region boundary is analogous to the transition rate across a cordon line in a transition rate diagram (see Chap. 5).

The following will refer to two types of diffusion models. The ***diffusion equation*** is a differential equation, first developed for molecular diffusion, but also applicable to queueing systems. A ***diffusion process*** (also called Brownian motion) is a stochastic process in which the interevent times are independent normal random variables. As applied to queueing, the fundamental assumption of the diffusion equation is that the number of arrivals and number of departures behave like diffusion processes and are mutually independent, whenever the queue size is positive. However, just because the arrival and departure processes behave like a diffusion process does not mean that they are diffusion processes. As we will see, other stochastic processes, including the Poisson process, can be approximated by the diffusion process.

The derivation of the diffusion equation, and its role in developing nonstationary results, are beyond the scope of this book. Interested readers are referred to Newell (1982). However, basic concepts can be illustrated with the following example, based on the diffusion process. In the example, it will be assumed that the probability of the queue being size zero is negligible. This might represent a situation where the queue size is initially large and the arrival rate exceeds the service rate. The following analysis is similar to that in Newell (1982).

First, consider the arrival process. Assume that at time 0, no arrivals have yet occurred. Then the time of the nth arrival, $A^{-1}(n)$, must equal the sum of n interarrival times, A_i:

$$A^{-1}(n) = \sum_{i=1}^{n} A_i \tag{6.A1}$$

From the central limit theorem, we know that if n is large and A_i are independent identically distributed, $A^{-1}(n)$ must (approximately) have a normal distribution. For large values of k, this implies that $A^{-1}(ki) - A^{-1}[k(i-1)]$ must also be approximately normally distributed. Because dependencies should be weak (at most) for large k, $A^{-1}(ki)$ should behave like a diffusion process. (This does not mean that $A^{-1}(n)$ behaves like a diffusion process.) A similar argument can be made for the departure process, $D_s^{-1}(ki)$, based on the assumption that the queue does not vanish. $D_s^{-1}(n)$ is the sum of n service times. If the service times are identically distributed, $D_s^{-1}(ki)$ will behave like a diffusion process. For an FCFS queue, the fact that $D_s^{-1}(n)$ and $A^{-1}(n)$ are approximately normal means that confidence intervals can be obtained for the waiting time of the nth customer.

Of greater interest than the processes $D_s^{-1}(ki)$ and $A^{-1}(ki)$ are the processes $D_s(t)$ and $A(t)$. The latter are related to the former. In fact, it is possible to describe events in either of two ways:

$$A^{-1}(n) \leq \tau \leftrightarrow A(\tau) \geq n \tag{6.A2}$$

The left side states that the arrival curve intersects the horizontal line, passing through the point n, at or to the left of time τ in Fig. 6.15. The right side states that the arrival curve

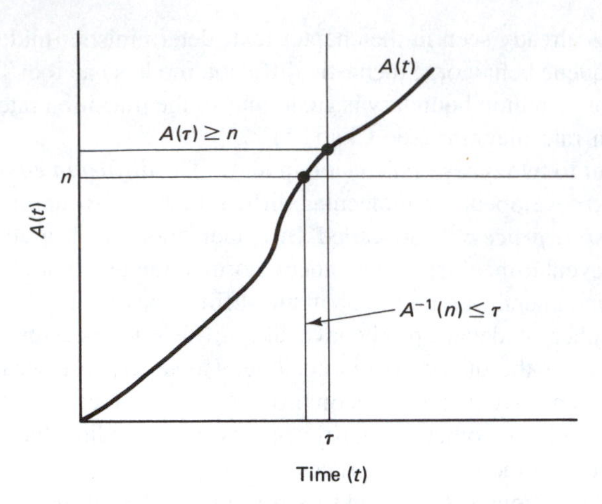

Figure 6.15 The event that $a^{-1}(n) \leq \tau$ is equivalent to the event $A(\tau) \geq n$.

intersects the vertical line, passing through the time τ, at or above the point n. As can be seen in Fig. 6.13, the statements are equivalent.

There is also an important relationship between the two points of intersection, $A^{-1}(n)$ and $A(t)$. Specifically, if $\tau\lambda \approx n$:

$$\lambda[\tau - A^{-1}(n)] + n \approx \lambda[2\tau - A^{-1}(n)] \approx \eta \tag{6.A3}$$

One can conclude (for large n) that if $A^{-1}(n)$ is approximately normally distributed, $A(t)$ must also be approximately normally distributed, with mean and variance defined by

$$E[A(t)] = \lambda t \qquad V[A(t)] \approx \lambda^2 V[A^{-1}(\lambda t)]$$
$$\approx \lambda^2(\lambda t)V(A) \qquad \text{(independent interarrival times)}$$
$$\approx \lambda t C^2(A) \tag{6.A4}$$

where $V(A)$ is the variance of the interarrival time and $C(A)$ is the coefficient of variation of the interarrival time. If the arrival process is Poisson, $V(A)$ is just $1/\lambda^2$, and $V[A(t)]$ equals λt, the variance of a Poisson distribution with mean λt. Similar relationships hold for $D_s(t)$, which is also approximately normally distributed:

$$E[D_s(t)] = ct \qquad V[D_s(t)] \approx c^2 V[D_s^{-1}(ct)]$$
$$\approx c^2(ct)V(S) \qquad \text{(independent service times)}$$
$$\approx ctC^2(S) \tag{6.A5}$$

where $V(S)$ is the variance of the service time distribution and $C(S)$ is the coefficient of variation.

It is now possible to derive the probability distribution for the number of customers in the system, $L_s(t)$. Because $L_s(t)$ is the difference between two (approximately) normal random variables, $A(t)$ and $D_s(t)$, it must also be normally distributed. Assuming service times are independent, that customers arrive by a Poisson process, and that arrival and departure processes are independent:

Poisson Arrivals, Independent Service Times

$$E[L_s(t)] = [A(0) - D(0)] + (\lambda - c)t \qquad (6.A6a)$$

$$V[L_s(t)] = t[\lambda + cC^2(S)] \qquad (6.A6b)$$

Equation (6.A6) can be used both to predict $L_s(t)$ and to obtain a confidence interval for $L_s(t)$, based on the normal distribution. It is valid for large values of t, provided that the probability that the queue vanishes is negligible. This is to say

Approximation Valid When

$$E[L_s(t)] - 2\sqrt{V[L_s(t)]} > 0 \qquad t \geqslant 0 \qquad (6.A7)$$

In reality, a queue will not instantaneously attain a large queue size, as assumed; it must first make the "transition through saturation." At or below saturation, the departure and arrival processes are certainly not independent. From time to time, the queue will vanish, and service must stop. Equation (6.A6a) will then *underestimate* queue size. Fortunately, there is an alternative model: the diffusion *equation*. Through the use of *boundary conditions*, the diffusion equation can be used to approximate the queue size distribution, even when the queue vanishes from time to time. The diffusion equation was the basis for Eq. (6.25), which gives an estimate for the queue size at the time when the queue makes the transition through saturation (that is, $\lambda = c$).

chapter 7

Reducing Delay through Changes in the Service Process

Queues are not merely an object of study. Queues are real problems faced by people, business, and government every day. So it is not enough to know how queues behave. One should also know how to eliminate queues, or at least how to reduce them to manageable size. Discussions of this topic begins in this chapter and continues throughout the remainder of the book.

The focus of this chapter is on reducing queueing delay through changes in the service process. Fundamentally, this means increasing the rate at which customers are served. How this is accomplished depends on the nature of the queue.

Queueing problems can be classified into three types: the *perpetual queue*, the *predictable queue*, and the *stochastic queue*.

The Perpetual Queue The perpetual queue is the worst type of queue. A perpetual queue is a queue for which *all* customers have to wait for service. All customers have to wait because the servers have insufficient capacity to handle the demand for the service.

The Predictable Queue A predictable queue occurs when the arrival rate is known to exceed the service capacity over finite intervals of time. The most familiar example of a predictable queue is the rush-hour traffic jam. But predictable queues also occur at restaurants at lunchtime, department stores prior to Christmas, and grocery stores before dinner.

The Stochastic Queue Stochastic queues are not predictable. They occur by chance when customers happen to arrive at a faster rate than they are served. Stochastic queues can occur whether or not the service capacity exceeds the average arrival rate.

Eliminating a perpetual queue is an imperative more than an option. Forcing all customers to wait for service is a pure waste of customers' time. Yet perpetual queues still exist as a way to ''ration'' scarce resources (prepaid health plans and many government services are examples). The mere existence of perpetual queues is a sign that the server disregards the customers' time and only seeks to minimize the cost of providing service. Such is the position of a monopolist, for if the customer had the option of patronizing another server he or she certainly would not wait in a perpetual queue. The solution to the perpetual queue is indeed simple: Either something must be done to increase the service capacity or something must be done to decrease the arrival rate. These ideas are discussed here and in the following chapter.

Eliminating a perpetual queue does not necessarily eliminate queues altogether. While the service capacity may exceed the average arrival rate, known variations may create capacity shortages over finite intervals of time. This creates predictable queues. Predictable queues can be eliminated by increasing the service capacity or decreasing the arrival rate over the intervals of time when queues are known to exist. This might be accomplished by hiring part-time employees to handle the peak demand. However, some servers, by their nature, must be available 100 percent (or nearly 100 percent) of the time. Roadways are one example. Eliminating a predictable queue may then require increasing the service capacity at all times, not just the times when queueing is a problem.

Just as eliminating perpetual queues does not necessarily eliminate predictable queues, eliminating predictable queues does not necessarily eliminate stochastic queues. Because they cannot be predicted, stochastic queues are the most difficult to eliminate. Fortunately, stochastic queues also tend to be small and short-lived. So it is not always imperative that they be eliminated. The three basic solutions are to increase the overall service capacity, develop a demand responsive service strategy, or reduce service time variations.

One demand responsive strategy was presented in Chap. 5: When queue length reaches a critical value, add another server; when queue length declines to zero, remove servers. This and similar strategies are common in grocery stores, banks, and other organizations where servers can alternate between direct customer service and ancillary activities. Demand responsive strategies cannot be used when servers cannot be employed in ancillary activities. Such may be the case when the server is a specialized piece of equipment rather than a person, or perhaps when union contracts prohibit job rotation.

The three queue types form a hierarchy, with perpetual queues the most insidious and stochastic queues the most benign. All three types can be addressed by increasing the service capacity or decreasing the arrival rate. But it is usually better to eliminate predictable queues by only increasing capacity or decreasing arrivals during the times when queues are a problem. And it is better to eliminate stochastic queues by only increasing capacity in response to changes in the arrival rate.

This chapter examines how to eliminate queues through changes in the service process. First, methods are presented for increasing the average service rate, with emphasis on reducing the service time. Second, methods are presented for selectively increasing the service rate at the times when queues are known to exist. A final section examines bulk service as a way to increase the service rate and reduce delay.

The emphasis of this chapter is on predictable queues, because of the importance of the problem and because of the wide range of solutions. However, many of the ideas presented, especially those in Sec. 7.1, apply to all three types of queues.

7.1 DECREASING THE SERVICE TIME AND INCREASING THE SERVICE RATE

The very first place to look for a solution to a queueing problem is in the existing servers. One should always ask whether customers can be served faster with the existing servers without sacrificing service quality. For example

Are Servers Devoting 100 Percent of Their Time to Customers When Customers Are Waiting? The simultaneous presence of queues and idle servers is likely the most common, and preventable, cause of excess delay. Are employees conversing with each other instead of with the customers? Is equipment adequately maintained so that it does not malfunction? Are servers distracted by secondary activities? Are some servers simply idle while customers are waiting? Are customers unaware that there are servers available? Here are some solutions.

Team service. Many queueing systems strictly adhere to a one server per customer rule. So when there are only three customers in the system and six servers available, three of the servers are idle. To reduce service time, the three idle servers should be teamed with the three busy servers. The objective is to make a multiserver queue behave like a high-capacity single server queue. That is, the system should work at full capacity whether there is one customer in the system, two customers in the system, or 100 customers in the system, just as a single server queue operates.

Flexible server assignment. Instead of forcing customers to use a specific server, customers can be allowed to use whichever server happens to be available. For example, patients at a medical clinic could be served by any one of a team of physicians, rather than a permanently assigned doctor. Flexibility in server assignment can then reduce waiting time. Flexible server assignment is the norm for most multiserver queues because it reduces waiting time. However, distance sometimes prevents it from being used to the fullest extent. For example, if the servers are not adjacent, the customer may not know whether an alternative server is free. Flexible assignment requires that similar servers be grouped together (discussed further in Chap. 11). Unfortunately, flexible assignment might also make the service less personal. For instance, many people would prefer to have a permanently assigned doctor, even if it meant long waits.

Are Servers Operating at Peak Efficiency? Inefficiency not only wastes the server's time; it wastes the customer's time. Are employees properly trained so that they can perform at maximum efficiency? Has management communicated the attitude that customer service is a top priority? (See Zemke 1986.) Is equipment properly maintained so that it works at maximum rate? Or, perhaps, do some customers receive *too much* service? If someone wants the spark plugs changed in his car, he shouldn't be given a tune-up. If someone wants to have her teeth cleaned, she shouldn't get a full dental examination with X-rays. Providing a service beyond what is called for can lead to delay. When delay is sufficiently large, the service process might change completely. Ninety percent of lawsuits are settled out of court, and many are now settled through private arbitrators, largely because of the length of time needed to bring a lawsuit to trial.

Solutions to queueing problems might include the following:

Automate the Service Process. Many simple tasks can be performed more quickly by computers or machines than by humans. Laser scanners in supermarkets have eliminated the need for cashiers to key in prices. Computer terminals at car-rental agencies have reduced the time needed to process a transaction. Automated cash registers at fast-food outlets allow an order to be rung up with the press of a single button. Today, most service processes involve humans working in concert with machines.

Reduce the Customer's Part of the Service Time. The time it takes to serve a customer is not the sole responsibility of the server; it also depends on the customer. In making a bank deposit, the tasks of endorsing checks and filling deposit slips belong to the customer. If these tasks are not completed prior to reaching the server, the service time will increase. In paying a toll on a highway, part of the service time is the time taken by the customer to hand the money to the toll taker. If a customer does not have the money out before reaching the server, the service time will increase.

These examples may seem minor, but consider trial court delays, a persistent problem in the United States. The length of a trial depends as much (or more) on the lawyers' (that is, the customers') actions as on the judges' actions. If the lawyer is unprepared, calls long lists of witnesses, or is a bombastic orator, the trial might drag on for months, or even years. See American Bar Association (1984), Church et al. (1978), Ebener (1981) for ideas on reducing trial court delays.

One of the responsibilities of the server is to ensure that the customer is prepared before reaching the front of the queue. Part of this is a simple matter of communication. For example, signs can be placed on turnpikes to inform drivers of the proper toll well in advance of the toll booth. Part of the solution may be in queue design, which may allow the customer to complete part of the service while waiting in line (for example, by filling out forms). And part of the solution may come in regulations and enforcement. For example, courts may assign penalties to lawyers for frivolous motions or delaying tactics.

Shift Some of the Service Responsibility to the Customer. Some tasks, by their very nature, belong to the customer (endorsing checks); some tasks belong to the server (collecting payment), and other tasks can be performed by either the customer or the server. Shifting some of this service responsibility to the customer can reduce the service time. In grocery stores, we load our groceries onto the checkout counter. In service stations we pump our own gas. In cafeterias we serve ourselves.

The key to effective customer participation is to convert idle time into productive time. Customer and server tasks should be performed simultaneously (a customer loads groceries *while* the checker rings them up) not sequentially (a customer loads groceries, *then* the checker rings them up). Neither should the server be idle waiting for a customer, nor should the customer be idle waiting for a server.

Unfortunately, customer participation not only reduces the service time; it changes the nature of the service. People may be less inclined to patronize businesses that place too large a burden on the customer. Customer participation should provide tangible benefits, either through reduced waiting time (that is, customer participation should not be used solely to reduce the number of servers) or reduced prices. See Bowen (1986) and Lovelock and Young (1979) for further ideas on how to tap customer participation.

Serve Customers Simultaneously. Many service tasks involve back-and-forth interaction between the server and the customer: A customer walks into a department store and asks to look at a camera. After answering a few questions, the clerk brings out the latest model and shows it to the customer, who proceeds to examine it. While the customer picks over the minute details of the camera, the server can move on to the next customer. The server can then serve two or more customers simultaneously and effectively increase the service rate.

Serve Similar Customers in Batches. The service time for a customer sometimes depends on which customer preceded it. For example, many manufacturing processes require machine setups for each type of item produced. If items of the same type are produced in batches, then the service time per item will decline. (Batching is examined in greater detail at the end of this chapter.)

Alternate Different Customer Types. In converse to the previous example, sometimes a customer will have a shorter service time when following a different type of customer. For example, automated warehouses operate more efficiently when a placement move is followed by a picking move. This makes productive use of both the move out from the storage/retrieval station and back to the station. The incentive here is to alternate a pick with a place, rather than grouping all the picks and all the places together.

The ideas presented in this section and summarized in Table 7.1 are just a few of the low-cost or no-cost ways to increase service capacity and reduce waiting time. More ideas are provided by Sasser (1976). No matter what type of system is considered, eliminating the queue should always begin with identifying ways to increase the service capacity without adding servers, provided that service quality does not suffer.

TABLE 7.1 WAYS TO REDUCE SERVICE TIME AND INCREASE SERVICE RATE

Ensure that servers devote 100% of time to customers during queues	
Team Service:	When a server is idle, have it help other busy servers.
Flexible Assignment:	Do not force customers to use a specific server. Let them choose among all available servers.
Ensure that servers operate at peak efficiency	
Automate the Service Process:	Purchase machines, such as bar code scanners, to speed up the service.
Reduce Customer's Part of Service:	Make sure that customers are prepared before joining the queue.
Shift Responsibility to Customer:	Let customer participate in the service; when the server is busy, the customer can be busy too.
Serve Customers Simultaneously:	If servers must wait for customers to complete a portion of the service, they should move on to other customers, and serve customers simultaneously.
Serve Customers in Batches:	Serving similar customers as a group can reduce setup time and increase service rate.
Alternate Customer Types:	Automated warehouses are more efficient when a pick is alternated with a place.

7.1.1 Impact of Reduced Service Time

We have discussed eight ways to reduce the service time and increase the service capacity. Most of the solutions come with little or no cost to the server. Yet the benefits can be enormous. The relationship between service capacity and waiting time will now be evaluated for a given arrival pattern that produces a *predictable* queue. Chapter 5 already examined this relationship for stochastic queues. Although the relationship for perpetual queues will not be evaluated, the impact of an increase in capacity will be at least as great as for a predictable queue.

Assumptions Figure 7.1 shows an example arrival curve along with a series of departure curves representing different service capacities. To make the analysis as clear as possible, this arrival curve is assumed to be deterministic, ($A(t) = \Lambda(t)$). The service time

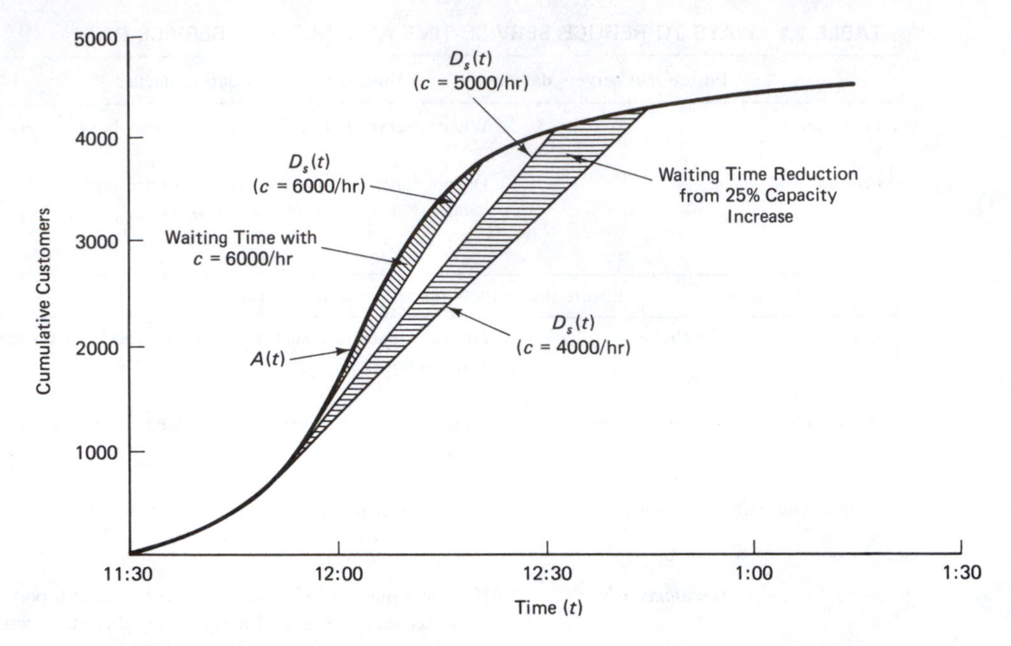

Figure 7.1 Cumulative arrival and departure curves for different service capacities, for a deterministic fluid model (short service time). Increasing the service capacity decreases both the duration and the magnitude of the queue.

is assumed to be small, so that $D_q(t) \approx D_s(t)$ (similar relationships hold for long service times). The service capacity is assumed to be constant over time.

Figure 7.1 demonstrates that an increase in service capacity both decreases the *magnitude* of the queue and the *duration* of the queue. This means that the relationship between total waiting time (the area between the arrival and departure curve) and the service capacity is highly nonlinear. Small changes in capacity create large changes in waiting time when queues are large. In the example, an increase in service capacity of 25 percent, from 4000 customers per hour to 5000 customers per hour, reduces waiting time by 55 percent (Fig. 7.2).

The figures show why it is essential to reduce queues to a manageable size. When a queue is large, small increases in capacity can have a tremendous impact on waiting time. One should not overlook these small changes.

7.1.2 Adding Servers

Adding servers has a similar impact to reducing the service time, except the cost is larger. One should not consider adding servers until all the possibilities for reducing service time have been exhausted. And one should not consider adding a full-time server until the option of adding a part-time server has been examined (discussed in Sec. 7.2). Why have a server operate continuously when it is only needed on occasion?

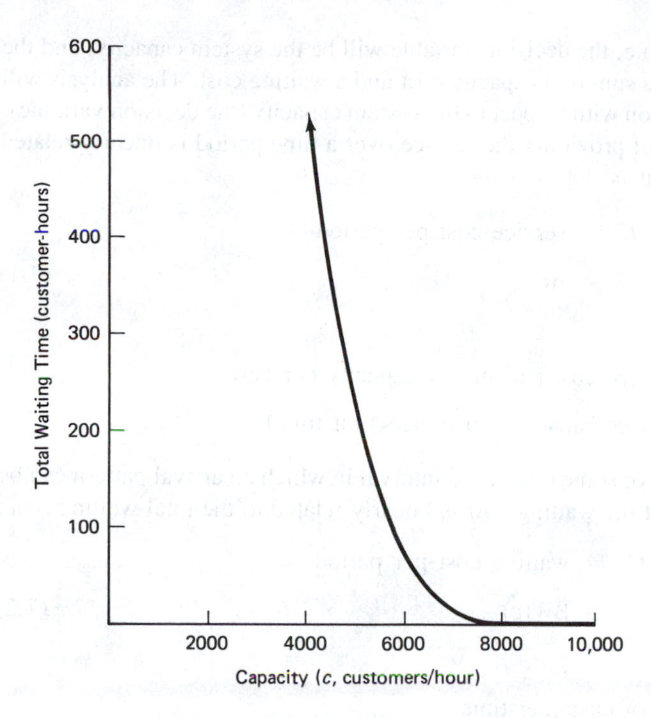

Figure 7.2 Total time in system versus service capacity. Waiting time increases at an increasing rate as the service capacity declines (based on arrival curve in Fig. 7.1).

Despite these caveats, sometimes there is no option other than to add full-time servers. If the server is a piece of equipment, a piece of land, or perhaps a lane of freeway, it may well be available 24 hours a day, whether or not it is needed. These are called ***all-or-nothing servers*** because their acquisition is an all-or-nothing proposition.

7.1.3 Balancing a Cost Trade-Off Through Graphical Analysis

For the all-or nothing server, eliminating queues may amount to balancing a trade-off between the cost of providing the service and the value of customer time. This trade-off will be analyzed graphically by examining the marginal change in cost due to small changes in the capacity of the system. The optimal capacity will be defined by the point where the marginal change in cost equals zero. ***Graphical analysis***, a general technique for analyzing queueing systems (Hall 1986), is used several times in this book.

Definitions 7.1

The ***decision variable*** is a quantity that can be controlled to improve a system's performance

The ***objective function*** measures the system's performance as a function of one or more decision variables.

In the following example, the decision variable will be the system capacity, and the objective function will be the sum of a capacity cost and a waiting cost. The analysis will optimize the objective function with respect to the system capacity (the decision variable).

Suppose that the cost of providing the service over a time period is linearly related to the service capacity. That is, let

$$\begin{aligned} C_s &= \text{service cost per period} \\ &= \alpha c \end{aligned} \tag{7.1}$$

where

α = service cost per unit of capacity per period

c = service capacity (customers/unit time)

and a period is a day, week, or some other time interval in which an arrival pattern can be predicted. Also suppose that the waiting *cost* is linearly related to the total waiting *time*:

$$\begin{aligned} C_w &= \text{waiting cost per period} \\ &= \beta W(c) \end{aligned} \tag{7.2}$$

where

β = value of customer time

$W(c)$ = total waiting time per period when capacity equals c

The objective can now be stated: Minimize $C_s + C_w$ with respect to the decision variable c.

A necessary condition for optimality is that cost must not decrease if the capacity is changed by a small amount Δc. The change in cost, ΔC, if c is *increased* by Δc can be written as

$$\Delta C = \alpha \Delta c + \beta[W(c + \Delta c) - W(c)] \tag{7.3}$$

The first term is the change in service cost and the second term is the change in waiting cost.

The change in waiting time (the term within the brackets) can be calculated graphically. Assume that one predictable queue occurs per time period. Then, for small values of Δc, the change in waiting time can be approximated from the area of the triangle shown in Fig. 7.3. That is

$$W(c) - W(c + \Delta c) \approx \tfrac{1}{2}T(c)\ [T(c)\Delta c] \tag{7.4}$$

where

$T(c)$ = *queue duration* when capacity = c (time from when queue first forms until it vanishes)

(If there is more than one queue per period, $W(c + \Delta c) - W(c)$ would be the sum of the areas of several triangles.)

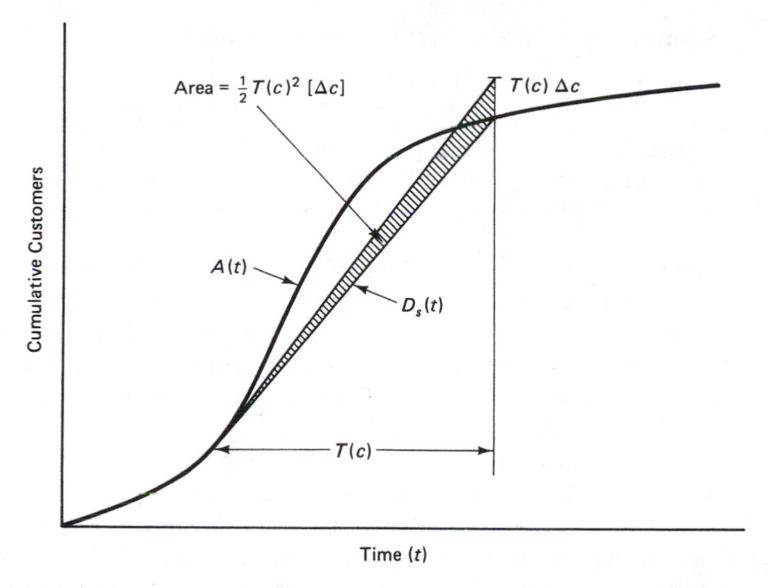

Figure 7.3 The change in total waiting time, due to a small change in service capacity, can be approximated by the area of a triangle.

ΔC represents the marginal change in cost, which must equal zero at the optimum (assuming that $W(c)$ is a continuously differentiable function). Substitution of Eq. (7.4) in Eq. (7.3) provides the following:

Optimality Criterion: Capacity of All-or-Nothing Server

$$0 = \alpha - (\beta/2)\, T^2(c) \rightarrow T^*(c) = \sqrt{2\alpha/\beta} \tag{7.5}$$

(The asterisk is used here and elsewhere in the book to denote optimality.) The interpretation of the result is that c should be set at a value that makes the queue duration equal $\sqrt{2\alpha/\beta}$. The larger the service cost, α, the longer the queue duration; the larger the waiting cost, β, the shorter the queue duration. Because $T(c)$ must be a decreasing function (larger capacity means shorter queue duration), Eq. (7.5) can be satisfied for only one value of c. Thus, it is *both a necessary and sufficient condition for optimality.*

In reality, the cost of providing the service is not a continuous function of the service capacity but a step function, with each step representing the cost of adding one more server. This means that the marginal graphical analysis only provides an approximate solution. The optimal solution is found by either rounding c up to the next integer multiple of a server's capacity or down to the next integer, whichever provides the least cost.

Example

A large company would like to know how many servers to hire at its company cafeteria. Each server is paid $150 per day, and employees are paid $20 per hour, on average. Servers can process 60 customers per hour. A single queue forms each day at lunchtime.

Solution The service cost per unit of capacity equals:

$$\alpha = [\$150/\text{day}] \ / \ [60 \text{ customers/hour}]$$
$$= \$2.50 \text{ per customer/hour per day}$$

Because $\beta = \$20/\text{hour}$, $\sqrt{2\alpha/\beta} = .5$ hour $= 30$ minutes. This solution is shown for the arrival pattern of Fig. 7.1. A capacity of 6000 customers per hour yields a queue duration of about 30 minutes. The capacity translates into 100 employees.

Just because the queue duration is 30 minutes does not mean that customers wait 30 minutes to be served. Notice in Fig. 7.1 that customers who arrive at the start of the queue are served well before the queue ends. In fact, most customers only wait a few minutes.

The example is somewhat unique in that customers are employees of the company providing the service. This made calculation of the customers' value of time straightforward. More commonly, the customer is not an employee, and one can only guess at the value of time. Further, customers usually are not captive and will renege when queues become large, leading to a loss in revenue (this topic is addressed in Chap. 8). An alternative form of analysis would be to determine the relationship between reneging and queue length and then derive an equation for profit as a function of service capacity. The profit equation would then be maximized with respect to c. Unfortunately, determining the reneging relationship is quite difficult, and one must often be content with a rough analysis based on assumed values of time.

A third approach is to set a service standard, such as maximum queue length or proportion of customers that have to wait, and provide the minimum capacity that meets the standard. This approach is common, and is discussed by Edie (1956), Linder (1969), and Kolesar et al. (1975). Edie's study is especially thorough and includes excellent discussions of data collection and modeling.

The example illustrates the expense of utilizing full-time servers when the arrival pattern is not stationary: Employees are paid a full day's wage when they are only needed a few hours. The example also illustrates why *some* queues should not be eliminated solely through changes in service capacity. If the arrival pattern is nonstationary and part-time servers cannot be used, adding sufficient capacity can be too expensive. This does not preclude eliminating queues through other means, such as altering arrival patterns, as is discussed in the following chapter.

7.2 ELIMINATING PREDICTABLE QUEUES: VARYING THE SERVICE CAPACITY

The ideal queueing system has neither customer delay nor server idle time. The cumulative departure curve for the ideal system exactly coincides with the cumulative arrival curve. Servers are added and removed in direct response to changes in the customer arrival rate, making the service capacity equal the arrival rate at all times. Customers never wait for servers and servers never wait for customers.

This depiction is not just a utopian dream. It describes how a queueing system would operate with effective use of part-time servers. What are some of the ways to utilize part-time servers? This depends on the nature of the queue. Chase (1978) classifies service systems by "extent of required customer contact in the creation of the service" on a scale from high customer contact to low customer contact, according to the following:

High Contact	Pure service (health center)
	Mixed service (banks)
	Quasi-manufacturing (home offices of banks)
Low Contact	Manufacturing (chemical plants)

(Chase uses the word "customer" in the conventional sense, that of designating the person requesting the service.)

High contact services must be performed in the presence of the person and, therefore, demand that waiting times be kept at a minimum. On the other end of the scale, low contact services are completely separated from the customer. Much longer waits are acceptable because a person is not present while service is performed. Whereas high contact waits rarely last more than an hour, low contact waits may last days, weeks, or months. These waits dictate the use of part-time servers. The former queue might be resolved by hiring permanent employees to work a few hours a day; the latter by hiring employees to work full time temporarily. In either case, the added capacity coincides with the duration of the queue.

In providing supplemental capacity, we must recognize that all servers are, to some extent, part-time. In the United States, employees normally work just 8 hours a day and 5 days a week, not 168 hours a week. Even equipment must be periodically shut down for maintenance. In eliminating predictable queues, the variable capacity should be used to the system's advantage, by effectively scheduling employee workhours and effectively scheduling equipment maintenance. With this in mind, solutions to predictable queueing problems include

Solutions to Predictable Queues

Queues of short duration (a few hours or less)

1. Hire part-time employees to work a few hours each day.
2. Schedule employee breaks (lunch, coffee, and so on) and equipment maintenance for periods of low demand.
3. Stagger starting and ending times of workshifts to match demand patterns.
4. Schedule days off for days of the week with low demand.

Queues of long duration (one or more days)

1. Hire temporary full-time employees or temporarily lease equipment.
2. Schedule employee vacations and major equipment maintenance for periods of low demand.
3. Pay employees to work overtime.

In addition, supplementary capacity might be provided by diverting servers from ancillary activities to direct customer service. And, in the case of mixed service, a portion of the service might be performed in advance of the peak arrivals. Restaurants, for example, partially prepare meals in advance so that the time from when a customer orders his or her meal until he or she receives it is minimized.

Many of these ideas, like the ones in Sec. 7.1, come with little or no cost. The key is to avoid idle time. So long as part-time servers are not idle, they should not be more expensive than full-time servers.

The following sections consider two aspects of adjusting capacity to match demand patterns. First, the impact of a variable capacity on waiting time will be evaluated, and second, techniques for scheduling employees to create a variable capacity will be presented.

7.2.1 Impact of a Variable Service Capacity

Before showing how to achieve a variable capacity, we will first look at the impact of a variable capacity. First, the impact of a temporary increase in capacity on waiting time will be examined, and second, the impact of a temporary decrease will be examined. The issues of when and how much to change capacity will be addressed through graphical analysis. Because the analysis is at times complicated, some readers may choose to proceed to Sec. 7.2.1.4, where general guidelines are summarized.

7.2.1.1 When to Add Capacity. First suppose that extra capacity is only available for a set length of time.

Definition 7.2 The ***supplemental period*** is the period of time in which extra capacity is provided.

Further, suppose that the supplemental period can be started at any time. (These characteristics represent a part-time employee who is restricted to working a set number of hours each day.) The issue is, *when is the best time to add the capacity*.

Figure 7.4 shows a hypothetical cumulative arrival curve along with three potential departure curves. In all three curves, customers are served at the rate of 2000 per hour during the supplemental period and at the rate of 440 per hour over the remainder of the day. The only difference between the departure curves is the starting time of the supplemental period.

It should be apparent that one departure curve creates less delay than any of the others. In the figure, this is the curve that adds capacity at the earliest time. The ''early start'' curve is constructed by shifting the diagonal segment, labeled A in Fig. 7.4, until it touches the arrival curve. That is, *the period should start sufficiently early that the queue vanishes sometime before the period ends*. Starting the supplemental period any later adds to delay, as indicated by the cross-hatched region. Put another way, waiting until a queue becomes a significant problem before adding capacity is a dangerous policy because it may be impossible to catch up with demand.

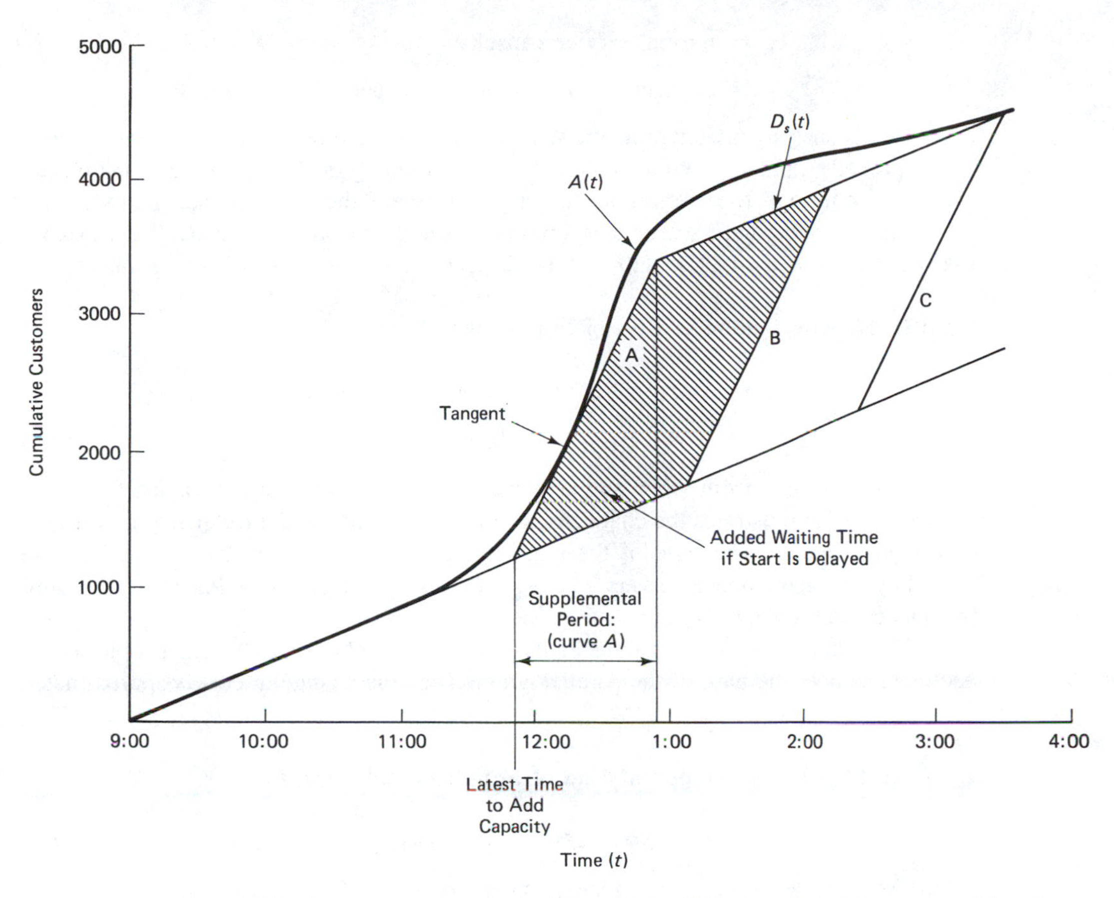

Figure 7.4 A capacity supplement should be added sufficiently early that the queue vanishes at least once during the supplemental period. Adding the capacity supplement any later will add to delay.

Under certain circumstances, it may be desirable to begin the supplemental period slightly earlier than indicated by curve A in Fig. 7.4. Determining the starting point would amount to evaluating a trade-off between reduced delay at the start of queueing and increased delay after the supplemental period ends. This analysis could be accomplished through a graphical analysis similar to that presented in Sec. 7.1.3.

7.2.1.2 Length of Supplemental Period Figure 7.4 invites an alternative solution: Provide supplemental capacity over a longer period of time. When work regulations do not prohibit it, extending the supplemental period can reduce waiting time.

It is impossible to determine how long the supplemental capacity should be provided without assigning costs. These are defined as follows:

γ = cost of supplementing capacity, per unit time

β = value of customer time

c_1 = normal service capacity

c_2 = capacity during supplemental period

In this analysis, assume that the starting time for the supplemental period is defined by the procedure in the previous section. Then determining the length of the supplemental period is equivalent to determining the ending time for the supplemental period.

First, assume that a queue exists when the supplemental period ends. Then extending the period by a small length of time Δt changes cost by the following amount:

Case I: ΔC with Queue at End of Supplemental Period

$$\Delta C = \text{change in service cost} + \text{change in waiting cost}$$

$$\Delta C \approx \gamma(\Delta t) - \beta[(c_2 - c_1)(\Delta t)][t_\ell] \tag{7.6}$$

where t_ℓ is the time from when supplemental period ends until queue vanishes.

Figure 7.5 illustrates the change in waiting time, which is approximated by the area of a parallelogram. The base of the parallelogram is the number of additional people served by the supplemental servers, $(c_2 - c_1)\,\Delta t$, and the height is the length of time until the queue vanishes, t_ℓ.

If a queue does not exist at the ending time, the number of additional people served declines, as does the base of the parallelogram. Then the change in cost is approximated by

Case II: ΔC with No Queue at End of Supplemental Period

$$\Delta C \approx \gamma(\Delta t) - \beta[\tilde{\lambda}_\ell - c_1)(\Delta t)]\,[t_\ell] \qquad \tilde{\lambda}_\ell \geqslant c_1 \tag{7.7}$$

where $\tilde{\lambda}_\ell$ is the arrival rate when supplemental period ends.

If $\tilde{\lambda}_\ell < c_1$, the second term is zero and ΔC is positive, meaning there is nothing to gain by extending the supplemental period further.

To find the optimal length of the supplemental period, Eq. (7.6) and (7.7) should be set equal to zero (as in Sec. 7.1.3) and solved. This leads to the following results:

Optimality Criteria: Ending Time of Supplemental Period

$$t_\ell^* = \frac{\gamma}{\beta} \; \frac{1}{(c_2 - c_1)} \quad \text{(case I)} \tag{7.8a}$$

$$t_\ell^* = \frac{\gamma}{\beta} \; \frac{1}{(\tilde{\lambda}_\ell - c_1)} \quad \text{(case II)} \tag{7.8b}$$

Most often, case II will define the optimum, meaning that the supplemental period should end some time after the queue vanishes.

The key point of the analysis is that the length of the supplemental period depends on a trade-off between the cost of providing the capacity and the cost of customer waiting

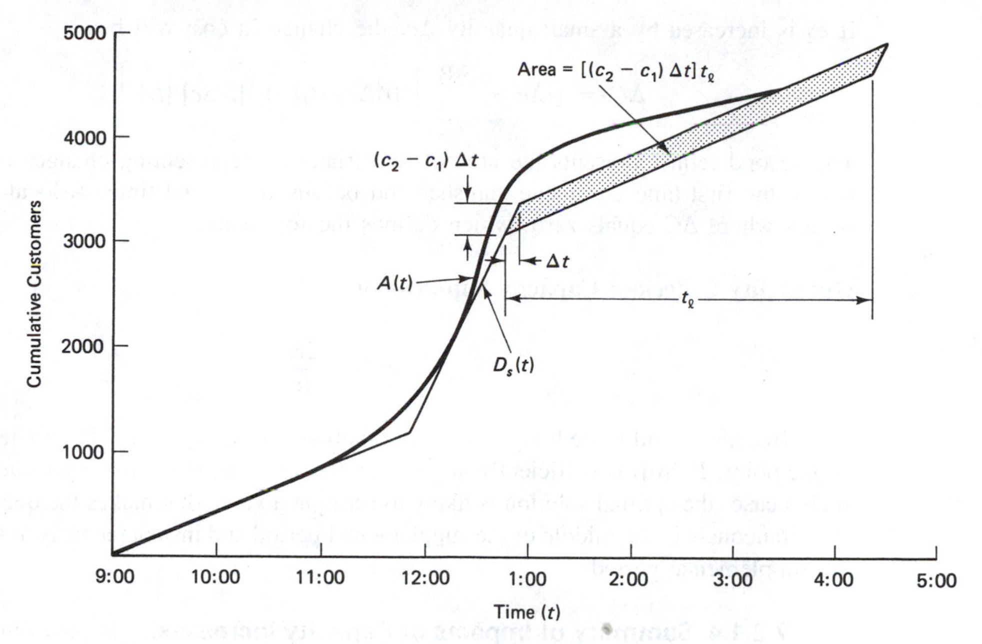

Figure 7.5 Extending the length of a supplemental period reduces waiting time until queue vanishes. Decrease in waiting time is approximated by a parallelogram.

time. The supplemental period should be longer when the ratio of γ (cost of providing the capacity) to β (cost of customer waiting) is small and when $c_2 - c_1$ is large (remember, the supplemental period gets shorter at t_ℓ increases).

7.2.1.3 Service Capacity. Consider one final question: Suppose that the capacity supplement is available over a set period of time, but the size of the supplement can be set at any value. For example, a restaurant hires part-time employees from 11:30 to 1:30 P.M. to handle the lunchtime crowds. How many extra employees should it hire?

For simplicity, we will assume that the queue vanishes twice during the supplemental period: once toward the middle and once at the end. Each time the queue vanishes, there may be a period in which there is excess capacity. The more capacity increases, the longer these periods will be, meaning there will be a drop in server efficiency. Nevertheless, waiting time will decline, and the added servers may still be worthwhile.

In the analysis, the following symbols are used:

α = cost per unit capacity (for example, cost per employee divided by service rate per employee)

t_1 = duration of the queue in the supplemental period

t_2 = duration of the second queue in the supplemental period

If c_2 is increased by a small quantity Δc, the change in cost will be

$$\Delta C = \alpha \Delta c - \frac{\beta}{2}\left[[t_1 \Delta c]\,[t_1] + [t_2 \Delta c]\,[t_2] \right] \tag{7.9}$$

The second term represents the area of two triangles, representing changes in waiting before the first time the queue vanishes and before the second time. A local optimum occurs where ΔC equals zero, which defines the following:

Optimality Criterion: Capacity Supplement

$$t_1^2 + t_2^2 = \frac{2\alpha}{\beta} \tag{7.10}$$

Because t_1 and t_2 are both decreasing functions of c_2, Eq. (7.10) is satisfied at most at one point. If $2\alpha/\beta$ is sufficiently large, it might not be satisfied for any value of c_2. In such a case, the optimal solution is likely to set c_2 at a value that makes the queue vanish instantaneously in the middle of the supplemental period and instantaneously at the end of the supplemental period.

7.2.1.4 Summary of Impacts of Capacity Increases.

In providing supplemental capacity, the following guidelines should be taken into consideration:

- The increase should occur early in the queue, to ensure that the queue vanishes sometime before the added capacity is removed (curve *A* in Fig. 7.4, for example).
- The duration of a supplemental period depends on a trade-off between service costs and waiting costs, as does the size of the capacity supplement.
- High service cost is an incentive for reducing the duration and size of the supplemental capacity; high waiting cost is an incentive for increasing the duration and size of the supplemental capacity (refer to Eqs. (7.8) and (7.10)).

These guidelines presume that supplemental capacity is only limited to set periods of time (Secs. 7.2.1.1 and 7.2.1.3) or a set number of supplemental servers (Sec. 7.2.1.2). If both the number of servers and the duration of the supplemental period are completely adjustable, then service capacity should be varied continuously, exactly matching the arrival rate. This would be an ideal situation: no waiting and no excess capacity.

7.2.2 Temporary Decrease in Capacity

Neither employees nor machines can work continuously. Both need breaks. Because breaks reduce system capacity, they should be carefully scheduled to avoid affecting customer waiting time. Sometimes all that need be done is to schedule breaks during times when there is a capacity surplus. But if a queue persists long enough, there is no way to avoid scheduling breaks during the queue and adding to customer delay.

Figure 7.6 shows an arrival curve along with three departure curves. Each departure curve includes a 25-minute break, but the breaks occur at different times. The timing of

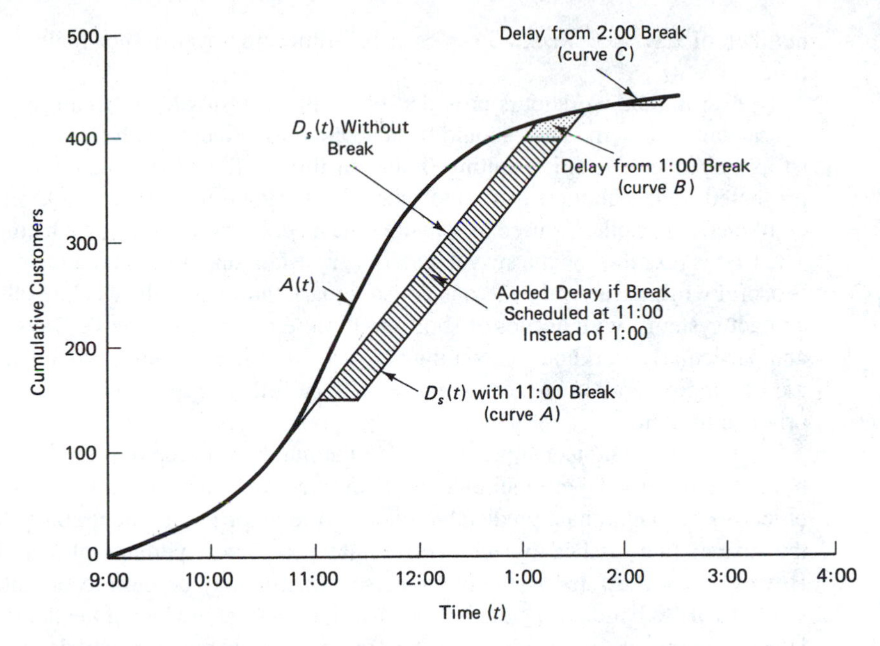

Figure 7.6 A server break delays all subsequent customers until the queue vanishes. Therefore, breaks should be scheduled as late as possible in the queue.

the break affects the waiting time for some customers, but not all. The break in curve C has the smallest impact because it occurs *after* the predictable queue has dissipated, when the arrival rate has fallen well below the service capacity. It does not take long for the system to recover from the loss in capacity.

If the break cannot be scheduled during low demand (for instance, if employees must have a break every so many hours), then it should be scheduled as late as possible. As shown in Fig. 7.6, a break not only affects customers who would have been served during the break; *all* subsequent customers are delayed until the queue vanishes. Curve A delays more than 250 customers by 25 minutes each, but curve B delays less than 30 customers. The break should be scheduled as late as possible in order to delay the fewest customers.

Figure 7.6 is just another illustration of the importance of providing a large service capacity toward the start of the queue. Queues should not be allowed to grow beyond control, either because capacity is supplemented too late or because breaks are scheduled too early. Breaks, or breakdowns (for example, an accident on a freeway or an ill employee), can have a devastating impact if they occur toward the start of a queue's duration. Quick response to correct such problems is essential.

7.3 TECHNIQUES FOR SCHEDULING TEMPORARY CAPACITY INCREASES

Employees do not and cannot operate on a continuous basis. Employees require breaks, are limited to working a set number of hours per day, and are limited to working a set number of days per week. So even a full-time employee is not really available "full time."

The limited workhours provided by employees should be used to the advantage of the queueing system. They should be scheduled to coincide with the peaks in arrival rate so as to reduce customer waiting times. In this section, two scheduling techniques are presented. One technique is used to schedule starting times for the workshifts of full-time employees. The other is used to schedule the days of the week that each employee works. The first is useful when the arrival pattern is nonstationary throughout the day. The second is useful when the arrival pattern is nonstationary throughout the week. Both apply to high contact systems, with queues of short duration (a few hours or less). Days-off scheduling and particularly workhour scheduling are of less use in low contact systems, where queues are of longer duration (where hiring temporary full-time employees and overtime are of prime importance).

The following techniques minimize the number of employees while fulfilling labor requirements. The labor requirements are derived from the cumulative arrival curve. If the objective is to eliminate predictable queues, the labor requirement at any time would be the arrival rate divided by the service rate per server, perhaps plus a safety margin. However, as illustrated in the previous sections, it may be more economical to allow a small queue to build during the peak demand, effectively reducing the labor requirements. Due to complexity, no method will be presented for simultaneously determining the best requirement pattern and employee schedule.

7.3.1 Workshift Scheduling

One way to provide a temporary increase in capacity is to overlap workshifts during the period of peak demand. For instance, if a business operates from 9:00 A.M. to 8:00 P.M., half the employees might work from 9:00 to 5:00 and another half from 12:00 to 8:00, providing twice as many workers between 12:00 and 5:00 as during the remainder of the day. If this time period happens to coincide with a peak in the customer arrival rate, then the added capacity can help eliminate queues. But one cannot rely on luck to eliminate queues. Workshifts are partly dictated by the starting and ending times of the business day. Had the business operated from 9:00 A.M. to 9:00 P.M., and the peak period been lunchtime, from 12:00 to 1:00, the overlap would not provide capacity when it is needed, but only between 1:00 and 5:00.

Browne (1979) describes a scheduling rule that minimizes the number of employees to achieve a labor requirements schedule. The rule, called the ***first-hour principle***, will be adapted here.

Definitions 7.3

$R(t)$ = labor requirement at time t (employees)

$P(t)$ = number of employees working at time t

$B(t)$ = cumulative number of employees who have begun their workshift by time t.

$E(t)$ = cumulative number of employees who have ended their workshift by time t

τ = length of a workshift

T = length of the business day

The objective is to minimize the total number of employees $[B(T)]$, subject to the following constraints:

$$P(t) \geq R(t) \qquad 0 \leq t \leq T \tag{7.11}$$

where:

$$E(t) = B(t - \tau) \tag{7.12}$$

$$P(t) = B(t) - E(t) = B(t) - B(t - \tau) \tag{7.13}$$

Equation 7.11 states that the number of employees working at time t must equal or exceed the requirement at time t; Eq. 7.12 states that the number of employees who have completed their day by time t equals the number who have started by time $t - \tau$, and Eq. 7.13 states that the number of employees working at time t equals the number that have started by time t minus the number that have ended by time t.

The first-hour principle works as follows:

Start an Employee at Time t if $R(t)$ Would Otherwise Exceed $P(t)$

An employee might be started for either of two reasons: $R(t)$ increases to $P(t)$ and beyond, or $P(t)$ drops below $R(t)$ because a workshift ended. The optimality of the rule follows from simple logic. Consider any feasible schedule that does not obey the first-hour principle. Then there must be at least one time t when $P(t) - R(t) \geq 1$. If the starting time of the last employee started before time t is delayed until $P(t) = R(t)$, the schedule will still be feasible, and no more employees will be needed (it may actually be possible to reduce the number of employees later). Repeating this process will produce the "first-hour schedule." Hence, the first-hour principle is optimal (though it is not necessarily the only optimal schedule).

Example

A supermarket would like to create a schedule for its employees, who work eight-hour shifts. The requirement curve, along with $P(t)$ and $R(t)$, is shown in Fig. 7.7a. Employees begin work at 8:00 (3), 10:40, 11:00, 11:20, 4:00 (3), 4:50, 5:30, 6:40, 7:00, 7:20.

Two important observations can be made from the solution. First, the first-hour principle does not guarantee that the ending times for workshifts coincide with the end of the business day, particularly when the peak demand occurs late in the day. When this happens, the schedule should be manually adjusted, as in Fig. 7.7b, by starting some of the workshifts earlier, perhaps allowing employees to be used in ancillary activities during the off-peak hours. The schedule might also be adjusted in order to schedule employee breaks during off-peak periods. The second observation is that providing sufficient

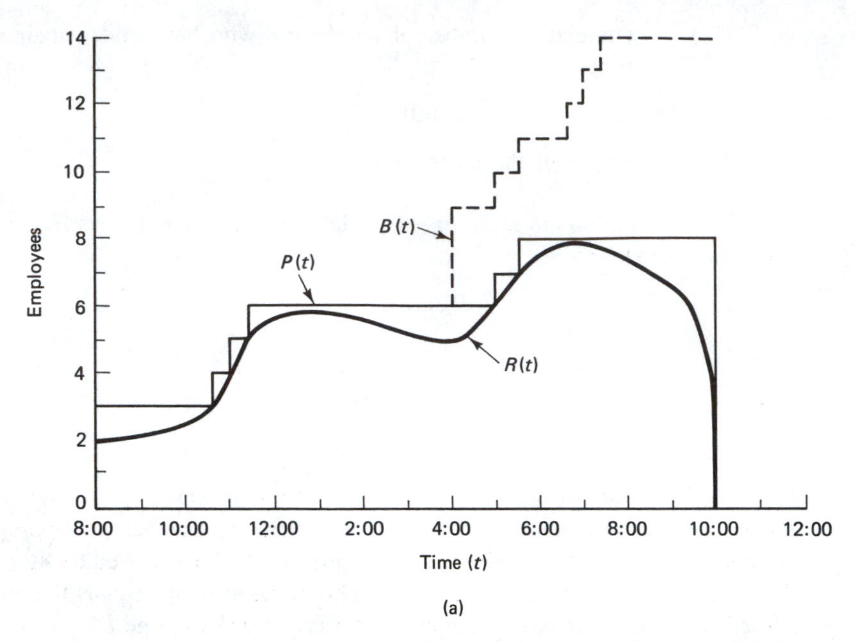

Figure 7.7a Workshifts created by first-hour principle. An employee is added when the labor requirement falls below the available supply of employees.

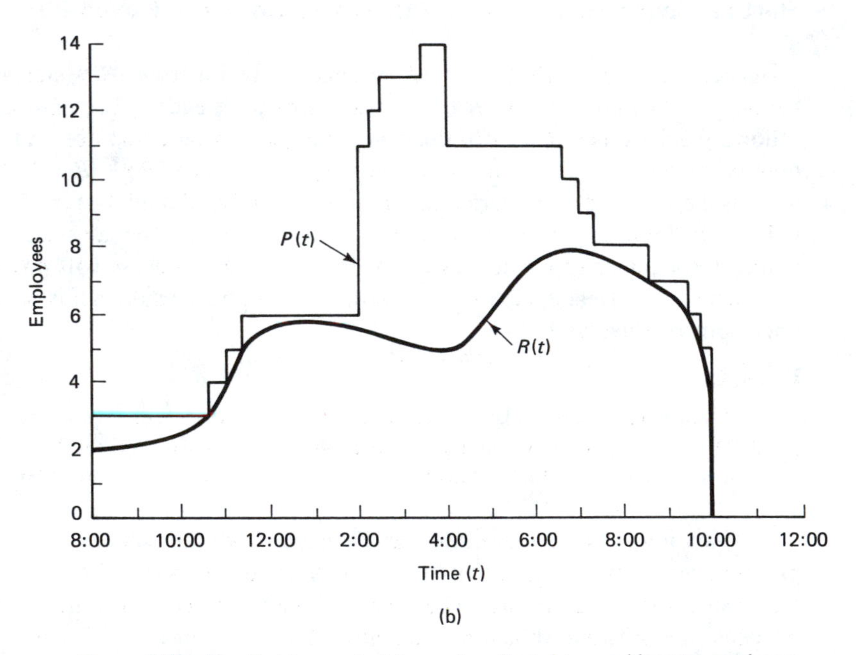

Figure 7.7b The first-hour schedule can be adjusted to provide extra employees during the middle of the day, when they can be diverted to ancillary tasks.

capacity to meet the peak demand can be expensive. In the example, one of the employees is needed for only an hour each day. Again, this points to the importance of using part-time servers to meet peak demand.

7.3.2 Days-Off Scheduling

In the United States, as well as in other countries, employees work five days per week. But most services are offered six or seven days a week. So even if the business is open just eight hours a day, no employee can be available all business hours. Furthermore, the demand for service usually varies throughout the week. Grocery and department stores typically have their busiest day on Saturday; firms that primarily cater to other businesses (for example, photocopy outlets) are busiest in the middle of the week.

Day-to-day variations can be accommodated through scheduling employee days off to coincide with low points in demand. The process might work like this. The first-hour principle is used to create workshifts and determine how many employees are needed on each day of the week. Based on the number of employees needed on each day, employees would then be assigned workshifts, with each employee receiving exactly five workshifts (except the last employee assigned, who might work less than five). So far so good—a schedule has been created. In fact, the schedule requires the minimum possible number of employees. But there is a problem. Some employees find that their days off are not consecutive—perhaps one is Sunday and the other is Wednesday. The employees are upset.

Tibrewala et al. (1972) developed an algorithm for scheduling days off that minimizes the number of employees, given that all employees' days off are consecutive. Although satisfying this constraint may mean hiring more employees, providing consecutive days off is usually essential to employee morale. The scheduling algorithm is iterative and schedules days-off pairs one employee at a time based on a daily requirements schedule. After each employee is scheduled, the daily requirements are adjusted and the process is repeated for the next employee.

The steps performed at each iteration follow:

1. Identify the day (or days) of the week with the smallest requirement.
2. Examine the days adjacent to the day (or days) with the smallest requirement. Among these days, select the one with the least requirement.
3. The day selected in step 2 along with its adjacent day from step 1 is the days-off pair.
4. Decrement the requirement by one on every day except the days off. Return to step 1 if a requirement still exists.

Ties can be broken in any manner. After the days-off pair is selected, the requirements on the other five days of the week are reduced by one, and the process is repeated for the next employee.

Example

A store has specified the following employee requirements:

M	T	W	T	F	S	Su
3	3	2	2	3	5	2

Iter. 1:
1. Wednesday, Thursday, and Sunday have the minimum requirement: 2.
2. Among the adjacent days, Thursday (adjacent to Wednesday) and Wednesday (adjacent to Thursday) have the minimum requirements.
3. Wednesday/Thursday is the days-off pair for employee 1.

After scheduling employee 1, the requirement curve becomes:

M	T	W	T	F	S	Su
2	2	2	2	2	4	1

Iter. 2:
1. Sunday has the minimum requirement.
2. Among the adjacent days, Monday has the minimum requirement.
3. Sunday/Monday is the days-off pair.

After assigning employee 2:

M	T	W	T	F	S	Su
2	1	1	1	1	3	1

Iter. 3: There are many possibilities here. Tuesday/Wednesday is selected.

M	T	W	T	F	S	Su
1	1	1	0	0	2	0

Iter. 4: Select Thursday/Friday (Note that this provides one more employee than needed on Sunday.)

M	T	W	T	F	S	Su
0	0	0	0	0	1	−1

Iter. 5: At this point, the only day with remaining requirement is Saturday. Sunday/Monday are selected as days off.

M	T	W	T	F	S	Su
0	−1	−1	−1	−1	0	−1

The complete schedule is summarized below:

Employee	1	2	3	4	5
Days Off	W/Tu	Su/M	Tu/W	Th/F	Su/M

Five employees were needed to meet a requirement for 20 days of work. Had there been no constraint on consecutive days off, one fewer employee would have been needed. Also, had the requirement on Saturday been reduced by one, one fewer employee would be needed.

As an alternative solution, one might have stopped the algorithm after iteration 3. At this point, no more full-time employees can be fully utilized. This might be the motivation for an alternative schedule in which three employees work full time, one employee works part-time four days a week (Monday, Tuesday, Wednesday, and Saturday), and a fifth just works on Saturday.

By no means is consecutive days off the only work constraint to consider. Employee contracts might specify that one of the days off must be Saturday or Sunday. Contracts might specify that workhours must be identical on consecutive days of the week. They might also specify that workshifts cannot change from week to week. These constraints can add considerable complications to the scheduling process and can, in some cases, lead to greater work force sizes.

7.3.3 Workshift Scheduling: Continuous Operation

Although most services have limited operating hours, some operate around the clock. Telephone directory assistance, police, and utility repairs are just a few examples. Scheduling employees for these services is complicated by the fact that there are no precise starting and ending times of each day. Although we have addressed the workhours and days-of-week questions sequentially, continuous operations might require them to be addressed simultaneously.

A reasonable heuristic for approaching the problem is first to identify the time of day with the least requirement and build up daily schedules from this point with the first-hour principle. The process can then be repeated for each subsequent day based on the solution for the previous day. But more elaborate methods may be called for.

Buffa et al. (1976) describe an operator scheduling system developed in 1973 for the General Telephone Company of California. The system accounts for nonstationarities in demand with respect to yearly patterns (with peak demand on Christmas Day and minimum demand in July), weekly patterns (with minimum demand on Saturday and Sunday), and daily patterns (with peak demand from 10:00 to 11:00 A.M. and minimum demand from 2:00 A.M. to 6:00 A.M.). The system begins with each day's demand forecast, which is found by adjusting the demand on the same day of the previous week to account for weekly growth in phone calls. The daily demand forecast is then converted into half-hour arrival rates, based on historical call distributions. Next, the arrival rates are translated into operator requirements, using the $M/M/m$ queueing model and set response time standards. Finally, workshifts are scheduled with a heuristic that attempts to minimize the absolute deviation between operator requirements and available operators, drawing from an approved set of workshifts.

7.3.4 Other Worker Scheduling Studies

Employee scheduling has been studied extensively in the industrial engineering and operations research literature. Some of the applications include scheduling: nurses in hospitals (Miller et al. 1976; Stimson and Stimson 1972; other studies are reviewed in Fries 1981); tellers in banks (Mabert and Raedels 1977; Matta et al. 1987); collectors at tollbooths (Brown 1984; Edie 1956); police cars (Kolesar et al. 1975); and telephone operators (Buffa et al. 1976; Sze 1984). The techniques developed in these papers can be extended to many other applications.

7.4 BULK SERVICE QUEUES

We now turn to a special category of queueing systems, ***bulk service queues***. In manufacturing, certain machines, such as stamping presses, require setups for each type of item produced. If items of the same type are produced in batches, then the service time per item declines. Bulk service also occurs in transportation: in freight transportation when groups of shipments are transported in the same vehicle, in mass transportation when groups of passengers are transported in the same vehicle, and in automobile transportation when groups of vehicles are served in the same traffic signal cycle. The fundamental characteristic of all of these examples is that the service process is faster, or cheaper, when similar customers are served in batches.

Bulk service effectively increases the capacity of the queueing system. But it also has a negative consequence. Consider how a bulk service process operates. First one type of customer is served, then a second, perhaps a third, and so on. Eventually, the server will return to the first customer type and the cycle might be repeated. But during the intervening time, when the second, third, and so on, customer types are served, none of the first type is served. The queue grows and does not shrink until the server returns. This cyclic, up-and-down, queue is a by-product of bulk service.

Setting up a bulk service system often amounts to balancing a trade-off between the benefit of increased capacity and the penalty of cyclic queues. This trade-off is elaborated on in the following sections.

7.4.1 A Bulk Service Queue with Two Arrival Streams

As a simple illustration of the trade-off, suppose that a queueing system serves two types of customers and that each type arrives at a constant rate (these may be different). The following analysis is readily extendable to multiple traffic streams.

Definitions 7.4

λ_1 = arrival rate of type 1 customers

λ_2 = arrival rate of type 2 customers

$\lambda = \lambda_1 + \lambda_2$ = total arrival rate

The service pattern can be represented by a two-phase cycle, one customer type being served per phase.

Definitions 7.5

T_1 = length of the service phase for customer type 1

T_2 = length of the service phase for customer type 2

T = cycle length

$= T_1 + T_2$

Each service phase has two parts. First, the server is set up for a customer type and, second, the individual customers are served. Suppose that the setup time and service rate are the same for both customer types.

Definitions 7.6

a = setup time

c = service capacity after the setup

Figure 7.8a shows cumulative departure and arrival curves for two customer types, one with a much larger arrival rate than the other. The waiting time within any cycle is the sum of the areas of two triangles, one for each customer type, as shown. The optimal values of the decision variables T_1 and T_2 are the values that minimize the average waiting time per customer. That is, T_1 and T_2 should minimize the waiting time per cycle *divided by the number of customers served per cycle*:

Minimize

$$W = \text{waiting time per customer}$$

$$= \frac{\text{Total wait per cycle, type 1} + \text{Total wait per cycle, type 2}}{\text{Customers served per cycle}}$$

$$= \frac{\dfrac{T_2 + a}{2}\left[\dfrac{(T_2 + a)\lambda_1 c}{c - \lambda_1}\right] + \dfrac{T_1 + a}{2}\left[\dfrac{(T_1 + a)\lambda_2 c}{c - \lambda_2}\right]}{[T_1 + T_2]\,[\lambda_1 + \lambda_2]} \tag{7.14}$$

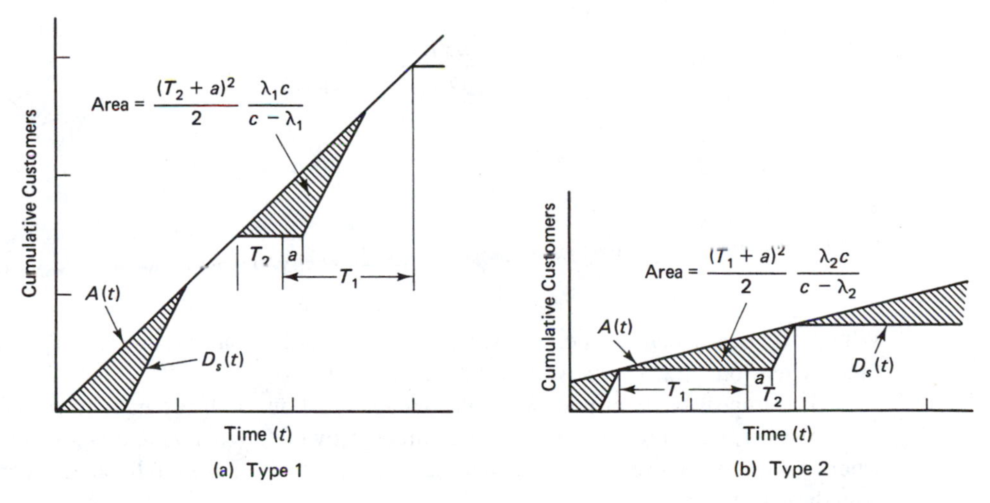

Figure 7.8 Cumulative arrival and departure diagrams for a bulk server. In part a, queue vanishes before end of service phase. In part b, service phase ends when queue vanishes.

The waiting time per customer must be minimized subject to the constraint that T_1 is sufficiently long to accommodate all the type 1 customers who arrive in a cycle and T_2 is sufficiently long to accommodate all of the type 2 customers who arrive in a cycle:

Subject to

$$(T_1 - a)c \geqslant T\lambda_1 \tag{7.15a}$$

$$(T_2 - a)c \geqslant T\lambda_2 \tag{7.15b}$$

Equations (7.14) and (7.15) can be simplified through substitution of the following symbols:

Definitions 7.7

r_1 = traffic intensity for type 1 customers
$= \lambda_1/c$

r_2 = traffic intensity for type 2 customers
$= \lambda_2/c$

r = combined intensity
$= r_1 + r_2$

$$K_1 = \frac{\lambda_1}{c - \lambda_1} = \frac{r_1}{1 - r_1}$$

$$K_2 = \frac{\lambda_2}{c - \lambda_2} = \frac{r_2}{1 - r_2}$$

which leads to the following:

$$\text{min:} \quad W = \frac{(T_1 + a)^2K_2 + (T_2 + a)^2K_1}{2\,(T_1 + T_2)r} \qquad r \leqslant 1 \tag{7.16}$$

$$\text{s.t.} \quad T_1 \geqslant \quad Tr_1 + a = \frac{a}{1 - r_1} + K_1T_2 \tag{7.17a}$$

$$T_2 \geqslant \quad Tr_2 + a = \frac{a}{1 - r_2} + K_2T_1 \tag{7.17b}$$

When r is greater than 1, then the server is oversaturated and the queue will grow over time. Therefore, Eq. (7.16) would not be applicable.

It is impossible to provide a simple general equation for the optimal values of T_1 and T_2. However, one important special case can be provided. When r_1 and r_2 are similar, *or* when r is large (close to 1), the optimal values of T_1 and T_2 fall on the constraint boundaries, and equal

r_1 and r_2 Similar or r Large

$$T_1^* = \frac{(1 - r_2 + r_1)}{1 - r} a \tag{7.18a}$$

$$T_2^* = \frac{(1 - r_1 + r_2)}{1 - r} a \tag{7.18b}$$

$$T^* = \frac{2}{1 - r} a \tag{7.18c}$$

Equation (7.18) specifies that each service phase ends exactly when the queue dissipates, as illustrated in Fig. 7.8b. In all cases, the cycle length, T^*, is directly proportional to the setup time, a, and inversely proportional to the excess capacity, $1 - r$. Either a large value of a or small value of $1 - r$ causes T^* to be large. T^* must be large in order to attain a service rate commensurate with the customer arrival rate. These factors also cause average waiting time to increase:

$$W^* = a \left[\frac{r - r_2^2 - r_1^2}{r(1 - r)} \right] \tag{7.19}$$

When r is very small, W^* is simply a, the setup time. As r becomes large, the excess capacity $(1 - r)$ approaches zero and W^* grows without bound.

Example

Two one-way streets cross at an intersection controlled by a traffic signal. The arrival rates on the streets are 1000 and 1500 vehicles/hour. The service capacity is the same on both streets, and equals 3000 vehicles/hour. The loss time (i.e., set-up time) at each signal change is 4 seconds. Find the optimal cycle and phase lengths.

Solution λ_1 and λ_2 are 1000 and 1500, and c is 3000, so r_1, r_2, and r are .333, .5, and .833, respectively. Substitution of these values in Eq. (7.18) results in $T^* = 48$ seconds, $T_1^* = 20$ seconds, and $T_2^* = 28$ seconds. From Eq. (7.19), $W^* = 13.6$ seconds.

Equation (7.19) points to another idea for reducing the waiting time (an idea exploited by Japanese manufacturers): Reduce the setup time. This has been accomplished through significant changes in the production technology. Setup times on large sheet metal stamping presses have been reduced from a matter of hours to a matter of minutes. Changes such as these allow items to be manufactured in smaller batches, greatly reducing waiting time and waiting cost.

If r_1 and r_2 are not similar *and* if r is not large, the expressions for T_1^* and T_2^* are more complicated. But Eq. (7.17) will always be an equality for the customer with the smallest arrival rate. That is, its service phase should end when the queue dissipates, as shown in Fig. 7.8b. The service phase for the customer type with the largest arrival rate may end *after* the queue dissipates (Fig. 7.8a), in effect providing priority to the customers with the highest arrival rate.

Figure 7.9 provides general results for two customer types and identical capacities. It illustrates the relationship between cycle length (expressed as T^*/a) and traffic intensity for different values of r_1/r_2 ($r_1 \geq r_2$). The figure shows that Eq. (7.18) applies to all values of r when $r_1/r_2 = 1$, to all values of r above .6 when $r_1/r_2 = 2$, and to all values of r above .9 when $r_1/r_2 = 4$. But no matter what this ratio is, cycle length increases at an increasing rate as r increases. Cycle length also increases as r_1/r_2 increases, because the service phase for the customer type with the larger intensity is extended beyond the time the queue dissipates (Fig. 7.8a). For larger values of r, when Eq. (7.21) applies, the cycle length is not affected by r_1/r_2; it is simply $2a/(1 - r)$.

Figure 7.10 provides a waiting time analysis, where waiting time is also expressed as a ratio to a, the setup time. All curves are similar in that the waiting time per customer increases at an increasing rate as r increases. Waiting time *decreases* as r_1/r_2 increases. It is better for all arrivals to be of a single type than to be evenly distributed among types.

7.4.2 General Principles for Bulk Service

There is no reason why the number of customer types should be limited to two, there may be three, four, even dozens of types. Such systems can be evaluated explicitly through the framework of Eq. (7.14). However, explicit analysis requires considerable effort. Moreover, the simple two-customer analysis illustrates most of the important principles:

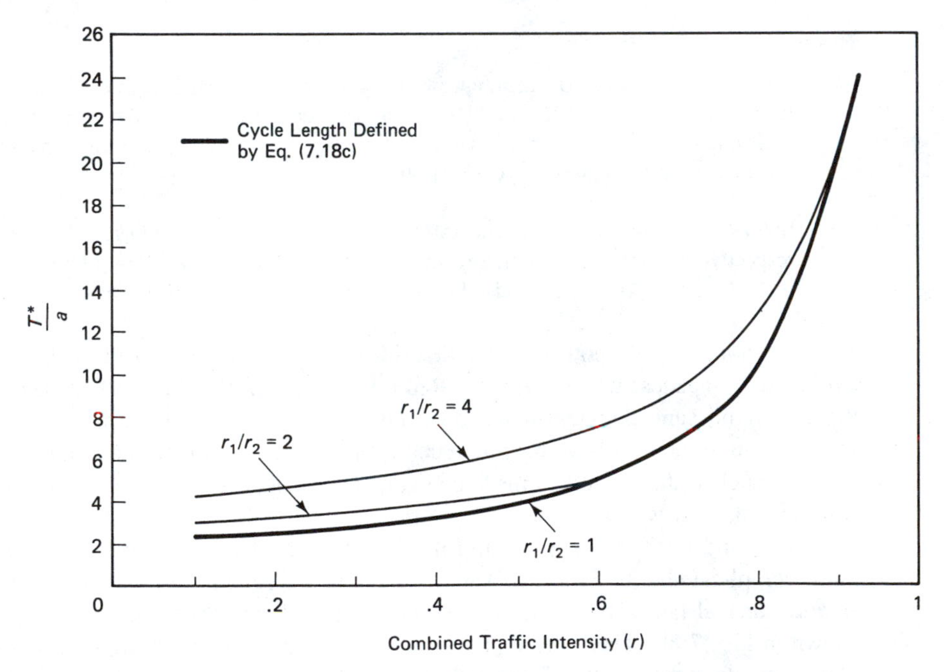

Figure 7.9 The optimal cycle length increases as the intensity increases. Cycle length is somewhat longer when the arrival rates of two customer types differ by a large amount.

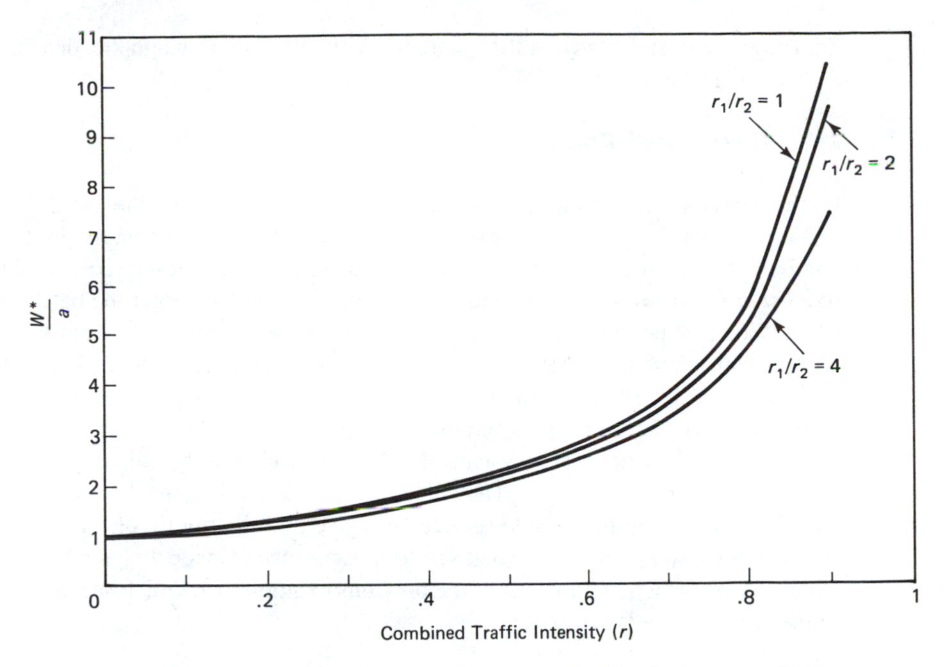

Figure 7.10 Waiting time per customer increases as the intensity increases. Waiting time per customer is somewhat smaller when the arrival rates of two customer types differ by a large amount.

1. A service phase should not end *until* all waiting customers of a given type are served.
2. If traffic intensities are similar, each service phase should end *exactly* when its queue vanishes.
3. If one of the customer types has a much larger intensity than the others, its service phase should be extended beyond the time its queue dissipates.
4. The length of a cycle should increase proportionately as the setup time increases and decrease as the excess capacity increases.

Finally, when r and a are sufficiently small, there is really no point to batching customers. Customers might as well be served on a first-come, first-served basis, without regard to type.

These principles are quite robust, and even apply to systems with time varying arrival and service rates. However, in such instances, the principles might be implemented in a different fashion. Instead of employing a set cycle length, the cycle length might vary depending on the demand. For instance, if the intensities are similar for all customer types, each phase should end as soon as the corresponding queue dissipates. If intensities are not similar, the phase for the customer(s) with the largest intensity could be extended a set proportion of time beyond the time the queue dissipates. These strategies are often used at isolated traffic signals, with the aid of magnetic detection devices embedded in the roadway.

7.4.3 Cost-Based Analysis

In the previous application, bulk service was motivated by the need for increased capacity. Another motivation for bulk service is that the service cost might depend on the number of customers served in a batch. In particular, the service cost might contain a fixed ***setup cost***, which is independent of the batch size. The larger the batch, the lower is the service cost per customer. This phenomenon is most prevalent in transportation. The cost of dispatching a bus between point a and point b is nearly independent of the number of passengers carried. The cost of dispatching a truck between point a and point b is nearly independent of the number of shipments carried.

A fixed cost per service means that bulk service may be called for, *even when all customers are the same type*. The following two sections consider two aspects of this question: the timing of a single service per cycle and the timing of multiple services. A simplifying assumption will be that services occur instantaneously (as when a bus departs from a stop), which contrasts with the previous examples, in which service occurs over a finite phase.

7.4.3.1 Single Service Time. Suppose that the customer arrival pattern is cyclic and that $\Lambda(t)$ is the expected number of customers to arrive by time t in any cycle (with time $t = 0$ being the start of the cycle). Because of the fixed cost per service, exactly one service occurs per cycle, at which time all of the waiting customers are served *simultaneously*. What is the best time to schedule this service? Oliver and Samuel (1967) examined a question of this type. In their paper, a cycle represented a day, $\Lambda(t)$ represented the expected number of letters received by time t at a postal box, and a service represented a mail pickup.

The answer can be found through a simple marginal analysis. With the objective of minimizing waiting time and with the number of services limited to one per cycle, no cost parameters are needed. Let

$$T = \text{length of the cycle}$$

$$\tau = \text{time that service occurs}$$

If τ is delayed by a small amount, $\Delta\tau$, two things will happen. First, all of the customers that arrived before time τ in the current cycle or after time τ in the previous cycle will be delayed by the time $\Delta\tau$. Second, all of the customers that arrived between time τ and time $\tau + \Delta\tau$ will be served a time T earlier (the length of one cycle), because they will not have to wait until the next cycle to be served. The net change in waiting time, ΔW, is then

$$\begin{aligned} \Delta W &= \text{time lost for customers arriving before } \tau - \text{time saved} \\ &\quad\ \text{for customers arriving between } \tau \text{ and } \tau + \Delta\tau \\ &= \Lambda(T)\Delta\tau - [\lambda(\tau)\Delta\tau]T \end{aligned} \tag{7.20}$$

A necessary condition for optimality is that W does not decrease if τ is changed by any small amount. Assuming that $\lambda(t)$ is a continuous function:

Optimality Criterion: Dispatch Time

$$\lambda(\tau^*) = \frac{\Lambda(T)}{T} = \text{average arrival rate over cycle} \tag{7.21}$$

The rule is indeed simple. The service should occur when customers are arriving at the average rate. However, because the arrival rate will vary over the cycle, Eq. (7.21) may be satisfied for several values of τ, some of which are local minima (occurring when $\lambda(t)$ is *declining*) and others of which are local maxima (occurring when $\lambda(t)$ is *increasing*). The minimum will ordinarily occur toward the end of a cycle, after the arrival rate has peaked (the maximum will occur before the arrival rate has peaked).

Example

Figure 7.11 shows $\Lambda(t)$ for a mail collection box. $\lambda(t)$ satisfies Eq. (7.24) at two times: 1:00 P.M. and 8:30 P.M. Because the arrival rate is still increasing, the first time is a local maximum (the worst time to schedule a dispatch). Because the arrival rate has passed its peak, the second time is a local minimum. Because it is the only local minimum, it must also be the global optimum.

The important thing to recognize is that the time of the service must balance the increased delay to customers that arrive before τ against the decreased delay to customers

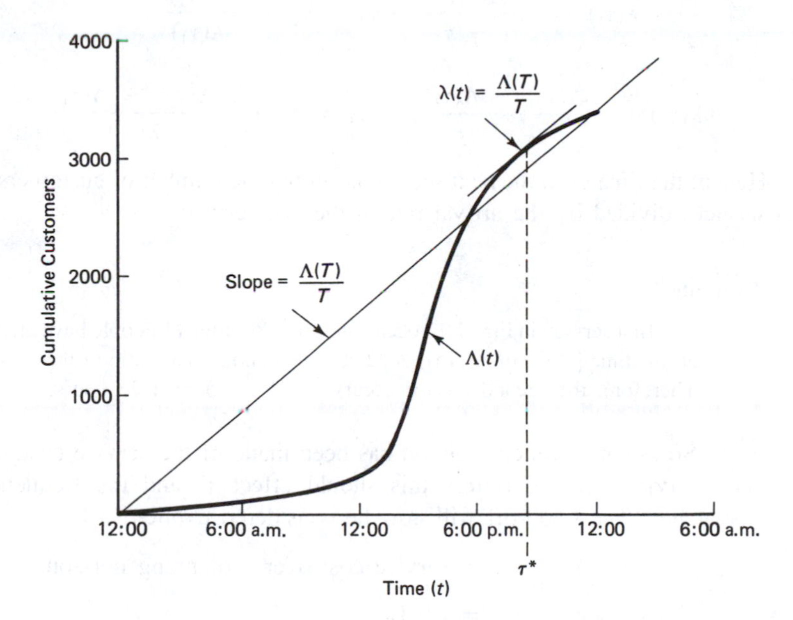

Figure 7.11 The optimal time for a single service within a cycle is when the arrival rate equals the average arrival rate over the cycle.

that arrive immediately after time τ. A similar trade-off applies to multiple services, as explained in the following section.

7.4.3.2 Multiple Services. Suppose now that the arrival pattern is defined by the general function $\Lambda(t)$, which may or may not be cyclic. Let τ_i be the time of the ith service, at which time all waiting customers are served simultaneously. The following optimization of τ_i relies on work by Newell (1971) and Bisbee et al. (1968).

From marginal analysis, the optimal time for any intermediate service can be derived from the times of the surrounding services. Without showing the derivation

Optimality Criterion: Service Time *i*

A. First service: $$\lambda(\tau_1) = \frac{\Lambda(\tau_1)}{\tau_2 - \tau_1} \tag{7.22a}$$

B. Second, third, . . . service: $$\lambda(\tau_i) = \frac{\Lambda(\tau_i) - \Lambda(\tau_{i-1})}{\tau_{i+1} - \tau_i} \tag{7.22b}$$

In either case A or B, a service occurs when the number of waiting customers, divided by the time until the *next* service, equals the arrival rate. These equations can be rewritten in a manner that allows all of the service times to be solved sequentially from any trial value of τ_1. For example

$$\lambda(\tau_1) = \frac{\Lambda(\tau_1)}{\tau_2 - \tau_1} \rightarrow \tau_2 = \tau_1 + \frac{\Lambda(\tau_1)}{\lambda(\tau_1)} \tag{7.23a}$$

$$\lambda(\tau_2) = \frac{\Lambda(\tau_2) - \Lambda(\tau_1)}{\tau_3 - \tau_2} \rightarrow \tau_3 = \tau_2 + \frac{\Lambda(\tau_2) - \Lambda(\tau_1)}{\lambda(\tau_2)} \tag{7.23b}$$

Hence, the time until the next service is simply the number of customers served at the last service, divided by the arrival rate at the last service.

Example

The first service in Fig. 7.12 occurs at time .75, after 11 people have arrived. The arrival rate at this time [slope of $\Lambda(.75)$] is 22 customers/hour. The ratio of these two values is .5 hour. Therefore, the second service occurs at .75 + .5 = 1.25 hours.

So far, no explicit mention has been made of the service cost, the motivation for bulk service here. Certainly this should affect τ_1 and the frequency of subsequent dispatches. The trade-off will now be considered explicitly. Let

$$\begin{aligned} C_s &= \text{service cost over a planning horizon} \\ &= F \cdot I \end{aligned} \tag{7.24}$$

where

F = fixed cost per service

I = number of services over a planning horizon (which could either be a single cycle or consecutive cycles)

As before, suppose that waiting cost is linearly related to the total waiting time:

$$C_w = \beta \int_0^T [A(t) - D_s(t)]dt \tag{7.25}$$

where T is the ending time of a planning horizon. Then the dispatch times $\tau_1, \tau_2, \ldots, \tau_I$ should be selected in a way that minimizes the sum of Eqs. (7.24) and (7.25). The optimal solution will be considered in two parts: first, for the special case where the arrival rate is constant, and second, for a more general arrival rate.

7.4.3.3 Dispatch Frequency, Constant Arrival Rate. Suppose that the customer arrival rate is constant; hence, $\Lambda(t) = \lambda t$. Then Eq. (7.22b) would reduce to

$$\lambda = \frac{\lambda(\tau_i - \tau_{i-1})}{\tau_{i+1} - \tau_i} \rightarrow \tau_{i+1} - \tau_i = \tau_i - \tau_{i-1} \tag{7.26}$$

To no surprise, when the arrival rate is constant, the interval between dispatches should also be constant. So, instead of minimizing cost over the entire planning horizon, one can equivalently minimize the cost *per customer* over a single cycle, where a cycle is the time between two services, of length τ.

$$\min_{\tau} \frac{\beta\left[\frac{\tau}{2}\right]\tau\lambda + F}{\tau\lambda} = \frac{\beta\tau}{2} + \frac{F}{\tau\lambda} \tag{7.27}$$

The first cost component, the waiting cost (β times the area of a right triangle bounded by τ and $\tau\lambda$), increases as τ (the decision variable) increases: The longer the time between services, the longer customers must wait. The second cost component, the service cost, decreases as τ increases: The longer the time between services, the less frequently the service cost is incurred. The minimum is found by setting the derivative of Eq. (7.27) with respect to the decision variable, τ, to zero. This yields the following:

$$\tau^* = \sqrt{\frac{2F}{\lambda\beta}} \tag{7.28a}$$

$$C^* = (C_s + C_w)^* = \sqrt{2F\lambda\beta} \tag{7.28b}$$

Equation (7.28) is a variation of the *Wilson economic-order-quantity equation*, a formula dating to the early twentieth century. The model was first applied to inventory planning with respect to balancing a trade-off between ordering costs and inventory costs. But the result is equally applicable to queues. Inventory costs are analogous to waiting costs, and ordering costs are analogous to service costs.

7.4.3.4 General Arrival Pattern. Newell provided a simple adaptation of Eq. (7.28a) that gives an approximately optimal solution when the arrival rate varies slowly over time:

$$(\tau_{i+1} - \tau_i)^* \approx \sqrt{\frac{2F}{\lambda(\tau_i)\beta}} \qquad i = 1, 2, \ldots \tag{7.29}$$

Substitution of Eq. (7.22b) allows Eq. (7.29) to be rewritten as follows:

$$\beta \left[\frac{[\Lambda(\tau_i^*) - \Lambda(\tau_{i-1}^*)]^2}{2 \cdot \lambda(\tau_i)} \right] = F \tag{7.30}$$

As shown in Fig. 7.12, the quantity in brackets is the area of a triangle approximately equal in size to the waiting time for customers who arrive between τ_{i-1} and τ_i. With this in mind, you can interpret Eq. (7.30) as stating that the waiting cost should equal the

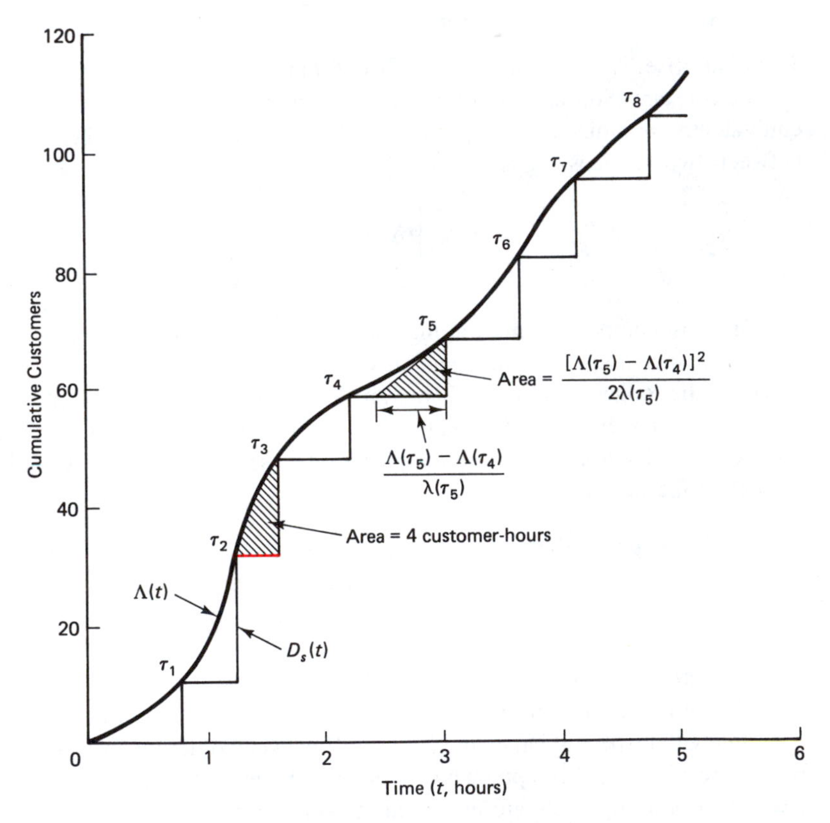

Figure 7.12 With multiple bulk services, the total waiting time per cycle should be constant.

dispatching cost. Substitution of the actual area for the approximate area leads to the following alternative scheduling rule:

Optimality Criterion: Dispatch Interval

$$\beta \cdot \text{total waiting time over interval} = F$$

$$\beta \int_{\tau_{i-1}}^{\tau_i} \Lambda(t) - \Lambda(\tau_{i-1})dt = F \tag{7.31}$$

Example

A bus company would like to determine a daily dispatching schedule, based on expected passenger arrivals. The cost of dispatching a bus is $40, and the value of a passenger's time is $10 per hour. From Eq. (7.31), the optimal waiting time per cycle, F/β, is 4 customer-hours. The solution is shown for a hypothetical curve $A(t)$ in Fig. 7.12.

The service strategy can be implemented in many different ways. The simplest is to create a schedule specifying exact service times (a bus schedule is one example). As an alternative, a dynamic strategy might be followed. Instead of initiating services at set times, services might be initiated as soon as the queue size reaches set quantities. The advantage of this latter strategy is that it is more responsive to random fluctuations in arrival rate. On the other hand, it makes service planning more difficult. If the exact service times are not known in advance, extra servers might have to be kept in reserve until needed. These two considerations must be balanced in selecting the best implementation.

7.4.4 Further Generalizations

Section 7.4.2 examined a bulk service queue with multiple customer types and Sec. 7.4.3 examined a bulk service queue with a single customer type and a fixed cost per service. Many applications are hybrids of these two models: multiple customer types *and* fixed cost per service. In the operations research literature, these hybrids are referred to as *joint replenishment* or *multiproduct lot scheduling* problems (see Doll and Whybark, 1973, and Kaspi and Rosenblatt, 1983, for example).

The simplest strategy is to follow a "rotation cycle" in which each customer type is served exactly once per cycle. In the case where service is noninstantaneous (as in Sec. 7.4.1), the optimal length of a rotation cycle is approximated by

$$T^* = \max\left\{ \sqrt{\frac{2\Sigma F_i}{\beta\Sigma\lambda_i(1 - r_i)}}, \ \frac{\Sigma a_i}{1 - \Sigma r_i} \right\}, \quad \Sigma r_i \leq 1 \tag{7.32}$$

where

F_i = setup cost for customer type i

λ_i = arrival rate for customer type i

$$r_i = \lambda_i/c$$

$$a_i = \text{setup time for customer type } i$$

The cycle length is defined by a trade-off between setup costs and waiting costs, as accounted for in the first part of Eq. (7.32). The second part of the equation is invoked if the first part provides insufficient capacity to serve all demand. T^* is then increased to provide the minimum capacity needed. Though a rotation cycle is simple, it is not necessarily best. If the arrival rates differ greatly between customers, a better strategy would be to allow some customers to be served every other cycle, others every third cycle, and so on. This issue is studied in the references cited earlier.

Another consideration in bulk service is that the number of customers served in any batch might be constrained by a machine capacity. This is particularly important in transportation, where the batch size cannot exceed the vehicle size. In such cases, $\tau_{i+1} - \tau_i$ would be constrained so that $\Lambda(\tau_{i+1}) - \Lambda(\tau_i)$ is less than or equal to the vehicle size. Consequently, the spacing between services must be reduced below the specified values, and the waiting cost will fall *below* the dispatching cost.

These are but two of many possible modifications to the bulk service model. Yet, in all cases, bulk service is motivated by the same two factors: need to obtain greater service capacity and/or need to reduce fixed setup costs. In the former case, batches should be sufficiently large to provide the needed capacity. In the latter case, batch sizes should optimize a trade-off between waiting costs and fixed service costs.

7.5 CHAPTER SUMMARY

Queues are not merely objects of study. Queues are real problems demanding real solutions. The nature of the solution depends on the type of queue. *Perpetual* queues can be eliminated by reducing the average arrival rate or increasing the service capacity. *Predictable* queues can be eliminated by decreasing the arrival rate or increasing the service capacity, during the duration of the queue. *Stochastic* queues can be eliminated by decreasing the variation in service time and interarrival time or through involving the ancillary activity policy discussed in Sec. 5.7.

No matter what the type of queue, the first place to look for a solution is in the existing servers. Inefficiency is costly to the queueing operator and the customer alike. The queueing operator should attempt to

Ensure that servers devote 100 percent of their time to customers when customers are in the system.

Increase server efficiency.

Automate the service process.

Increase customer efficiency.

Shift some service responsibility to the customer.

Serve customers simultaneously.

Serve similar customers in batches, when there are setup costs.

All of these are either no-cost, low-cost, or money-saving ideas. And even small increases in capacity can provide substantial reductions in waiting time. One should not overlook small changes.

Predictable queues can be eliminated through adding service capacity during the time of the queue. In *high contact* systems, such as medical clinics, waits tend to be short, rarely lasting more than an hour. Short duration queues can be eliminated by hiring part-time employees, scheduling breaks and maintenance for periods of low demand, staggering workshifts, and scheduling days off when demand is low. In *low contact* systems, such as manufacturing plants, queues tend to be long, lasting days, weeks, or months. These can be eliminated by hiring temporary full-time employees or leasing equipment, scheduling vacations and major maintenance for periods of low demand, and paying employees to work overtime during peak demand.

Ideally, these strategies should result in a service capacity that matches the variations in customer arrival rate—never is there excess capacity; never is there deficient capacity. Failing this, increases in capacity should be provided sufficiently early that the queue vanishes at least once before the supplemental capacity is removed. Adding capacity too late may cause the queue to grow beyond control. Server breaks, on the other hand, should occur as *late* in the queue as possible. Any break will delay all subsequent customers, until the queue dissipates. The later the break is added, the fewer customers that are affected.

This chapter was written within the context of eliminating queues. But this is not the only kind of queueing problem. Another potential problem is that too many servers are idle—either because scheduling is poor or because there are just too many servers. Even in this situation, the techniques presented in this chapter can be used to create more effective schedules. On the other hand, having too many servers is not as common because queue operators are usually more aware of their own costs than of their customers' costs: They tend to underserve rather than overserve their customers. Certainly, if servers are idle most or all of the time their number should be reduced. And whether or not queueing is a problem, everything possible should be done to maximize server efficiency.

FURTHER READING

AMERICAN BAR ASSOCIATION. 1984. *Attacking Litigation Costs and Delay.*

BISBEE, E. F., J. W. DEVANNEY, D. E. WARD, R. J. VON SAAL, and S. KURODA. 1968. "Dispatching Policies for Controlled Systems," *Fourth International Symposium on the Theory of Traffic Flow*, ed. W. Leutzbach and P. Baron. Bonn, West Germany: Herausgegeben vom Bundisminister für Verkehr.

BOWEN, D. E. 1986. "Managing Customers as Human Resources in Service Organizations," *Human Resources Management*, 25, 371–383.

BROWN, J. 1984. *Management and Analysis of Service Operations.* New York: North-Holland Elsevier.

BROWNE, J. J. 1979. "Simplified Scheduling of Routine Work Hours and Days Off," *Industrial Engineering*, 11, 27–29.

Buffa, E. S., M. J. Cosgrove, and B. J. Luce. 1976. "An Integrated Work Shift Scheduling System," *Decision Sciences*, 7, 620–630.

Chase, R. B. 1978. "Where Does the Customer Fit in a Service Operation?" *Harvard Business Review*, 56, 137–142.

Chaudry, M. L., and J. G. C. Templeton. 1983. *A First Course in Bulk Queues*. New York: John Wiley.

Church, T. W., J. Q. Lee, T. Tan, A. Carlson, and V. McConnell. 1978. *Pretrial Delay, A Review and Bibliography*. Publication R0036. Williamsburg, Va.: National Center for State Courts.

Curson, C., ed. 1986. *Flexible Patterns of Work*. London: Institute of Personnel Management.

Doll, C. L., and D. C. Whybark. 1973. "An Iterative Procedure for the Single-Machine Multi-Product Lot Scheduling Problem," *Management Science*, 20, 778–784.

Ebener, P. A., 1981. *Court Efforts to Reduce Pretrial Delay, A National Inventory*, R-2732-ICJ. Santa Monica, Calif.: Rand Institute for Civil Justice.

Edie, L. C. 1956. "Traffic Delays at Toll Booths," *Operations Research*, 2, 107–138.

Fries, B. E. 1981. *Applications of Operations Research to Health Delivery Systems*. Lecture Notes in Medical Informatics. Berlin: Springer-Verlag.

Hall, R. W. 1986. "Graphical Techniques for Manpower Planning," *International Journal of Production Research*, 24, 1267–1282.

Kaspi, M., and M. J. Rosenblatt. 1983. "An Improvement of Silver's Algorithm for the Joint Replenishment Problem," *IIE Transactions*, 15, 264–267.

Kolesar, P. 1984. "Stalking the Endangered CAT: A Queueing Analysis of Congestion at Automatic Teller Machines," *Interfaces*, 14, 16–26.

———, K. L. Rider, T. B. Crabill, and W. Walker. 1975. "A Queueing-Linear Programming Approach to Scheduling Police Patrol Cars," *Operations Research*, 23, 1045–1062.

Larson, R. C. 1972. *Urban Police Patrol Analysis*. Cambridge, Mass.: MIT Press.

Linder, R. W. 1969. "The Development of Manpower and Facilities Planning Methods for Airline Telephone Reservations Offices," *Operational Research Quarterly*, 20, 3–21.

Lovelock, C. H., and R. F. Young. 1979. "Look to Consumers to Increase Productivity," *Harvard Business Review*, 57, 168–178.

Mabert, V. A., and A. R. Raedels. 1977. "The Detail Scheduling of a Part-Time Work Force: A Case Study of Teller Staffing," *Decision Sciences*, 8, 109–120.

Matta, K. F., J. M. Daschbach, and B. N. Wood. 1987. "A Non-Queueing Model to Predict Teller Requirements in Retail Bank Branches," *Omega*, 15, 31–42.

Miller, H. E., W. P. Pierskalla, and G. J. Rath. 1976. "Nurse Scheduling Using Mathematical Programming," *Operations Research*, 24, 850–870.

Morse, P. M. 1958. *Queues, Inventories and Maintenance*. New York: John Wiley.

Newell, G. F. 1971. "Dispatching Policies for a Transportation Route," *Transportation Science*, 5, 91–105.

Oliver, R. M., and A. Samuel. 1967. "Reducing Letter Delays in Post Offices," *Operations Research*, 10, 839–892.

Sasser, W. Earl. 1976. "Match Supply and Demand in Service Industries," *Harvard Business Review*, 54, 133–140.

Stimson, R. H., and D. H. Stimson. 1972. "Operations Research and the Nurse Staffing Problem," *Hospital Administration*, 17, 61–69.

SZE, D. Y. 1984. "A Queueing Model for Telephone Operator Staffing," *Operations Research*, 32, 229–249.

TIBREWALA, R. K., D. PHILIPPE, and J. J. BROWNE. 1972. "Optimal Scheduling of Two Consecutive Idle Periods," *Management Science*, 19, 71–75.

ZEMKE, R. 1986. "Contact! Training Employees to Meet the Public," *Training*, August, pp. 41–45.

PROBLEMS

1. Describe two examples of each of the following queueing systems:
   (a) Perpetual queue
   (b) Predictable queue
   (c) Stochastic queue

2. Select an example of a low contact system with which you are familiar. Give examples of how to increase the service rate by
   (a) Automation
   (b) Reducing customer portion of service time
   (c) Shifting work to the customer
   (d) Serving customers simultaneously
   (e) Serving customers in batches

3. For the system that you selected in Prob. 2, describe how you might supplement capacity during times of predictable queues.

4. Repeat Probs. 2 and 3 for a high contact system.

***5.** The state highway department would like to determine how many lanes to provide on one of its highway segments. Each lane can serve 1800 vehicles per hour, and the cost of building each lane is $10 million. Amortized, the cost amounts to $6000/day/lane. The agency believes that vehicle time should be valued at $10/hour. The agency also believes that the only way to reduce waiting time is to build highway lanes.
   (a) Assuming that Fig. 6.14 is the cumulative arrival curve for the morning rush hour, that this is the only period of concern, and that the number of lanes can be approximated by a continuous variable, determine the optimal duration of the queueing period.
   (b) Round off your solution to part a to obtain an integer number of lanes. Along with the cumulative arrival curve, draw the cumulative departure curve on a piece of graph paper.
   (c) Estimate the total cost of your solution (per day).
   (d) Can you think of a less expensive way to serve vehicles? Describe an alternative to building extra lanes.

***6.** Suppose that the highway department has reassessed the value of time to $6/hour. Repeat Prob. 5 with this new number.

7. The arrival curve in Prob. 5 has changed to the one shown in Fig. 7.13. Repeat part a of Prob. 5. (Hint: Because there are two queueing periods, Eq. 7.5 cannot be directly applied.)

8. A national fast-food franchise has hired five extra employees to work 2 hours during lunchtime. With the extra employees, customers can be served at the rate of 500/hour; otherwise, the rate is 350/hour. Copy the cumulative arrival curve in Fig. 6.14 on a piece of

*Difficult problem

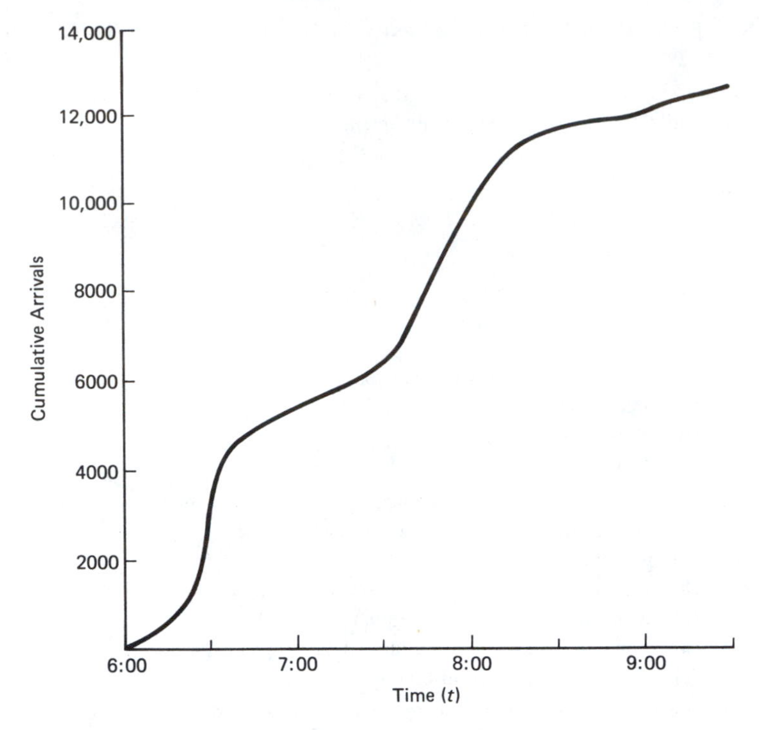

Figure 7.13 Example cumulative arrival curve.

graph paper, rescale the horizontal axis from 11:00 A.M. to 2:30 P.M., and rescale the vertical axis to 1200 total customers (service time is short).

(a) Using an "early start," determine when the extra employees should begin their shift, and draw the cumulative departure curve.

(b) In words, describe how the following changes would affect the firm's hiring policy:

i. Value of customer time increases 30 percent.

ii. Employee wages drop 10 percent.

iii. An increase in customer arrival rate of 20 percent at all times.

iv. Employee productivity drops by 15 percent.

9. Figure 7.14 shows the arrival rate of customers at a supermarket checkout counter. The owner would like to provide a very high level of service by ensuring that the number of employees equals or exceeds $\rho(t) + \sqrt{\rho(t)}$ at all times. Employees can serve 20 customers/hour, and work full time, 8 hours per day.

(a) Draw the labor requirement curve, $R(t)$, on a piece of graph paper.

(b) Following the first-hour principle, draw the optimal $P(t)$ (employees provided) curve.

(c) How many employees will the supermarket need?

(d) Discuss how you could adjust your solution to allow for worker breaks.

10. Repeat Prob. 9 with lower service standard. That is, ensure that the number of employees equals or exceeds $\rho(t) + .2\sqrt{\rho(t)}$. What are the disadvantages and advantages of this solution over your solution to Prob. 9?

11. The engineering library would like to have the following number of employees working on each day of the week:

Day	M	Tu	W	Th	F	Sa	Su
Employees	3	4	4	2	7	5	5

(a) Assuming that all employees must work five days a week, with consecutive days off, develop a minimum cost work schedule. How many employees must be hired?

(b) The employees have come to the owner with an offer. They will agree to nonconsecutive days off if the pay is increased from $5/hour to $5.50/hour. Should the owner accept the offer?

(c) The employees have returned with a further offer. They will also agree to work less than five days per week, if necessary. With this additional concession, should the owner increase the wage?

12. A stamping press is used to produce two types of parts. The first part is used at the rate of 60/hour and the second is used at the rate of 90/hour. After setup, the machine produces at the rate of 225/hour. The setup time is 30 minutes (setup cost is zero).

(a) Calculate, exactly, the length of the production run for each part. (Hint: Examine Fig. 7.9 to see whether Eq. (7.18) applies.)

(b) What is the average number of finished parts in inventory, given a constant demand rate.

(c) In an effort to achieve just in time, the firm has reduced the setup time to 10 minutes. How will this change your answers to parts a and b?

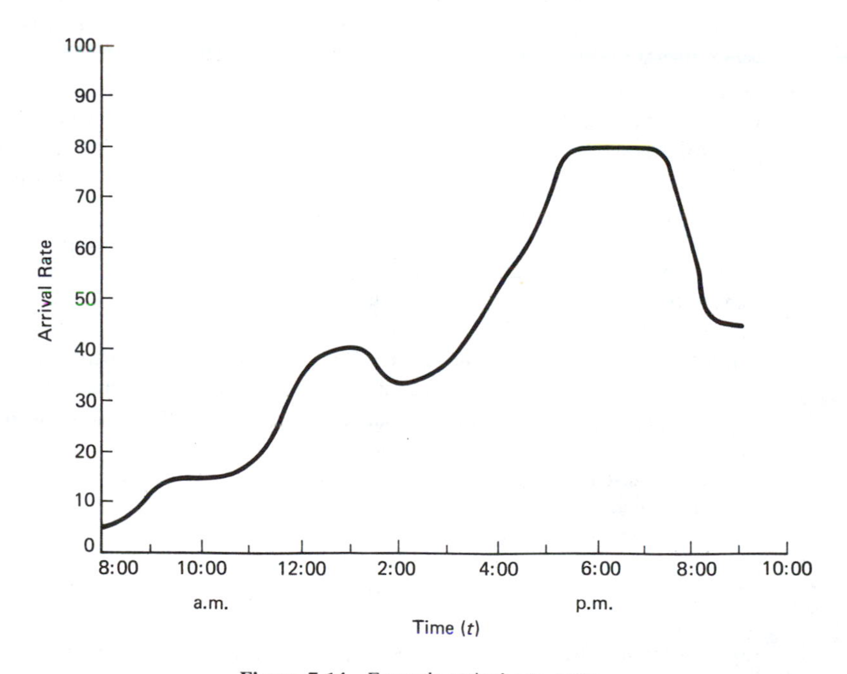

Figure 7.14 Example arrival rate curve.

*13. Traffic has been found to arrive at constant rates at an isolated signalized intersection, in quantities shown below:

	Vehicles/15 minutes	
	Street 1	**Street 2**
7:00–7:15	250	60
7:15–7:30	400	100
7:30–7:45	600	150
7:45–8:00	350	90

The street can serve 3300 vehicles/hour, in each direction. The setup time is 5 seconds.

(a) For each period, determine the optimal cycle length (derive from Fig. 7.9).

(b) From your answer to part a, calculate the optimal lengths of the green phases in each direction for the third period.

(c) How much space should be provided on street 1 to accommodate the maximum queue size?

14. A clerical worker at a health insurance company processes four types of forms. The forms are received at the rates of 24, 26, 28, and 33/day. Forms are processed at the rate of 120/day. However, each time the employee shifts from one form to another, a setup time of 1/8 day is incurred. Following the guidelines in Sec. 7.4.2, select the cycle length that minimizes waiting time.

15. Each time a grocery store places an order for a brand of corn flakes, it incurs a fixed cost of $10. Corn flakes are sold at the constant rate of 100 boxes per day, and travel time is one week. Inventory holding cost amounts to $.25/carton per week, with 24 boxes per carton. Determine the optimal order quantity and the optimal cost per week.

16. Suppose that demand for cartons in Prob. 16 follows the pattern in Fig. 7.13, except that the vertical axis is rescaled downward by a factor of 100 and the horizontal axis is rescaled from 1-hour intervals to 2-week intervals. Using the approximate solution technique in Sec. 7.4.3.4, determine when orders should be received. Draw the curves representing cumulative shipments and cumulative receipts.

***17.** An integrated automobile manufacturer produces components at one plant for assembly in another. Parts are produced and used at the constant rate of 1000 per day. The cost of transporting a shipment is $200, travel time is one day, and inventory cost is $.05 per part per day.

(a) Suppose that the order quantity is 5000 parts. Draw the following curves for a 15-day period: cumulative part production at the component plant, cumulative part shipments from the component plant, cumulative part receipts at the assembly plant, and cumulative part usage at the assembly plant.

(b) Calculate the inventory cost and transportation cost per day.

(c) Determine the optimal order quantity and optimal cost per day (this will involve a slight modification of Eq. (7.28)).
2

(d) How might your solution be modified if plants operated 16 hours per day, 5 days per week?

(e) How could you incorporate a vehicle capacity into your solution to part c?

*Difficult problem

18. A chemicals plant is used to refine four grades of oil. Each time a new grade is produced, the equipment must be purged, at a cost of $200 and a time of 1 hour. The equipment can process 1000 gallons per hour, and operates 24 hours/day. Inventory holding cost amounts to $.01/gallon per day. Demand rates are shown below:

A: 100 gal/hour B: 50 gal/hour C: 75 gal/hour D: 150 gal/hour

(a) Following a rotation cycle, what should the cycle length be?

(b) Suppose that grade B is only produced every other cycle. How would this change Eq. (7.32) and change the cycle length?

(c) Do you believe that your solution to part b is better than your solution to part a? Explain.

CASESTUDY

THE NEED FOR SPEED

Brian Quinton and Susie Stephenson

Call it the First Law of Foodservice Acceleration: The faster food moves, the more an operator will sell.

Today's breakfast and lunch customers have more options and less time than ever before. The two-paycheck home means that either partner, or both, may spend the noon hour running errands. A drive for greater business productivity has made lunch-time clock watching a career survival tactic. And when they do fit lunch into their schedules, people undoubtedly have more places to get it. In addition to restaurants and company cafeterias, options include phone-in delivery, street vendors, supermarket takeout and microwaveable meals at home.

All foodservice segments, especially those dependent on heavy breakfast or lunch traffic, must find ways to stay top-of-mind with the dashing diner. One way to do this is to tailor the meal occasion to give him what he needs most—time.

"Breakfast and lunch are becoming convenience-driven meals," says Robert Goldin, a senior principal with Technomics Inc. "Consumers have little time available, especially in quick-service and midscale segments, and they don't want to spend it waiting in lines."

Speed is of the essence, but it is not a cure-all. Quick service is only one part of a value equation in the customer's mind; food quality, consistency, service and price also play a part in the choice—and probably a larger part. Even the fastest feeder has to recognize that simply pouring on the speed will not lift traffic counts very high for very long if the customer does not think the meal is good value for the money.

The Fast-Food Tradition

Some type of quick foodservice has been around since the first stagecoach travelers straggled into the inn with a 15-minute layover between horses. Lunch wagons, cafeterias, automats and drive-ins have pared the extraneous elements of foodservice down to the bare minimum for customers in a hurry. Ray Kroc, stopwatch in hand, and the legions who followed him, brought the tradition of speed down to the present day.

But something has happened to fast food: It stopped being quite so fast. In an attempt to appeal to a broad customer base, chicken chains began offering spareribs, burger places sold fish and desserts, and everyone tried breakfast. The new rollouts gave their vendors a higher profile against the competitive background. But they also began to drag down order execution times.

Occasionally the new products simply took too long to cook. Wendy's dropped breakfast omelets because customers had to wait several minutes to get them.

When the legendary Nathan's Famous hot dog chain was bought in 1985, the new owners reduced by half the 150-item menu the stores were asked to carry. "These were items the Coney Island store had been known for—frogs' legs, crab sandwich, clams on the half shell," says Operations Director Bob Varma. "But we eliminated all but two of the seafood items and concentrated on hot dogs, french fries and drinks, which made up 70% of our sales anyway." The result was greatly increased counter speed and no complaints about the restricted menu.

Among the major players, McDonald's appears to enjoy a speed advantage in the mind of consumers. This reputation is partly a legacy of Ray Kroc's stopwatch and concern for time-motion efficiency, but it also stems from the company's effort to keep menu growth within easily managed bounds.

Burger King and Wendy's won years of market visibility with their offer to "personalize" their hamburgers on request. Now, however, "having it your way" may be less important than having it the quickest way, and analysts think both chains would like to downplay the customized-burger option. One reason Wendy's rolled out its Big Classic burger, the chain's first sandwich with set condiments, was to cut service time.

Combination meals also build speed into an overabundant menu. Usually available only at the pick-up window, where time is really at a premium, a typical meal consists of a regular-sized sandwich, fries and drink for a slightly reduced price. More importantly, the combination meal helps shift customer demand to certain basic menu items. This lets the kitchen crew prepackage or at least prep more efficiently.

Still, the conflict between quicker service and wider choice threatens to get worse, if anything, now that fastfeeders are taking food bars from dinner houses, adding promotional items through their menus for a matter of weeks, selling tiny burgers and chicken nuggets and testing ice cream and dessert items. These strategies may make good marketing sense, but they also add to the average time spent in line, particularly at peak hours. "The food is great," said one Wendy's SuperBar customer recently. "But you can get pretty frustrated waiting in a long line for nothing but a plastic plate."

Wendy's is also the last major chain to use the single-point counter service line, with one cashier rather than four or five. Most chains use multiple lines with banks of registers. Burger King switched to this configuration in 1984; McDonald's has used it since it began franchising.

Denny Lynch, vice president of corporate communications at Wendy's, maintains that the single-line system can actually assemble an order faster. "We only make two orders at a time—one at the pick-up window and one in the dining room," he says. "McDonald's makes 50 at once. And though you may say it takes longer to make the sandwich, we say we can get it made in the time it takes one person to go around to four stations and get a whole order put together. We've proven that we can get fast. We can beat the clock."

Accelerating the Drive-Thru

Whatever its standing in the counter-service sweepstakes, Wendy's is arguably doing the most right now to accelerate the drive-thru. That is appropriate, since Wendy's helped boost the popularity of drive-thrus by putting a window on its first free-standing unit in 1971. Today, one quarter of the stores use headsets for the drive-thru kitchen crew. When the customer gives his order at the outdoor order station, it is transmitted instantly to both the drive-thru cashier and the crew. Getting the order directly instead of via the cashier saves the crew 10 or 15 seconds and improves accuracy. Some units also use digital timers to make drive-thru crew members aware of how often they meet the 30-second deadline.

Wendy's is now testing several refinements to its drive-thru systems. At peak hours, some units replace the conventional outside order stations with employees with headsets moving along the line of cars. When these workers take an order, they rearrange it into Wendy's standard "sandwich, fries, drinks" format before transmitting it to the pick-up kitchen, saving a few seconds. And a dozen units are using manned outdoor kiosks as drive-thru order stations. This system has streamlined the drive-thru process so much that the kiosks are specifically signed and marketed as "Express Windows." Some use 30-second timers and are experimenting with a half-minute meal guarantee.

The perception that once-fast food is slowing down opened a window of opportunity for a slew of double drive-thru chains offering limited menus, low prices and speed. But the Second Law of Foodservice Acceleration says that nothing is ever "fast enough," and there are signs that a new generation of double drive-thrus has found it necessary to pick up an already swift pace.

For example, the Lexington, Ky.-based Grand Junction chain is offering a "Meal-in-a-Minute" guarantee to customers at its 15 double drive-thru units. If a diner cannot get a quarter-pound burger, regular fries and a drink for $2.39 in 60 seconds, the order is free. To fulfill that promise, says President Burt Spinks, Grand Junction made some crucial design changes in its prototype. The new design establishes a totally separate makeup area for the walk-up window. By adding five feet to the width of the building, it also enlarges the central drive-thru work space. This eases congestion and keeps the cashier and order taker out of the production flow at peak times.

The new speed program has meant putting one more person on duty in the store at all times. Grand Junction also found it necessary to trim even its limited menu a bit; a popular grilled chicken sandwich was cut because it could not be prepped quickly. But even without it, the one-minute guarantee produced a 64% sales increase in its first quarter. Average service time under the program is under 50 seconds, Spinks says; giveaways amount to 6% of sales.

"This kind of speed is something that our industry has to get back to," Spinks says. "We just can't go around calling ourselves fast food and not be fast."

Full-Service Speed Takes Training

If fast food had to regain its speed, full-service restaurants never had any to begin with. But competitive pressures and the need for a share of the quick-lunch market have forced them to find it. This has meant sharpening service techniques, rethinking systems and inventing new menu items—all to remove the "veto power" of the customer in a hurry.

Six years ago, for example, customers expected to wait 20 to 30 minutes for a fresh pizza, an impression that cut most full-service pizza outlets out of the lunch rush. Pizza Hut had tried an all-you-can-eat pizza-and-salad smorgasbord, but had been disappointed by its profit potential.

The answer was rolled out in 1983: the Personal Pan Pizza, a thick-crust pie that cooked in eight minutes instead of 20. Suddenly Pizza Hut could promise customers a pie five minutes after the order reached the kitchen. Within a year, same-store lunch sales went from 7% of revenue to 30%, even 40% in some units.

Numerous complex and costly factors had to fall into place to make this quick-lunch program possible. A new recipe had to be developed for a pizza small enough to suit one individual, one that would make for only minimal waste if it should be held too long. The company made a substantial investment in conveyorized ovens; these allowed the kitchen crew to start cooking slightly ahead of the 11 o'clock start and then just keep feeding in pies at specified intervals as the supply diminished. The personal pizzas could be held for up to five minutes after leaving the oven, so no one got a pie more than 13 minutes old. And since people are just as important as machines in the full-service sector, Pizza Hut hired 20,000 new servers.

Quick-lunch traffic has fallen off a bit since Pizza Hut stopped promoting it heavily; the typical Pizza Hut's lunch business has dropped back to 25% to 28% of revenue. But the program will remain.

The Third Law of Foodservice Acceleration: Speed is contagious. If one player in a segment gets faster at lunch, most of the others will try to keep up. When Pizza Hut began building its lunch business, other full-service chains went full tilt for speed, too.

Bennigan's decided to break out of its bar-with-food niche. The casual dining chain now promises to serve diners anything on the 100-item menu in 15 minutes or it is free—anytime during business hours. Many other full-service chains are now making similar, if more restricted, offers. Stuart Anderson's Fast-Track Lunch promises 10 minutes from order to table or the meal is free. Perkins Family Restaurants hands out "Oops!" coupons for a free comparable meal if the food takes longer than 10 minutes. Friendly's gives away ice cream sundaes if it misses the 5-minute deadline. That is also the time Marriott Hotel coffee shops will take to get breakfast to their customers.

"No Secret" to Speed

"There's no big secret to speed, no breakthrough restaurant technology," says Dick Monroe of Red Lobster. "You just have to have your fundamentals down. You can't forget the cocktail forks or the salad dishes. In my opinion, if you have to spend a lot of money to get into an express-meal program, you'd better look at your whole operation. You're probably doing something wrong anyway."

All these guarantees employ a canny marketing strategy. They impress consumers with the restaurant's confidence in its ability to deliver. At the same time, they maneuver many diners into secretly betting *against* the restaurant and being more tolerant of any delays. Finally, as Mike Brandt of Stuart Anderson's points out, "It's good advertising to have someone go back to the office and say they just won a free lunch. Once in a great while, of course."

But while they're rushing the food, full-service restaurants have to take pains not to rush the customers. Some people in this world are not under time pressures, even at lunch, and servers have to be sensitive to cues that indicate the pace at which a diner wants his meal to proceed.

Chains that respond most successfully to their customers' needs will be rewarded with market share taken from their competition. When Stuart Anderson began advertising the Fast-Track Lunch, lunch receipts jumped 42%; Red Lobster raised its noontime sales by 20%. Those figures mean better cost-efficiency, which has become increasingly important in this time of single-digit chain expansion. "The price of real estate, labor and basic operation mean that you can't make it with just one daypart anymore," says analyst Richard Pyle of Alex, Brown & Sons. "Pizza Hut going after lunch and McDonald's doing breakfast are just ways of trying to maximize existing assets."

Many Students, Little Time

Institutional foodservice is faced with that same necessity, thanks to rising costs, declining subsidies and increased competition from other meal options. If they are more fortunate than many commercial operations, it is because they often have long experience in serving people quickly.

Speed is probably most critical in school foodservice. Although changes in serving methods, menu items and service times have pretty much alleviated the long lines that once irritated institutional customers, school foodservice directors still face the challenge of feeding thousands of children in lunch periods that can be as short as 25 minutes.

Directors have confronted the problem by implementing a variety of new programs and procedures. "Organizing a line well is critical to speed," says Betty Bender, foodservice director at the Dayton, Ohio, public schools. "We've found that students won't stand in a line for more than five to eight minutes. Any longer than that and they'll go somewhere else or skip the meal. We also try to supply them with information before they actually get to a line. For example, students should, when possible, know what's on the menu and where they must go to get it."

To ensure this, Bender recommends that foodservice directors and dietitians post menus prior to service, announce menus over public-address systems and make sure the different service areas are well marked. "Students should know as soon as they walk into a cafeteria where to go to pick up a special, if that's what they want," she says. "If that means posting a big sign over that area, then that's what you do. You shouldn't have customers wandering around a dining area. It's distracting to everyone and is a frustrating experience for the student." Placing vending machines away from serving lines helps to keep down congestion. Salad bars and other self-serve items also minimize the bunching up of students in particular spots.

But as in commercial quick-service restaurants, there is a gap between theory and practice. Thelma Becker found, to her surprise, that lines at the Souderton, Pa., public schools moved *faster* if students made their lunch choices when they got to the cafeteria lines. Before implementation of the new choice system, students made their selections early in the morning—without benefit of looking at the food. According to Becker, students seem to make their lunch selections more quickly when the actual food is before them.

But, as is so often the case, even the most sophisticated equipment and well-planned serving areas are of little use in improving efficiency if employees are not well trained. Simply, they must know how to operate equipment.

Employees should also know how to make use of their entire bodies. "It's amazing how few of us effectively use both our hands," says another school foodservice director. "There's absolutely no reason line workers can't serve with their left *and* right hands, but very few do."

Schools are also cutting down on lines by serving food in locations other than cafeterias. In the Miami schools, for instance, students can purchase meals from outside sites and window units; many schools have also picked up on the mobile carts/kiosks concept initiated in San Diego. The rationale is that if students refuse to come into the cafeteria—because of long lines or some other reason—the school will take the food to them.

Other Institutional Segments

The other institutional segments—colleges, universities and employee feeding (both hospital and business and industry)—generally lack the same mealtime pressure as school foodservice directors. Many of the following examples can easily be adapted to the needs of any institutional segment.

"A student's schedule is much less restricted than when I went to school," comments Michigan State's Ted Smith. "Today, kids have much more control over when they want to be in and out of classes. They're also much less likely to eat at a specific time. I can remember when at 12 noon, a bell would ring and we'd have every kid on campus lined up outside a dining area and wanting to eat right now." The same is true of customers in hospitals and business and industry accounts.

"We just don't have a cattle call the way we used to," says a business and industry director in the Midwest. "Yes, the majority of people are still prone to come down to lunch at 12 noon, but many more people float in here at different times of the day."

Offering seconds has helped to speed lines, too. According to a college foodservice director from the South, students used to mull over their decision because they knew once they left the serving area, they would not be allowed back in. Now that they know they can have seconds, they move more quickly. Placing side dishes on separate plates has also made the agony of decision quicker and more bearable for the chooser and those behind him.

Biology has determined that cafeteria lines that flow from left to right move more quickly than those structured to flow the other way. Studies have shown that most right-handed people (and there are a lot more righties than lefties) will almost always take items with their right hands. If the line is going to be left, they will push the tray with the right hand, too. If it is moving to the right, they will push the tray with their left hand—making use of both left and right, and moving down the line more quickly.

Commercial quick-service has given institutional speed a boost by getting customers used to self-busing. Institutional customers expect to clear their own trays now, which makes for cleaner and more attractive dining. Customers do not have to wait for a busperson to clear and wipe tables before they sit down.

Texas A&M University has met the need for speed in an impressive way. At the College Station campus, Foodservice Director Lloyd Smith and recently retired director Fred Dollar helped design and implement a new feeding system for the 2,200 cadets attending the school. The renovated Duncan Dining Center will enable all 2,200 cadets to be fed an entree, salad, dessert and drink in under 14 minutes.

The logistics are truly awesome. The cadets march up the quadrangle outside the dining area in formation. The doors to Duncan are opened and the cadets enter in formation—16 across. They pick up their trays and move to soda stations—126 heads in all powered by four 5-horsepower compressors (each module has eight carbonators and four transfer pumps).

Then they proceed to the entree stations, each with an illuminated overhead menu that tells exactly what is offered at that station. If even one item is different, the meal is plated and served at a different station. For instance, eggs, bacon and *hashbrowns* might be served at stations one and two. Eggs, bacon and *grits* are served at stations three and four.

Each station is serviced by two adjacent lines (the front of the serving station in these two lines is the same color). The corridor between the two stations is a different color. When a cadet is served, he turns toward the empty area to walk back out to the main dining area without running into anyone else on line.

"One of the beauties of this system," Smith continues, "is that each segment of the meal is finished in its entirety before the next begins. So the people who serve move over to help in the dishroom because when it's time for the cadets to bus their trays, they've all already been served their food."

Most commercial operators can only envy Texas A&M, its new equipment and its customers' deportment. But as an example of a well thought-out system, the college shows that speed of service depends at least as much on finding the logical way to operate as it does on technological bells and whistles.

And, like Smith's operation, commercial restaurants that want to succeed at speed must be fully committed to satisfying more than just the desire for a quick bite: They must deliver quick quality.

"People have so many options open to them that you have to offer good food fast, with good service and at a good price all at once," says Pizza Hut's Jenkins. "The restaurants who can do that all at once and do it consistently are going to benefit the most. The rest are going to have to settle for taking less of the market."

Source: B. Quinton and S. Stephenson, "The Need for Speed," *Restaurants and Institutions*, Sept. 30, 1988, pp. 40–49. Reprinted by permission of the publisher.

CASESTUDY

AN INTEGRATED WORK SHIFT SCHEDULING SYSTEM

Elwood S. Buffa, UCLA
Michael J. Cosgrove, General Telephone and Electronics Service Corporation
Bill J. Luce, General Telephone Company of California

An integrated work shift scheduling system is developed and applied in the scheduling of 2600 telephone operators in 43 locations of the General Telephone Company of California. The system involves the forecasting of calls on a half-hourly basis, the conversion to operator requirements, the scheduling of tours by a heuristic algorithm, the assignment of operators to tours, and the operation of the system.

One of the general problems which has emerged from the study of service and nonmanufacturing systems is work shift scheduling. The problem is essentially one of scheduling manpower to meet some variable demand pattern which is usually short-term, such as daily and/or weekly, but which also may be seasonal. The reason that the problem emerges in the context of service and nonmanufacturing systems is that they are characterized by variable short-term demand which cannot be backlogged and by an output which cannot be inventoried. This paper presents an integrated system of demand forecasting, conversion to operator requirements, shift scheduling, and operator assignment which is operating at the General Telephone Company of California.

The range of applications of work shift scheduling in the literature is an indicator of the importance of the problem. For applications to nurse scheduling see [1] [3] [21]. Other applications discussed in the literature are police protection, [7], airline reservation offices [16], supermarkets and retail stores [29], public transportation [5], bank tellers [20], the post office [25], and the telephone industry [8] [10] [11] [12] [17] [18] [19] [28]. A similar problem exists in scheduling manpower for continuous seven day manufacturing operations of employees who work only a forty hour week [4] [27] [29].

Demand for Service

The service offered is the telephone exchange by which operators are assigned to provide directory assistance, coin telephone customer dialing, and toll call assistance. The standard for service is supplied by the Public Utilities Commission in unusually specific terms: service must be provided at a resource level such that an incoming call can be answered within ten seconds, 89 percent of the time. The difficulty in implementing the response standard is in the severe demand variability of incoming calls.

Figures 1, 2, 3, and 4 show typical call variation during the year, the week, the day, and within a peak hour. Figure 1 shows the annual variation, highlighting the two sharp peaks which include Mother's Day and Christmas Day. The data indicate calls during the busiest hour in each of the 52 weeks. The minimum occurred in the 28th week (3200 calls) and the maximum during Christmas (4400 calls). The peak to valley ratio is 1.38. To accommodate seasonal variation, the company must provide about 38 percent more capacity at Christmas time than in the 28th week. Generally the summer months involve lighter loads than the rest of the year.

Figure 2 shows daily call load for January 1972. A pronounced weekly pattern emerges, showing the Saturday and Sunday call load to be only about 55 percent of the typical load for Monday through Friday. Although the telephone company offers lower weekend toll call rates to help smooth the load, the resultant weekly variation is still very large.

Figure 3 shows the half-hourly variation for a typical 24 hour period. Peak call volume is in the 10:30–11:00 A.M. period (2560 calls), and the minimum occurs at 5:30

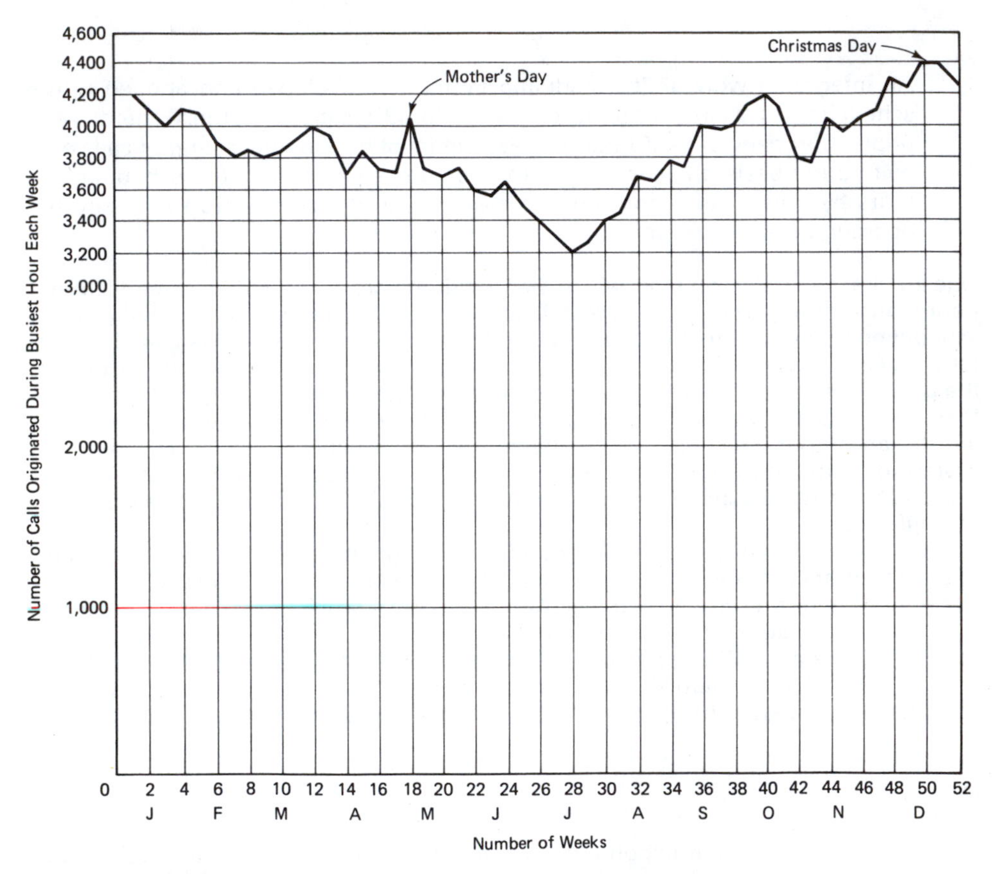

Figure 1 Typical distribution of calls during the busiest hour for each week during a year.

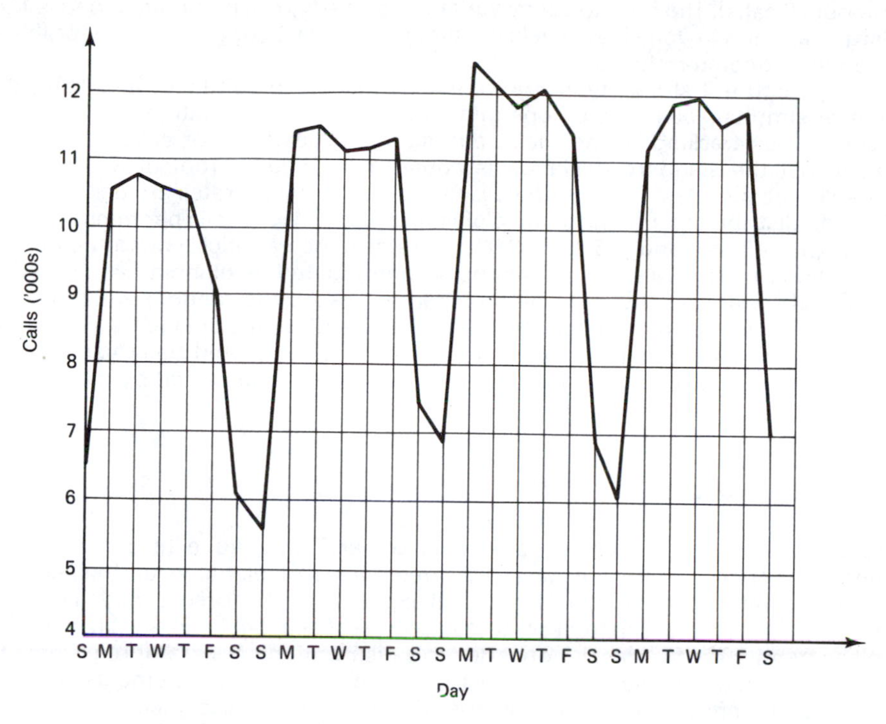

Figure 2 Daily call load for Long Beach, January 1972.

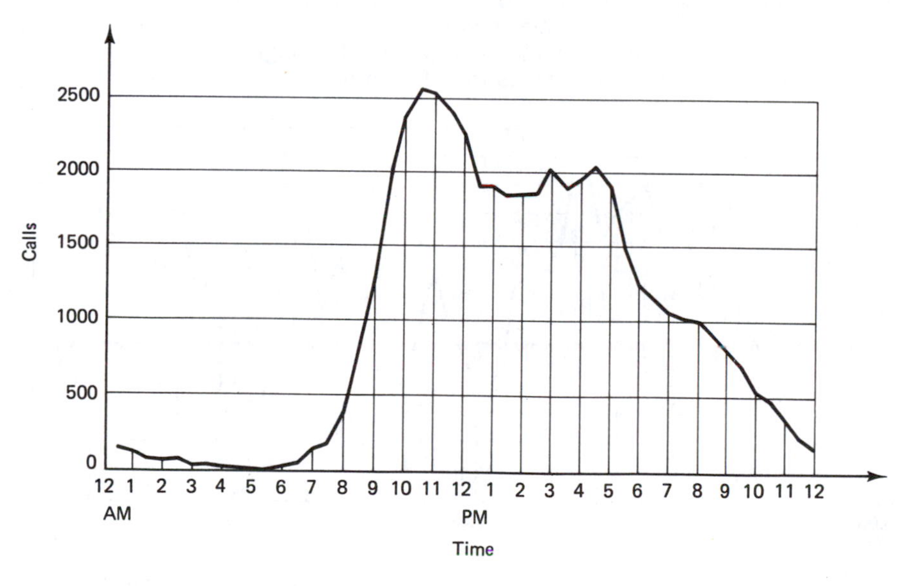

Figure 3 Typical half-hourly call distribution (Bundy D A).

A.M. (about 20 calls). The peak to valley variation for the typical half-hourly load is 128. The large variation in daily load levels apparent in Figure 3 suggests a daily problem of scheduling operator shifts.

Finally, Figure 4 shows typical intra-hour variation in call load, indicating the number of simultaneous calls by one-minute intervals. This variation is random. In fact, continuous tracking of the mean and standard deviation of calls per minute indicates that the standard deviation is equal to the square root of the mean, a reasonable, practical test for randomness, and that the arrival rates are described by the Poisson distribution. For the sample of Figure 4, $\bar{X} = 15.75$ calls per minute; $SD = 4.85$ calls per minute; and $\sqrt{15.75} = 3.99$. Therefore, the variation within the hour is taken as random. This variation cannot be accommodated by planning and scheduling, thus sufficient capacity must be provided to absorb the random variations.

The overall situation evidenced by the typical distributions of Figures 1, 2, 3, and 4 is that a forecastable pattern exists for seasonal, weekly, and daily variation. In addition, the call rate at any selected minute is adequately described by the Poisson distribution.

The Integrated System

Figure 5 indicates the system developed at General Telephone to accommodate demand for service. There are basically three cycles of planning and scheduling which involve information feedback of actual experience. The forecast of daily calls is the heart of the system. The forecast takes account of seasonal and weekly variation as well as trends. The forecast is converted to a distribution of operator requirements by half-hour increments. Based on the distribution of operator requirements, a schedule of tours or shifts is developed, and specific operators are assigned to tours. This sequence of modules is entirely computerized as indicated in Figure 5.

Besides the operator schedule, there are two additional cycles which operate on a normal basis. First, a schedule for "today" may be impacted by unintended events such as operator illness, an emergency increase in call load, and so on. Supervisors in local installations use the "Intraday Management Cycle" (Figure 5) to cope with such unintended events. In addition, there is the "Monthly Future Force Cycle" in which

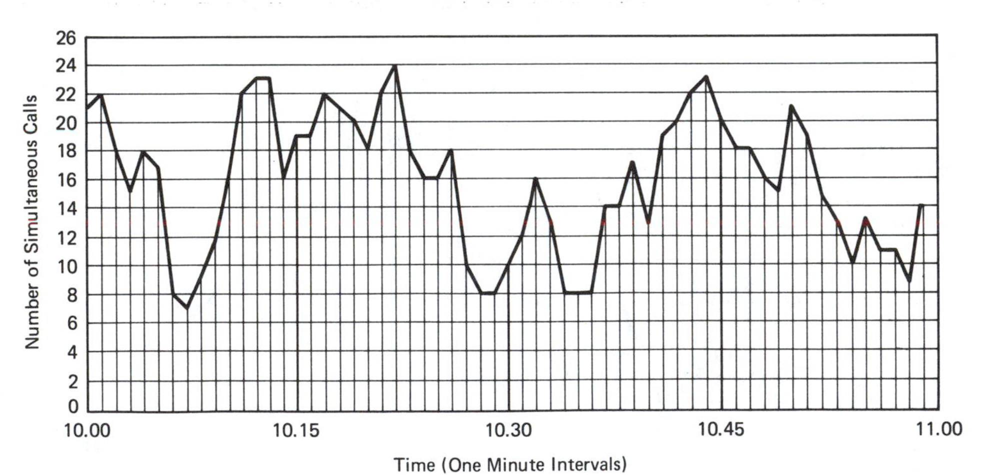

Figure 4 Typical intrahour distribution of calls, 10:00–11:00 A.M.

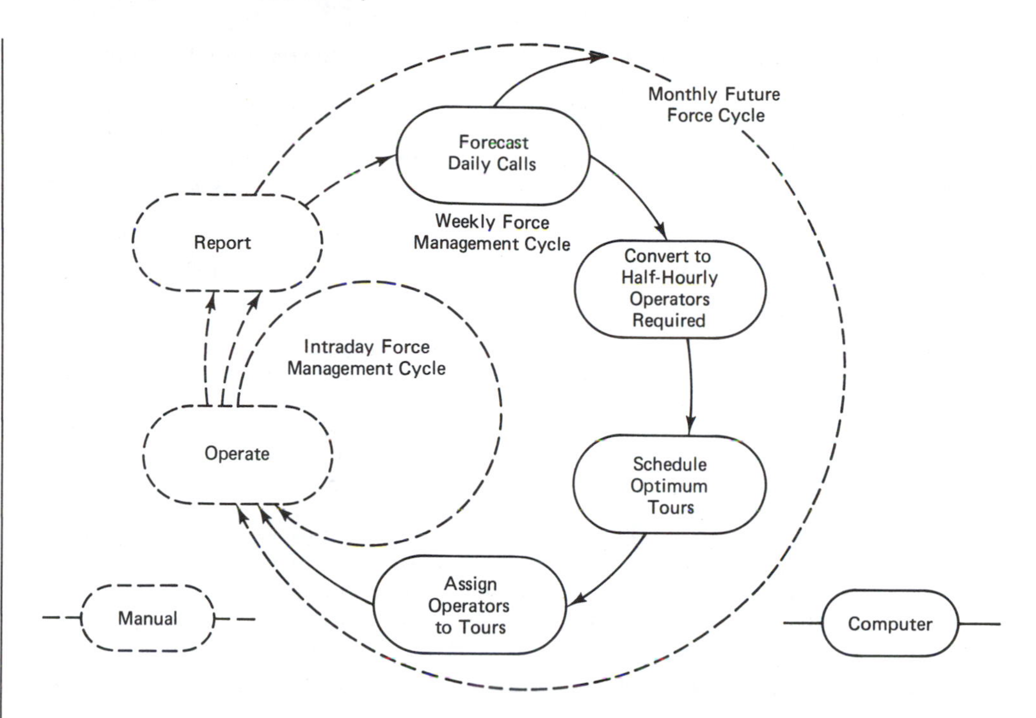

Figure 5 "Force Management System"

management can make higher level adjustments based on reports of actual operations or based on forecasts involving particular trend and seasonal factors. Hiring and training of operators is planned in the future cycle which projects up to 12 months forward [18].

Forecasting Demand

The demand forecasting system is based on a Box-Jenkins [6] model. The following major terms are used in forecasting the number of calls at a specific location for next Monday.

Calls next Monday = Calls last Monday
+ Weekly growth at this time last year (Monday_{-52} − Monday_{-53})
− error last week × θ
− error 52 weeks ago × ϕ
+ error 53 weeks ago × θ × ϕ.

θ is a non-seasonal moving average parameter, and ϕ is a seasonal moving average parameter.

In terms of actual operation, the computer inputs are last week's calls by day and type of service (toll, assistance, directory service), coefficients (work units per call) for the forecast week by day and type of service, and the board load (productivity) by day for the forecast week. The computer outputs are forecasts of daily calls up to five weeks in advance and a translation of the forecast into required operator hours by day up to five weeks in advance.

Figure 6 shows a typical record of comparison between forecast versus observed numbers of calls for Santa Monica. The uncanny forecast for day 151 is Thanksgiving when people are predictably more interested in dinner and family affairs than in communication. Average error for the forecasting system as a whole is 3.5 percent.

Conversion to Half-hourly Operator Requirements

The total day operator hours must be distributed to the individual half hours of the work day to provide a required operators' profile for the shift scheduling stage. The daily distribution of operator requirements is developed by exponential smoothing of current and historical percent of total day requirements by half hours. These requirements are derived from actual counts of operators staffed each half hour and achieved speed of answer (response time). The M-M-C queueing model is used to adjust the actual operators staffed to reflect staffing required to meet the constraint of the response time standard.

The conversion process from estimate of calls to operators required is shown in Figure 7. The parameters which are required in the process are average call duration (coefficient) based on studies of actual times, operator efficiency (work units per hour), and the response time standard. The final product is a profile of operators

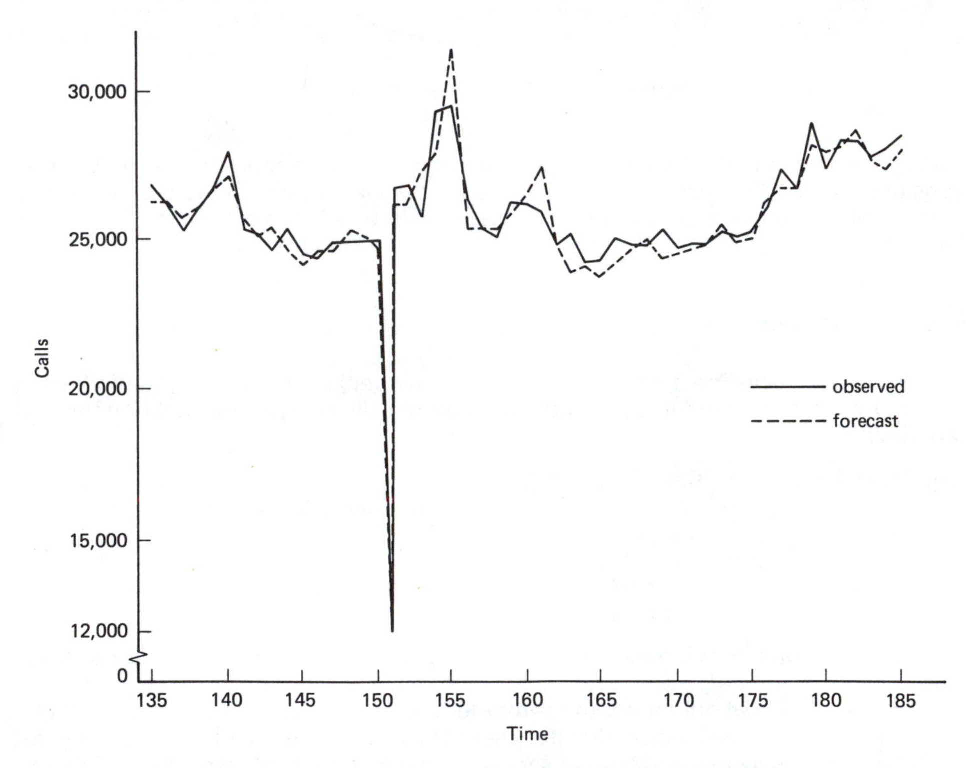

Figure 6 Sample of forecast versus observed numbers of calls at Santa Monica.

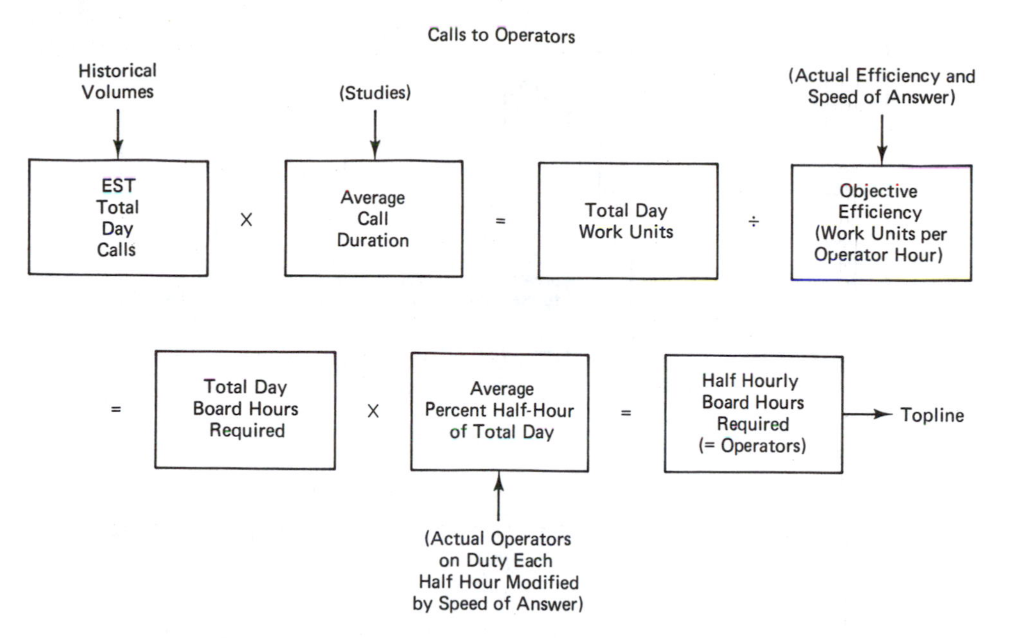

Figure 7 Model for conversion of calls to half-hourly operator requirements (topline).

required for each half hour of the day to process the call-load within the specified response time constraint.

Scheduling of Shifts

The graphical representation of an operator profile is shown in Figure 8. The problem in assigning tours or shifts is one of fitting in shifts so that they aggregate to the operator profile, as is also shown in Figure 8.

In order to be able to build up shifts so that they aggregate to the topline profile, flexibility is required in the types of shifts. Flexibility is provided by the shift length and the positioning of lunch hours and rest pauses. The set of shifts is constrained by state and federal laws, union agreements, company policy, and practical considerations. Shifts in the set are actually selected based on California State restrictions, company policy, and local management input concerning the desirability of working hours by their employees.

Each shift consists of two working sessions separated by a rest pause which may be the lunch period. Each working session requires a 15 minute rest pause near the middle of the session. The admissible shift set is limited by the following constraints.

1. Shifts are 8, 7, or 6.5 hours.
2. Work sessions are in the range of 3 to 5 hours.
3. Lunch periods are either a half-hour or an hour.
4. Split work periods are in the range of 3.5 to 5 hours (split work periods are separated by substantial non-work periods).

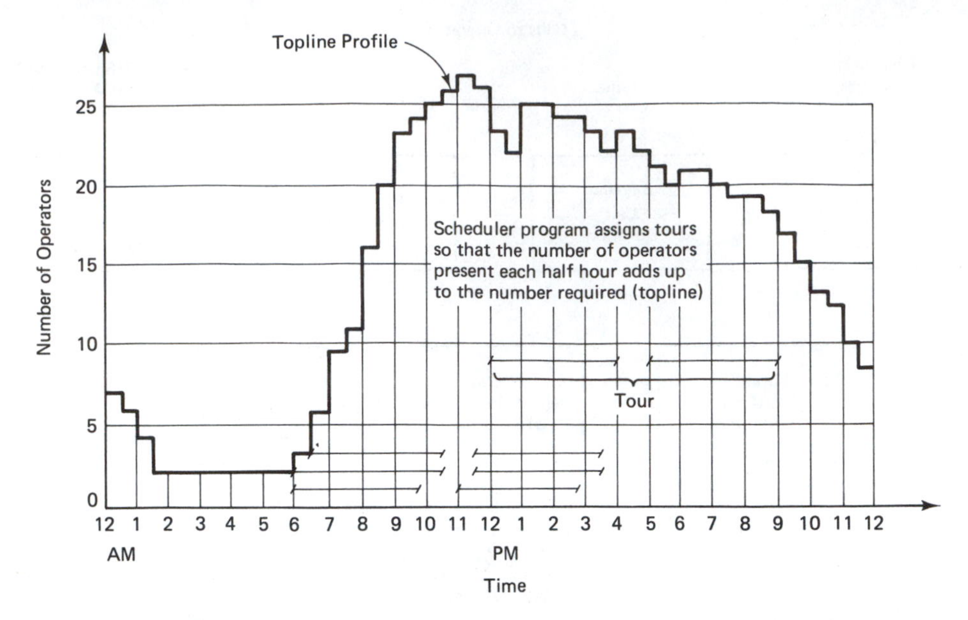

Figure 8 Topline profile and concept for assigning tours to aggregate to the topline.

5. Eight hour shifts end before 9 P.M.
6. Seven hour shifts end from 9:30 to 10:30 P.M.
7. Six and one-half hour shifts end at ll:00 P.M. or later.
8. Earliest lunch period is at 10:00 A.M.

Luce [17] developed a heuristic algorithm for choosing shifts from the approved set in such a way that the absolute differences between operators demanded by the topline profile in period i, D_i, and the operators provided, W_i, is minimized when summed over all n periods of the day, that is

$$\text{Minimize} \sum_{i=1}^{n} | D_i - W_i |. \tag{1}$$

The strategy is to build up the operator resources in the schedule, one shift at a time, drawing on the universal set of approved shifts. The criterion stated in (1) is used to choose shifts at each step. Conceptually, as the schedule of W_i values is built up, the distance is minimized between the demand and operators supplied curves as illustrated by Figure 9.

At each stage in building up the schedule, there exists some remaining distance between D_i and W_i. The criterion for choice of the next shift is the following test on each alternate shift: add the contributions of the shift to W_i (1 for all working periods and 0 for idle periods such as lunch and rest pauses), and recalculate expression (1). Choose the shift which minimizes (1). In order to negate the preference of the preceding choice rule to favor shorter length shifts, the different shift lengths are weighted in the calculation. The longest shift is given a weight of 1.0, and shorter shifts are weighted by the ratio of the working times. Thus, if the longest shift is 8 hours, then, a 7 hour shift would be weighted 8/7 = 1.14.

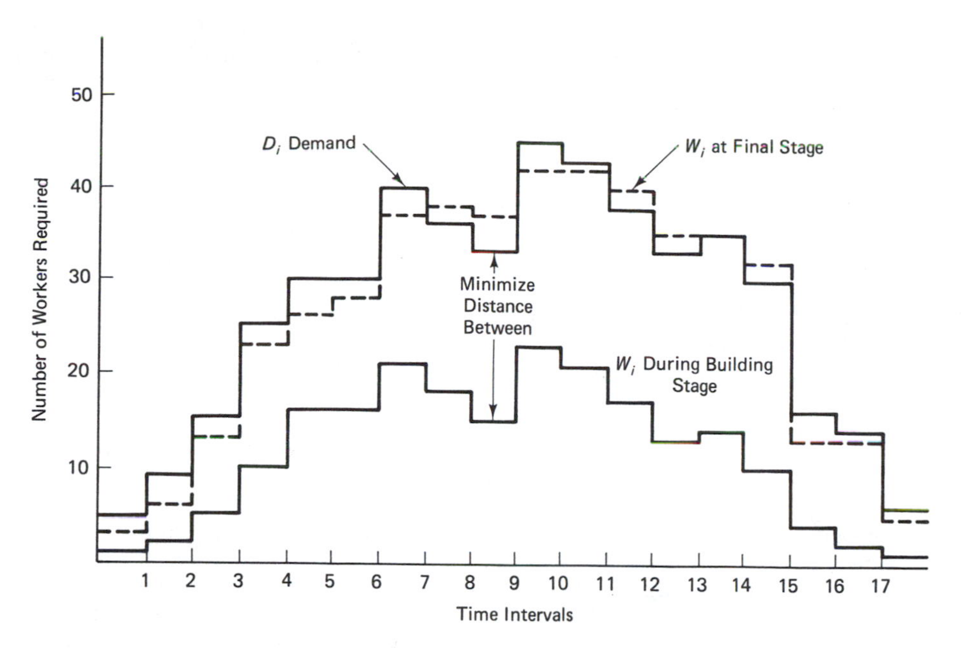

Figure 9 Concept of the schedule building process, using the criterion, minimize

$$\sum_{i=1}^{n} |D_i - W_i|.$$

As the number of time intervals and shift types increase, the computing cost increases. Luce states that computing costs are moderate when the number of time intervals is less than 100 and the number of shifts is less than 500 [17].

As indicated in Figure 9, the final profiles for D_i and W_i do not coincide perfectly in any real case. Operators provided by the algorithm will be slightly greater or less than the demand, and the aggregate figures are a measure of the effectiveness of a given schedule. For example, in a typical output the aggregate statistics were total hours required = 295 and total hours scheduled by half hours = 296.50.

For each 1/2 hour of the day, the computer output indicates the operators required and provided. The excess or shortage of operators provided is reported by half hours.

Also printed in the output is a list of shifts actually used. The shift length is shown, as well as the positioning of lunch and rest periods. The system also provides absentee relief (AR) allocations to be used as needed, based on experience factors.

Assigning Operators to Shifts

Given a set of shifts which meets the demand profile, the next step is to assign operators to shifts. There are many complications to this process which arise because of the 24 hour per day, 7 day per week operation. There are important questions of equity regarding the timing of days off and the assignment of overtime work which carries extra pay. There are also employee shift preferences and seniority status to be taken into account.

Luce [19] presents a computing algorithm which makes "days off" assignments:

1. At least one day off in a week;
2. Days off are one or two;
3. Maximize consecutive days off;
4. If days off cannot be consecutive, maximize the number of work days between days off;
5. Weekends are treated separately on a rotational basis in order to preserve equity because
   a. Overtime pay for weekend work and
   b. Weekends are the most desirable days off;
6. Requests for additional days off are honored on a first come-first assigned basis.

The days off procedure must be carried out to assure that a final feasible schedule will result. Employee trading of days off is allowed. The actual assignment of operators to shifts takes into account employee shift preferences. Each operator makes up a list of shifts in ranked order. The list can have different preferences for each day of the week. Seniority is the basis for determining the order of satisfying preferences, and assignments are made to the highest ranked shift available for each operator.

The final employee schedule for each day is a computer output which specifies for each operator the beginning and end of two work periods (lunch between), and the time for each of two rest periods.

Concluding Remarks

The integrated work shift scheduling system described has been in operation at the General Telephone Company of California since 1973. The company routinely schedules approximately 2600 telephone operators in 43 locations using the system. The size of the work force at each location ranges from approximately 20 to 220 operators.

During 1974, the company realized a net annual savings in clerical and supervisory costs of over $170,000, as well as achieving a 6 percent increase in work force productivity, savings which were attributable to this scheduling system.

CASE STUDY REFERENCES

ABERNATHY, W. J., N. BALOFF, and J. C. HERSHEY. "The Nurse Staffing Problem: Issues and Prospects." *Sloan Management Review*, Vol. 13, No. 1 (Fall, 1971), pp. 87–109.

ABERNATHY, W. J., N. BALOFF, J. C. HERSHEY, and S. WANDEL. "A Three-Stage Manpower Planning and Scheduling Model–A Service-Sector Example." *Operations Research*, Vol. 21, No. 3 (May–June, 1973), pp. 693–711.

AHUJA, H., and R., SHEPPARD. "Computerized Nurse Scheduling." *Industrial Engineering*, Vol. 7, No. 10 (October, 1975), pp. 24–29.

BAKER, K. R. "Scheduling a Full-Time Workforce to Meet Cyclic Staffing Requirements." *Management Science*, Vol. 20, No. 12 (August, 1974), pp. 1561–1568.

BENNETT, B. T., and R. B. POTTS. "Rotating Roster for a Transit System." *Transportation Science*, Vol. 2, No. 1 (February, 1968), pp. 25–34.

BOX, G. E. P., and G. M. JENKINS. *Time Series Analysis, Forecasting, and Control*. San Francisco: Holden-Day, 1970.

BUTTERWORTH, R. W., and G. T. Howard, "A Method of Determining Highway Patrol Manning Schedules." ORSA, 44th National Meeting, November, 1973.

CHURCH, J. G. "Sure Staff: A Computerized Staff Scheduling System for Telephone Business Offices." *Management Science*, Vol. 20, No. 4 (December, 1973), pp. 708–720.

HEALY, W. E. "Shift Scheduling Made Easy." *Factory*, Vol. 117, No. 10 (October, 1969).

HARVESTON, M. F., B. J. LUCE, and T. A. SMUCZYNSKI. "Telephone Operator Management System—TOMS." ORSA/TIMS/AIIE Joint National Meeting, November, 1972.

HENDERSON, W. B., AND W. L. BERRY. "Heuristic Methods for Telephone Operator Shift Scheduling: An Experimental Analysis." Working Paper No. 20, Center for Business and Economic Research, College of Business Administration, The University of Tennessee, February, 1975.

HENDERSON, W. B., and W. L. BERRY. "Determining Optimal Shift Schedules for Telephone Traffic Exchange Operators." Paper No. 507, Institute for Research in the Behavioral, Economic, and Management Sciences, Krannert Graduate School of Industrial Administration, Purdue University, April, 1975.

HILL, A. V., and V. A. MABERT. "A Combined Projection—Casual Approach for Short Range Forecasts." Paper No. 527, Institute for Research in the Behavioral, Economic, and Management Sciences, Purdue University, September, 1975.

JELINEK, R. C. "Tell the Computer How Sick the Patients Are and It Will Tell How Many Nurses They Need." *Modern Hospital*, December, 1973.

LARSON, R. C. "Improving the Effectiveness of New York City's 911." In A. W. DRAKE, R. L. KEENEY, and P. M. MORSE, eds. *Analysis of Public Systems*, Chapter 9. Cambridge, Mass.: M.I.T. Press, 1972, pp. 151–180.

LINDER, R. W. "The Development of Manpower and Facilities Planning Methods for Airline Telephone Reservations Offices." *Operational Research Quarterly*, Vol. 20, No. 1 (1969), pp. 3–21.

LUCE, B. J. "A Shift Scheduling Algorithm." ORSA, 44th National Meeting, November, 1973.

LUCE, B. J. "Dynamic Employment Planning Model." ORSA/TIMS, Joint National Meeting, April, 1974.

LUCE, B. J. "Employee Assignment System." ORSA/TIMS, Joint National Meeting, April, 1974.

MABERT, V. A., and A. R. RAEDELS. "The Detail Scheduling of a Part-Time Work Force: A Case Study of Teller Staffing." Paper No. 531, *Institute for Research in the Behavioral, Economic, and Management Sciences*," Purdue University, Sept. 1975.

MAIER-ROTH, C., and H. B. WOLFE. "Cyclical Scheduling and Allocation of Nursing Staff." *Socio-Economic Planning Sciences*, Vol. 7, 1973, pp. 471–487.

MONROE, G. "Scheduling Manpower for Service Operations." *Industrial Engineering*, August 1970.

MURRAY, D. J. "Computer Makes the Schedules for Nurses." *Modern Hospital*, December 1971.

PAUL, R. J., and R. E. Stevens. "Staffing Service Activities With Waiting Line Models." *Decisions Sciences*, Vol. 2, No. 2, April 1971.

RITZMAN, L. P., L. J. KRAJEWSKI, and M. J. SHOWALTER. "The Disaggregation of Aggregate Manpower Plans." *Management Science*, Vol. 22, No. 11, pp. 1204–1214.

ROTHSTEIN, M. "Hospital Manpower Shift Scheduling by Mathematical Programming." *Health Services Research*, Spring 1973, pp. 60–66.

ROTHSTEIN, M. "Scheduling Manpower by Mathematical Programming." *Industrial Engineering*, April 1972, pp. 29–33.

SEGAL, M. "The Operator-Scheduling Problem: A Network-Flow Approach." *Operations Research*, Vol. 22, No. 4, July–August 1974, pp. 808–823.

WALSH, D. S. "Computerized Labor Scheduling: Supermarkets Jumping on the Bandwagon." 25th Annual Conference and Convention, *American Institute of Industrial Engineers*, May 1974.

Source: E. S. Buffa, M. J. Cosgrove, and B. J. Luce, "An Integrated Work Shift Scheduling System," *Decision Sciences*, 7, 620–630 (1976). Reprinted by permission of the Decision Sciences Institute.

CASE STUDY QUESTIONS

The Need for Speed

Answer the following in terms of the concepts outlined in Chap. 7.

1. Wendy's dropped breakfast omelets because they take too long to cook. Does a long cooking time necessarily mean a small service rate per cashier? Explain why or why not.
2. What was the incentive for Nathan's to reduce its menu size?
3. Denny Lynch is quoted as saying, "We only make two orders at a time—one at the pick-up window and one in the dining room." Compared to McDonald's, which method would have the shorter service time and which would have the higher service capacity (among all servers)?
4. Most drive-thru restaurants use a single cashier to process cars. Why is this the case, and what limitations does it put on the restaurant?
5. In a drive-thru restaurant, how does the distance from the point where the order is placed to the point where the order is received influence the service capacity of the system? Develop a model.
6. Give examples of how restaurants have reduced their service time through (a) automation, (b) reducing customer portion of service time, (c) shifting work to the customer, (d) serving customers simultaneously, and (e) serving customers in batches. Also, how do restaurants increase their capacity to accommodate periods of peak demand?

An Integrated Work Shift Scheduling System

1. Describe the relationship between arrival rate and (a) time of year, (b) time of week, and (c) time of day.

2. What do the authors mean when they say the variation in calls within the hour "cannot be accommodated by planning and scheduling"?
3. What procedure was used to convert the estimate of calls to operators required?
4. How is the topline profile converted into actual workshifts?
5. The authors state that their procedure is heuristic, and not optimal. What is it about their solution that is nonoptimal? Can you think of any way of improving on their solution?

chapter 8

Reducing Delay through Changes in the Arrival Process

The total time spent by customers waiting equals the area between the cumulative arrival and the cumulative departure curves. This area can be reduced through either, or both, of two means: by making the service process more closely match the arrival process or by making the arrival process more closely match the service process. The former idea was presented in Chap. 7; the latter is the focus of this chapter.

The sequence of Chaps. 7 and 8 is not accidental, and reflects the order in which queueing problems should be attacked. First, the existing servers should be scrutinized to see whether it is possible to reduce the service time without sacrificing quality. Failing this, if queues are predictable, capacity should be supplemented during the times when queues are a problem, through use of part-time servers. In most instances, these two approaches are sufficient to eliminate predictable queues. But there is an exception. Some servers, by their nature, cannot be used on a part-time basis. A freeway lane, an airport runway, an expensive piece of machinery: these may all be available full time whether or not they are needed full time. All three are examples of *all-or-nothing* servers; they are either available all of the time or not at all—there is no middle ground. Adding all-or-nothing servers can be an expensive way to eliminate queues.

The alternative to adding servers is to alter the arrival process. A price discount might induce customers to arrive during periods of low demand. A reservation system might be implemented so that customers arrive at constant intervals. Or service hours might be changed to spread out customer arrivals. These are mechanisms for encouraging customers to arrive at the best possible times.

Consider a situation where the arrival rate varies over time, as in Fig. 6.6:

If each customer is induced to change its arrival time to the time it departs from queue (that is, if $A(t) = D_q(t)$), the queue would disappear. The queue would disappear *without changing the times customers depart from the system.*

In a sense, one can say that predictable queues are caused by customers arriving at the wrong times. Why should customers arrive any earlier than the time they begin service? Why not arrive when a server is available? Queues are something like lifeboats. Just as boats have capacities, so do servers. Going beyond capacity benefits no one and, perhaps, hurts everyone. Allowing customers to arrive at a faster rate than they can be served delays every customer in the queue.

One should keep in mind that altering the arrival process does not come without cost. Reservations and pricing pose some inconvenience to customers, though the inconvenience is usually preferable to the waits that would otherwise result. Nevertheless, because customers are inconvenienced, one should not consider altering the arrival process until all of the ideas presented in Chap. 7 for altering the service process have been exhausted.

This chapter covers three ways to alter the arrival process: reservation and appointment systems, pricing, and changing the underlying arrival process. In addition, reneging and its impact on system performance is examined. Like Chap. 7, the emphasis is on eliminating queues rather than increasing server utilization, the reason, as before, being that it is the more prevalent problem.

8.1 RESERVATION AND APPOINTMENT SYSTEMS

Reservations and appointments are two names for the same idea—a system whereby the customer contacts the server in advance to schedule a time to perform the service. Reservations and appointments are common for medical offices, restaurants, and transportation, to name a few examples. They are especially common for queueing systems with long service times (10 or more minutes). It is difficult, though not impossible, to use reservations for short service times (customers may not want to go to the time and effort of scheduling an appointment). Reservations are unheard of at automobile toll plazas, for example, even though waits are sometimes enormous (service time is only a few seconds).

Reservations should encourage customers to arrive at a constant rate matching the rate of service. Customers arrive at a constant rate because they are scheduled to arrive at a constant rate. Thus, the contribution of reservations is the elimination of predictable queues. To a lesser extent, reservation systems can also help eliminate stochastic queues by reducing the variation in customer interarrival times. But this depends on customer punctuality. If customers do not arrive precisely on time, particularly if the spacing between appointments is small, then the arrival process will resemble a Poisson process and the reservation system will not eliminate stochastic queues.

There is an important danger in reservations: They tend to mask queueing problems rather than eliminate them. The presence of a customer waiting in the office provides a constant reminder to the server that queues are a problem. A customer waiting at home for an appointment does not. But whether in an office or at home, the customer is still part of a queue. Worthington's 1987 article on the United Kingdom's health service is informative. He states: "Waiting lists have been an unsatisfactory feature of the National Health Service (NHS) since its inception in 1948. In part, they are a necessary evil in that they provide a pool of patients to ensure that expensive health-service resources do not lie idle for the want of suitable patients. However, waiting lists are usually much bigger than is necessary for this purpose."[1] From this statement, it is apparent that the NHS reservation system has led to a perpetual queue—a queue in which *all* customers wait for appointments. Reservations reduce the cost of providing the service (by reducing the number of servers) but do not improve the quality of the service.

This is not to say that reservations are a bad idea. On the contrary, they can be very effective at eliminating queues. The message is that reservations should not be used with the purpose of achieving 100 percent server utilization. Further, the appointment queue should be constantly monitored to ensure that customers do not have to wait excessive lengths of time for their appointments.

8.1.1 Reservation Systems

The two types of reservation systems that have been studied the most are medical appointments and airline reservations. In the medicine application, private clinics commonly operate as single server queues, even when several doctors practice at the same location. This means that patients are given appointments for their own personal physicians. Public clinics sometimes operate as multiple server queues, meaning that patients are not given appointments for specific doctors. In either case, servers operate more or less on a continuous basis for periods up to 8 hours, the length of a workday. Airlines, along with railroads and other forms of transportation, are examples of bulk service queues. Groups of customers are served simultaneously by the same vehicle. Customers are not served on a continuous basis.

Due to the natures of the service processes, the main issues in reservations are different for medical clinics and airlines. For clinics, the issues are the spacing between appointment times and the number of patients scheduled for each appointment time. For airlines, spacing between appointment times is not an issue. The "appointment times" are defined by the flight schedule. The issue for airlines is the number of customers to schedule for each "appointment" (that is, each flight), which depends on the likelihood that customers show up on time.

Restaurants are another important, though less studied, application of reservation systems. In restaurants the timing of appointments is crucial due to the accentuated peak in the arrival pattern. The following three sections examine reservations for continuous service queues (for example, medical clinics), peaked arrival patterns (restaurants), and bulk service (transportation).

8.1.2 Reservations for Continuous Operation

Soriano classifies appointment systems into four types: ***pure block*** appointment systems, ***individual*** appointment systems, ***mixed block-individual*** appointment systems, and other systems.

> A pure block system assigns a common appointment time at the beginning of the clinic session for all the patients scheduled to be treated on any given day. An individual appointment system is one in which each patient is assigned a different appointment time, and these times are equally spaced throughout the clinic session. A mixed block-appointment system arranges for an initial group of patients to arrive at the beginning of the clinic session with others scheduled to arrive at equally spaced intervals.[2]

From the clinic's perspective, the advantage of the pure block system is that doctors are kept busy from the beginning of the clinic session until the last patient is served. But from the patient's perspective, the pure block system is a disaster. It forces patients to arrive much earlier than they could possibly be served. Fortunately for patients, the pure block system is no longer common. The individual appointment system is more sensible, considering that most clinics operate as single server queues. In essence, the doctor's schedule for the day is determined in advance by the appointment schedule. Customers are scheduled to arrive at the same rate that they are served and, one hopes, neither waiting time nor idle time is large. The motivation for the mixed block-individual system is that queueing systems do not instantly enter steady state at the beginning of the day. It takes time for the system to "warm up." The warm-up phase might be eliminated by scheduling extra appointments at the start of a clinic session. This, in turn, can eliminate doctor idle time.

Bailey (1952), Jackson (1964), and White and Pike (1964) studied the impacts of changing the ***appointment interval***, the time separation between appointments. In all of these studies, the major concern was balancing a trade-off between server idle time and patient waiting time. For example, Jackson collected data on doctor service times in the United Kingdom, observing a mean of 4.55 minutes. By way of simulation, he then applied the data to varying appointment intervals, assuming that customers arrive on time over a clinic session lasting 2 hours (doctors continue to work until the last patient is served).

Figure 8.1 gives Jackson's result. It demonstrates that shortening the interval between appointments reduces server idle time and increases patient waiting time. This relationship is highly nonlinear: Small changes in the appointment interval have a large impact on waiting time when the intervals are small. Jackson writes: "It is worth noting that the reason the patients' average waiting time does not become infinite when the appointment interval is less than the average consulting time is that it is not an infinite queueing system, each surgery being limited to approximately 2 hr."[3] Patient waiting time depends on the length of the clinic session. A short appointment interval may be satisfactory when the session is relatively short, but not if it is long (for example, if it lasts the entire day). A short session provides an earlier opportunity for the server to "catch up" with arrivals. (It is worth noting that the system may never approach steady state. Therefore, *the steady-state models developed in Chap. 5 are not applicable*.)

Figure 8.1 is expressed as waiting time per customer. At least as important is the *total* customer waiting time. For an appointment interval of 6 minutes, a wait per customer of 2 minutes translates into a total customer waiting time of 40 minutes over 2 hours, a value that exceeds the server idle time. An appointment interval of 5 minutes creates more than ten times as much waiting time as server idle time. This suggests that

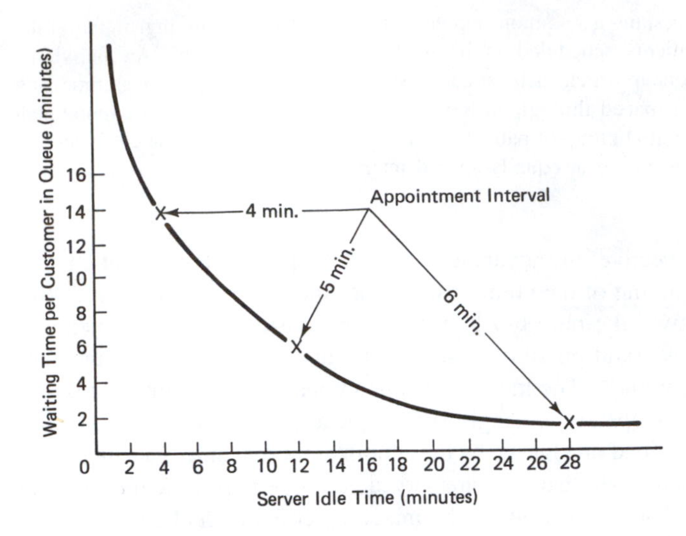

Figure 8.1 Increasing the interval between appointments increases server idle time but decreases customer waiting time. *Source:* Jackson 1964.

the appointment interval should be somewhat longer than the mean service time. Nevertheless, a 2- to 5-minute wait *per customer* is probably tolerable, given that customers are unlikely to arrive at their precise appointment times anyway.

Welch's 1964 study of the mixed block-individual system provided results on the impact of scheduling multiple patients at the start of a clinic session (Fig. 8.2). Welch assumed that both the appointment interval and the mean service time are 5 minutes, and that 25 patients are served. As might be expected, the more patients that are scheduled at the start, the longer is the patient waiting time and the shorter is the doctor idle time. Both Welch and White and Pike suggest scheduling two patients at the start of the session to achieve a balance between doctor idle time and patient waiting time. However, this conclusion is premised on equating doctor idle time to waiting time *per patient*. A fairer comparison would be to *total* patient waiting time, as in Fig. 8.2. In Welch's study, scheduling two patients at a session's start created 40 times as much waiting time as doctor idle time!

It is unfortunate that patient waiting time far exceeds doctor idle time in most clinics. The evidence for this is obvious—just compare the average number of waiting patients to the average number of idle doctors. Whereas somewhat more waiting time than idle time is justified (due to value of a doctor's time and the willingness of customers to endure short waits), typical waits far exceed reason. (I have been told by one expert in this field that some doctors do not believe that they should *ever* have to wait for patients.)

8.1.2.1 Setting practical appointment times. One of the practical considerations in setting up a reservation system is that appointment times should be rounded to useful numbers—for example, to 5-, 10-, or 15-minute increments. Little is gained by setting an appointment at 8:07, say, because the customer will round the time to 8:05 or 8:10 anyway.

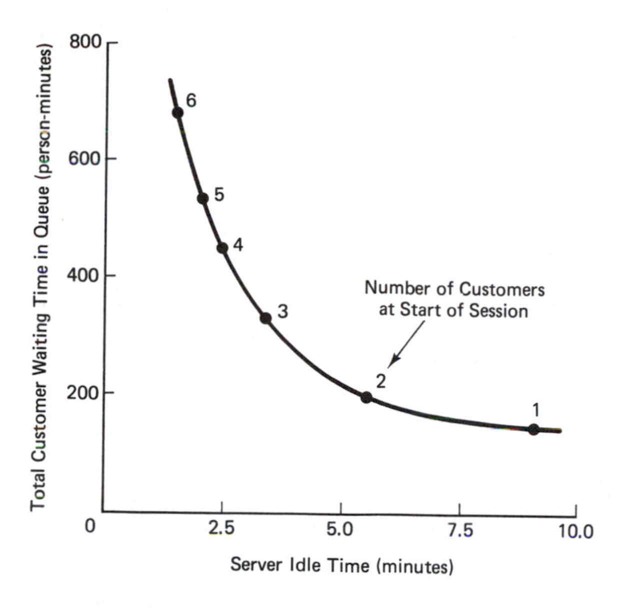

Figure 8.2 Increasing the number of customers at the start of a session decreases server idle time but increases customer waiting time. *Source:* Welch 1964.

Example

A doctor has determined that the mean time to serve a patient is 6 minutes. Her appointment system works as follows:

8:00, 8:05, 8:10 — 8:20, 8:25, 8:30 — . . .

This provides an average appointment interval of 6.7 minutes, slightly larger than the mean service time, and provides shorter intervals at the start of each 20 minute cycle.

A second practical consideration is that the service time can sometimes be predicted in advance, from the type of service requested. In a doctor's office, a physical will take longer than a simple treatment, for example. The reservationist should assess the nature of the patient's service and assign him or her to an appropriate appointment interval.

Example

In another office, three types of treatment are performed, treatment A taking 8 minutes on average, B taking 13 minutes, and C taking 26 minutes. The treatment times are first rounded to "practical time intervals": 10 minutes, 15 minutes, and 25 minutes. Then appointments are scheduled in 5-minute blocks, A using two blocks, B three blocks, and C five blocks.

The appointment system is still not complete. A system is needed for assigning customers to time blocks. One approach is simply to assign each customer to the available time slot, of sufficient length, nearest to the time requested. This might result in a pattern like the Gantt chart in Fig. 8.3. But note a problem. Because of the way appointments are filled, it is impossible to schedule any more 25-minute appointments, while various 5-minute intervals cannot be used at all.

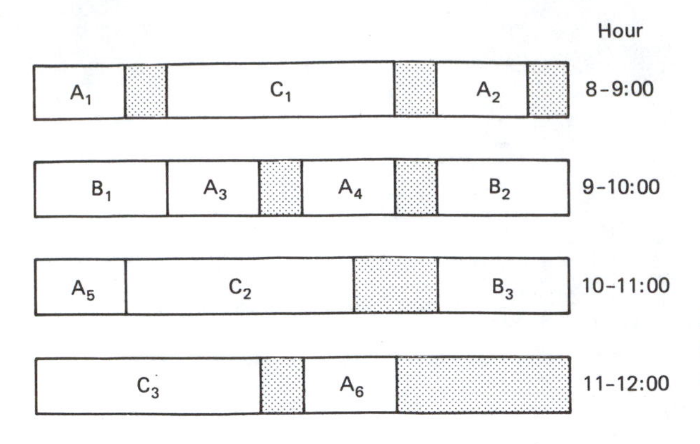

Figure 8.3 Gantt chart showing customer appointment times, when customer is allowed to set appointment at any available time.

Figure 8.3 is an example of a ***packing problem***—a problem of fitting a set of elements (in this case, appointments) within a fixed space (the length of the clinic session). Packing problems are greatly simplified when two features are maintained:

1. All appointment intervals are multiples of the same value (10 minutes, 20 minutes, 30 minutes, and so on).
2. Some time slots are reserved for longer appointments.

Figure 8.4 possesses these features. Instead of allocating 15 minutes for treatment B, just 10 minutes is allocated. This puts a greater strain on the system, but because treatment B is less common than treatment A, the system still has some excess capacity. As partial compensation, treatment C is rounded up to 30 minutes. Each hour block is divided into two parts. The second part is reserved for treatment C, and is not used for A or B unless the first half of the hour is fully scheduled, and the patient cannot be persuaded to take an appointment in another hour. Compared to Fig. 8.3, the schedule provides much more flexibility for scheduling further appointments, even though the total idle time is exactly the same (1 hour). Customers have slightly less discretion in choosing appointment times, but they are still given appointments within an hour of the desired time.

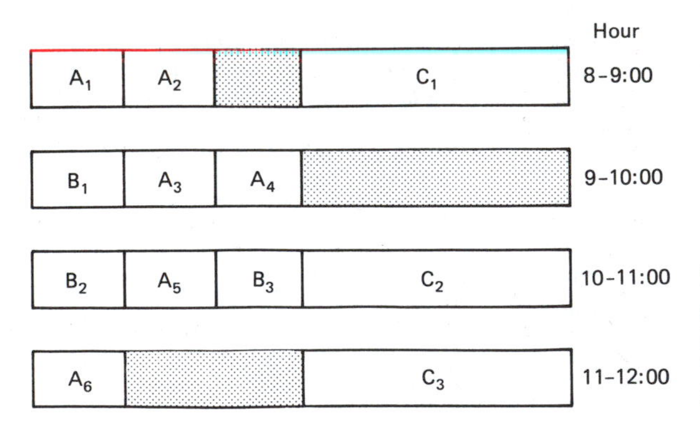

Figure 8.4 Gantt chart showing customer appointment times, when different types of treatment are assigned to time blocks. This approach converts small time blocks into more useful long time blocks.

A final practical issue is that service times are influenced by the size of the appointment intervals. That is, servers tend to work faster when customers are scheduled to arrive at a faster rate. Some queue operators reduce the appointment interval with the explicit intention of forcing greater output from the servers. This is a dangerous practice, for it might be detrimental to service quality. On the other hand, there is also a danger in setting the appointment interval too long: It encourages the server to take more time than necessary. This is an important dilemma, but it is also a dilemma that is best addressed with techniques that fall outside the traditional realm of queueing (for example, the methods of industrial psychology and organizational behavior). The thing to recognize is that service times are not necessarily hard-and-fast numbers, and that they are influenced by the reservation system.

8.1.2.2 Filling appointments. There are two important parts to a reservation system. One, setting the appointment times, has been discussed. The other part, filling appointments, is now presented. Rising, Baron, and Averill's 1973 study of the outpatient clinic at the University of Massachusetts is instructive. The authors divided customers into two categories, controllable and uncontrollable. "The *controllable* component of demand was defined as the patients who made (or could be induced to make) an advance appointment for their physician use. The *uncontrollable* component of demand, the walk-in patients, was defined as the patients who arrive without notice. This latter category included both emergencies and patients whose need for medical care possibly could be postponed, but was not."[4]

Prior to instituting the new appointment system, the number of patients served varied from 253 on Wednesdays to as many as 351 on Mondays, with a particularly large number of walk-ins on Monday. Arrival times also varied within days, with a peak during the hour from 2:00 to 3:00 P.M. The new system equalized the number of patients throughout the day and week. The number of advance appointments was set to complement the pattern of "uncontrollable" arrivals. Advance appointments varied from 96 on Monday to 116 on Thursday. By hour, the number of advance appointments varied from 34 in the 4–5:00 hour to 105 in the 9–10:00 hour (partly reflecting doctor availability). This left extra "open" appointments to handle the uncontrollable demand, when needed.

An important aspect of this reservation system, as with any medical reservation system, is the distinction between customer types. Nonurgent appointments can be delayed, while emergencies must be taken immediately. Because the arrival pattern of emergencies cannot be controlled, the arrival pattern of nonemergencies should be controlled to complement the emergency pattern, yielding an overall constant arrival rate.

When the patient phones for an advance appointment, the reservation clerk assigns him or her to one of the available advance appointment times. Some of these times are inherently more attractive than others (for example, Monday afternoons) and naturally will be filled far in advance. To maintain high service quality, a good reservation system should provide the customer with a number of alternative appointment times. And, when possible, the reservation clerk should try to persuade the customer *not* to select a popular appointment time. The popular times should be left open as long as possible for the customers who have no choice but to arrive at these times. Further, if the clerk sees that many appointments in the upcoming days have not yet been filled, he or she might persuade the patient to take an early appointment. On the other hand, if the next few days or weeks look heavily booked, then the clerk should encourage the patient to delay his or her appointment. Simple persuasion can smooth out the arrival times. It may be that some customers are truly indifferent as to when they arrive. If they are indifferent, why not encourage them to arrive when the system has excess capacity?

How long should a customer have to wait for an appointment? At a minimum, every week should begin with a few open appointments. That is, if the customer is completely flexible about appointments, he or she should never have to wait more than a week. Unfortunately, some medical clinics have earned reputations for much longer waits, sometimes booking all appointments for weeks into the future. This is a misuse of reservation systems. Reservations should not preclude the necessary acquisition of additional servers.

8.1.3 Peaked Arrival Patterns

What if all customers arrive over a very short period of time, so short that it barely exceeds the length of one or two services. In all parts of the world, the arrival pattern for restaurants is markedly peaked. Hour-long intervals sometimes encompass all of the arrivals.

The two key issues in queues with peaked arrival times are being able to serve customers at a time close to their desired arrival time and maximizing ***turnover*** (or, simply, "turns": the number of parties served per table per meal). These objectives are clearly conflicting, and force some form of compromise. Suppose that an average dinner lasts 1½ hours, and all of the customers wish to arrive between 7:00 and 8:00. It would then be impossible to have a turnover greater than one and still serve customers at their desired times. A table used at 7:00 would not become available until 8:30, 30 minutes after the last customer arrived. Thus, if the restaurant desires to increase its turnover, it must somehow spread out arrivals.

According to Chiffriller (1982), "There are two ways in which reservations can be handled. One allows parties to come at *staggered* intervals. The other has everyone come at the same time; this is sometimes called *sittings*."[5] The staggered approach is advantageous in that it more directly accommodates customer needs. On the other hand, sittings greatly simplify scheduling; it ensures that all tables become available for the next group of customers at more or less the same time. Sittings are also attractive because of the nature of the "table queue." With multiple servers and long service times, $D_q(t)$, the cumulative departures from the queue, should not be a straight line but "rippled" (as was illustrated in Fig. 6.9). First one party (a party, a group of diners, is a "customer") is served, then another, and so on. These ripples should be mimicked in the appointment system.

Ideally, the appointment system should be a hybrid of the staggered system and the sittings system. Here is an approach:

1. Based on the expected customers per meal (N) and number of tables (M), determine the number of turnovers per table per meal: $T = N/M$.

2. Convert the turnovers into integer values. For example, if $T = 1.6$, 60 percent of the tables should turn over twice ($p_1 = .6$, $T_1 = 2$) and 40 percent of the tables should turn over once ($p_2 = .4$, $T_2 = 1$).
3. For each turnover value (T_1 and T_2), determine sitting times in a manner that minimizes the deviation between *cumulative reservations* and *cumulative desired arrivals*.
4. For each sitting, stagger reservations slightly to smooth the peaked arrival rate.

These steps are not precise but, nevertheless, provide a reasonable scheme for developing a schedule.

Example

In a large 150-table restaurant, the average mealtime has been found to be 1.25 hours. The restaurant expects 250 parties, whose desired arrival times are represented by the cumulative curve in Fig. 8.5.

Solution T equals 250/150, or 1.67 turns. This is converted to $p_1 = .67$ and $T_1 = 2$; $p_2 = .33$ and $T_2 = 1$. Thus, 100 tables turn over twice and 50 turn over once. The initial reservation pattern is a cumulative curve with three steps: The first step is the initial sitting for tables that turn over twice, the second step is the sitting for tables that turn over once, and the third step is the second sitting for tables that turn over twice. The first and third steps are separated by 1.25 hours, the service time.

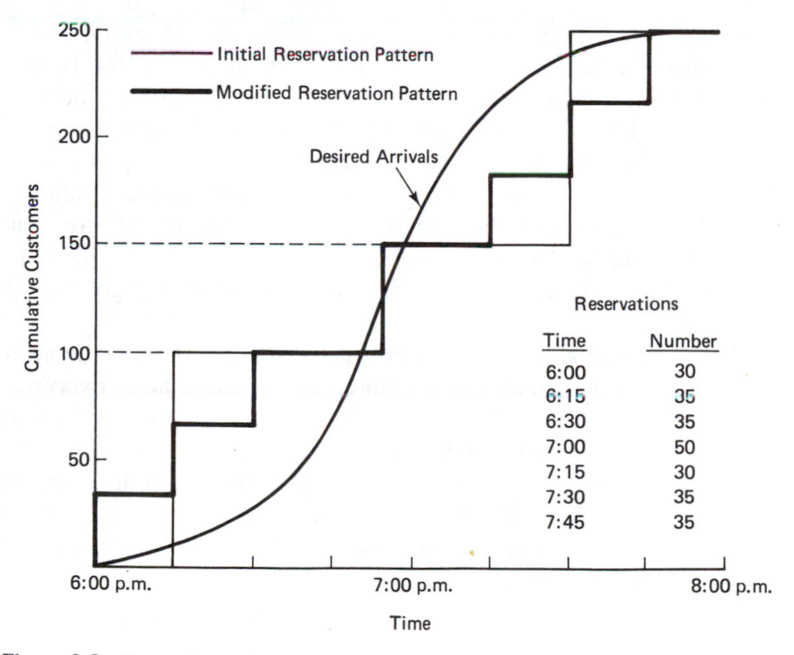

Figure 8.5 Comparison of reservation times to desired arrival times. Objective is to minimize deviation between curves, given available servers and service time.

The modified solution involves shifting portions of the first and third steps of the reservation curve right or left in unison to minimize deviation from desired arrivals. In addition, reservation times are rounded to 15-minute increments.

Note that it is impossible to serve all diners at their desired times; some must be served early and others must be served late.

There comes a point where the service time is so large that it is impossible to have two sittings, in which case the reservation system is quite simple. With only one party per meal per table, there can be no conflict between customer appointments. Customers may, more or less, be scheduled at the time of their choosing. While this certainly simplifies operations, it also effectively cuts revenue in half. Hence, there is considerable incentive to reduce the service time and increase the number of turnovers to two or more.

8.1.4 Bulk Service Queue

Reservations are the norm in long-distance transportation, whether the mode be train, ship, or plane. Reservations are the norm because of the long spacing between services. If a customer were not given a seat on one plane, he or she might have to wait hours, or even days, until the next. Worse yet, if reservations were not used during the busiest travel months, a customer might have to wait weeks for an available seat.

Travel reservations provide customers an assurance that space will be available on a desired departure. (They do not provide assurance that passengers will not have to wait at other airport queues, such as at the ticket counter, which is a separate issue.) Because "appointment times" are dictated by flight departures, one might think that airline reservations is a simple process. It is not. The number of reservations issued depends on the probability that customers show up for their flight. Unless the airline overbooks its seats, revenue will be lost. But if it books too many passengers, the plane may not be able to accommodate all who show up, and the airline will have to compensate passengers for their inconvenience. Adding to this issue is the great range of fares offered, with varying restrictions. Airlines continue booking high-fare passengers long after the low-fare tickets are fully issued. Further, the airline must consider "spillovers" from one flight to another as flights are filled. That is, just because a flight is full does not mean that the airline loses revenue; the customer might simply switch to another flight (see Powell, 1983, for example).

Developing a procedure for booking transportation reservations (called ***yield management***) is far too complicated a process to undertake here. Every year, several papers are written on this subject for the AGIFORS (Airline Group of the International Federation of Operational Research Societies) conference, and the state of the art is constantly evolving. However, the essence of virtually all models in the literature is balancing the cost of compensating "bumped" passengers against lost revenue for empty seats. This problem resembles the classic "newsboy" problem found in the inventory literature. As a simple illustration, let

M = number of reservations booked

R = revenue per seat

C_b = cost per bumped passenger

The variable M should be set at a value that provides maximum profit, given R, C_b, and the size of the vehicle. A necessary condition for optimality follows from marginal analysis:

$$R \cdot [1 - P(\text{full plane} \mid M^*)] - C_b \cdot P(\text{full plane} \mid M^*) = 0 \tag{8.1}$$

That is, the marginal change in profit from increasing M^* by the value one should equal zero. Equation (8.1) can also be written in the following form:

$$P(\text{full plane} \mid M^*) = \frac{R}{R+C_b} \tag{8.2}$$

Example

An aircraft has a seating capacity of 200 people, and all tickets are sold at the price of $150. Passengers bumped from a flight are compensated in the amount of $100, in addition to receiving a ticket refund.

From Eq. (8.2), $P(\text{full plane} \mid M^*) = 150/(150 + 100) = .6$. Reservations should be taken until there is a 60% probability that the plane will be full.

Note that M^* is represented as a probability. This value should be applied to the probability distribution function for the number of "shows" given M reservations. The exact shape of the distribution function will vary from situation to situation and must be determined through data analysis. Also note that M^* is the maximum number of reservations that should be taken; there may or may not be sufficient demand to reach this limit.

8.1.5 General Issues in Reservation Systems

"No-shows" (and, to a lesser extent, cancellations) is an important issue for all types of queues, conventional or bulk service. In fact, no-shows is probably the biggest weakness of reservation systems. If the customer does not show up, the server might be left with a block of time that cannot be used productively. Effort should be taken to ensure that customers do show up, through reminder calls or postcards, or possibly charging no-shows a fee. Beyond this, the best way to ensure that customers show up is to guarantee prompt service. Why should the customer be obliged to arrive on time if the server is unable to do the same? If no-shows are a problem, perhaps the solution is not to overbook customers. The solution may be to reduce customer bookings and guarantee prompt service for customers.

A related issue is customer punctuality. Again, the best way to ensure punctuality is to guarantee that the server is punctual. If the server is habitually late, customers will have little motivation to show up on time. But some servers take the opposite tack, and actually keep "double books" for their reservations: One time is given to the customer and a later time is used internally. This is the wrong attitude, one obviously motivated by a desire to eliminate all server idle time at the expense of long customer waits.

Servers must also contend with "walk-in" demand, customers who arrive without an appointment. Many servers view walk-ins as a problem, mostly because they have not established procedures to deal with them. In fact, walk-ins are attractive because they can fill empty slots in the schedule. Walk-ins are usually more than willing to wait, understanding that customers with appointments should receive priority. For the server, this means that a high level of service can be maintained for customers who went to the effort of making appointments, while still achieving a high server utilization.

8.2 PRICING

Reservation systems provide a mechanism for coping with variations in customer arrival rate. But they do not remove the underlying variations. Desirable appointment times are filled in advance; unpopular times only become filled when the wait for desirable times becomes long. Although this process reduces the size of queues at the server, it creates another queue at home, waiting for the appointment.

The main problem with reservations is the way that customers are "entitled" to arrive at desirable times. Customers earn this right through enduring long waits until appointments. This, in a way, is not too different from queues without reservations, for which customers earn the right to be served at desirable times by enduring long waits in queue. But why should the receipt of a service depend on the willingness of a customer to endure a queue (or any other form of self-immolation)? Time spent in queue is, to a great degree, time wasted. The server does not benefit from the time and neither does the customer.

An alternative "entitlement" mechanism is pricing. The service price is varied in relation to the popularity of the arrival time—high prices for desirable times, low prices for undesirable times—or the popularity of the server. Prices are adjusted until *predictable* queues are eliminated. So customers are entitled with their money rather than their time.

The strength of pricing is that it merely transfers money from one pocket (the customer's) to another (the server's). In general, this is more efficient than waiting, which demands that productive time (the customer's) be lost forever.

Absence of efficient pricing is probably the most frequent cause of *perpetual* queues. In capitalist countries, pricing plays a key role in the distribution of goods and services. When goods become scarce, prices rise; queues do not. In communist countries, prices are controlled. When goods become scarce, prices do not rise; queues may rise instead. Customers are entitled to scarce goods based on their waiting rather than their money. This is inefficient because it wastes customer time. Of course, this problem is not unique to communism. Government agencies are reluctant to raise prices, in capitalist and communist countries alike (post offices are notorious worldwide for their long waits). For instance, the U.S. government revoked a pricing plan used at Boston's Logan Airport to reduce congestion caused by small aircraft. Prepaid health plans also use queues to ration free services.

Although the principle that the price of a good is related to its scarcity is ingrained in the capitalist system, it is not accepted among all people. Many believe that everyone should have the right to basic commodities and basic services. Whereas queues treat customers equally (one customer's time is not different from the next), pricing hits the poor hardest. Thus, some say, it is preferable to charge a low price for basic commodities (and accept waiting) than to charge a market price for basic commodities and eliminate the waiting.

The debate between pricing and waiting is a debate between ***efficiency*** and ***equity***. But keep in mind that charging low prices is not the only way to achieve equity. Progressive income taxes and benefit programs are other mechanisms. The poor person may be more than willing to pay higher prices for basic commodities if it came with a commensurate decrease in income tax. In this way, everyone might be better off.

8.2.1 Principles of Congestion Pricing

The idea of using prices to eliminate queues (often called "congestion pricing") is not new. It has been a subject of academic interest since the 1920s and Pigou's *Economics of Welfare* (1920). A key point in Pigou's work is that prices should be used to encourage people to make decisions that are good for society as a whole, not just good for them as individuals. A further subject, primarily of academic interest, is whether prices in capitalist economies actually achieve this objective. There is actually some evidence that market prices are "efficient"; that is, prices encourage people to make decisions that are good for society as a whole. However, this philosophical issue will not be studied here.

Congestion pricing is a scheme for charging customers for the delay that they impose on other customers as a way to achieve efficiency. Congestion pricing has been studied in greatest depth as a way to eliminate delay on transportation networks. The motivations for this include the following:

Transportation is often publicly owned or regulated. Hence, it is not subject to the usual market forces.

Adding capacity often requires the addition of expensive all-or-nothing servers (the construction of freeway lanes, for example).

Arrival patterns are highly nonstationary, on a daily basis (morning and evening rush hours) and also throughout the year (vacation travel before holidays).

The second and third factors imply that predictable queues cannot be easily eliminated by the means presented in Chap. 7. Furthermore, reservation systems are impractical for some forms of transportation, such as highway travel. This leaves pricing as the basic alternative.

The fundamental principle of pricing (defined by Dupuit in 1844, according to Hotelling, 1938) is that:

The price charged to the user should equal the *marginal cost* imposed by the user on the system.

The ***marginal cost*** is the change in total cost (to all customers) due to the *addition* of one customer to the system (marginal means "of the edge"). A customer joining a queue will endure a wait. This wait is, in a sense, a "price" charged to the user. But this price alone does not equal the marginal cost imposed on the system. A customer joining a queue will also cause subsequent customers to endure a *longer* wait. The extra wait endured by others is the additional marginal cost imposed by the user on the system. The principle of marginal cost pricing states that the user should be monetarily charged a price equivalent to the additional marginal cost.

Example

During a period of heavy congestion, it has been determined that each additional customer will increase total waiting time by 20 customer-minutes. However, the average waiting time per customer is just 8 minutes. Therefore, the additional delay imposed on other customers is $20 - 8 = 12$ minutes, and the price should be commensurate with the value of 12 minutes' time.

The following two sections illustrate the principle of marginal cost pricing with two examples. The first addresses the decision of which queue to join. The second addresses the decision of when to arrive.

8.2.2 Selecting a Queue to Join

In transportation, customers often have two or more servers to choose from, one of which may be more attractive than the others (for example, one choice might be a fast freeway and the other might be a slower arterial road). Under ordinary circumstances, the customer would choose the superior server. But if the arrival rate at the superior server is sufficiently large to cause queueing, the second choice could be preferred. For any given arrival rate, there would exist an optimal proportion of customers to use the superior server. Does this proportion match customer choices?

As a simple illustration, suppose that customers select between two single server queues prior to inspecting the states of the systems. At both queues, customers arrive by a Poisson process, and service times are exponential random variables. Server 1 is very convenient and takes less time for customers to reach than the inconvenient server, server 2. The following symbols will be used:

λ_1 = arrival rate at the convenient server
λ_2 = arrival rate at the inconvenient server
$\lambda = \lambda_1 + \lambda_2$ = total arrival rate
μ_1 = service rate at server 1
μ_2 = service rate at server 2
τ = extra time needed to reach server 2

The total arrival rate, λ, is a fixed value.

When customers are presented with two options, a fair behavioral assumption is that they will select the option with the minimum cost. In the absence of pricing, the cost is merely the customer's time (waiting plus access). So, if one server requires less time than the other, customers will use it in preference to the other. However, if too many customers use the server, delay will increase until, eventually, both servers require the *same* time. In ***equilibrium***, *the total time (waiting plus access) per customer will be the same for each server used.*

Because customers are unaware of the state of the system prior to choosing, they will select on the basis of expected time in system. Equation (5.23) gave the expected time in system per customer for the $M/M/1$ queue. Adding in the access time and substituting λ/μ for ρ, Eq. (5.23) defines an equilibrium as follows:

Equilibrium (No Pricing)

$$\text{Wait per customer server 1} = \text{Wait per customer server 2}$$

$$W_1 = W_2$$

$$\frac{1}{(\mu_1 - \tilde{\lambda}_1)} = \tau + \frac{1}{(\mu_2 - \tilde{\lambda}_2)} \tag{8.3}$$

where the ˜ symbol denotes equilibrium. For any given value of λ, Equation (8.3) can be solved algebraically to find $\tilde{\lambda}_1$ and $\tilde{\lambda}_2$. It can also be solved graphically by plotting W_1 and W_2 versus λ_1 (or λ_2) and identifying the point of intersection, as shown in Fig. 8.6. In the figure, $\mu_1 = 50$ customers/hour, $\mu_2 = 80$ customers/hour, $\lambda = 50$ customers/hour, and $\tau = .06$ hour. The equilibrium is $\tilde{\lambda}_1 = 36.5$ and $\tilde{\lambda}_2 = 13.5$, and the wait per customer is .075 hour, which must be the same at both servers.

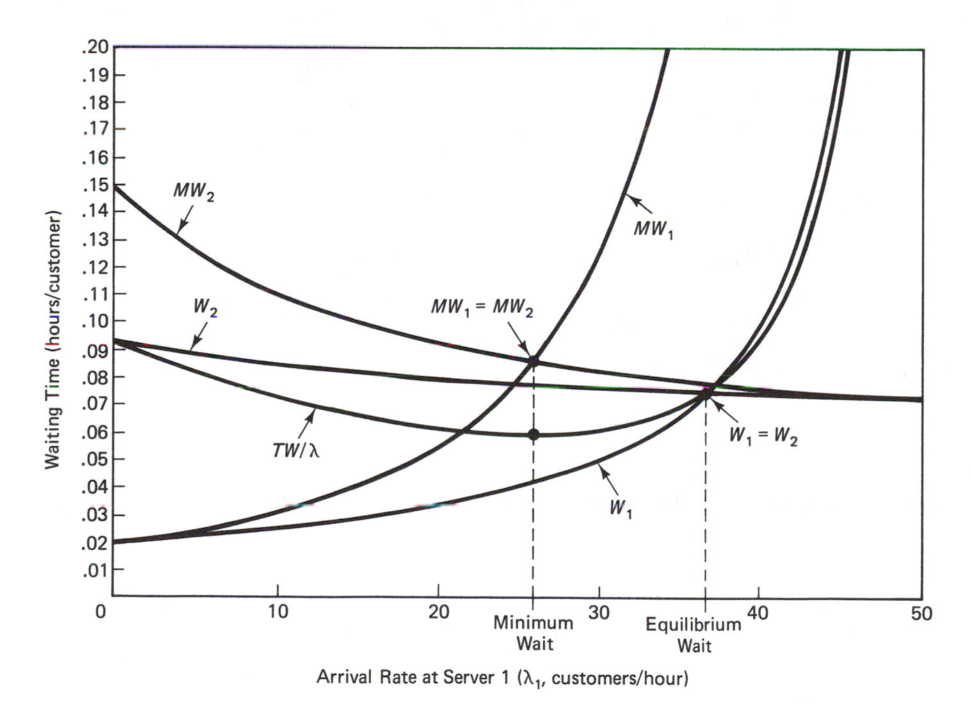

Figure 8.6 Minimum total waiting time occurs when the marginal waiting times are equal. This solution is preferred to the equilibrium solution, in which average waiting times are equal.

It is not difficult to see that the *equilibrium* solution is not the *optimal* solution. The optimal solution should minimize the total waiting time among all customers:

$$TW = \text{total wait}$$
$$= \lambda_1 W_1 + \lambda_2 W_2 = \lambda_1 W_1 + (\lambda - \lambda_1) W_2 \tag{8.4}$$

The minimum occurs where the derivative of TW with respect to λ_1 equals zero. This is equivalent to the following:

$$\frac{\partial TW}{\partial \lambda_1} = \frac{\partial(\lambda_1 W_1)}{\partial \lambda_1} - \frac{\partial(\lambda_2 W_2)}{\partial \lambda_2} = 0$$
$$\frac{\mu_1}{(\mu_1 - \lambda_1)^2} - \left[\frac{\mu_2}{(\mu_2 - \lambda_2)^2} + \tau\right] = 0$$
$$MW_1 - MW_2 = 0 \tag{8.5}$$

where MW_1 and MW_2 are marginal waiting times. That is, the optimal solution occurs when the *marginal* waiting time (the change in total waiting time if the arrival rate is increased slightly) is the same for both servers, not when the waiting time itself is the same for both servers. This solution can also be found graphically, by plotting MW_1 and MW_2 and identifying the point of intersection, as shown in Fig. 8.6. For the example, $\lambda_1^* = 26$ and $\lambda_2^* = 24$, which yields a waiting time per customer of just .059 hour (derived from Eq. (8.4), which is also plotted in Fig. 8.6). Substitution of the solution back into Eq. (8.3) for W_1 and W_2 provides the following:

$$W_1^* = .042 \text{ hour/customer} \qquad W_2^* = .078 \text{ hour/customer}$$

For the optimal solution, the time required to use server 2 is nearly twice as large as the time required to use server 1. Therefore, the optimal solution is not an equilibrium solution: Customers would prefer to switch from server 2 to server 1.

The role of pricing is to encourage customers to make the right choice. This is accomplished by making the cost to the user equal to the cost of the system (by marginal cost pricing). Let

$$P_i = \text{surcharge for using server } i$$
$$\beta = \text{value of a customer's time}$$

Then, following the principle of marginal cost pricing

$$P_i = \beta(MW_i - W_i) \tag{8.6}$$

That is, the surcharge should be commensurate to the difference between the cost borne by the user (βW_i) and the total cost incurred by the system (βMW_i).

Example

Substitution of λ_1^* and λ_2^* in Eq. (8.4) yields $MW_1 = MW_2 = .086$ hour. The average waits are .042 and .078 hour. So the prices should be $P_1 = \beta \cdot .044$ and $P_2 = \beta \cdot .008$.

When an additional customer arrives at server 1, it waits .042 hour on average, and other customers are caused to wait an additional .044 hour. To encourage customers to make optimal decisions, a price commensurate with .044 hour of time should be charged to customers who use server 1. Similarly, a price commensurate with .008 hour of time should be charged to customers who use server 2.

Note that the difference between the surcharges is equivalent to .036 hour of time, which is also the difference between the waiting times at the two servers. With these prices, we would expect the equilibrium distribution of customers and the optimal distribution to be identical. There would be no incentive for customers to switch from one server to the other because the costs would be the same.

In this example, the marginal cost pricing principle leads to 22 percent less waiting time. Hence, it promotes efficiency. It is also equitable in the sense that customers pay for the additional waiting time imposed on others.

8.2.3 Time-Based Pricing

Prices can be used to encourage customers to select the best server. Prices can also be used to encourage customers to arrive at the best times. The focus of this section is on the use of prices to change arrival times and eliminate predictable queues. To make the concepts as clear as possible, stochastic effects will be ignored.

In Chap. 7, we saw that any delay in serving customers during a predictable queue delays all subsequent customers (until the queue vanishes). Now we can see that any additional customer who arrives during a predictable queue also delays all subsequent customers (until the queue vanishes). Specifically

The change in total waiting time due to the addition of one customer equals the time from when the additional customer arrives until the time the queue vanishes. This change represents the marginal "cost" to the system.

This principle is illustrated in Fig. 8.7. The marginal customer that arrives at 6:50 increases the height of the cumulative arrival curve by one, over the period from 6:50 to 8:00. Therefore, the total waiting time increases by the amount of 70 customer-minutes.

Figure 8.8 shows how marginal waiting time and waiting time for the customer depend on the customer's arrival time, assuming that customers are served first-come, first-served. The curve is based on the arrival and departure curves in Fig. 8.7. Marginal waiting time starts large and declines to zero, while waiting time for the customer increases gradually. Toward the end of the queue's duration, the two values are nearly the same.

To encourage customers to arrive at optimal times, they should be charged a price commensurate with the *difference* between the two curves. This means that the price should be largest toward the beginning of the queue's duration and negligible toward the end. This pattern is not new. It again reflects the fact, seen in the previous chapter, that events occurring toward the start of the queue have the biggest impact on total waiting time. A key to controlling waiting is to control arrivals at the queue's beginning.

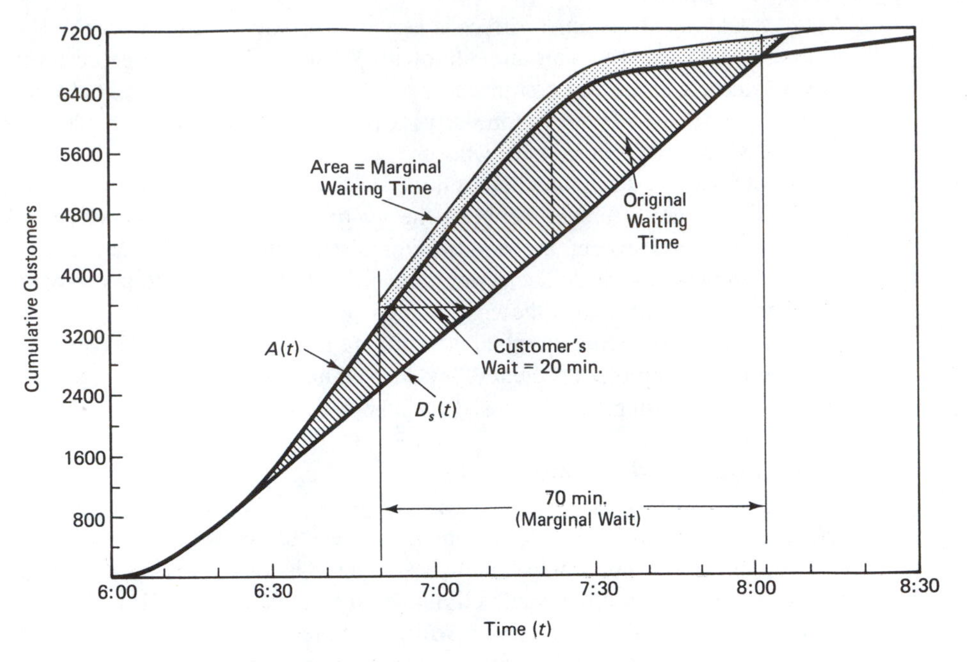

Figure 8.7 The marginal waiting time created by an additional customer during a predictable queue equals the time from when the customer arrives until the queue vanishes.

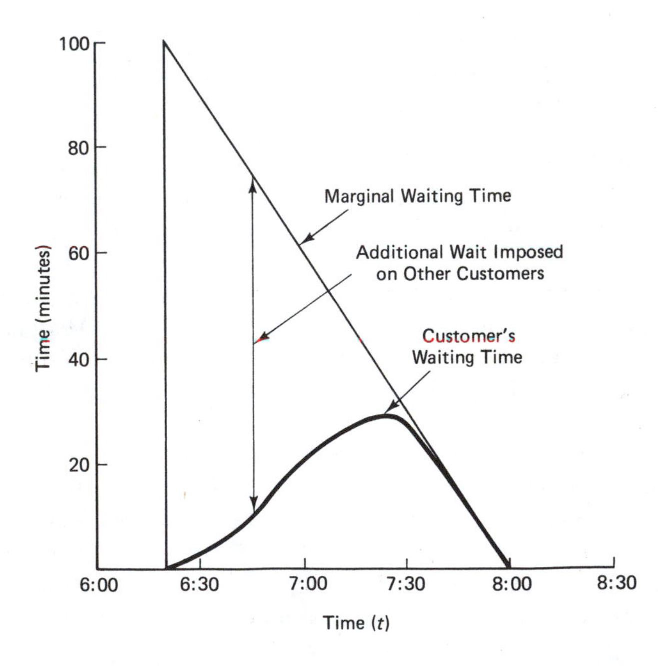

Figure 8.8 The marginal waiting time declines at a constant rate during a predictable queue. The customer's waiting time increases gradually, then declines at nearly the same rate as the marginal waiting time. Prices should be commensurate with the difference between the curves and be largest toward the queue's beginning. (Figure based on arrival and departure curves in Fig. 8.7.)

The objective of pricing should actually be to eliminate predictable queues completely. Prices should encourage customers to arrive at a steady rate over a "congested period," the congested period being a period of time when the system operates at or near capacity. The price should be large at the beginning of the congested period (equivalent to the cost of waiting through the entire period) and decline until the period ends, at which time the price should be zero (see Hendrickson and Kocur, 1981, Newell, 1987, and Vickrey, 1969, for more details).

8.2.4 Setting Practical Prices

The theory and practice of pricing are quite different, particularly when it comes to public enterprises. Time-based pricing is virtually unheard of in highway transportation, even though it should be one of its best applications (see Elliott, 1986, for example). Some of the obstacles include the public's perceptions that

People should not be charged for government services.

Congestion pricing unfairly hurts workers, who have no choice other than to drive during peak hours.

People who endure queues should not be charged more for the service; they should be charged less.

Collection of tolls causes queues rather than eliminates them.

Time-based pricing has also been proposed, but not widely accepted, for mass transit systems. Vickrey's 1955 article on setting fares for the New York City subway system is an example. In his conclusions, Vickrey proposed establishing a "higher fare for rush-hour trips over congested sections," but also noted that "popular acceptance of such a fare structure, or for that matter, even serious consideration by responsible public officials, seems at the present to be rather a long way off."[6] Today, it still seems a long way off.

In addition to public attitude, technology is an obstacle to time-based pricing. On highways, collecting fees traditionally has meant building tollbooths, with all their concomitant costs. Recent technological developments may change this. For example, Hong Kong has experimented with an electronic road-pricing system that automatically records vehicles passing road segments, and bills them through the mail (Dawson and Catling 1986). Singapore has achieved some success by allowing only drivers who purchase monthly permits in the central business district. In subways, automated ticketing systems, such as the ones used in Washington, D.C., and San Francisco, make it feasible to charge different prices at different times of day (this is actually done in Washington).

From the practical point of view, prices cannot be changed continuously, as suggested by the theory. They must, instead, be changed in steps. The most familiar example of this is in telephone price structures, which, in the United States, provide a 40 percent discount between 5:00 and 11:00 P.M. and a 60 percent discount between 11:00 P.M. and 8:00 A.M. Simplified price structures are important to public understanding and acceptance. Prices should not appear random or arbitrary, and the public should be aware of the prices before they decide when to arrive. Otherwise, prices will not have the desired effect of changing customer decisions.

A practical problem in collecting fees is that the time when a customer arrives and the time when the customer departs from queue are not necessarily the same. This can create havoc if the price changes in the intervening time. For instance, if a surcharge is added between the hours of 8:00 and 9:00, a customer might arrive at 7:58 and not reach the toll collector until, say, 8:02. The customer could then rightly argue that he or she should not have to pay the surcharge. One solution would be to collect the fee upon arrival, but this would require an additional service process to collect the fee, and possibly create another queue. A better solution would be to change the price only when there are no queues. But even this solution is not perfect, for it may lead to a rush of customers arriving immediately prior to the imposition of a surcharge, creating an artificial queue.

Though pricing is not a perfect solution, and may not eliminate queues in entirety, it can still have an enormous impact. Though the theory says that prices should vary throughout the queueing period, a flat fee should ordinarily achieve good results. And though the price structure might have to be changed from time to time, to reflect changing arrival patterns, the adjustments do not have to be significant to achieve desirable results.

Pricing provides one other important benefit. It is a means for assessing the effectiveness of expanding service capacity.

Example

Adding one server at a congested facility will cost $200 per day and be capable of serving 100 customers over the congested period. Currently, customers pay $1 to arrive during the congested period. If the server is added, revenue will increase by no more than $100 (100 customers served × $1 per customer), and cost will increase by $200. The added server is not justified.

Revenue would not increase by more than $100 because the new server cannot accommodate more than 100 customers, and the revenue per customer after a capacity expansion would not increase (it may actually decrease). This upper bound on the revenue generated provides a quick estimate for whether the project is worthwhile. Vickrey, writing on congestion tolls in transportation, states

> In the absence of the information that would be provided by the charging of appropriate tolls, planning of investment in expanded transportation facilities is half blind, and resort is sometimes had to arbitrary rules of thumb such as that of providing capacity adequate to handle the traffic during the thirtieth heaviest hour of traffic out of the year. The capriciousness of such a rule should be fairly obvious.[7]

This principle applies to all types of queues. How does one know whether an investment in capacity is justified if one cannot assess the benefit of reduced customer waiting? There must be some way to convert hours of waiting into a dollar cost. Otherwise, adding capacity will be half blind.

8.3 CHANGING THE UNDERLYING ARRIVAL PROCESS

Instead of attacking the symptom of nonstationary arrivals, queues might be eliminated by attacking the causes of the variations. This strategy has been promoted, to varying degrees of success, as a way to remove rush-hour traffic congestion under the name of "transportation system management" (TSM). The concept can also be applied elsewhere.

Consider the problem of traffic queues. Congestion is most prevalent during just a few hours of the day (7:00–9:00 A.M. and 4:00–6:00 P.M.) when large numbers of employees begin and end work. TSM attacks the problem of congestion, not through reservations or pricing, but by promoting alternative workhours—encouraging employers to institute ***flextime***. Flextime allows employees to begin and end work over wider ranges of time (Martin and Jones 1980; Nollen 1982). The TSM philosophy is that, under flextime, employees will choose to travel during uncongested periods, effectively smoothing out the arrival pattern.

Commuting is not unusual in that the nonstationarity is caused by a form of deadline (in this case, the time work begins). Deadlines for filing tax forms or for filing college admissions applications also create nonstationarities: The arrival rate peaks immediately before the deadline. Figure 8.9, for instance, displays a cumulative arrival curve for applications submitted to the University of California at Berkeley, 50 percent of which are received within a ten-day span. Queues like this can be controlled by staggering the deadline. Customers might be categorized, and each category given a different deadline.

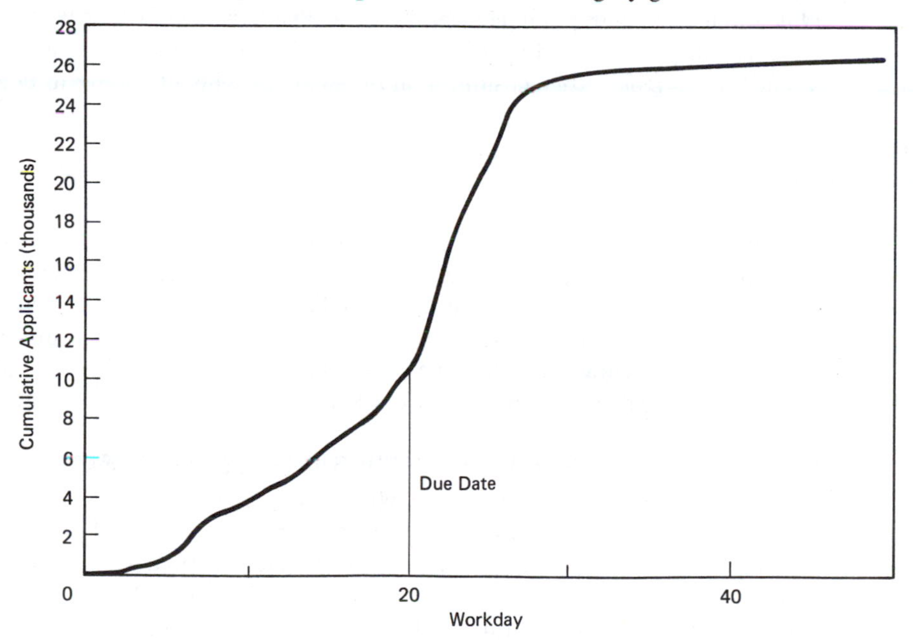

Figure 8.9 Cumulative arrivals of applications at the University of California at Berkeley reveals nonstationary process. Staggering deadlines might smooth out arrivals.

The starting time for an event—football game, play, concert—is another form of deadline. While it may be impossible to stagger the start of an event, arrivals might still be spread out by providing incentives for customers to arrive early (or leave late after the event is completed). In the United States, it is customary for bands to play before and after college football contests, for instance. This secondary entertainment provides an incentive for some customers to arrive early and leave late, which smooths out the process.

Another cause of nonstationarity is server operating hours. If the hours are cut too short, queues may be created either at the beginning or end of the day. Most people are familiar with the term "banking hours" because banks were notorious for their short hours. Banking hours created heavy demand for service immediately prior to 3:00, the traditional closing time. Fortunately, banks are open longer today, which alleviates the rush.

Arrival processes can also be altered through advertising. It may be that customers are not aware of the best times to arrive. Provided with this information, they may adjust. Advertising is also a good way to promote business during the lean months of the year. For example, automobile garages might remind their customers to bring their cars in for periodic maintenance when demand is low.

A final strategy for altering the arrival process is to change or expand the service. Personal accountants are sure to have peak demand immediately prior to tax deadlines. To smooth out the demand, the accountant might expand his or her practice to serve small businesses, which have more of a year-round demand. Better yet would be to expand into a line of business with a complementary demand pattern—one with least demand prior to tax time and peak demand at some other time of year. A smoothed pattern of business would lead to more efficient utilization of resources without adding to queueing.

8.4 RENEGING

Human behavior is a mysterious thing. We are all different, and because we are all different we react to situations in different ways. It is difficult to predict how customers will respond to changes in queue design. Will the customer arrive at a different time? Will the customer renege upon arrival? If the customer reneges, will the customer return later or never again? How will queueing affect future business? These are all important questions concerning customer behavior. For the most part, they are also unanswered questions.

It is generally accepted that the likelihood that a customer reneges (leaves a queue before being served) increases as the length of the queue increases, both in terms of time and customers. Little is known about the exact form of this relationship. Even if this form were known, it would certainly vary from situation to situation. To gain a general feel for the qualitative aspects of reneging, a simple model will now be presented.

Suppose that when a customer arrives at a queue, it inspects the queue's length and decides whether or not to wait. If the customer decides to wait, then it will wait until served. If the customer reneges, it will never return. As an illustration, suppose that the probability that a customer reneges is described by the following function:

$$P[L_q(t)] = \begin{cases} L_q(t)/a & L_q(t) \leq a \\ 1 & L_q(t) > a \end{cases} \tag{8.7}$$

Hence, the probability that a customer reneges increases at a constant rate as the queue length increases, until $L_q(t) = a$, at which point all customers renege.

Figure 8.10 illustrates the impact of reneging through cumulative arrival and departure curves. These curves are based on the same arrival pattern used in Chap. 6 to create Fig. 6.6. However, in Fig. 8.10, adding capacity not only reduces waiting time but also increases the number of customers served. It affects two measures of performance. Note that reneging tends to moderate the arrival rate as queues become large, until a form of equilibrium, where the arrival rate equals the service rate, exists. Reneging is the moderating force that keeps queues from growing even larger when the arrival rate exceeds capacity. In fact, most perpetual queues maintain this sort of equilibrium, with customers always forced to decide whether it is worth waiting for the service.

Whether or not reneging is good or bad depends on one's perspective. The server may lament the loss in business. On the other hand, the customers who remained are pleased because queues, and waits, have been shortened. Finally, while the reneging customers may resent the time they wasted traveling to the server, they are at least happy they did not waste even more time in the queue. On the whole, it is probably a good thing

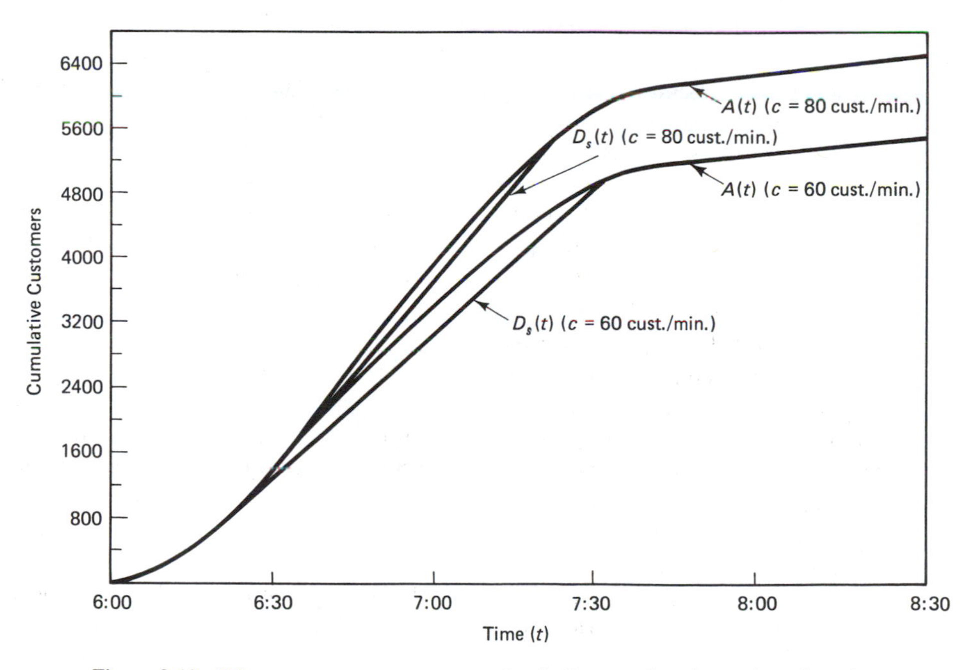

Figure 8.10 When customers renege, queue size declines, as does the total number of customers served. Because increasing service capacity will increase arrivals, queues may not decline as much as expected.

that customers renege, if only to remind the server that queues are a problem. But one should not look at reneging as an alternative to pricing. Why allow the queue to grow large in the first place? Why not use prices to reduce arrivals at the outset? Furthermore, with reneging, the wrong customers are induced to leave the system. Few customers will renege during the critical period of queue formation because the queue has not yet grown to intolerable length. Customers will not begin reneging until much later, when their impact on the system is greatly reduced.

Reneging is just one sign that the server does not, and cannot, control every aspect of the system. The server cannot order customers to arrive at set times. The server can merely provide incentives. Reneging is also one of the least understood aspects of queueing systems. Predicting customer reaction to changes in queue design is imprecise at best.

8.5 CHAPTER SUMMARY

There are two basic ways to eliminate queues: Change the service process and change the arrival process. Changing the service process is the preferred action. However, queues can also be eliminated by inducing customers to arrive at a steady rate, at times when they would otherwise have departed from queue. Altering the arrival process in this manner eliminates predictable queues without delaying departures.

The mechanisms for changing the arrival process, summarized in Table 8.1, are reservations, pricing, and altering the underlying causes of variation. Reservations are common for medical offices, restaurants, and other systems with long service times. Reservations are also common in transportation, particularly air travel. An important consideration in establishing a reservation system is the length of the interval between appointments. The shorter is this interval, the longer customers will have to wait and the less time servers will be idle. In general, the appointment interval should be slightly larger than the mean service time. Appointment times should also be rounded to practical values (5-, 10-, or 15-minute increments) and, when possible, different types of services should be given different length appointments.

Reservations is a mechanism for coping with variations in arrival rate, but it does not remove the underlying variations. Customers are "entitled" to arrive at desirable times on the basis of how long they are willing to wait for an appointment. An alternative way to entitle customers is pricing. High prices are charged for popular arrival times; low prices for unpopular times. Whereas waiting requires that productive time (the customer's) be wasted, pricing merely transfers money from one pocket (the customer's) to another (the server's).

Prices should be set according to the principle of marginal cost and account for all costs imposed on the system (including other customers). Marginal cost pricing promotes efficiency by eliminating predictable queues. It also provides a means for evaluating the benefit of increased service capacity. The willingness of customers to pay for expansion is an appropriate yardstick.

TABLE 8.1 STRATEGIES FOR CHANGING THE ARRIVAL PROCESS

Reservations:	Change the appointment interval. Provide different appointment intervals for different services. Adopt an individual system instead of a block system. Fill fewer advance appointments at times when a large "uncontrollable" demand is expected.
Pricing:	Adopt marginal cost congestion pricing to Encourage customers to use underutilized servers Encourage customers to arrive at uncongested times
Underlying Arrival Process:	Give commuters flexibility in setting workhours. Stagger deadlines. Provide secondary service as incentive for customers to arrive early for events (and leave late). Extend operating hours. Advertise during periods of low demand. Expand into a complementary business.

A final way to eliminate predictable queues is to attack the underlying causes of variations. For example, rush-hour traffic queues are caused by employees beginning and ending work at the same time. Merely changing prices does not address the cause. Instead, alternative workhours should be promoted, perhaps through use of flextime.

Combined, Chaps. 7 and 8 provide a range of options for eliminating queues. As stated at the outset of this chapter, the sequence of topics prescribes the order in which queueing problems should be attacked. However, queueing problems might also be attacked through a combination of approaches, taking ideas from both chapters. As an example, airlines might view the passenger reservation system and the flight scheduling system as an integrated process. That is, both service and arrivals are planned simultaneously (as seats are booked, more flights are added). Integrated approaches can achieve results that go beyond individual mechanisms.

FURTHER READING

BAILEY, N. T. J. 1952. "A Study of Queues and Appointment Systems in Hospital Out-Patient Departments, with Special Reference to Waiting Times," *Journal of the Royal Statistical Society*, 14, 185–199.

CHIFFRILLER, T. F., Jr. 1982. *Successful Restaurant Operation*. New York: Van Nostrand Reinhold.

DAWSON, J. A. L., and I. CATLING. 1986. "Electronic Road Pricing in Hong Kong," *Transportation Research*, 20A, 129–134.

ELSE, P. K. 1986. "No Entry for Congestion Taxes," *Transportation Research*, 20A, 99–107.

ELLIOTT, W. 1986. "Fumbling toward the Edge of History: California's Quest for a Road-Pricing Experiment," *Transportation Research*, 20A, 151–156.

Hendrickson, C., and G. Kocur. 1981. "Schedule Delay and Departure Time Decisions in a Deterministic Model," *Transportation Science*, 15, 62–77.

Hotelling, H. 1938. "The General Welfare in Relation to Problems of Taxation and of Railway and Utility Rates," *Econometrica*, 6, 242–269.

Jackson, R. R. P. 1964. "Design of an Appointment System," *Operational Research Quarterly*, 15, 219–224.

Martin, V., and D. W. Jones. 1980. *A Guide to Flextime*. Berkeley, Calif.: Institute of Transportation Studies.

Newell, G. F. 1987. "The Morning Commute for Nonidentical Travelers," *Transportation Science*, 21, 74–88.

Nollen, S. D. 1982. *New Work Schedules in Practice, Managing Time in a Changing Society*. New York: Van Nostrand Reinhold.

Pigou, A. C. 1920. *The Economics of Welfare*. London: Macmillan.

Powell, B. 1983. "A Stochastic Passenger Loading Model of Airline Schedule Performance," *Transportation Research*, 17B, 399–410.

Rising, E. J., R. Baron, and B. Averill. 1973. "A Systems Analysis of a University-Health-Service Outpatient Clinic," *Operations Research*, 21, 1030–1047.

Soriano, A. 1966. "Comparison of Two Scheduling Systems," *Operations Research*, 14, 388–397.

Vickrey, W. S. 1955. "Revising New York's Subway Fare Structure," *Journal of the Operations Research Society of America*, 2, 38–68.

———, 1969. "Congestion Theory and Transport Investment," *American Economics Review*, 59, 251–261.

Welch, J. D. 1964. "Appointment Systems in Hospital Outpatient Departments," *Operations Research Quarterly*, 15, 224–232.

———, and N. T. J. Bailey. 1952. "Appointment Systems in Hospital Outpatient Departments," *Lancet*, 262, 1105–1108.

White, J. J., and M. C. Pike. 1964. "Appointment Systems in Out-Patients' Clinics and the Effect of Patients' Unpunctuality," *Medical Care*, 2, 133–142.

Worthington, D. J. 1987. "Queueing Models for Hospital Waiting Lists," *Operational Research Society Journal*, 38, 413–422.

NOTES

1. D. J. Worthington, "Queueing Models for Hospital Waiting Lists," *Journal of the Operational Research Society*, 38, (1987) 413.

2. A. Soriano, "Comparison of Two Scheduling Systems," *Operations Research*, 14, (1966), 389.

3. R. R. P. Jackson, "Design of an Appointments System," *Operational Research Quarterly*, 15 (1964), 222.

4. E. J. Rising et al., "A Systems Analysis of a University-Health-Service Outpatient Clinic," *Operations Research*, 21, (1973), 1032.

5. T. F. Chiffriller, Jr., *Successful Restaurant Operation* (New York: Van Nostrand Reinhold, 1982), p. 173.

6. W. S. Vickrey, "Revising New York's Subway Fare Structure," *Journal of the Operations Research Society of America*, 2 (1955), 68.

7. W. S. Vickrey, "Congestion Theory and Transport Investment," *American Economics Review*, 59 (1969), 260.

PROBLEMS

1. Give three examples of systems where it is difficult, or impossible, to dynamically change the service *capacity* in response to changes in the arrival rate.

2. For each of the following, give two examples of systems for which the specified reservation system is appropriate (if possible):

(a) Individual appointment times

(b) Multiple appointments assigned to each time interval (that is, customer can arrive at any time within the interval)

(c) Multiple appointments assigned to the same time (that is, customer must arrive before specified time to receive service)

***3.** A high school counselor is developing an appointment system for students. The length of a visit has been found to have a uniform distribution, with a range from 6 minutes to 12 minutes. The counselor would like to schedule 25 appointments each day, all in one session.

(a) Assume that students arrive precisely on schedule. Write a program to simulate the queueing system. Simulate the system ten times for each of the following appointment intervals: 8 minutes, 10 minutes, and 12 minutes. Determine average student waiting time and average counselor idle time per session for each appointment interval.

(b) Based on your answer to part a, what do you believe the appointment interval should be? Justify your answer.

(c) Suppose that all students arrived early for their appointments, with a mean earliness of 3 minutes and standard deviation of 3 minutes. Do you believe this should change your answer to part b? Explain.

***4.** Simulate the system described in part c of Prob. 3 with an exponential earliness distribution. For this system, answer the questions posed by parts a and b of Prob. 3.

5. A doctor's office has classified patients according to type of treatment and service time:

Arrival Rate	Service Time
Average 4/hour short:	7 minutes on average, range of 5–10 minutes
Average 1/hour medium:	14 minutes on average, range of 10–20 minutes
Average 1/3 per hour long:	23 minutes on average, range of 20–30 minutes

Describe a practical system for scheduling appointments.

6. Based on past requests, a restaurant has determined that dinner parties prefer to arrive at the following times:

Time	Parties	Time	Parties
6:00–6:29	5	8:00–8:29	4
6:30–6:59	8	8:30–8:59	8
7:00–7:29	15	9:00–9:29	5
7:30–7:59	10	9:30–9:59	5

Assuming that the restaurant has 40 tables, and sittings last 3 hours, develop a practical reservation system for the restaurant.

***7.** Suppose that the length of a dinner in Prob. 6 has a normal distribution, with mean 3 hours and standard deviation .5 hour. Also, suppose that the arrival time of each dinner party has a

*Difficult problem.

normal distribution with mean equal to the reservation time and standard deviation of 5 minutes. Simulate the queueing process over ten evenings, with reservations specified by your solution to Prob. 6. From the simulation, estimate mean waiting time. Also, for each evening, determine the maximum waiting time. Based on your results, do you see any obvious way to improve the reservation system?

8. A rock-and-roll promoter uses an open seating plan for events (that is, customers are not offered reserved seats). His theater can accommodate 1000 customers, but, because many customers do not show up, he often sells extra tickets. If a ticket holder arrives after the theater is full, he or she is refunded the $10 ticket price, plus given an additional $20 to compensate for the inconvenience. If ticket holders show up for the event independently, with probability .95, how many tickets should be sold? (Hints: Approximate the binomial distribution with the normal distribution and use the quadratic formula.)

9. Give three examples of where imposition of congestion prices could reduce waiting or delay.

***10.** The arrival process at two parallel servers is stationary and Poisson with a rate of 15/hour. The service time distribution has a mean of 5 minutes at each server, but the first server has a coefficient of variation of 1, and the second has a coefficient of variation of .5. The first server is somewhat inconvenient to use and requires walking 3 minutes out of the customer's way.

(a) Suppose that an equilibrium exists between the servers, so that the average waiting + travel time is the same for both. To the nearest integer, determine the equilibrium arrival rate at each server and the average waiting time among all customers.

(b) Following the principle of identical marginal waiting time, determine the optimal arrival rate at each server (nearest integer) and the minimum average waiting time among all customers.

(c) If customers value their time at $10/hour, what prices should be charged to obtain the solution from part b?

11. Suppose that Fig. 7.13 represents the arrival of telephone calls at an information line and that customers can be processed at the rate of 4500 per hour.

(a) Draw the curves representing customer waiting time and marginal waiting as a function of arrival time for the 6:00 to 9:30 period.

(b) Based on your answer to part a, describe a practical pricing scheme that could eliminate some of the queueing.

12. Describe three systems for which queues might be eliminated by changing the underlying arrival process. Describe how you would change the arrival process.

***13.** Fans arriving at a basketball ticket window have a tendency to renege when the queue is very long. From past observation, the following relationship has been determined:

$$P(\text{renege upon arrival}) = \begin{cases} 0 & L_q(t) \leq 5 \\ .0025[L_q(t) - 5]^2 & 5 \leq L_q(t) \leq 25 \\ 1 & L_q(t) > 25 \end{cases}$$

Fans arrive at the constant rate of 50/hour from 9:00 to 12:00 A.M., the rate of 120/hour between 12:00 noon and 1:00 P.M., and the rate of 60/hour between 1:00 and 5:00 P.M. The service rate is constant at 60/hour. Ticket windows remain open until all customers in line at 5:00 are served.

(a) Write an equation giving the arrival rate of *non-reneging* customers, as a function of the queue length, for the 12:00 to 1:00 period.

*Difficult problem.

(b) Using a computer, and a time increment of one minute, plot cumulative arrivals of non-reneging customers, and cumulative departures, over the entire day. Determine the total number of customers served, and the average waiting time.

(c) Suppose that the service rate drops to 50/hour. Repeat Part (b) with this new rate.

(d) Compare your solution to Part (b) to your solution to Part (c). How would this comparison be different if customers did not renege?

14. A city library decided to install a computerized card catalog to replace its old microfiche system. According to its initial calculations, the new catalog would reduce the time required to look up an entry by 30%. However, to be on the safe side, it replaced its microfiche readers for terminals on a one-for-one basis.

To its surprise, waiting times were larger than ever after the new system was installed, even though the time to look up an entry was no more than predicted, and terminals did not break down. What could have happened?

CASE STUDY

KEEP PATIENTS WAITING? NOT IN MY OFFICE

William B. Schafer, M.D.
Pediatrician
La Canada, Calif.

Good doctor-patient relations begin with both parties being punctual for appointments. This is particularly important in my specialty—pediatrics. Mothers whose children have only minor problems don't like them to sit in the waiting room with really sick ones, and the sick kids become fussy if they have to wait long.

But lateness—no matter who's responsible for it—can cause problems in any practice. Once you've fallen more than slightly behind, it may be impossible to catch up that day. And although it's unfair to keep someone waiting who may have other appointments, the average office patient cools his heels for almost 20 minutes, according to one recent survey. Patients may tolerate this, but they don't like it.

I don't tolerate that in my office, and I don't believe you have to in yours. I see patients *exactly* at the appointed hour more than 99 times out of 100. So there are many GPs (grateful patients) in my busy solo practice. Parents often remark to me: "We really appreciate your being on time. Why can't other doctors do that too?" My answer is: "I don't know, but I'm willing to tell them how I do it."

Booking Appointments Realistically

The key to successful scheduling is to allot the proper amount of time for each visit, depending on the services required, and then stick to it. This means that the physician must pace himself carefully, receptionists must be corrected if they stray from the plan, and patients must be taught to respect their appointment times.

By actually timing a number of patient visits, I found that they break down into several categories. We allow half an hour for any new patient, 15 minutes for a well-baby checkup or an important illness, and either five or 10 minutes for a recheck on an illness or injury, an immunization, or a minor problem like warts. You can, of course, work out your own time allocations, geared to the way you practice.

When appointments are made, every patient is given a specific time, such as 10:30 or 2:40. It's an absolute no-no for anyone in my office to say to a patient, "Come in 10 minutes" or "Come in a half-hour." People often interpret such instructions differently, and nobody knows just when they'll arrive.

There are three examining rooms that I use routinely, a fourth that I reserve for teenagers, and a fifth for emergencies. With that many rooms, I don't waste time waiting for patients, and they rarely have to sit in the reception area. In fact, some of the younger children complain that they don't get time to play with the toys and puzzles in the waiting room before being examined, and their mothers have to let them play awhile on the way out.

On a light day I see 20 to 30 patients between 9 A.M. and 5 P.M. But our appointment system is flexible enough to let me see 40 to 50 patients in the same number of hours if I have to. Here's how we tighten the schedule:

My two assistants (three on the busiest days) have standing orders to keep a number of slots open throughout each day for patients with acute illnesses. We try to reserve more such openings in the winter months and on the days following weekends and holidays, when we're busier than usual.

Initial visits, for which we allow 30 minutes, are always scheduled on the hour or the half-hour. If I finish such a visit sooner than planned, we may be able to squeeze in a patient who needs to be seen immediately. And, if necessary, we can book two or three visits in 15 minutes between well checks. With these cushions to fall back on, I'm free to spend an extra 10 minutes or so on a serious case, knowing that the lost time can be made up quickly.

Parents of new patients are asked to arrive in the office a few minutes before they're scheduled in order to get the preliminary paperwork done. At that time the receptionist informs them. "The doctor always keeps an accurate appointment schedule." Some already know this and have chosen me for that very reason. Others, however, don't even know that there *are* doctors who honor appointment times, so we feel that it's best to warn them on the first visit.

Fitting in Emergencies

Emergencies are the excuse doctors most often give for failing to stick to their appointment schedules. Well, when a child comes in with a broken arm or the hospital calls with an emergency Caesarean section, naturally I drop everything else. If the interruption is brief, I may just scramble to catch up. If it's likely to be longer, the next few patients are given the choice of waiting or making new appointments. Occasionally my assistants have to reschedule all appointments for the next hour or two. Most such interruptions, though, take no more than 10 to 20 minutes, and the patients usually choose to wait. I then try to fit them into the spaces we've reserved for acute cases that require last-minute appointments.

The important thing is that emergencies are never allowed to spoil my schedule for the whole day. Once a delay has been adjusted for, I'm on time for all later appointments. The only situation I can imagine that would really wreck my schedule is simultaneous emergencies in the office and at the hospital—but that has never occurred.

When I return to the patient I've left, I say: "Sorry to have kept you waiting, I had an emergency—a bad cut" (or whatever). A typical reply from the parent: "No problem, Doctor. In all the years I've been coming here, you've never made me wait before. And I'd surely want you to leave the room if *my* kid were hurt."

Emergencies aside, I get few walk-ins, because it's generally known in the community that I see patients only by appointment except in urgent circumstances. A non-emergency walk-in is handled as a phone call would be. The receptionist asks whether the visitor wants advice or an appointment. If the latter, he or she is offered the earliest time available for non-acute cases.

Taming the Telephone

Phone calls from patients can sabotage an appointment schedule if you let them. I don't. Unlike some pediatricians, I don't have a regular telephone hour, but my assistants will handle calls from parents at any time during office hours. If the question is a simple one, such as "How much aspirin do you give a 1-year-old?" the assistant will answer it. If the question requires an answer from me, the assistant writes it in the patient's chart and brings it to me while I'm seeing another child. I write the answer in—or she enters it in the chart. Then she relays it to the caller.

What if the caller insists on talking with me directly? The standard reply is: "Doctor will talk with you personally if it won't take more than one minute. Otherwise you'll have to make an appointment and come in." I'm rarely called to the phone in such cases, but if the mother is very upset, I prefer to talk with her. I don't always limit her to one minute; I may let the conversation run two or three. But the caller knows I've left a patient to talk with her, so she tends to keep it brief.

Dealing with Latecomers

Some people are habitually late; others have legitimate reasons for occasional tardiness, such as a flat tire or "He threw up on me." Either way, I'm hard-nosed enough not to see them immediately if they arrive at my office more than 10 minutes behind schedule, because to do so would delay patients who arrived on time. Anyone who is less than 10 minutes late is seen right away, but is reminded of what the appointment time was.

When it's exactly 10 minutes past the time reserved for a patient and he hasn't appeared at the office, a receptionist phones his home to arrange a later appointment. If there's no answer and the patient arrives at the office a few minutes later, the receptionist says pleasantly: "Hey, we were looking for you. The doctor's had to go ahead with his other appointments, but we'll squeeze you in as soon as we can." A note is then made in the patient's chart showing the date, how late he was, and whether he was seen that day or given another appointment. This helps us identify the rare chronic offender and take stronger measures if necessary.

Most people appear not to mind waiting if they know they themselves have caused the delay. And I'd rather incur the anger of the rare person who *does* mind than risk the ill will of the many patients who would otherwise have to wait after coming in on schedule. Although I'm prepared to be firm with parents, this is rarely necessary. My office in no way resembles an army camp. On the contrary, most people are happy with the way we run it, and tell us so frequently.

Coping with No-shows

What about the patient who has an appointment, doesn't turn up at all, and can't be reached by telephone? Those facts, too, are noted in the chart. Usually there's a simple explanation, such as being out of town and forgetting about the appointment. If it happens a second time, we follow the same procedure. A third-time offender, though, receives a letter reminding him that time was set aside for him and he failed to keep three appointments. In future, he's told, he'll be billed for such wasted time.

That's about as tough as we ever get with the few people who foul up our scheduling. I've never dropped a patient for doing so. In fact, I can't recall actually billing a no-show; the letter threatening to do so seems to cure them. And when they come back—as nearly all of them do—they enjoy the same respect and convenience as my other patients

Source: W. B. Schafer, "Keep Patients Waiting? Not in My Office," *Medical Economics*, May 12, 1986, pp. 137–141. Copyright © 1986 and published by Medical Economics Company, Inc., Oradell, NJ 07649. Reprinted by permission.

CASE STUDY

A SYSTEMS ANALYSIS OF A UNIVERSITY-HEALTH-SERVICE OUTPATIENT CLINIC

E.J. Rising, R. Baron, and B. Averill

In the fall of 1970 the University of Massachusetts Health Service delivered primary health care to approximately 19,000 students on a compulsory prepaid basis. There were, in addition to the outpatient department that is of interest here, approximately 70 inpatient beds, a laboratory and x-ray facilities, an emergency room, a pharmacy, and a mental-health clinic that is housed separately. The University Health Service operates a health education program and an environmental health and safety program.

The outpatient department usually treats between 400 and 500 patients per day. About half of these patients see a physician on either an appointment or a walk-in basis; the rest visit clinics such as the nurse-practitioner clinic, where four nurses deliver primary care under the direct supervision of a physician, and special-purpose clinics operated by nurses for things such as immunizations, TB tests, allergies, warts, obesity, etc. During the fall semester of 1970, the Health Service had 12 full-time physicians on its medical staff. Because of duties in administration, the inpatient area, the nurse-practitioner clinic, 'on-call' periods during the evenings and weekends, and other tasks, only 260 physician hours per week were made available in the outpatient department during regular clinic hours. The rotating schedule meant that no more than seven physicians could be available at one time.

The outpatient department of the Health Service at the University of Massachusetts has many problems in common with other outpatient medical-care delivery systems. The rapid growth experienced over the past several years has resulted in conditions common to most overcrowded health-care facilities. The alleviation of the following conditions was identified as the immediate target of the study:

1. There was a long waiting time for patients.
2. The professional staff felt overworked and harassed.
3. There was much confusion and crowding in the waiting rooms at predictable times (Monday, Tuesday, and Friday afternoons).
4. The physicians were still seeing patients as long as an hour past closing time.

5. During the day, physicians were sometimes idle because patients did not always keep appointments scheduled several weeks in advance.
6. The current building was (and is) overcrowded, as it was designed for a student body of 10,000 and is currently serving a student body of over 19,000 students.

The system was conceptualized as a complex queueing system, and consequently, two types of data were required. Information was needed on the arrival patterns of patients and on the way they spent their time in the system. This latter information was broken down into waiting time, the routing of patients through the system, and the amount of time required to serve their needs at each of the places in the system they received service.

The data used to determine arrival patterns were taken from the encounter form that all patients fill out prior to any service they receive from the Student Health Service. These forms elicit information from each patient including name, symptoms, particular service desired, etc. After it is filled out by the arriving student, it is stamped with the date and time and placed with the medical record. The physician who sees the patient enters the tentative diagnosis; the various paramedical departments that serve the patient also record any services that are rendered. These forms were designed as part of a management information system and are in constant use. Data on the consultation time physicians spent with patients and time patients spent in the laboratory, x-ray, etc. were taken separately. Clerks were stationed near the entrance to physicians' examining rooms and other facilities, and were furnished with date-time stamping clocks. Special record sheets provided for the purpose were stamped as each patient entered and again when he left each service. These records were also time stamped and collected when the patient left the building. The information stamped on these forms gave an accurate account of the services the patient used and of the time necessary to provide the service in question, as well as all waiting times involved. During the data-taking periods, the number of these special records collected agreed with the number of medical encounter forms within about 7 percent of the total; in addition, less than 5 percent of the special forms provided to collect service times and routing data were unusable because of a missing arrival or departure stamp.

Step 1. Smoothing Patient Arrivals by Day

Enrollment projections and historical data on student usage were utilized to estimate the number of physician visits to be expected during the fall semester of 1970. It was estimated that there would be, on the average, approximately 1,000 physician visits per week.

The average numbers of patients arriving at the University Health Service to see either a doctor or a nurse concerning a health problem are shown in Table I for the fall semesters of 1969 and 1970; Table II shows the numbers of these total arrivals who actually saw physicians for the same periods. These data are broken down by day of the week, and the average number of arrivals that occur on each day of the week is displayed, the percentage of the semester-long daily average appearing in parentheses. The largest number of arrivals occurred on Monday, with Tuesday a close second. The smallest values occurred on Thursday, with Wednesday and Friday in an intermediate position.

Figure 1 shows the average total hourly patient arrivals over the day for Monday and Thursday. Monday and Thursday were chosen because they show the heaviest and lightest patient load, respectively, and because the academic calendar ordinarily schedules Monday-Wednesday-Friday or Tuesday-Thursday classes. Thus, a picture

TABLE I Numbers of Patients Arriving at the University Health Service to See Either a Physician or Nurse Concerning a Health Problem

	Average number of visits per week	Average number of visits per day	Average number of visits (percentages of average daily visits given in parentheses)					University enrollment (approx.)
			Mon.	Tue.	Wed.	Thu.	Fri.	
Fall semester 1969	1487	297.4	351 (118%)	341 (114%)	264 (88%)	253 (84%)	288 (96%)	17800
Fall semester 1970	1640	328.0	404 (123%)	342 (104%)	306 (93%)	290 (88%)	298 (91%)	19125

The data are selected to include only arrivals between 8 A.M. and 5 P.M. on Mondays through Fridays for weeks in which there are no holidays, and also excludes data for Fridays before holidays occurring on the following Monday.

TABLE II Numbers of Patients Who Saw Physicians at the University Health Service

	Average number of visits per week	Average number of visits per day	Average number of visits (percentages of average daily visits given in parentheses)				
			Mon.	Tue.	Wed.	Thu.	Fri.
Fall semester 1969	889	177.8	219 (123%)	190 (107%)	170 (93%)	162 (84%)	169 (92%)
Fall semester 1970	1008	201.6	216 (106%)	202 (100%)	198 (98%)	194 (94%)	204 (101%)

The data are selected to include only arrivals between 8 A.M. and 5 P.M. on Mondays thru Fridays for weeks in which there are no holidays, and also excludes data for Fridays before holidays occurring on the following Monday.

of the arrival patterns on these two days should reveal any strong bias by volume or day of the week. Figure 2 illustrates that the patient interarrival times over the entire day have a negative exponential distribution. From this fact it was assumed that this form of distribution could be used throughout the day to generate walk-in arrivals, while the mean value of the distribution was changed hour by hour to correspond to the observed values.

The data for Fig. 1 were obtained from the arrival date-time stamp made on all medical encounter forms. The similarity of the pattern between Monday and Thursday data demonstrates that there was little biasing effect of class hours, which tend to be scheduled at the same hours on Monday, Wednesday, and Friday, or at the same time on Tuesday and Thursday. Although the 1970 data were not available when the analysis was performed, it is presented here to show the stability of the pattern.

The consultation times (service times) that physicians spent with patients were measured in three separate categories: for appointment patients, walk-in patients, and the time required for a second service. These distributions are shown by the

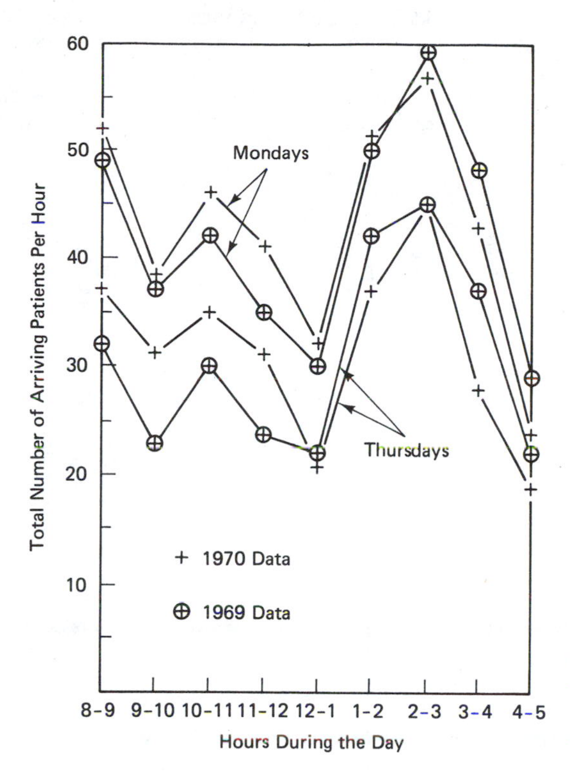

Figure 1 Hourly arrivals at the Student Health Service (Monday and Thursday averages for the fall semesters in 1969 and 1970).

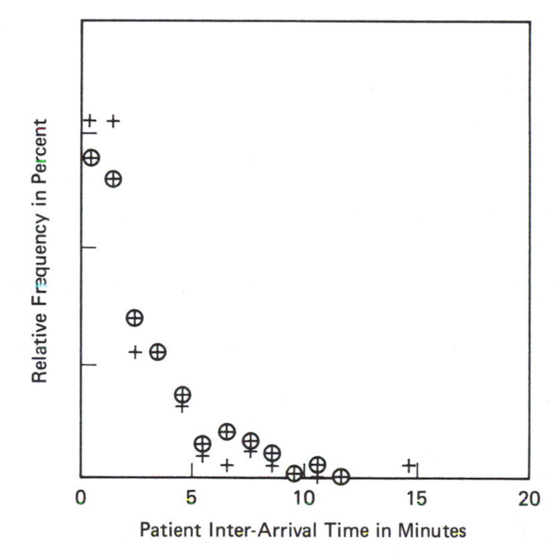

Figure 2 The frequency distribution of patient interarrival times. + Monday, April 6, 1970; $\bar{x} = 2.167$, $s = 2.402$, $n = 237$. ⊕ Thursday, April 9, 1970; $\bar{x} = 2.626$, $s = 2.838$, $n = 202$.

histograms in Fig. 3. The sample mean and sample standard deviation for appointment service times were 12.74 minutes and 9.56 minutes, respectively; for walk-in service times they were 9.61 minutes and 4.91 minutes. No published data were found for comparison purposes.

In the actual operation of the clinic, the physicians see patients in a sequence governed by three priority considerations. First priority is given to emergency pa-

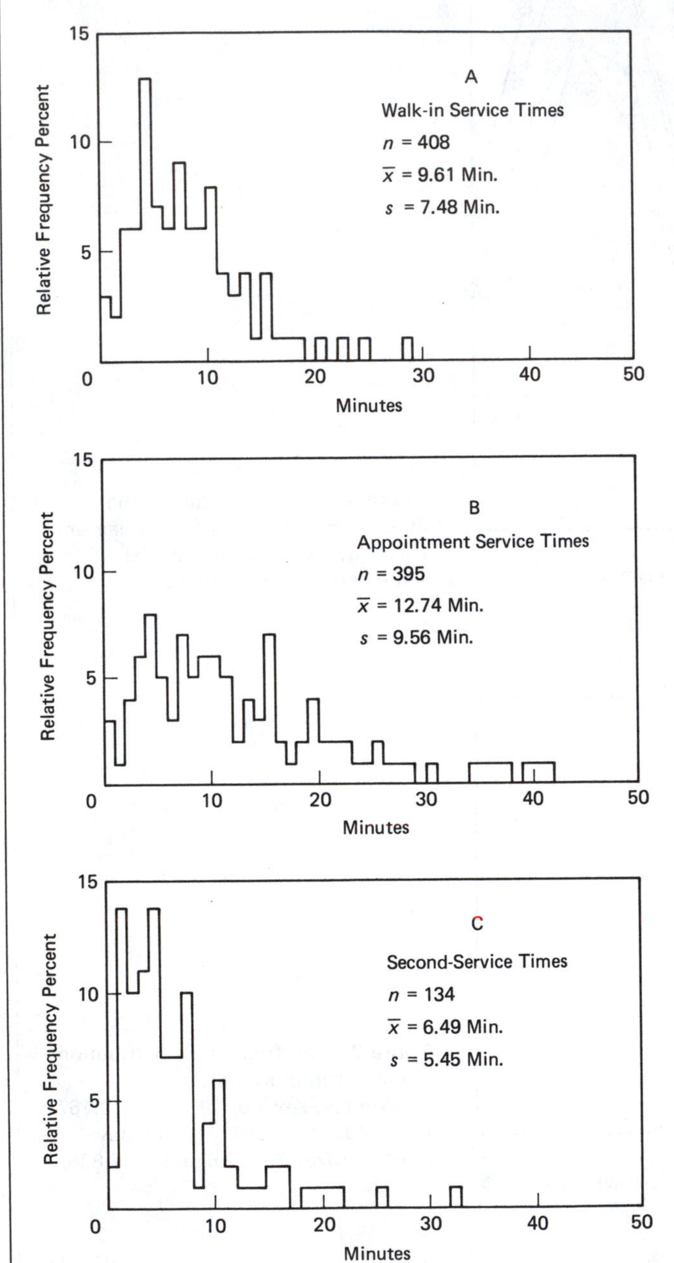

Figure 3 The histograms of service times.

tients entering the system and patients who are returning from a visit to the laboratory, x-ray, etc., to see the physician they have already seen earlier on the same day. Second priority is given to patients who have made an advance appointment with a specific physician. Last priority is given to walk-in patients who are then seen on a first-come, first-served basis. The walk-in patients are seen by any physician as soon as he becomes free of higher priority work. Most physicians use two examining rooms, which means a patient is being seen in one room while the next patient is being prepared in another. When a physician finishes with one patient, the priority system is used to select a patient for the examining room just vacated. This priority system is administered by a nurse who controls the flow of patients through the release of medical records to individual physicians in the proper order. The implication of this priority system is to eliminate immediately and completely lost physician time due to no-show appointment patients.

Source: Excerpted from E. J. Rising, R. Baron, B. Averill, "A Systems Analysis of a University-Health-Service Outpatient Clinic," *Operations Research*, 21:5, 1030–1047 (1973). Reprinted by permission of the publisher.

CASE STUDY QUESTIONS

Keep Patients Waiting? Not in My Office

1. Why is it important to give patients a specific appointment time?
2. The author states that he reduces waiting by using multiple examining rooms. Justify this policy in terms of a queueing model.
3. The author likes to reschedule appointments when he is interrupted by an emergency. If the interruption occurs between 1:00 and 2:00 P.M., how might this policy benefit patients who arrive later, say, 4:00 to 5:00 P.M.?
4. Is the policy of not seeing walk-ins sound? Can you think of an effective way to serve this component of demand?
5. Given that patients sometimes arrive late or early, in what order should they be served: (a) according to appointment time; (b) according to arrival time; (c) some alternative system? Justify your answer.
6. Should a doctor be treated the same way when he or she is late as patients are treated when they are late?

A Systems Analysis of a University-Health-Service Outpatient Clinic

1. Describe the different components of demand. To what extent can each component be controlled?
2. In what ways does the University Health Service differ from the doctor's office described in the previous reading. Which system is more efficient? Which provides the better service, and why?
3. Given the information provided in the reading, develop a system for scheduling patient appointments. Discuss how you would fill appointments and set appointment length.

4. Do you believe that the introduction of a pricing scheme could improve the health service? Describe such a system.
5. What can the health service do to affect the underlying demand process? How might this reduce waiting time?
6. Suppose that a new appointment system eliminates all waiting after arriving at the clinic. Does this necessarily mean that the queueing system is providing excellent service and that all waiting is eliminated?

chapter 9

Queue Discipline

Chapters 7 and 8 present a variety of ways to eliminate queueing problems. One can alter the service process, perhaps through the use of part-time servers, and one can alter the arrival process, through reservations or pricing. So suppose that you've implemented these ideas and have eliminated the *predictable* queues. Yet from time to time, for unexplained reasons, queues still materialize and your customers are clamoring for service. These are the *stochastic* queues, caused by random variations in the arrival and service patterns. What do you do now?

While stochastic queues can be reduced through eliminating the causes of *random* variability, or possibly by altering servers with ancillary activities (see Chap. 5), they can seldom be eradicated. Fortunately, there are ways to cope with stochastic queues. The size of these queues, and the associated cost, can be controlled through the use of an effective queue discipline. As introduced in Chap. 1, the ***queue discipline*** prescribes the order in which customers are served. The most familiar is the first-come, first-served (FCFS) discipline. For reasons of equity, FCFS is the way people expect to be served in most circumstances. Nevertheless, from several perspectives, FCFS is not always the best way to serve customers, particularly when customers are not people but work orders instead.

Queue discipline, or ***scheduling***, has been studied in great depth within the context of manufacturing, especially job shop manufacturing. A job shop is a collection of machines and individuals capable of performing a variety of tasks in the completion of jobs. Two examples are machine shops (containing drills, lathes, and so on) and printing shops (containing presses, binding equipment, and so on). In an ***open job shop***, jobs (customers) arrive in the form of work orders, detailing tasks, production quantities, and usually due dates, and leave in the form of completed work. The queue discipline specifies the sequence in which orders are processed on the equipment. In a ***closed job shop***, production is not scheduled to meet specific customer orders. Instead, production is planned to stock inventories, which are later withdrawn to fill orders. Closed shops are common among firms selling a few standardized products, while open shops are common among firms providing customized service.

Job shops can be further classified into those using ***dynamic schedules*** and those using ***static schedules***. A dynamic schedule is a queue discipline that is continuously updated as jobs arrive and are completed. The next job to enter service is selected from the entire set of jobs in the queue. A static schedule is a periodic process. At regular time intervals (usually a day, week, or month), the queue is examined and a job sequence is formulated for the next interval. This sequence is usually locked in and not altered, even if higher priority jobs arrive during the interval.

Production scheduling is an important issue in both open and closed shops, with either dynamic or static schedules. However, only the *dynamic open job shop* precisely matches the characteristics of a queueing system, with streams of customers (jobs) arriving continuously over time. This is the focus of this chapter. Nevertheless, closed job shops and static schedules are closely related to queueing systems, and many of the results are transferable.

Though job shop scheduling and queue discipline are virtually two names for the same thing, each has developed its own terminology. For instance, in the job shop literature, ***flow time*** has the same meaning as time in system. ***Machines*** is used in place of servers, ***jobs*** is used in place of customers, ***processing time*** is used in place of service time, and ***completion time*** is used in place of departure time from system. For consistency, this chapter will continue to use the queueing terminology, with the exception that customers will sometimes be called jobs and departure time from system will sometimes be replaced by completion time. The term *job* will allow the work being completed to be distinguished from the person submitting the work, who will then be called the customer.

Unlike a queue of people, there are less compelling reasons to process jobs in FCFS order. Fairness is less of a concern. A job will not complain if it is served out of FCFS order, provided that it is completed on time. This opens up a realm of queue disciplines that are capable of outperforming FCFS according to one or more measures of performance.

In this chapter, queue discipline is examined as a way to reduce queueing delay and the cost associated with queueing delay. The perspective will be that of a single type of server, perhaps working in parallel, that receives work according to a random process and completes work with random service times. The primary application will be in processing work orders, though the principles will also apply to serving people. In Chap. 10, concepts are extended to networks of servers performing different functions.

The discussion in this chapter only scratches the surface of a rich area of research. Interested readers may wish to consult the classic reference on scheduling by Conway et al. (1967). Other references include the survey articles by Graves (1981) and Panwalkar and Iskander (1977) and the books by Baker (1974), Coffman (1976), French (1982), and Muth and Thompson (1963).

9.1 TRIAGE

No matter what system is used, the queue discipline must be tied to identifiable characteristics of the customers. Arrival time is certainly one of these characteristics. It is the basis for the FCFS discipline. But customers have many other identifiable characteristics, such as

1. Service time (or *job size*)
2. Promised delivery time (also called *due date*)
3. Requested delivery time
4. Urgency of work

These and other characteristics are the bases for queue disciplines. Jobs may be processed according to service time, shortest first; jobs may be processed in order of promised delivery time, earliest first. Or the queue discipline may account for combinations of characteristics, such as "shortest slack first," where slack is the time until delivery minus the service time. Most practical queue disciplines are in fact hybrid strategies that account for two or more characteristics.

Unfortunately, customer characteristics cannot always be precisely defined. Even worse, characteristics can sometimes be manipulated by the customer to gain higher priority than warranted. The service time, for instance, is not known with certainty until the service is completed. Prior to service, when the work is scheduled, it can only be estimated, usually from a measure of job size or difficulty. Delivery times also are not concrete. Customers frequently request that work be completed earlier than necessary, so as to provide them with a safety margin. Job urgency in particular is hazy, and is usually the product of the customer's persuasive skills.

Triage is the process by which the server assesses the customer characteristics and assigns priority. It is a term most often used in emergency medicine for prioritizing the treatment of patients in major accidents. Triage occurs before treatment begins, to ensure that the most critical injuries are served first. But triage is essential to any queueing system that employs a discipline other than FCFS. Among other things, this means that service time must be accurately estimated, and realistic due dates must be assigned. Triage should be a formal part of the queueing system, even though it necessarily adds some delay (triage is a queueing system in itself). All too often, the process of assigning priorities is ad hoc; every job becomes urgent, and one can only react to emergencies rather than plan ahead. Triage can add order to a queueing system.

Example

Students at an ivy-clad institution must wait in long queues to receive financial aid information. Some of their questions are simple and can be answered immediately, while others require detailed explanations. To reduce waiting, students could be asked the nature of their questions upon arrival. Simple questions are then answered on the spot. Otherwise, the student is referred to an appropriate line.

With these comments on triage in mind, the characteristics of some of the classic queue disciplines are presented in the following section.

9.2 CLASSIC QUEUE DISCIPLINES

In Sec. 9.2 and 9.3, the queueing system will be assumed to have a single server. Later in the chapter, this condition will be generalized to parallel servers.

9.2.1 Service Time Independent Disciplines: FCFS, LCFS, SRO

Suppose that it is impossible to differentiate between customers' service times prior to their completing service. For example, a customer phones an airline reservation line requesting information. Prior to the call, the reservation clerk does not know what type of information is needed, and cannot predict the service time. It is not an identifiable customer characteristic and cannot be the basis for a queue discipline. Examples of service time independent disciplines include the following:

Definitions 9.1

First-Come, First-Served (FCFS) Each time a server becomes available, select the customer with the earliest arrival time.

Last-Come, First-Served (LCFS) Each time a server becomes available, select the customer with the latest arrival time.

Service in Random Order (SRO) Each time a server becomes available, randomly select a customer.

Among service time independent disciplines, the following important property always holds:

Expected time in system and expected time in queue are independent of queue discipline.

Why is this true? Because the state of the system and the transition rates between states only depend on the number of customers in the system, not the number of customers of each type. (Note that this principle conforms to the results shown in earlier chapters, which give expected waits independently of the queue discipline.)

Even though queue discipline does not affect expected waiting time, it will affect the waiting time *variance*. Because certainty is usually preferred to uncertainty, minimizing the variance is a logical basis for selecting the queue discipline. In steady state

Processing customers according to the FCFS discipline minimizes waiting time variance (time in queue and in system) when queue discipline is service time independent.

While not a formal proof (see Kingman 1962), one can see why this statement is true by considering the variance for a discipline other than FCFS, as in Fig. 9.1b relative

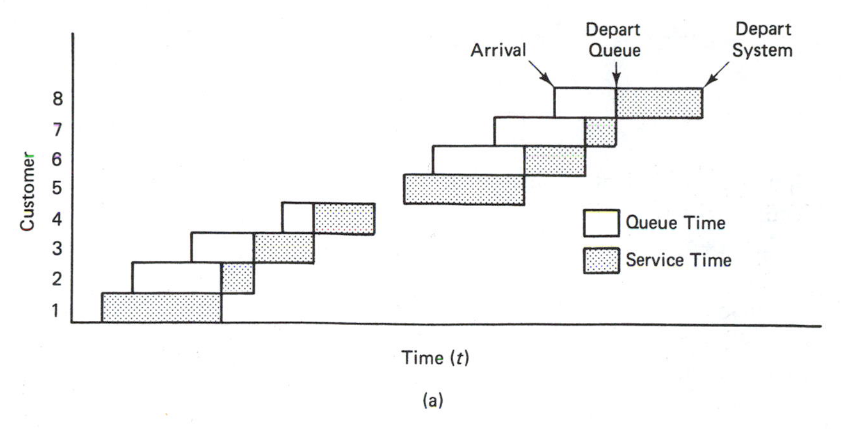

Figure 9.1a Gantt chart demonstrating first-come, first-served (FCFS) discipline.

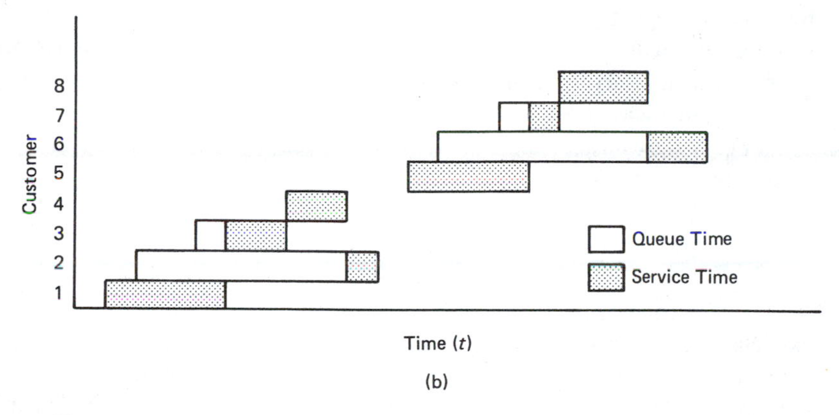

Figure 9.1b Gantt chart for same group of arrivals, with last-come, first-served (LCFS) discipline.

to an FCFS discipline (Fig. 9.1a). Given that expected waiting time is the same for all queue disciplines, the variance is minimized when the expectation of the *squared* waiting time is minimized. Let

t_n = arrival time of nth customer

$S(n)$ = service time of nth customer

$\Sigma \tilde{W}_{qn}^2$ = sum of squared times in queue, non-FCFS

ΣW_{qn}^2 = sum of squared times in queue, FCFS

Consider the change in squared time in system if two customers are served in a sequence other than FCFS. Suppose that at time T, the server is available to serve either customer n or customer $n + 1$:

$$\begin{aligned}\Sigma\tilde{W}_{qn}^2 - \Sigma W_{qn}^2 &= \text{difference in squared time in queue}\\ &= (T - t_2)^2 + (T - t_1 + S_2)^2 - [(T - t_1)^2 + (T - t_2 + S_1)^2]\\ &= 2(T - t_1)S_2 + S_2^2 - [2(T - t_2)S_1 + S_1^2] \qquad (9.1)\end{aligned}$$

S_1 and S_2 have identical distributions. Therefore, $E(S_1) = E(S_2)$ and $E(S_1^2) = E(S_2^2)$. This leads to the following:

$$E[\Sigma\tilde{W}_{qn}^2 - \Sigma W_{qn}^2] = 2(t_2 - t_1) \cdot E(S) > 0 \qquad (9.2)$$

Because $t_2 > t_1$ by definition, Eq. (9.2) must be greater than zero. Hence, processing customers in any order other than FCFS increases the waiting time variance. Not only is FCFS equitable, it optimizes an important measure of performance.

For various practical circumstances (explained in Chap. 11), it is not always possible to implement an FCFS discipline. Though undesirable, customers are sometimes served LCFS or SRO. While not changing expected waiting time, these disciplines necessarily provide greater variations in waiting time. With LCFS, a customer might be fortunate enough to arrive immediately prior to a service completion, in which case waiting time will be very small, or unfortunate enough to arrive immediately prior to a long sequence of arrivals, in which case waiting time will be very large (the subsequent arrivals being served first).

For a steady-state single server queue, with Poisson arrivals and a general service time distribution, Conway et al. (1967) showed the following:

$$\frac{E(W_{q,\text{LCFS}}^2)}{E(W_{q,\text{FCFS}}^2)} = \frac{1}{1 - \rho} \qquad (9.3)$$

where the second subscript denotes the queue discipline. Equation (9.3) indicates that the expectations of the waiting time squared are similar when ρ is small and very different when ρ is close to 1. In essence, the waiting time variability associated with LCFS increases as the system becomes more heavily utilized and queues become long.

If service times are also exponentially distributed, then

$$\frac{E(W_{q,\text{SRO}}^2)}{E(W_{q,\text{FCFS}}^2)} = \frac{1}{1 - \rho/2} \qquad (9.4)$$

As with LCFS, this ratio increases as ρ approaches 1. However, the ratio is never more than 2. Comparing all three disciplines in steady state, the LCFS discipline has been shown to maximize the variance of time in system, and we have already seen that FCFS minimizes the variance. The variance for SRO falls between (see Conway et al., 1967, for a review of these results).

9.2.2 Diagrams of Cumulative Work

A useful technique in comparing queue disciplines is to diagram cumulative arrival of work as a function of time.

Definition 9.2

Cumulative arrival of work, $AW(t)$ The total *system* time required to serve the customers that have arrived by time t.

$$AW(t) = \sum_{A^{-1}(n) \leq t} S(n)/m \tag{9.5}$$

As before, m is the number of servers. Figure 9.2 shows an example diagram of $AW(t)$. Like the curve $A(t)$, each step in $AW(t)$ represents the arrival of a customer; unlike $A(T)$, step sizes vary in proportion to service time. Figure 9.2 also shows ***cumulative departure of work***, $DW(t)$. So long as there is work in the system, and the system is operating at capacity, $DW(t)$ is a line of slope one. In other words, it takes 1 hour to complete 1 hour of work, 2 hours to complete 2 hours of work, and so on. Unlike departure of customers, $DW(t)$ is not a step curve.

Definition 9.3

Queue of work at time t, $LW(t)$ The total work remaining in the system at time t

$$LW(t) = AW(t) - DW(t) \tag{9.6}$$

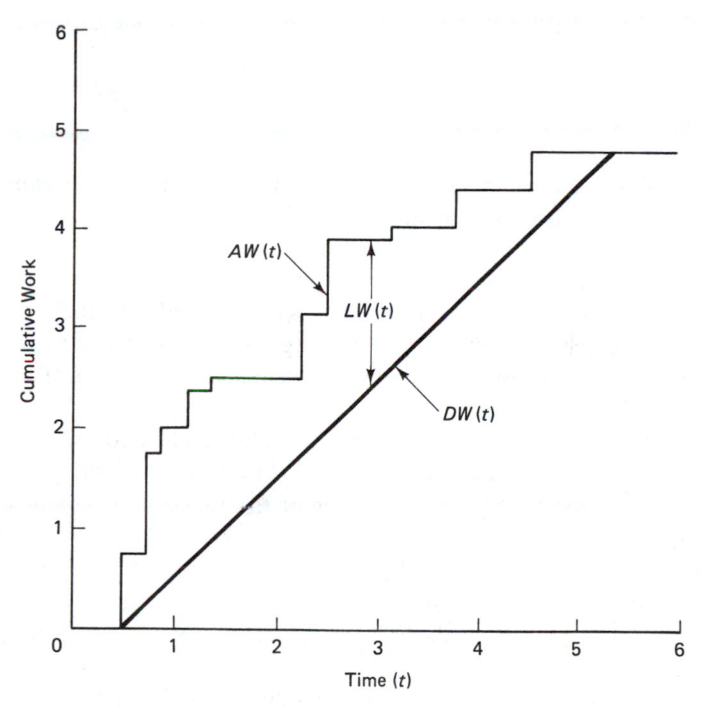

Figure 9.2 In a work-conserving queue, work in system does not depend on the queue discipline. Slope of departure curve equals one when work is in the system.

The queue of work is the time required to serve all of the customers remaining in the system at time t.

Many queueing systems possess the property of ***work conservation***. This means that $LW(t)$ does not depend on the order in which customers are served. The choice of a queue discipline then amounts to minimizing the cost associated with a fixed amount of work, as discussed in the following section.

9.2.3 Shortest Service Time First (SST)

Most often, the server can differentiate service times prior to service. In a store, service time can be estimated from the number of items being purchased; in a job shop, service time can be estimated from the production quantity. Almost always, this information should be incorporated in the queue discipline.

Definition 9.4

Shortest Service Time (SST) Each time a server becomes available, select the customer with the shortest service time.

If service times are not known with certainty, then jobs could be processed in order of *expected* service time.

If a single server system is work-conserving (for example, service times are not sequence dependent) and jobs arrive by a Poisson process, SST has the following important property:

Processing jobs according to the SST discipline minimizes expected time in system and expected time in queue.

Recall that $LW(t)$ is fixed in a work-conserving queue. Therefore, the total number of customers in the system at time t, $L_s(t)$, is $LW(t)$ divided by the average remaining service time of the jobs in system. By selecting short jobs first, SST maximizes the average service time of the jobs *remaining* in system and minimizes $L_s(t)$. From Little's law, this also minimizes time in system.

The optimality of SST can also be understood by considering a discipline in which jobs are not processed according to SST (Fig. 9.3b, for example). If the order of jobs 2 and 3 is reversed, note that the latter of the two jobs is completed at the same time. However, the first job is completed earlier with SST because it has the shorter service time. In general, if any pair of jobs is completed out of the SST order, exchanging their sequence will reduce expected time in system. Repeating this reversal process for all such pairs eventually produces the SST discipline (Fig. 9.3a). (A formal proof can be found in Phipps, 1956.)

Note in Fig. 9.3 that SST does not guarantee that the entire job sequence is SST. It only guarantees that the next job selected from the queue is shortest. Even though job 1 is longer than job 3, job 1 is processed first; at the time it begins service, job 3 has not yet arrived.

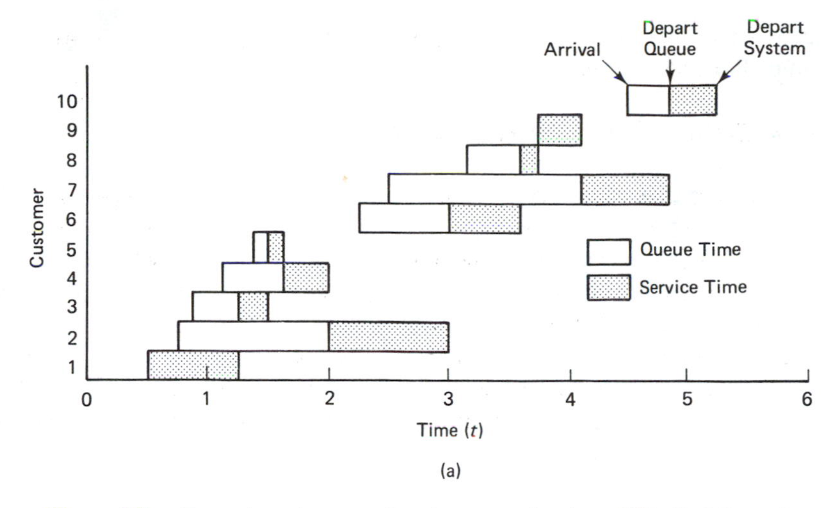

Figure 9.3a Gantt chart demonstrating shortest-service-time (SST) discipline (chart based on arrivals in Fig. 9.2).

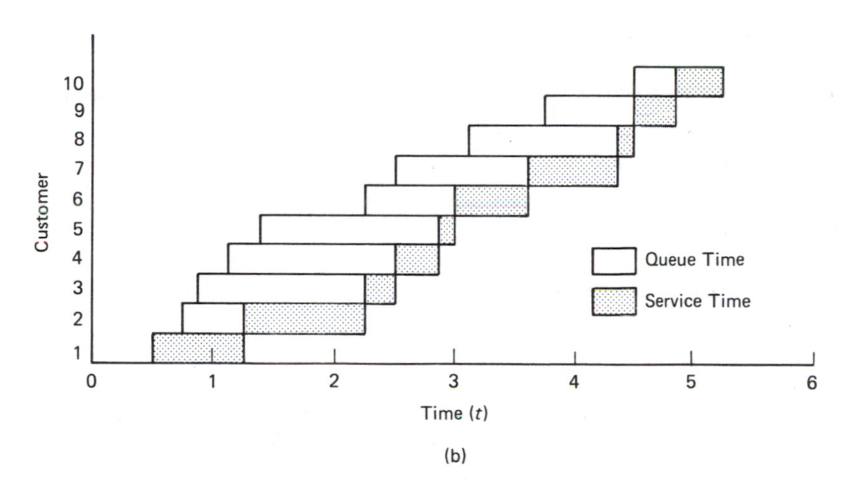

Figure 9.3b Gantt chart for same group of arrivals as Fig. 9.3a, with FCFS discipline.

Because it minimizes expected time in system, SST is an extremely robust rule that performs well relative to virtually all measures of performance—standard deviation of waiting time, completion of jobs on time, and so on. This has been confirmed in numerous simulation studies, such as the classic study of a nine-machine job shop by Conway (1965a, 1965b). As reviewed by Conway et al. (1967), the mean time in system for SST was over 50 percent less than FCFS. More surprising was that the *standard deviation* was over 30 percent less than FCFS. Of course, this does not mean that all jobs are completed at the earliest possible times. Some jobs, particularly the long ones, are necessarily delayed. However, the vast majority (90 percent or more) of jobs are completed earlier under SST than under FCFS.

9.2.4 Earliest Due Date First (EDD)

The due date may be a time requested by the customer, a time suggested by the server, or some combination of the two. The due date may be a point of negotiation, between the sales department of the server and a buyer for the customer. Whatever the case, the due date is a fixed time (not necessarily a day, but possibly even a specific hour and minute) to be compared against the job's completion time. Completing a job on time is good, whether it is completed one minute or one week before the due date. Completing a job late is bad, but completing it just a little late is better than completing it very late. Unfortunately, these qualitative statements are difficult to quantify, which makes scheduling jobs with due dates a rather subjective process.

Definitions 9.5

Lateness, L = completion time minus due date

Tardiness, T = $\max\{0,L\}$

When a job is late, the tardiness and the lateness are the same; when a job is early, the lateness is negative and the tardiness is zero.

With due dates, an appealing discipline is the following:

Definition 9.6

Earliest Due Date (EDD) Each time a server becomes available, select the customer with the earliest due date.

When all jobs are available for scheduling at the same time, EDD is known to minimize the *maximum* tardiness and the *maximum* lateness among a set of jobs. In general, research by Jackson (1961) indicates that EDD will tend to minimize maximum tardiness and lateness. More importantly, however, EDD *does not* minimize the more important objectives of expected tardiness or number of tardy jobs.

The problem with EDD is easy to recognize. If a few jobs are difficult to schedule (either because of long service time or tight deadline), it would be unwise to give them priority at the expense of making many other jobs late. But this is exactly what EDD does; it guarantees the best possible service for the most difficult jobs. Furthermore, EDD often invites abuse by giving too much discretion to the customer. Customers often can set or influence due dates without regard to true priority.

Expected lateness is minimized with the SST discipline; the objectives of minimizing expected time in system and expected lateness are equivalent. Simulation studies have also found that SST performs well (though not optimally) relative to the objective of minimizing number of jobs late and expected tardiness, particularly when the system is heavily utilized. A procedure for minimizing number of jobs late was developed by Moore (1968), but this relies on scheduling all jobs simultaneously, an impossible condition in most situations. Researchers have also experimented with a variety of hybrid disciplines, which combine the features of SST and EDD. These are discussed in Sec. 9.2.7. For a review of queue disciplines that account for due date, see Baker (1984).

9.2.5 Shortest Weighted Service Time (SWST)

The last of the classic queue disciplines is SWST, shortest *weighted* service time. Here, the weight is a measure of a job's importance. It may represent the cost per unit time for delaying a customer. The objective is to minimize an expression of the following type:

$$\min \quad \sum_n f_n E[W_s(n)] \tag{9.7}$$

where

$$f_n = \text{weight for job } n \text{ (cost per unit time)}$$

$$E[W_s(n)] = \text{expected time in system for job } n$$

The importance of a job is proportional to its weight, with a larger weight signifying a more urgent job.

According to the SWST rule, the order in which customers are served is defined by a priority index, which equals the ratio of a job's service time to its weight:

$$r_n = \text{priority index for job } n$$

$$= \frac{S(n)}{f_n} \tag{9.8}$$

Definition 9.7

Shortest Weighted Service Time (SWST) Each time a server becomes available, select the job with the smallest priority index r_n.

As is logical, SWST gives preference to jobs with small service times and large weights. SWST also minimizes Eq. (9.7) for work-conserving single serve queues with Poisson arrivals. That is, for a fixed amount of work in the system $LW(t)$, SWST minimizes the summed weights among the jobs remaining in the system.

To be successful, the weights used in SWST must not be arbitrary and should be tied to well-defined customer characteristics.

Example

Tour groups arrive at a major tourist site by a Poisson process. The service time for selling and collecting tickets has been found to be the following function of group size, N:

$$S = 1 + .1N$$

If groups were served according to SST, then the smallest groups would be served first. However, the site operator believes that a group's importance should be proportional to its size: $f(N) = N$. Hence, the priority ordering is defined by

$$\frac{S}{f(N)} = \frac{1}{N} + .1$$

With the weighted service time rule, the largest, not the smallest, group is served first.

This example is representative of a wide variety of queueing systems: Both service time and priority increase with job size. As is common, the service time increases less than proportionately with size, and highest priority goes to the large jobs, not the small. A common mistake in applying SST is not to consider job importance. *One should always be cautious of the possibility that large jobs are more important than small, and that SWST should be used instead of SST.*

SWST provides a framework for the creation of hybrid rules, as the weights can be derived from a variety of job characteristics. For instance, the weight might be inversely proportional to the schedule slack. Then, as the due date approaches, the weight would increase, providing higher priority. Weights might also be related to specific costs. A contract with the customer might specify a penalty for each unit of time that completion is delayed. This penalty would then become the weight in Eq. (9.8).

9.2.6 Other Scheduling Characteristics

While arrival time, service time, due date, and job weight are the most common characteristics used in queue disciplines, other possibilities include:

Makespan: Time required from the beginning of a set of jobs until the completion of the last job in the set

Setup time: A service time that depends on the type of job that immediately precedes service

Reservation time: Promised time for beginning service

Makespan and setup time are significant factors in many job shops. Makespan is most important in job shop networks, so its discussion will be delayed until Chap. 10. Setup times have already been discussed in Chap. 7 and will not be studied in detail here. However, one practical discipline would be to give priority to jobs that require no setup time relative to the job last completed. If no such job is in the queue, another discipline, such as FCFS, SST, EDD, or SWST, might then be invoked. Another discipline would be to select the job with shortest immediate service time, which would include the appropriate setup time relative to the job just completed. On the whole, however, setup time is a more significant factor for closed job shops than for open job shops (see Baker 1974; French 1982; Graves 1981).

Reservation time is somewhat analogous to a due date, except that it pertains to the time service is promised to *begin* rather than the time service is promised to be *completed*. It most often comes into play in queues of people, such as restaurants and medical offices (see Chap. 8). If customers do not arrive in the same order as their reservations, the server must decide whether to serve customers FCFS or, more likely, first-reservation, first-served (FRFS). As a rule of thumb, if the server is operating ahead of schedule or has some excess capacity, the FCFS order makes the most sense. No one is harmed by this sequence and the early customer has a shorter wait. On the other hand, if scheduling is tight, allowing a customer to be served early will likely delay other customers. Certainly it would be unfair to delay a customer who arrives on time in favor of someone else who arrives early.

Many queue operators give lowest priority to customers who arrive late (perhaps with a few minutes leeway), fitting them in only if there is time. This practice is fair so long as the server itself is punctual. But it would be unfair to penalize a customer for being late when the server is usually late. A habitually late server should not worry so much about queue discipline as in making reservations realistic (the server is the one who needs the discipline!).

9.2.7 Hybrid Scheduling Rules

Conflicting objectives usually make it impossible to stick completely to any of the pure queue disciplines. Hybrid rules, which account for two or more customer characteristics, are used instead. Typically these rules are based on a combination of the service time and arrival time characteristics or the service time and due date characteristics. The following examples, taken from Panwalkar and Iskander (1977), are representative:

FCFS/SST From jobs waiting more than a specified time, select according to FCFS. If no job has been in the queue longer than the specified time, select according to SST.

Alternating FCFS/SST Use SST for a certain time, then FCFS, and repeat as a cycle.

Prioritized FCFS Divide jobs into priority classes. Within each class, serve according to FCFS.

Min slack Select the job with the least slack (available time before due date minus remaining service time).

Though not all inclusive, Panwalkar and Iskander's article still presents well over 100 different disciplines.

Despite the plethora of possibilities, most simulation studies of queueing systems (such as Conway and Maxwell 1962) have found that SST is hard to beat with respect to nearly all measures of performance, particularly when the queueing system is heavily utilized. For lower levels of utilization (less than 90 percent), disciplines based on slack can sometimes outperform SST with respect to minimizing number of jobs late. But SST will always be best relative to the objective of minimizing time in system.

One of the ways to obtain the short time in system advantage of SST and still ensure that long jobs are completed on time is to ensure that due dates are set realistically. If due dates factor in an extra waiting time allowance for long jobs, there should be no problem with completing jobs on time. But more on this in Chap. 10, which includes a discussion of hybrid rules in the context of queueing networks.

9.3 INTERRUPTION AND INSERTED IDLE TIME

A dynamic schedule may operate with or without *interruption*.

Definition 9.8

Interruption (also called a *preemption*) means that a job that is already in service is interrupted to serve another job with higher priority.

Interruption is common in computer systems. Long, low priority jobs are interrupted by short, high priority jobs and resumed later. This prevents the long jobs from tieing up the system for extended periods of time. Whether or not jobs should be interrupted depends on the costs associated with resuming service. That is, it depends on how much of the attained service time is lost when a job is interrupted. In computer systems, the resumption time is negligible, but in job shops, interrupting a job may amount to starting from scratch. The former is called ***interrupt resume***, the latter ***interrupt repeat***. Clearly, interruption should be rare with interrupt repeat.

9.3.1 Interruption for New Arrivals

With interrupt resume, one does not have the dilemma of whether or not to begin a long job, for fear of higher priority jobs arriving soon after. If a more urgent job comes along, the job in service is stopped, the new job is begun, and the original job is resumed later. It may be that the new job is also interrupted when an even higher priority job arrives. No matter, two or more jobs can be resumed in order of priority.

The most important characteristic of interrupt resume is that in steady state, with known service times and a single server, average time in system is minimized when jobs are served according to the *shortest remaining service time* discipline (as proved by Schrage 1968):

Definition 9.9

Shortest Remaining Service Time (SRST) Each time a server becomes available, select the job with the shortest *remaining* service time (total service time minus attained service time). Each time a job arrives, compare the job's service time to the remaining service time of the job in service. Interrupt the job in service for the new job if the new job's service time is smaller (Fig. 9.4).

Example

A 2-hour job begins service at 1:00. At 2:00, the job is interrupted in favor of a 30-minute job. At 2:15, a 90-minute job arrives. When the 30-minute job is completed at 2:30, the 2-hour job is resumed, because its remaining service time is less than 90 minutes.

The SRST rule is exemplary of the fact that whatever discipline is used under interrupt resume, the jobs in service should receive no favoritism over jobs in queue. Nevertheless, as a practical matter, the server may still wish to avoid interruption, if

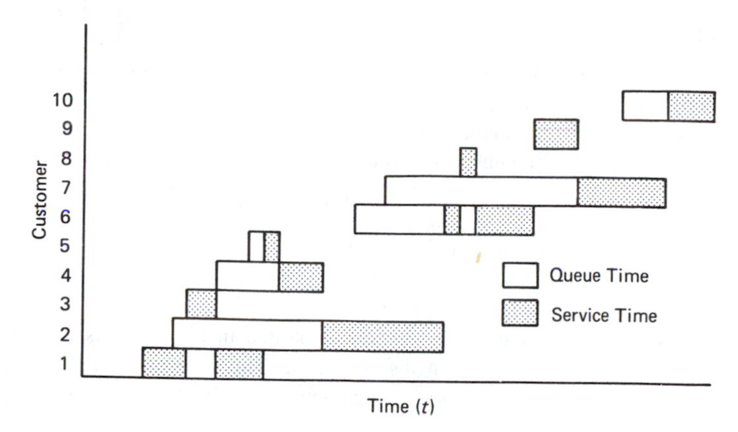

Figure 9.4 Gantt chart for same group of arrivals as Fig. 9.3, with shortest-remaining-service-time (SRST) discipline and interrupt-resume.

solely to eliminate confusion. While not a major issue in computer systems, job shops may find that the risks of losing a job or forgetting to perform an operation outweigh the benefits of interruption.

9.3.2 Interruption Due to Learning

Beyond new arrivals, a reason why a job might be interrupted is that information is learned about its service time characteristics *after* beginning service. Suppose that a person walks up to the counter at a camera store needing service. That person might be there to drop off film for developing, purchase a camera, or any number of things. Certainly some of the possibilities will demand more of the server's time than others, with the camera purchase likely demanding the most. After learning what the customer wants, the server may then choose to interrupt service to take care of shorter jobs first.

To a degree, the camera store is just another example of triage. But learning can occur in more subtle ways. As a job is being served, one can derive new estimates of the *expected remaining service time*. This value is easily derived from the laws of conditional probability, as a function of the attained service time:

$f(s)$ = probability density function for the service time

$R(x)$ = expected remaining service time for a job that has already attained a service time of x

$$= E(S - x \mid S > x) = \frac{\int_x^\infty sf(s)ds}{\int_x^\infty f(s)ds} - x \tag{9.9}$$

Although the expected remaining service time will ordinarily decline as a job is processed, in some instances it will increase. When this happens, it may be advantageous to interrupt the job in service for a job in queue which has not yet begun service.

Example

The service time for a machining operation falls into either the category of difficult jobs or easy jobs, randomly with equal probability. If the job is easy, the service time has a uniform distribution over [0,5] minutes, and if the job is difficult, the service time has a uniform distribution over [20,30] minutes. Therefore, the probability density function is the following:

$$f(s) = \begin{cases} 1/10 & 0 \leq s \leq 5 \\ 0 & 5 \leq s \leq 20 \\ 1/20 & 20 \leq s \leq 30 \end{cases}$$

The expected remaining service time is plotted in Fig. 9.5. As can be seen, $R(x)$ initially increases relative to the attained service time. This is because a job that has attained a service time of, say, 4 minutes without being completed is likely to be a difficult job and have a large remaining service time.

Given the above, a reasonable strategy would be to interrupt a job in service when $R(x)$ climbs by some critical value above the minimum of $R(x)$ among the jobs in the queue. The job with the minimum value of $R(x)$ would then enter service. In the example, it might be reasonable to interrupt jobs with attained service times between 2 and 10 minutes in favor of new jobs. Note that this interruption can occur at any point in time, not just when a new job arrives.

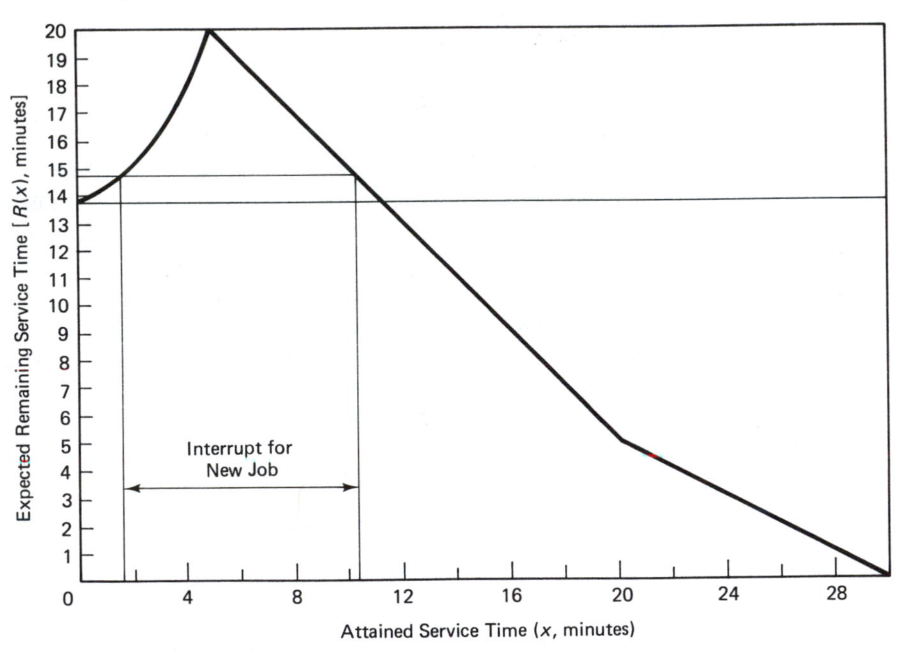

Figure 9.5 Expected remaining service time versus attained service time. Fifty percent of jobs are easy (uniform [0,5] distribution) and 50 percent of jobs are difficult (uniform [20,30] distribution). Interrupting jobs that have been in service 3 to 10 minutes can improve average waiting times.

While the attained service time provides a clue as to a job's total service time, in most instances better information is available—asking the customer a few questions or assessing the size of a job, for instance. An exception is computer systems, where there is little direct human involvement in the queue discipline. A variety of queue disciplines have been developed for this environment to exploit the fact that $R(x)$ might be an increasing function. For example, the ***forward-backward*** discipline selects the job that has so far received the least service. The idea is that if a job hasn't been completed after a certain length of time, it will likely be a long job and have a very long service time remaining. Hence, jobs that have already received a large amount of service should have lowest priority.

Speed is obviously a unique characteristic of computer systems. Whereas a job shop could not rapidly switch from one job to another, computer systems can. A computer can operate in the following manner. After each time unit, called a ***quantum***, the next job to enter service is selected according to the following rule:

Definition 9.10

Round robin (RR) If at the end of a quantum, the job currently in service is not yet completed, move it to the *end* of the queue. The job at the front of the queue then enters service, and all other jobs in the queue move forward by one place.

If the quanta are very small, the switching between jobs can occur so fast that the computer system behaves as though its capacity is equally divided among all of the jobs in the system. That is, if ten jobs are in the system, each job will be served at a rate equal to 1/10 of the server's undivided capacity.

When the quanta are very small, an interesting property of the single server round-robin discipline (Poisson arrivals) discovered by Sakata et al. (1971) is that the expected time in system is a linear function of a job's service time:

$$W_s(s) = \text{expected time in system as function of a job's service time, } s$$

$$= \frac{s}{1-\rho} \tag{9.10}$$

This result is independent of the service time distribution. Perhaps this should be no surprise, as a result of the fact that capacity is equally apportioned among all jobs, without regard to length.

The round-robin discipline is appealing from the standpoint of equity. There is something eminently fair about making the time in system proportional to job length. Relative to the FCFS discipline, it is also appealing because time in queue is not independent of service time; it is smaller for small jobs. With round robin, the expected time in system is simply

$$W_{s,\text{RR}} = \frac{1/\mu}{1-\rho} \tag{9.11}$$

Through simple algebra, one can show that $W_{s,\text{RR}}$ is less than $W_{s,\text{FCFS}}$ when the coefficient of variation for the service time is greater than one. This means that the variation must be very large, and greater than that of the exponential distribution, before RR provides a smaller time in system on average.

Readers interested in a further discussion of queueing disciplines used in computer systems should consult Kleinrock (1976).

9.3.3 Interrupt Repeat and Inserted Idle Time

Certainly interruption should be rare under interrupt repeat, if used at all. However, one might want to use another strategy instead: ***inserted idle time***. The server may choose not to begin a job if a higher priority job is expected to arrive in the near future. As a rule, inserted idle time should not be used for single server queues if the server has no knowledge of future arrivals; that is, if arrivals occur according to a stationary Poisson process. With multiple servers, a ''cutoff priority'' discipline can be instituted (Schaack and Larson, 1989). Under this policy, a subset of the servers is reserved for future arrivals when only low priority customers are in the queue. Inserted idle time should not be used at all with interrupt resume; it would only sacrifice productive time and not benefit high priority customers.

9.4 PARALLEL SERVERS

In recent years, ***parallel processing*** (that is, parallel servers) has been an important topic in computer science. What parallel processing means is that a computer program is divided into tasks and completed on several processors simultaneously. If processors can operate just as efficiently in parallel as individually, this should reduce time in system. The concept of parallel processing goes well beyond computers and applies to all types of servers. In the job shop literature, a job that can be served in parallel is called ***preemptive***. That is, a preemptive job is divisible into subtasks that can be performed separately and simultaneously.

If jobs are preemptive, with no loss in efficiency, a system with n parallel servers behaves the same as a single server, whose capacity equals the sum of the n servers' capacities. Jobs would be served one by one simultaneously on the n machines. Hence, the disciplines developed for a single server would apply.

More often than not, jobs are not preemptive. They are assigned to individual servers and processed until completed. If the servers have identical properties, then the single server disciplines generally need only be modified slightly. Whenever a server becomes available, the customer with highest priority begins service. Under this scenario, with Poisson arrivals in steady state, the SST discipline is known to minimize expected time in system. Among service time independent disciplines, with Poisson arrivals in steady state, FCFS minimizes the time-in-system variance. However, SWST is not guaranteed to minimize expected weighted time in system, though simulation studies have shown it to provide near-optimal results (Amar and Gupta, 1986, for example).

If servers have different service rates, scheduling becomes more complicated. The next job served would not necessarily have highest priority if the available server does not have the largest service rate. It may be preferred to hold the job until the better server becomes free. Results are difficult to generalize, and effective scheduling likely requires algorithmic procedures, such as the one developed by Horn (1973).

A common practice in stores and banks is to restrict certain servers to high priority customers. The express checkout line in supermarkets is the most familiar example. Express servers effectively give priority to smaller jobs, but not so much as the SST rule. Small jobs do not automatically move to the head of the queue; they only have greater choice as to which queue to join. This is a practical compromise between the objective of keeping queues orderly (it may create havoc to allow customers to bypass each other in queue) and the objective of minimizing time in system. From the standpoint of the second objective alone, SST is always preferred to express servers.

9.5 EVALUATION OF QUEUE DISCIPLINES

The ideal way to evaluate a discipline is to apply it to a system, measure its performance, and compare its performance to other disciplines. Unfortunately, industry is usually unable or unwilling to participate in such experiments. Most evaluations found in the scheduling literature are theoretical. Further, most rely on simulation. This is sad indeed, for theoretical studies cannot possibly capture all of the facets of a real queueing system.

Simulation studies can be further classified into the hypothetical and the applied, the latter being based on empirical data from real job shops and the former being based on theoretical data. Several important applied studies, based on work at General Electric, Hughes, Texas Instruments, and Western Electric, are reviewed by Elmaghraby and Cole (1963), Moodie and Novotny (1968), Buffa (1966), and Muth and Thompson (1963). The studies by Conway et al. (1967) are largely hypothetical. Amar and Gupta (1986) compared applied to hypothetical simulations and concluded that applied problems are easier to schedule than the hypothetical and that the SWST discipline performs better on the applied problems than the hypothetical (and close to optimal). This is certainly good news for practitioners.

Simulating either a single server or parallel server queue with priorities is not a difficult matter and follows directly from the methodology presented in Chap. 4. Without interruption, the only difference from the FCFS queue is that when a job is selected to enter service, all of the jobs in the queue must be examined to see which has highest priority. With interruption, an additional difference is that whenever a job arrives, it should be compared against the job (or jobs) in service to see whether it has higher priority, in which case an interruption occurs. Finally, if priorities change dynamically over time, then an activity scanning simulation must be used. After each advance of the simulation clock, the queue must be scanned to check for interruption. All in all, these changes are minor and do not warrant detailed coverage here (though the topic of simulation is brought up again in Chap. 10).

The disciplines of certain types of single server queues, or queues that behave like single server queues, can be analyzed without use of simulation, through analytical and graphical models. The general assumption is that customers are classified into n categories. Category 1 is given highest priority, category 2 given second highest priority, and so on, and the next job selected from the queue is taken from the highest priority category present. Within each category job, jobs are selected FCFS. Disciplines of this type are often called ***head of line*** because one can think of the customers as being arranged in sequence (determined by the priorities and arrival times) and the next customer selected always comes from the head of the line. The classification may be made according to any static criterion, such as SST. The model cannot evaluate dynamic criteria, such as slack time or remaining service time.

An analytical model for a steady-state queue will be presented first, then a graphical model for deterministic queues with varying arrival rate. Both are more complicated mathematically than the previous topics in this chapter.

9.5.1 Steady-State Single Server Without Interruption

In 1954, Cobham developed the equations that define the expected time in queue for head-of-line disciplines, with Poisson arrivals and general service times. We will consider first the case without interruption and then move on to the case with interruption.

Cobham observed that the time spent in queue can be divided into three parts. First, a customer must wait until the customer currently being served completes service. Next, a customer must wait until all customers of *equal or higher* priority that were in the queue when it arrived are served. Finally, a customer must wait until all customers of *higher* priority that arrived while it was waiting are served. These statements can be expressed in terms of the following symbols:

W_{qj} = expected time in queue per type j customer

λ_j = arrival rate of type j customers

s_j = expected service time of type j customers

s_j^2 = expected squared service time of type j customers

p_j = proportion of customers that are type j

W_r = expected remaining service time for the customer in service at time of arrival

E_{ji} = expected number of type i customers in the queue when a type j customer arrives

H_{ji} = expected number of type i customers to arrive while a type j customer is in queue.

The word description is then

$$W_{qj} = W_r + \sum_{i=1}^{j} s_i E_{ji} + \sum_{i=1}^{j-1} s_i H_{ji} \tag{9.12}$$

To finalize the result, E_{ji} and H_{ji} must be expressed in terms of the waiting times. Further, W_r must be expressed in terms of the service time distribution and the arrival rate. Finally, a system of equations (one equation for each priority class) must be solved.

To show the result, the symbol σ_j will be defined. It represents the sum among the utilizations for priority classes 1 to j:

$$\sigma_j = \begin{cases} \sum_{i=1}^{j} \rho_i = \sum_{i=1}^{j} \lambda_i s_i & j = 1, 2, \ldots \\ 0 & j = 0 \end{cases} \tag{9.13}$$

By simple laws of averages, the expected service time squared and the expected service time, among all categories, are the following:

$$E(S^2) = \Sigma\, p_j s_j^2 \tag{9.14}$$

$$E(S) = \Sigma\, p_j s_j \tag{9.15}$$

The expected time in queue per type j customer is then

$$W_{qj} = \frac{\lambda E(S^2)}{2(1 - \sigma_{j-1})(1 - \sigma_j)} \tag{9.16}$$

Equation (9.16) resembles the Pollaczek-Khintchine formula (for the $M/G/1$ queue, see Chap. 5). In the special case where the number of classes equals one, the equations are identical.

Equation (9.16) indicates that waiting time increases as the variation in service times increases and as the utilization (among classes 1 to j and 1 to $j - 1$) increases. W_{qj} does not depend on lower priority classes, except to the extent that these classes influence $E(S^2)$. Further, as is logical W_{qj} increases as j increases; lower priority customers have to wait longer on average. The expected time in queue among all customers is

$$W_q = \sum_j p_j W_{qj} \tag{9.17}$$

Other measures of performance, such as expected time in system and expected customers in queue, can be derived in the usual manner from Eq. (9.15) and (9.17) (as we saw in Chap. 2).

Equations (9.16) and (9.17) can be used as a model for evaluating the performance of the ***shortest-expected-service-time (SEST)*** discipline (customers are sequenced in order of expected service time).

Example

A warehouse retrieval system serves two types of customers, representing small orders and large orders (signified by the subscripts 1 and 2). Customers arrive by a Poisson process, with rates $\lambda_1 = 20$/hour and $\lambda_2 = 10$/hour. Each type of order has an exponential service time distribution, defined by the rates $\mu_1 = 60$/hour and $\mu_2 = 20$/hour ($s_1^2 = 2/60^2$, $s_2^2 = 2/20^2$).

FCFS discipline If the discipline is FCFS, the expected time in queue is defined by the Pollaczek-Khintchine formula:

$$E(S) = (20/30) \cdot (1/60) + (10/30) \cdot (1/20) = .028 \text{ hour}$$

$$E(S^2) = (20/30) \cdot [2 \cdot (1/60)^2] + (10/30) \cdot [2 \cdot (1/20)^2] = .0020 \text{ hour}^2$$

$$\rho = (20 + 10) \cdot E(S) = .84$$

$$W_{q,\text{FCFS}} = \frac{1}{\lambda}\left[\frac{\rho^2}{1-\rho}\right]\left[\frac{1 + C^2(S)}{2}\right] = \frac{\lambda E(S^2)}{2(1-\rho)} = .19 \text{ hour}$$

SEST discipline The expected time in queue is now defined by Eq. (9.17):

$$\sigma_1 = \lambda_1(1/\mu_1) = (20)(1/60) = .333$$

$$\sigma_2 = \sigma_1 + \lambda_2(1/\mu_2) = .333 + (10)(1/20) = .833$$

$$W_{q1} = \frac{\lambda E(S^2)}{2(1-\sigma_1)} = .045 \text{ hour}$$

$$W_{q2} = \frac{\lambda E(S^2)}{2(1-\sigma_1)(1-\sigma_2)} = .270 \text{ hour}$$

$$W_{q,\text{SEST}} = (20/30)W_{q1} + (10/30)W_{q2} = .12 \text{ hour}$$

For the example, the SEST discipline reduces expected time in queue by 36 percent.

The advantages of the SEST discipline over FCFS are illustrated in Fig. 9.6, which gives the ratio $W_{q,\text{SEST}}/W_{q,\text{FCFS}}$ for a single server queue with Poisson arrivals. In drawing the figure, each of two classes was assumed to have an exponential service time distribution, and the classes were assumed to have the same arrival rate. Further, the following parameter is used:

$$K = s_2/s_1 \tag{9.18}$$

As expected, SEST reduces time in queue for all parameter values, especially when ρ is close to 1 and K is very large. This is indicative of the general relationship that SEST *is most advantageous when the system is heavily utilized and when service times differ greatly among priority classes*.

9.5.2 Steady-State Single Server with Interruption

We now consider interruption/resume within the context of a head-of-line discipline for a single server with Poisson arrivals and general service times. With interruption, the job in service must always have the highest priority among all of the customers in the system. An interruption occurs when a new job arrives with higher priority than the job currently in service. The interrupted job then moves to the head of the line (possibly ahead of other interrupted jobs from lower priority classes) and is resumed later, without loss in service time. Note that even if priority classes are arranged in order of expected service time, this

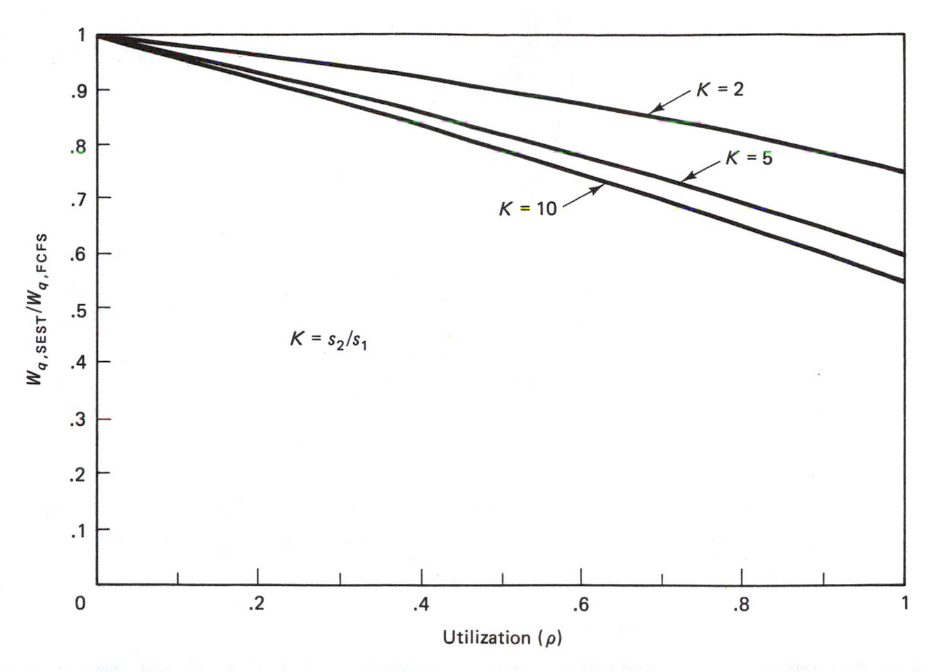

Figure 9.6 SEST discipline provides a shorter expected time in queue than the FCFS discipline, especially when the utilization is large and there is a large variation in service times. (Figure assumes that each of two service classes has Poisson arrivals, and service times for each class are exponential.)

discipline is not a shortest-expected-remaining-service-time (SERST) discipline. Higher priority jobs retain their priority, even when lower priority jobs are very close to completion. Nevertheless, the performance characteristics are similar to such a discipline.

The equation for expected time in queue is more complicated when interruption is allowed. It was also derived by Cobham, in a manner similar to that described for the case without interruption. Using the terminology from the previous section

$$W_{qj} = \text{expected time in queue per type } j \text{ customer}$$

$$= \frac{1}{1 - \sigma_{j-1}} \left[s_j + \frac{\sum_{i=1}^{j} \lambda_i s_i^2}{2(1 - \sigma_j)} \right] - s_j \tag{9.19}$$

And, as before

$$W_q = \sum_j p_j W_{qj} \tag{9.20}$$

Example

Suppose that the warehouse studied in Sec. 9.5.1 operates under interrupt-resume. Then expected time in queue is calculated as follows:

$$W_{q1} = \frac{\lambda_1 s_1^2}{2(1 - \sigma_1)} = .0083 \text{ hour}$$

$$W_{q2} = \frac{1}{1 - \sigma_1}\left[s_2 + \frac{\lambda_1 s_1^2 + \lambda_2 s_2^2}{2(1 - \sigma_2)}\right] - s_2 = .30 \text{ hour}$$

$$W_q = (20/30) \cdot W_{q1} + (10/30) \cdot W_{q2} = .106 \text{ hour}$$

For the example, interruption reduces expected time in queue by 12 percent.

Using the parameters defined earlier, Fig. 9.7 compares the interruption case to the no interruption case for a single server queue with Poisson arrivals and two customer types (each with exponential service times and 50 percent of total arrivals). As can be seen, interruption reduces time in queue for all parameter values, especially when K is large and ρ is *small*. This is indicative of the general relationship that *interruption is most advantageous when the system is lightly utilized and when service times differ appreciably among priority classes*. Interruption is rare when ρ is large, so it provides little added benefit. Large mean queue sizes provide SEST with a large selection, allowing it to pick jobs with small service times at the outset.

9.5.3 Graphical Models

Though usually not the best solution to predictable queues, adopting SST can still help reduce delay. This section uses a fluid model to analyze graphically the change in waiting time when the SST discipline is adopted over a period when arrival rate exceeds service capacity.

As in the previous section, suppose that customers are categorized according to service time and that customers with short service times receive priority. Also suppose that the arrival and service processes are deterministic and can be approximated with a fluid model (see Chap. 6).

Let

$\lambda_j(t)$ = arrival rate of type j customers at time t

$\omega_j(t)$ = departure rate of type j customers at time t

c = server capacity

$L_j(t)$ = queue size of type j customers at time t

High priority customers are allocated capacity first; lower priority customers are only served when capacity remains:

$$\omega_1(t) = \begin{cases} c & L_1(t) > 0 \\ \lambda_1(t) & L_1(t) = 0 \end{cases} \tag{9.21a}$$

$$\omega_j(t) = \begin{cases} c - \sum_{m<j} \omega_m(t) & L_j(t) > 0, j > 1 \\ \lambda_j(t) & L_j(t) = 0, j > 1 \end{cases} \tag{9.21b}$$

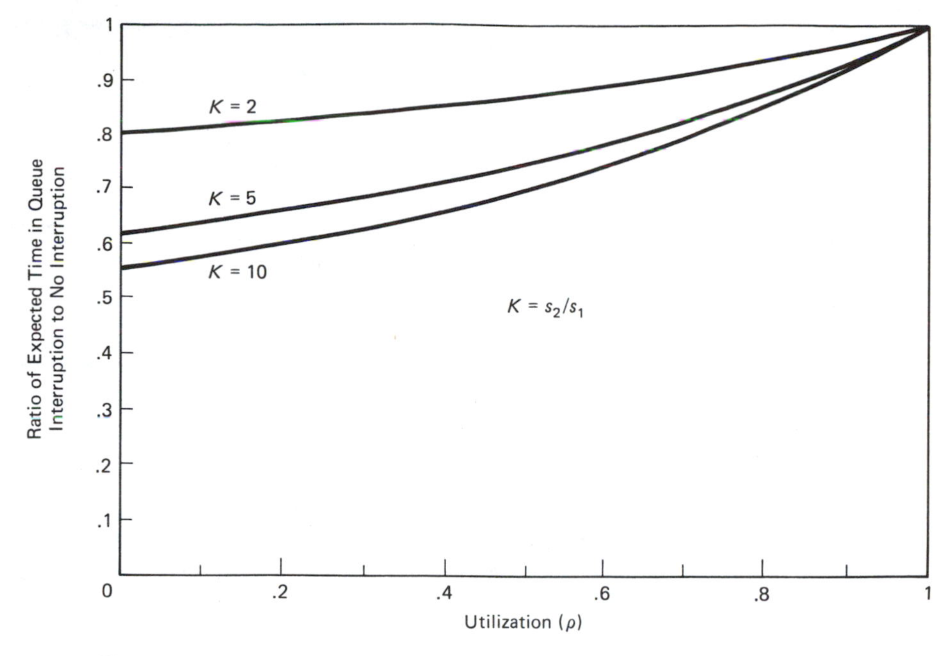

Figure 9.7 Interrupting long jobs for short jobs will reduce expected time in queue, especially when utilization is low and there is a large variation in service times. (Figure assumes that each of two service classes has Poisson arrivals, and service times for each class are exponential.)

$$L_j(t) \quad = \quad \int_0^t [\lambda_j(\tau) - \omega_j(\tau)]d\tau \qquad (9.21c)$$

High priority customers tend to be served at the same rate as they arrive, and therefore do not queue. Lower priority customers are only served when capacity is available. When the system is overloaded, their service rates may be zero.

Example

A freeway on-ramp is controlled by a traffic signal that restricts the total flow of vehicles to 800 per hour. Ordinary vehicles must wait for a green signal before entering, while car pools (carrying 2.5 people, on average) can bypass the queue and enter without delay (provided that their arrival rate does not exceed 800 vehicles/hour). The arrival rate of people in ordinary vehicles is $\lambda_r(t)$ and the arrival rate of people in car pools is $\lambda_c(t)$, as shown in Fig. 9.8a.

The service capacity is 800 vehicles per hour, which translates into 800 people/hour for ordinary vehicles and 2000 people/hour for car pools. Using Eq. (9.21) to allocate capacity, Fig. 9.8a shows the cumulative departure curves. Throughout the entire period, car pools never have to queue. Ordinary vehicles begin queueing at about time 30, and the queue vanishes at about time 100.

Note that as the arrival rate of car pools increases, the departure rate of ordinary vehicles declines, but the overall departure rate (people/hour) increases. The overall departure rate increases because a higher percentage of the vehicles served are car pools.

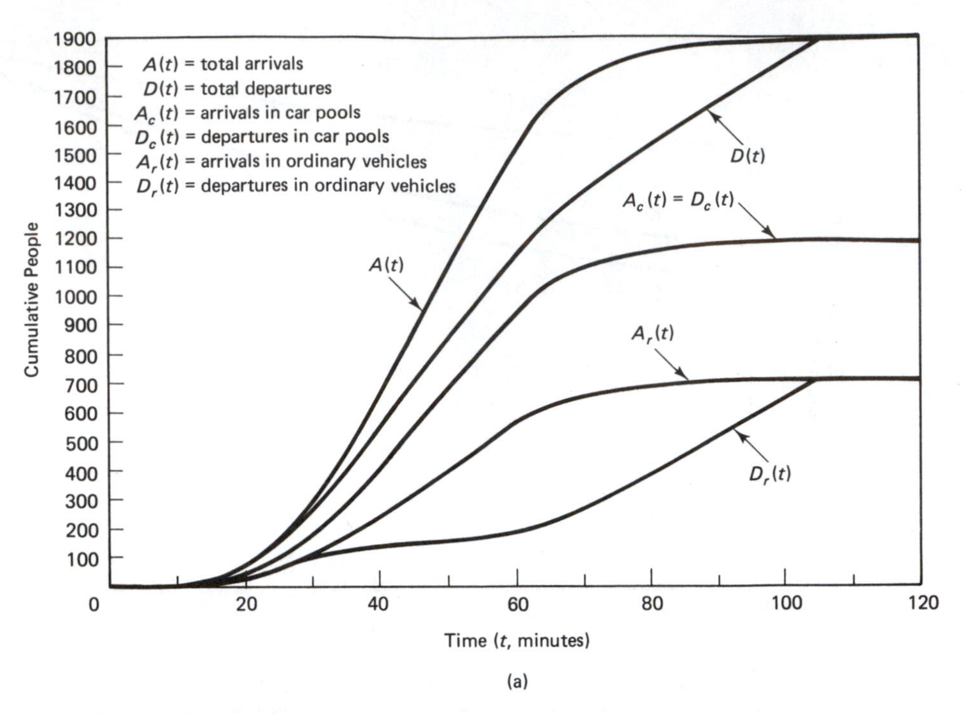

Figure 9.8a Cumulative arrival and departure curves showing the impact of giving priority to car pools at a freeway entrance. In the example, only customers in single-occupant vehicles wait in queue.

For comparison, Fig. 9.8b shows the cumulative departure curves without car-pool priority. It should be evident that total person delay increases substantially (though total vehicle delay has stayed the same).

In traffic, giving car pools priority is like implementing the SST discipline: It takes less time to serve a person in a car pool than a person in a single-occupant vehicle. Not only does favoring car pools reduce delay in the short run, but it should also, in the long run, encourage more car pools and reduce delay further. This effectively increases the capacity of the system, as shown in Fig. 9.8b.

9.6 CHAPTER SUMMARY

If one message is to be gleaned from this chapter, it is that service time should be a key part of virtually all practical queue disciplines. It is the one way to reduce average time in system. Consequently, disciplines such as shortest service time (SST) also tend to reduce expected tardiness, reduce waiting time variability, and ensure that more jobs are completed on time. SST is especially attractive when the system is heavily utilized and when there are large and predictable variations in service time. If service time cannot be

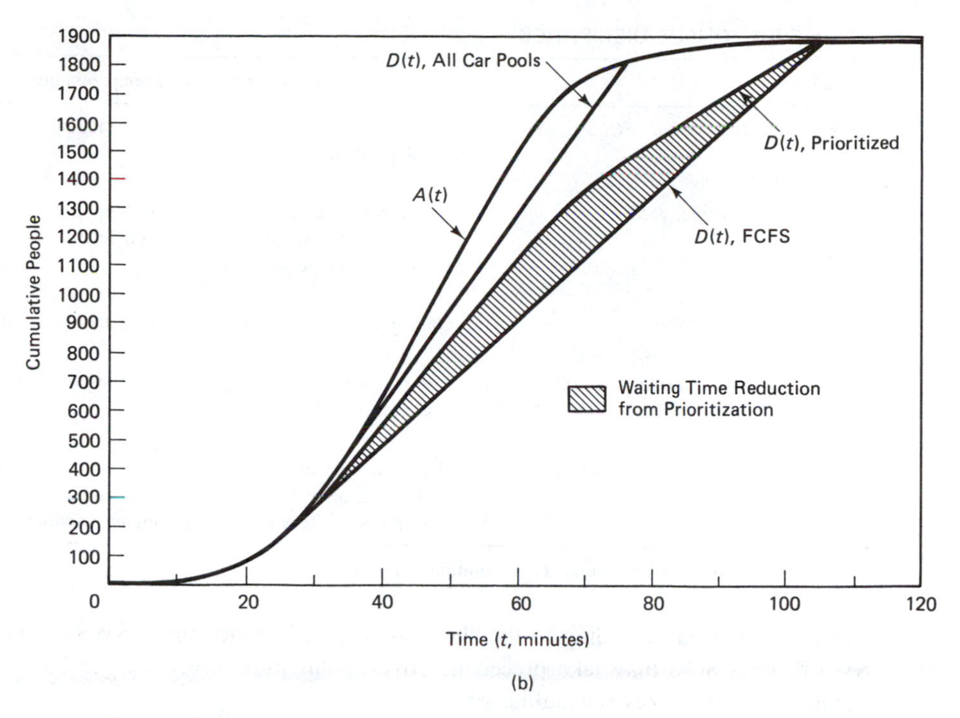

Figure 9.8b Giving priority to car pools reduces average waiting time per person because people in car pools are served at a faster rate than people in single-occupant vehicles.

predicted in advance, then first-come, first-serve (FCFS) is a good choice. It would minimize the standard deviation of time in system and make waiting time more predictable, especially when the system is heavily utilized.

Depending on the situation, one may incorporate other customer characteristics in a hybrid scheduling rule. In job shops, due date might be used in combination with service time. Short jobs would receive priority, but long jobs would increase in priority as their due dates approach. (It is unwise to prioritize strictly on the basis of due date, for it does not optimize any important measure of performance.) In pure service systems (where a person waits for service) arrival time might be combined with service time and implemented through a head-of-line discipline.

Jobs should be interrupted for higher priority customers, so long as they can be resumed later without a service time penalty. An interruption might result from the arrival of a higher priority customer or possibly through learning that the job in service has a large remaining service time. In the case of interrupt/repeat, interruption should be used sparingly. Not only is the job in service delayed, but subsequent jobs are delayed due to the sacrifice of system capacity.

One should be aware that all customers may not have the same value of time. It may be more costly to delay a large job by 1 minute than a small job. In such instances, jobs

TABLE 9.1 QUEUE DISCIPLINES

Discipline	Features of single server/nonpreemptive*
First-come, first-served (FCFS)	Minimizes variance of time in system among service time independent disciplines.
Shortest service time (SST)	Minimizes expected time in system among work-conserving systems (performs well relative to most measures of performance).
Earliest due date (EDD)	Only minimizes maximum tardiness (static schedule).
Shortest weighted service time (SWST)	Minimizes expected *weighted* time in system (work-conserving, Poisson arrivals).
Round robin (RR)	Expected time in system is linear function of service time (Poisson arrivals). Shorter expected time in system than FCFS when service times are highly variable.

*Some properties can be generalized to multiple server/preemptive

can be prioritized according to the shortest-weighted-service-time (SWST) rule. Under SWST, large jobs may take precedence over small jobs, reversing the SST ordering. Table 9.1 summarizes scheduling rules.

No matter what the discipline, priorities should be assigned in as objective a fashion as possible, in a manner that truly measures the urgency of the work. One should guard against manipulations of the system, through unrealistic due dates or perhaps through customers splitting jobs into smaller sizes. This objective is the ideal. In reality, the myriad considerations that come into play in setting priorities, and particularly the politics, make the ideal difficult to achieve. The case studies at the end of this chapter, one taken from medicine and the other from emergency services, are exemplary.

FURTHER READING

AMAR, A. D., and J. N. D. GUPTA. 1986. "Simulated versus Real Life Data in Testing the Efficiency of Scheduling Algorithms," *IIE Transactions*, 18, 16–25.

BAKER, K. R. 1974. *Introduction to Sequencing and Scheduling*. New York: John Wiley.

______. 1984. "Sequencing Rules and Due-Date Assignments in a Job Shop," *Management Science*, 30, 1093–1104.

BUFFA, E. S., ed. 1966. *Readings in Production and Operations Management*. New York: John Wiley.

COBHAM, A. 1954. "Priority Assignment in Waiting Line Problems," *Operations Research*, 2, 70–76.

COFFMAN, E. G., ed. 1976. *Computer and Job-Shop Scheduling Theory*. New York: John Wiley.

CONWAY, R. W. 1965a. "Priority Dispatching and Work-in-Process Inventory in a Job Shop," *Journal of Industrial Engineering*, 16, 123–130.

———. 1965b. "Priority Dispatching and Job Lateness in a Job Shop," *Journal of Industrial Engineering*, 16, 228–237.

———, and W. L. MAXWELL. 1962. "Network Dispatching by the Shortest Operation Discipline," *Operations Research*, 10, 51–73.

CONWAY, R. W., W. L. MAXWELL, and L. W. MILLER. 1967. *Theory of Scheduling*. Reading, Mass: Addison-Wesley.

ELMAGHRABY, S. E., and R. T. COLE. 1963. "On the Control of Production in Small Job Shops," *Journal of Industrial Engineering*, 14, 186–196.

FRENCH, S. 1982. *Sequencing and Scheduling, An Introduction to the Mathematics of the Job-Shop*. Chichester, U.K.: Ellis Horwood.

GRAVES, S. C. 1981. "A Review of Production Scheduling," *Operations Research*, 29, 646–675.

HORN, W. A. 1973. "Minimizing Average Flow Time with Parallel Machines," *Operations Research*, 21, 846–847.

JACKSON, J. R. 1961. "Queues with Dynamic Priority Disciplines," *Management Science*, 8, 18–34.

KINGMAN, J. F. C. 1962. "The Effect of the Queue Discipline on Waiting Time Variance," *Proceedings of the Cambridge Philosophical Society*, 58, 163–164.

KLEINROCK, L. 1976. *Queueing Systems, Vol. 2: Computer Applications*. New York: John Wiley.

MCKAY, K. N., F. R. SAFAYENI, and J. A. BUZACOTT. 1988. "Job-Shop Scheduling Theory: What Is Relevant?" *Interfaces*, 18:4, 84–90.

MOODIE, C. L., and D. J. NOVOTNY. 1968. "Computer Scheduling and Control Systems for Discrete Part Production," *Journal of Industrial Engineering*, 19, 336–341.

MOORE, J. M. 1968. "An n-Job, One-Machine Scheduling Algorithm for Minimizing the Number of Late Jobs," *Management Science*, 15, 102–109.

MUTH, J. F., and G. L. THOMPSON, eds. 1963. *Industrial Scheduling*. Englewood Cliffs, N.J.: Prentice Hall.

PANWALKAR, S. S., and W. ISKANDER. 1977. "A Survey of Scheduling Rules," *Operations Research*, 25, 45–61.

PHIPPS, T. E. 1956. "Machine Repair as a Priority Waiting-Line Problem," *Operations Research*, 4, 76–85.

SAKATA, M., S. NOGUCHI, and J. OIZUMI. 1971. "An Analysis of the *M/G/*1 Queue under Round-Robin Scheduling," *Operations Research*, 19, 371–385.

SCHAACK, C. and R. C. LARSON. 1989. "An N Server Cutoff Priority Queue Where Arriving Customers Request a Random Number of Servers," *Management Science*, 35, 614–634.

SCHRAGE, L. E. 1968. "A Proof of the Optimality of the Shortest Remaining Service Time Discipline," *Operations Research*, 16, 687–690.

PROBLEMS

1. Describe two types of businesses that operate as a closed shop and two types that operate as an open shop. Referring to these examples, what are the incentives for operating as a closed shop and what are the incentives for operating as an open shop?

2. Give two advantages to following a dynamic schedule and two advantages to following a static schedule. For which type of shop, open or closed, is a dynamic schedule most useful?

3. For each of the following examples, describe a triage process that could be used to prioritize customers:

(a) Customers at a camera sales counter
(b) Tax forms at the Internal Revenue Service
(c) Patients at a hospital emergency room

***4.** An *M/M/1* queue, with $\rho = .9$ and $\lambda = 15$/hr, currently operates FCFS.

(a) Calculate the variance of time in queue $(\rho^3(2 - \rho)/\lambda^2(1 - \rho)^2)$.
(b) Calculate the variance of time in queue for an LCFS discipline.
(c) Calculate the variance of time in queue for an SRO discipline.
(d) Why does the LCFS discipline have the largest variance?

5. The following gives the arrival times, service times, and due dates for a series of jobs:

Job	Arrival	Service	Due	Job	Arrival	Service	Due
1	5	11	50	5	31	2	60
2	7	6	60	6	32	10	65
3	13	15	30				
4	30	21	62				

(a) Draw a diagram of cumulative work.
(b) Using dynamic scheduling, without interruptions, create the following schedules. Show the time each job enters service and the time each job completes service:
i. FCFS ii. SST iii. EDD iv. minimum slack
(c) Draw the Gantt chart for each solution from part b.

6. Repeat Prob. 5 for the shortest remaining service time discipline with interruptions.

7. For each of the solutions in Prob. 5, calculate

(a) Mean and maximum tardiness
(b) Number of jobs late
(c) Mean time in system

In your opinion, which rule performed best?

8. A print shop has decided to follow a SWST discipline, where the weight is equivalent to the price of the job. Currently, the following jobs are waiting to be printed:

Job	1	2	3	4	5
Price ($)	21	105	5	12	130
Originals	1	5	1	2	10
Total copies	1000	5000	200	500	6000

The printing time for a job equals .1 hour per original, plus .0002 hour per copy.

(a) Determine the job priority numbers and order the jobs.
(b) By looking at the data, describe a simple rule of thumb for prioritizing jobs that achieve the same objective as SWST.

***9.** Jobs submitted to a campus computer center fall into either the category of short jobs or the category of long jobs with equal probability. Short jobs have an exponential distribution with mean 5, and long jobs have an exponential distribution with mean 20.

(a) Write the equation for expected remaining service time.
(b) Through use of a computer, plot the equation from part a.
(c) Based on your answer to part b, which of the queue disciplines described in Chap. 9 is equivalent to giving priority to the job with the shortest expected remaining service time?

*Difficult problem.

10. Using the Pollaczek-Khintchine formula, prove that the round-robin discipline will have a shorter time in system than the FCFS discipline when the coefficient of variation for the service times is greater than one (single server with Poisson arrivals).

11. Give an example of when it would be desirable to use inserted idle time.

12. Jobs arrive at a computer center according to a Poisson process with rate 3000 per hour. Jobs can be classified as follows:

A:	Service time = .1 second	(50% of jobs)
B:	Service time = .5 second	(25% of jobs)
C:	Service time = 1 second	(15% of jobs)
D:	Service time = 2 seconds	(8% of jobs)
E:	Service time = 5 seconds	(2% of jobs)

(a) If the round-robin discipline is followed, what would be the expected time in system for each type of job?

(b) What is the expected time in system among all jobs?

(c) If jobs were served FCFS, what would be the expected time in system among all jobs?

(d) For this example, which discipline do you prefer, FCFS or RR?

***13.** A lathe operator receives two types of jobs according to a Poisson process:

Job A: $\lambda_A = 1/4$ per hour, $s_A = 1$ hour, $s_A^2 = 1.5$ hours^2

Job B: $\lambda_B = 1/8$ per hour, $s_B = 4$ hours, $s_B^2 = 20$ hours^2

(a) Using an FCFS discipline, what is the expected time in queue for each job type, and what is the expected time in queue among all jobs?

(b) Repeat part a for a SEST discipline.

(c) Repeat part a for a head-of-line discipline that gives priority to job type A, with interruption.

14. For the data in Prob. 13, what would happen to the ratio of $W_{q,\text{SEST}}$ to $W_{q,\text{FCFS}}$ if λ_A and λ_B both increased 20 percent? (Describe in words.) What would happen to the ratio of W_q for part c to $W_{q,\text{SEST}}$? Describe in words.

***15.** Suppose that a third customer is added to the set in Prob. 13.

Job C: $\lambda_C = 2$/hour, $s_C = .1$ hour, $s_C^2 = .0125$ hour^2

Repeat parts a–c from Prob. 13.

***16.** Write a program to simulate the system described in Prob. 13. For each discipline, simulate the system for a 100-hour period, estimate W_q, and derive a 95% confidence interval for W_q.

17. A bridge is being redesigned so that vehicles carrying two or more people will have priority when entering. Currently, 20 percent of the vehicles carry two people and 5 percent carry three people (the remaining 75 percent only carry the driver). Vehicles arrive at the rate of 3000/hour between 6:00 and 7:00, 5000/hour between 7:00 and 8:00, and 2500/hour between 8:00 and 9:00. The two highway lanes are capable of serving 2000 vehicles per hour each.

(a) Draw the curve representing cumulative arrivals of people.

(b) Assuming that all vehicles receive equal priority, draw cumulative departures of people. Also, calculate total waiting time among all people (this can be done exactly, with trigonometry).

*Difficult problem.

(c) Now assume that car pools have priority. Again, draw the cumulative departures of people and calculate total waiting time among all people.

(d) Which solution will minimize waiting time of vehicles, prioritized or nonprioritized?

(e) In words, describe how the priority scheme might encourage people to behave in a way that further reduces delay.

CASE STUDY

TRANSPLANT GIFT IN SHORT SUPPLY

Ronald Kotulak

Organ transplants may be a miracle of modern medicine, but they involve a cruel irony. Though transplants have become a vital lifesaving procedure, thousands of people die each year waiting for donor organs that never arrive.

In desperation, some resort to media blitzes. A father's plea for his dying daughter pulls at the nation's heartstrings, and Jamie Fiske gets a liver. A tearful Phil Donahue show wins "Baby Jesse" a new heart.

Even President Reagan has made appeals so that dying patients may live, and he has offered use of his presidential plane to ferry the vital cargo. Nevertheless, his administration has steadfastly resisted paying for transplants with federal insurance programs.

The system for allocating donor organs is disorganized, inequitable and morally disturbing. This major medical procedure, with all of its lifesaving promise, is being rationed in full public view.

"Modern technology has provided us with a double-edged sword," said Dr. Gary Friedlaender, chief of orthopedics at Yale University. "On the one hand, organ transplants are miracles that will save the lives of many.

"On the other hand, we are faced with thousands of people whose lives potentially could be saved with a transplant but who will die because they are not available."

The hard truth is there are nowhere near enough donor organs and probably never will be. In addition, transplant surgery entails extreme expense: A heart transplant costs $95,000 for the first year alone; a liver, $110,000; and a kidney, $35,000.

Nationwide, more than 36,000 heart, liver and kidney transplants a year could be justified right now, but fewer than 11,000 organs are likely to be available, one study indicates.

More than 14,000 Americans die each year who might have been saved by heart transplants, had the organs been available, said Roger Evans, a medical technology assessment specialist at Battelle Human Affairs Research Centers in Seattle.

But each year, only 500 to 1,300 donor hearts can be expected to be available, said Evans, who recently ended a four-year study of heart transplants for the federal Department of Health and Human Services.

Liver transplants could save 9,500 patients a year, but only 1,800 donor livers can be expected, Evans said. And 10,000 to 12,000 kidneys could be transplanted, though the supply may peak at 8,000.

Though hundreds of thousands of people die each year, few are candidates for organ donations. Physiological and legal restrictions severely limit such donations.

With few exceptions, organs must come from patients who are brain dead but maintained on life-support systems. From 20,000 to 22,000 people, most of them young adults killed in accidents fall into this category.

But for many reasons, including a cutoff age of 35 for heart donors and lack of consent from the relatives of dying patients, the number of available organs is much smaller.

These critical issues—the donor shortage and the extreme expense—raise a fundamental ethical question: Should society limit such care?

"Should we say the hell with it, we're not going to make available to everybody everything that the imagination of man can come up with?" said George Annas, professor of health law at Boston University School of Medicine.

"Should we tell people . . . there are some things that are either too expensive or simply don't make sense for society to underwrite for everybody."

Most people believe organ transplants should be done, even if they can't be given to everyone who needs them. But serious questions arise in deciding who gets the organs. So far, organs have been given to the person who, first of all, is the most ill and, second, most likely to benefit from them.

For example, a patient whose life is threatened by liver disease is more likely to get a donor organ than one who has liver disease plus a bad heart or other damaged organ.

Age is another factor. Some centers will not implant hearts into patients older than 50; other centers stretch the age limit to 65.

Though society does not exclude those who are unable to pay, it does not provide a uniform, institutionalized way of funding transplants for them—and this sometimes makes it very difficult for poor people and those without adequate insurance to obtain transplants.

Eighty-five percent of private medical insurers offer some transplant coverage, especially for hearts, but whether payment is made is decided on a case-by-case basis, Evans said. And one out of three Americans is not covered for such surgery at all, he noted.

Recently a federal task force recommended that the government underwrite such transplants, but the Reagan administration has been opposed to this.

U.S. officials recently announced that Medicare, the federal insurance program for the elderly and disabled will pay for some transplants. But because older or disabled people are seldom eligible for transplants, the move is not expected to make much of a difference.

Transplant coverage under Medicaid, the federal-state insurance program for the poor, would make a big difference. But given efforts to cut back on this program, such coverage is considered unlikely.

"The poor and uninsured will have to scramble as best they can to find some hospital that will provide them with a transplant," Evans said.

In 1979, a medical breakthrough enabled heart, liver and other transplants to become more than just experimental surgery. Scientists discovered cyclosporine, a drug that dramatically reduced the threat that donor organs would be rejected.

Four years later, the federal Food and Drug Administration approved the drug for general use, spurring a fourfold increase in heart and liver transplants between then and 1985.

Last year, there were 719 heart transplants, up from 164 in 1983. There were 602 liver transplants, up from 164; and 7,800 kidney transplants, up from 6,112.

At the same time, transplants have become the most glaring example of health-care rationing in this country.

"Organ transplants have really put massive health-care rationing on the public agenda for the first time," said Annas, of Boston University. "It's always been there, but in the past it has always been done covertly."

Hidden rationing has long been one of medicine's dark secrets, but most Americans have never had to deal with it. The victims have traditionally been too weak to make their complaints heard. But the transplant situation is making the issue more visible and urgent.

As with other issues of high-technology medicine, organ transplants have raised serious moral questions.

Will society allow only the rich or insured to get donor organs? Will the poor in need of transplants be condemned to die? Will there be a black market for donor organs? Will the media play an even greater role in deciding who gets transplants?

Americans have shown no stomach for overt rationing. They got a taste of it in Seattle during the late 1960s, after development of the kidney dialysis machine.

At the time only a few machines were available, and so local hospitals formed committees to anonymously decide which patients would get dialysis—and which would be turned down to die.

But it was disclosed that these "god squads" had denied dialysis to patients because they were prostitutes or playboys or for other "social worth" reasons. Society reacted with repugnance.

To avoid the agony of such decisions, Congress in 1972 approved federal payments for all dialysis and kidney transplants.

But that decision only sidestepped the issue and, in doing so, created a heavy financial burden. The program, originally estimated at $200 million to $300 million a year, now costs $2.5 billion annually and cares for 80,000 patients.

In the Great Society programs of the 1960s, rationing kidney dialysis was considered intolerable.

"The political climate has changed since then," said Arthur Caplan, associate director of the Hastings Center, a New York ethics think tank. "I think we could and would tolerate some explicit rationing today."

Many transplant centers require potential patients to deposit $100,000 or more just to be placed on eligibility lists for donor organs. The enormous deposits require those without insurance coverage or with inadequate coverage to make dramatic public pleas for money.

Some large institutions have been severely chastised for using donor organs from deceased Americans in the cure of wealthy foreigners—who are willing to pay much more—while U.S. patients are passed over.

The federal National Organ Transplant Task Force recently recommended that no U.S. hospital implant hearts or livers into foreigners or immigrant aliens until all the needs of American citizens have been met.

Given the demand in this country, the recommendation, if adopted by states or the federal government, would in effect prohibit foreigners from getting those organs here.

The task force also proposed a quota for kidney transplants, urging that no center be allowed to perform more than 10 percent of these procedures on foreigners.

An international "gray market," based in Bombay, has sprung up for the buying and selling of kidneys. An impoverished Indian can sell one of his kidneys, which will be made available to wealthy kidney patients, usually from other countries. Normal life is possible with one kidney.

The issue may be complicated even more by Evans' study, which asserts that heart transplants have joined kidney transplants in being cost effective. Liver transplants, though they continue to improve, have not approached the long-term success of hearts and kidneys.

With cyclosporine warding off rejections, more than 80 percent of the patients receiving new hearts at the best centers will survive a year, and more than half will survive longer than five years. The average yearly cost of the procedure over a five-year period is now $23,478.

Heart transplants are less expensive than kidney dialysis, cancer therapy and AIDS treatment, Evans said.

"We can't say that heart transplantation is experimental anymore, because it is not," he said. "If we're saying we can't afford it, then we better reconsider some of the other things we are doing, like kidney dialysis, and maybe discontinue them, because we can't afford them either."

For livers, the one-year survival is about 80 percent, and 50 percent of the recipients survive three to five years. Children have the best success with new livers.

Ninety-five percent of those receiving kidneys from living related donors are alive after a year, and 91 percent at three years. For kidneys obtained from unrelated donors, the figures are 86 and 78 percent, respectively.

For all of its medical magic, cyclosporine has drawbacks. It can have serious side effects, not the least of which is an increased risk of cancer, and a patient must take it for the rest of his life, at an annual cost of $6,000.

Time is a major factor in limiting the availability of organs. Surgeons have just three to four hours to take a beating heart out of one body and place it into another. Livers last about 12 hours, and kidneys about 72. Because of this, hearts seldom can be shipped beyond a metropolitan area, and livers can go only to nearby cities.

As scientists look for ways to relieve the organ shortage, they must try to answer incredibly difficult questions, both medical and ethical.

For example, would society consider removing life support systems and using organs from some of the 10,000 irreversibly comatose patients on artificial maintenance? Would it consider using organs from aborted fetuses or anencephalic newborns most of whose brains are missing?

Society might accept the standard of brain death in the former case, but using organs from abortions or anencephalic newborns would involve a dramatic shift in thinking.

It also might be possible to obtain organs from people who are dead-on-arrival at hospitals, but whose bodies might be resuscitated long enough to remove a viable kidney or liver.

Far less controversial are "required request" laws, which mandate that hospitals ask relatives of dying patients to consider donating their organs. Eleven states passed such laws this year, and 11 more, including Illinois, are considering them.

But even if all these measures were used, there still wouldn't be enough organs to meet the need.

Another possibility is the use of animal organs, if that can be made technically feasible, but it undoubtedly would raise additional moral questions. And it, too, probably would demand a serious shift in society's view of medicine and its hold over life.

A workable, fully implantable artificial heart is probably 15 to 20 years down the road, some experts say. And this raises its own ethical questions. For example, with a potential demand of up to 100,000 artificial hearts a year, society may say it can't afford them for everyone.

Another problem is the number of transplant centers, which has grown as steadily as the procedure's success rate.

Nationwide, 80 centers are vying to perform heart surgery—and more are getting ready to open their doors—even though the number of available donor organs justifies no more than 24 to 62 centers, Evans said.

To ensure such centers do their best work, they should perform at least 12 heart transplants, 15 liver transplants and 25 kidney transplants a year, according to the federal task force.

"There's no reason a center should be doing two or three heart transplants a year," Evans said. "That's crazy. Those people are not going to be providing the same quality of service as an institution doing 10 or 15.

"We can't afford to be doing poorly, because donor organs are such a scarce commodity."

Source: R. Kotulak, "Transplant Gift in Short Supply," *Chicago Tribune*, July 16, 1986, pp. 1, 11.

CASE STUDY

HOT LINE

Jerry Carroll

Nights are no place for the faint of heart on the fourth floor at the Hall of Justice. The squeamish should steer clear, too.

There, you hear about it first when knives cut and blood spurts, when gun muzzles flash and bullets tear into flesh, when cars deliver death and injury to an intersection.

You know when women are beaten and children abused, when people come home and find front doors kicked open and their things gone, when drunks pass out on the street and are lying in their own filth. If it's bad and ugly enough, you get the word.

The fourth floor is where police communications is. It is a shabby, blue-lit room where the phones ring and computers record the social pathology of the city's streets the way the brain registers pain from the body's ganglia.

When you work on the fourth floor you begin worrying after a while about getting callouses on your feelings and losing your sensitivity to the people who call for help—that endless stream of victims who were mugged, raped, robbed, beaten or swindled. You worry that the time will come when their pain doesn't register on you anymore.

The fourth floor is where Lisa Befera, 26, puts in 3-to-11 p.m. shifts answering the 911 emergency telephone calls. When she isn't doing that she is sending officers to find a sniper. Or telling somebody to turn the radio down. The pendulum swings like that. Moments of high stress interrupting long spells of tedium.

Befera is one of 78 dispatchers, a number stretched thin by a hiring freeze earlier this year and a ban on overtime. "The hardest-working civilians in the police department," says her boss, Lieutenant Chris Weld. "Ideally, we should have 110."

Befera has been on the job only two years, not long enough to become desensitized or a victim of burnout, conditions she sees in some of her co-workers.

Blond, brown-eyed and articulate, the former college literature major is enthusiastic, full of pride in her work.

"I'm helping people. I'm not just sitting there watching the world go to hell in a hand basket."

In a busy 24-hour period, there will be something like 3400 telephone calls coming into police communications, 40 percent of them to 911. But sometimes there is an inexplicable surge.

"We had 4073 calls last Monday," Weld said. "I didn't believe it myself."

When the telephone calls come in hot and heavy, not all can be answered. So those coming in on the 553–0123 non-emergency line ring and ring.

But when the 911 calls don't get picked up, a red light goes on and a buzzer sounds until they are.

A lot of times the calls coming in on that line are a far cry from critical.

"That's no emergency," dispatchers say abruptly. "Call 553–0123." They hang up.

Although it's a Friday night when people are more apt to cut loose and get into trouble, it has been a quiet one for Befera, who is manning one of the non-emergency lines just now.

A woman calls to complain about an abandoned car that has occupied a priceless parking space for weeks on Russian Hill. It's high on the Top 10 of police complaints. "The car is slowly being picked apart," the woman say peevishly.

She had called Northern Station and was told by the cop who answered to tell it to police communications. Befera recognizes this as a dodge by somebody trying to avoid work.

"Call back," she advises. "This time talk to the sergeant, not just anyone who answers the phone."

The relationship between the police dispatcher and the cop in the field is a complex one. Mostly it consists of crisp coded exchanges over the radio in neutral voices. Sometimes there is banter.

But other times it gets snappish when an officer gets the wrong information or thinks it took too long to get a question answered about some nullity with zeroes for eyes he stopped and wants a warrant-check on.

When the answers come slowly it's usually because the dispatcher is juggling requests from several other cops needing the same kind of rap sheet on the bad actor they have leaned up against his car, legs spread.

"It's not unusual for a female dispatcher to mutter an obscenity after ending a radio transmission to a policeman who gets surly.

"Dispatchers are not appreciated by the people in the field," laments Deputy Chief Victor Macia. "They have no idea of the stress that goes with the job."

When Befera answers the non-emergency line, she says, "Police communications." Her voice is warm and interested.

When she answers the 911 line, her voice is charged with urgency. "Police emergency!" The telephone number and address of the caller automatically appear on her computer screen when she picks up the telephone.

Her fingers fly over the computer keyboard in front of her when Befera takes a complaint. She sends the information to the screen of a dispatcher a few feet away. The dispatcher relays it to the nearest patrol cars as fast as Befera sends it.

Most people calling in with an emergency are unaware the exchange is this swift. Some are infuriated by what seems like a leisurely questioning by 911 operators about picky details.

Not Just Wasting Time

"People assume I'm taking notes and will dispatch somebody when I get good and ready," Befera says. She leans back in her chair to stare at her nails to give an impression of vacant languor.

But, with a unit already en route, what she and the others are doing is gathering the sort of detail that saves lives if the police arrive when the criminal is still on the scene. Or it allows them to collar suspects when the trail is still hot.

What does the robber look like? What is he wearing? Is he with someone? Is he armed? What kind of weapon is it? The callers reply impatiently, their voices rising.

Befera had to take a stress reduction course to deal with the job pressures. One of the techniques was how to stop replaying things in her mind when she got home.

"On the sad stories that happened, I'd ask myself, did I do well? If it happened again, what would I do differently?"

She's thinking tonight about a call she got a couple of nights earlier. It was a man who was deeply depressed and crying. Befera sensed he was suicidal and sent a cop out to him. But what happened?

"I wonder," she says.

"You never know. There was a man once who slashed his throat and both wrists on the sidewalk and went staggering off. He was never found. Just vanished."

That's a war story. Police dispatchers collect them like cops. "Every day there is something new." When they gather after work, that's all they talk about, the "runs" they handled during the shift and the strange things that happened.

Red Alert

"Two or three weeks ago we got a call from a bartender who said an off-duty policeman had his gun drawn on two people and needed assistance."

That is a Code 1025, which means everybody in the area drops whatever they are doing and hauls butt with lights flashing and siren screaming to where the cop is in trouble.

"Everybody went flying over there. We had 15 officers and six dispatchers talking on the radio," Befera said. The off-duty cop kept the suspects under control until help arrived.

It is when this kind of hell breaks loose that a dispatcher proves her mettle. The officer in the field sees only a small bit of the developing picture.

He has to be guided by the dispatcher, who is analyzing the information flooding in and making rapid fire decisions. She moves cars around the city like chess pieces, scrambles paramedics if necessary, notifies supervisors, dispatches speciality units like the crime lab.

"I'm probably OK in situations like that, but there are women here who are like surgeons. They hear everything and know where every unit is and they have absolute control. They leave your mouth open with awe. I can't even imagine being that good."

The women sound like a lynch mob when a child molester is taken into custody. The room rings with their angry shouts. "These women get furious, and I love being part of that," Befera says.

Besides the civilians, there are two lieutenants, eight sergeants and 27 patrolmen working in police communications. Some of the patrolmen are there because they got in trouble on the street and are awaiting disciplinary proceedings.

"This is the department's penal colony," Weld says.

Women greatly outnumber men among police dispatchers. "I think it's 10-to-1," said Befera.

Tense Relationship

The relationship between the dispatchers and the officers working in communications is not without its tension.

Some of the cops aren't too handy at the job because they don't type fast enough to interact efficiently with the computer. But, at $33,000 a year at top scale, they earn a lot more than the $25,320 a dispatcher does at the same pay grade.

Befera gets along well with the cops, chiefly because she admires them so much.

"It irks me to hear people talking about lazy cops eating doughnuts and drinking coffee. People's perceptions are so distorted. What are they doing when those officers are fishing a body out of the bay the crabs have been eating or a kid has been run over in an intersection?"

But she would never be a cop or marry one.

"The stories they tell. The bodies that have been lying for two weeks and have maggots in them. The domestic beefs they have to break up day after day where the wife has been beaten up and the kids are crying. Terrible things.

"I always wonder when I send them out on a run what will happen, like when there's a crazy man with a gun on a corner. They just say 10-4 and go."

Blood and a Hang-glider

The night is slow. An officer calls in on his hand radio as he approaches an apartment where neighbors think there has been suicide. "There's blood all over," he says. But it's dried blood, and nobody's there.

An off-duty policeman reports seeing a hang-glider fall into the bay near Candlestick Point. Nothing is ever found.

Ten men battle four others outside the Greyhound bus station. One man is injured and one arrested.

Two passers-by see a man lying apparently dead in the street near the Holiday Inn on Seventh Street. He's merely drunk.

Security firms regularly call to report silent alarms have been triggered. "Our sensors pick up voices inside," says one.

A shoplifter is arrested at a drugstore on Market Street, families have fights. "When they call and you can hear the children crying in the background, that's when it hurts," Befera says.

An old woman collapses in a restaurant. A woman says smoke is coming from under the hood of her car. A man beats up his girlfriend and vandalizes her car. Drunks call and so do crazies. People call 911 when they mean to call 411.

Sitting there with her Starset I headset on, Befera takes it all in. Sometimes she wishes she still smoked. She wonders whether she will still be able to do this job five years from now. But she knows now she likes it.

"I like to be there when people are hurting. I go home at the end of the day with a feeling of accomplishment. People feel better after talking to me. What more could you ask from a job?"

She lives on the Peninsula. The reason is simple. At the end of her day, Befera says, "I have to get out of the city."

Source: J. Carroll, "The Hot Line," San Francisco Chronicle, Oct. 6, 1986, pp. 38, 41.

CASE STUDY QUESTIONS

Transplant Gift in Short Supply

1. Describe how the allocation of transplant organs behaves like a queueing system. Describe the server and the service process. Describe the customer and the arrival process. What might a renege represent?

2. In what ways has modern technology affected the arrival and service processes for the queueing system?
3. Describe the current "queue discipline" for allocating liver and heart transplants. Do you believe this is a good system? How might the system be improved?
4. It has been suggested that transplant organs be made available for sale, both as a way to encourage more donors and to ensure that transplants are given to the most worthy patients. Is this a good idea?

Hot Line

1. Describe the triage process used by the dispatcher. How does triage improve the performance of the 911 system?
2. The same dispatchers answer calls on the emergency 911 number and the nonemergency 553–0123 number. How does the system ensure that more critical calls receive priority?
3. When a call is received, the closest officer to the incident may be busy with a previous call. Under what conditions do you believe this officer should be interrupted, instead of sending a more distant officer? Describe the information you would like to know before making this decision.
4. Police officers sometimes follow beats within set "turfs." That is, each officer is assigned his or her own zone (turf) to serve, and only leaves the zone in an emergency. How would "turfing" affect the queue discipline? How would "turfing" affect the quality of customer service?
5. From the reading, it is apparent that the queue discipline cannot be described with a simple formula. That is, considerable discretion is given to the dispatcher. Do you believe that this situation is unique to emergency dispatching or that it might occur in other sorts of queueing systems? Give examples.

chapter 10

Queueing Networks

The analysis of a queueing system, as with the analysis of most any type of system, should follow a building-block approach. To understand the operation of the whole, one must first understand the operation of the parts. This has been the goal of the last nine chapters. Rather than taking on complicated queueing systems, only simple systems, consisting of individual servers or identical servers working in parallel, have been studied.

The purpose of this chapter is to extend the concepts already introduced to ***queueing networks***. Queueing networks occur when the service process, for reasons of efficiency, is divided into separate tasks (a concept going back to Adam Smith's pin factory and a hallmark of industrialization). Consequently, queueing networks require the coordinated effort of two or more servers, as in the following examples:

> A data file is sent across the country between two computers. Along the way, the file is routed via several processors, each of which performs a portion of the transportation service.
>
> Work orders processed in a machine shop require the execution of various operations on drills, lathes, and presses. Each machine acts as a server.
>
> A patient undergoes a major physical at a medical clinic. The physical entails a battery of tests on different equipment and examination by several specialists.

It is tempting, when studying queueing networks, to examine queues individually, without considering the impacts on other parts of the system. But only under rare circumstances will such a localized approach be successful. Jackson (1957) established sufficient conditions under which a queueing network can be decomposed to individual *M/M/*1 servers: Customers must arrive from outside the network by a Poisson process, service times must be independent and exponential, queue discipline must be independent of routing and service time, and routing must depend on a fixed probability transfer matrix. While these results are noteworthy in the theoretical sense, they are quite limited in application, for queueing systems rarely possess the required properties. (For further background, see Disney and König 1985.)

It is also tempting, when studying queueing networks, to treat the entire network as an individual server. This is especially true when the inner workings of the network are hidden from view, as is the case for computer and communications systems. But this too can be misleading, for it becomes impossible to decipher time spent in queue and time spent in service. If the only information available are the times the customer entered and exited the network, then it will be impossible to determine whether the network is very congested (most of the time spent in queue) or not (most of the time spent in service). Further, to improve the network's performance, one must know which servers within the network are causing delay, something that cannot be detected without looking at the inner workings.

The most important concept in queueing networks is that of the ***bottleneck***, a concept that will receive considerable attention in this chapter. In most queueing networks, queueing delays are greatly influenced by the performance of one or more bottleneck servers. To improve the performance of the entire system, one must identify the bottlenecks and improve their performance. The bottleneck will be an important concept in each of the two basic types of network service: ***serial service*** and ***parallel service***. With serial service, customers are served one task at a time in *consecutive* order. With parallel service, tasks are performed *simultaneously* by different servers. For instance, in a job shop, a lathing operation might be performed on a component at the same time as a drilling operation on another. Of course, many networks contain both serial and parallel elements.

There is an extensive literature on queueing networks that this chapter will only begin to touch. Much of this literature is application specific. Prime examples include computer/data communications networks (Bertsekas and Gallager 1987; Kleinrock 1976), traffic networks (Haight 1963), and production networks (Hax and Candea 1984; Johnson and Montgomery 1974). Despite the specificity of this literature, there is much that all of the applications hold in common. This chapter highlights these common factors, emphasizing the most fundamental results.

The complexity of queueing networks makes it extremely difficult to develop analytical models of their stochastic behavior. It is not, however, difficult to develop graphical models of their deterministic behavior under varying levels of demand, and this is the emphasis here. Models are also provided for estimating the ***throughput*** (that is, output rate) of the queueing network, an important measure of system performance. The chapter concludes with a discussion of how to use computer software to diagnose and resolve problems in queueing networks.

10.1 QUEUES IN SERIES

In the simplest type of queueing network, queues and servers are organized in a series, and every customer passes through the entire series in the exact same sequence (see Fig. 10.1a, a special case of serial processing. In production this is sometimes called a *flow shop*). For example, imagine a two-step service process for returning clothing at a store. First, a customer goes to the return window where a credit slip is obtained, and second, the customer goes to a cashier, who reimburses the customer. This contrasts with Fig. 10.1b, which shows parallel service.

The two serial servers do not operate independently of each other because the arrival of customers at the cashier is tied to the service of customers at the return window. While

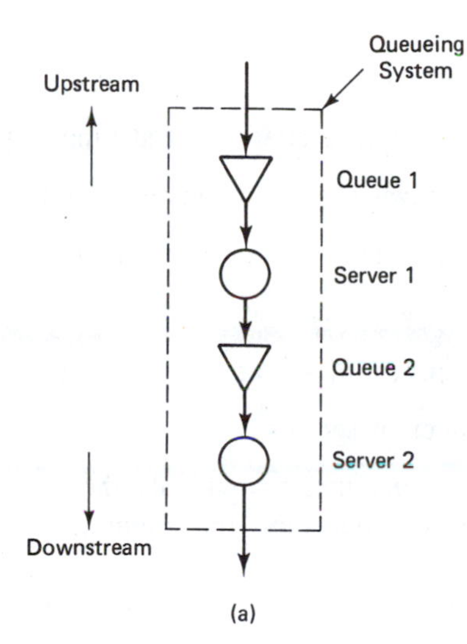

Figure 10.1a Diagram of two queues in series showing upstream and downstream directions.

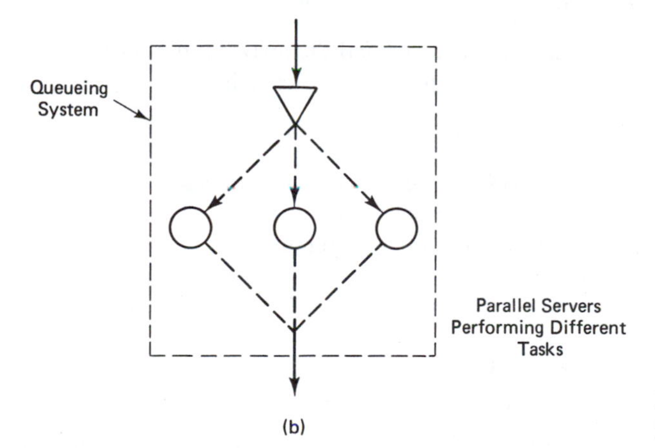

Figure 10.1b Diagram of three servers in parallel, fed by a single queue.

the arrival process at the first server may be Poisson, it should no longer be Poisson at the second server. More generally, *one server's departure process dictates the next server's arrival process*. This statement is the crux of the issue. It is why servers cannot be examined in isolation.

10.1.1 Fluid Model

As an illustration, consider a system in which customers are served continuously (non-batch), with short service times. Further, suppose that the arrival rate varies deterministically over time, and that a fluid model is valid. Servers are numbered 1 to N, in the order visited, and the following definitions are used:

$A^n(t)$ = cumulative arrivals at server n up to time t

$D^n(t)$ = cumulative departures from server n up to time t

$L^n(t)$ = customers in queue at server n at time t = $A^n(t) - D^n(t)$

τ_n = free-flow travel time from server $n-1$ to server n

$\omega_n(t)$ = departure rate at server n at time t

$\lambda_n(t)$ = arrival rate at server n at time t

c_n = service capacity of server n

N = total number of servers

The free-flow travel time is the time that it would take the customer to travel from server $n - 1$ to server n in the absence of queueing (this might also incorporate the service time at server $n - 1$).

The cumulative arrivals at any secondary server are tied to the cumulative departures at the previous server:

$$A^n(t) = D^{n-1}(t - \tau_n) \qquad 2 \leq n \leq N \tag{10.1}$$

If there is no queue at server n, the departure rate and the arrival rate will be the same. Once there is a queue, the departure rate is the server's capacity:

$$\omega_n(t) = \begin{cases} \lambda_n(t) & L^n(t) = 0 \\ c_n & L^n(t) > 0 \end{cases} \tag{10.2}$$

$D^n(t)$ can be derived from $\omega_n(t)$ in the usual manner, through integration, and $\lambda_n(t)$ can be derived in the usual manner from $A^n(t)$, through differentiation.

Definitions 10.1.

Time in system for a queueing network is the difference between the arrival time at the first queue and the departure time from the last server.

Customers in system for a queueing network is the total number of customers in queue or in service in the network: $A^1(t) - D^n(t)$.

Time in service is the sum of the free-flow travel times.

Time in queue is the time in system minus the time in service.

Server A is **upstream** from server B if the customer is served at A before B.

Server B is **downstream** from server A if the customer is served at B after A.

As can be seen, the fluid analogy carries over to the upstream and downstream terminology of queueing networks.

Figure 10.2 illustrates the behavior of a simple two-server system. A large queue forms behind server 1, beginning at time 8:20. Once the first queue forms, the arrival rate at server 2 becomes constant, equaling the capacity of server 1. Consequently, a somewhat smaller queue forms behind server 2, beginning at time 8:35. The free-flow travel time, the separation between the departure curve at server 1 and the arrival curve at server 2, is 15 minutes.

Based on Fig. 10.2, we would like to answer the following: If we could increase the capacity at either one of the servers by some small amount, which capacity should we increase? The naive answer would be to look at the two queue sizes, note which is larger, and increase the capacity of the corresponding server. This approach is wrong. Note in

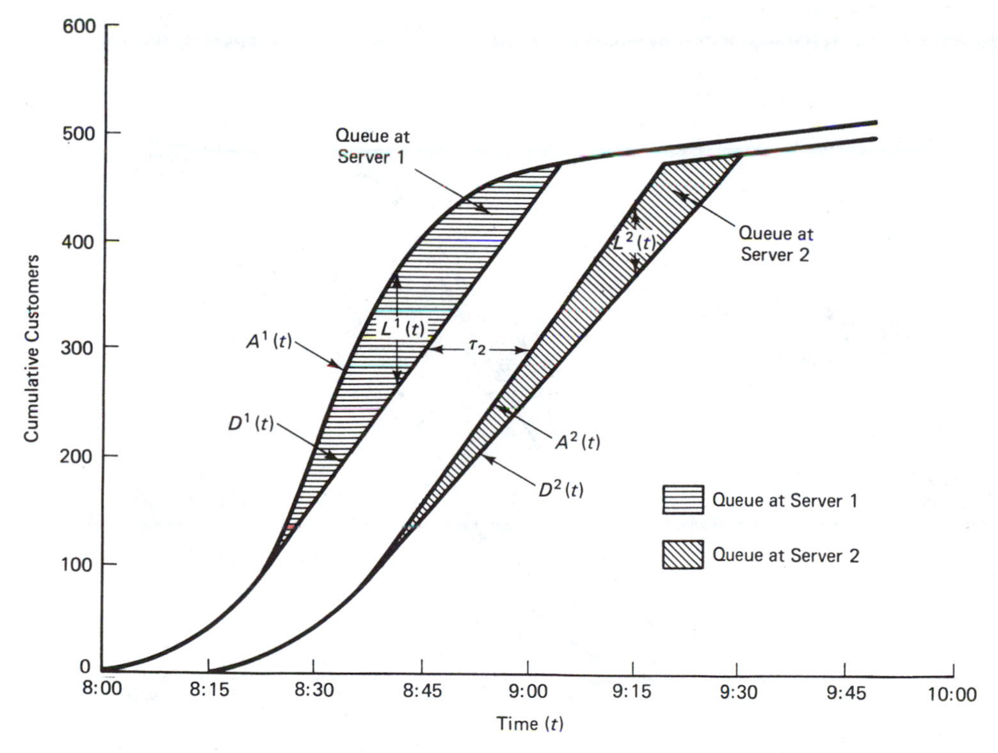

Figure 10.2 Cumulative arrivals and departures for a deterministic-fluid model of two queues in series. The second server is the bottleneck because it is the last place where the customer encounters a queue.

Fig. 10.3a the impact of increasing the capacity of the first server. The delay at the first server has decreased, as one might expect. But the delay at the second server has *increased*—by an equivalent amount—because of the increase in its arrival rate. The total time in system is not reduced, *it is merely shifted from the first server to the second.*

The correct answer for the example would be to increase the capacity of the second server, the server with the smallest capacity. The impact of this action is shown in Fig. 10.3b, in which the capacity has been increased to that of the first server. The queue at the second server has vanished and the total time in system has declined.

The second server is the bottleneck in Fig. 10.3.

Definitions 10.2

Bottleneck server For deterministic queues in series, offering continuous service, the server (or servers) with the smallest service capacity.

It is possible to have more than one bottleneck, but these must all have the same minimum capacity. In series queues where there is just one bottleneck, another important property holds:

The bottleneck is the last server where the customer encounters a queue. The bottleneck is *not* necessarily the server with the largest queue.

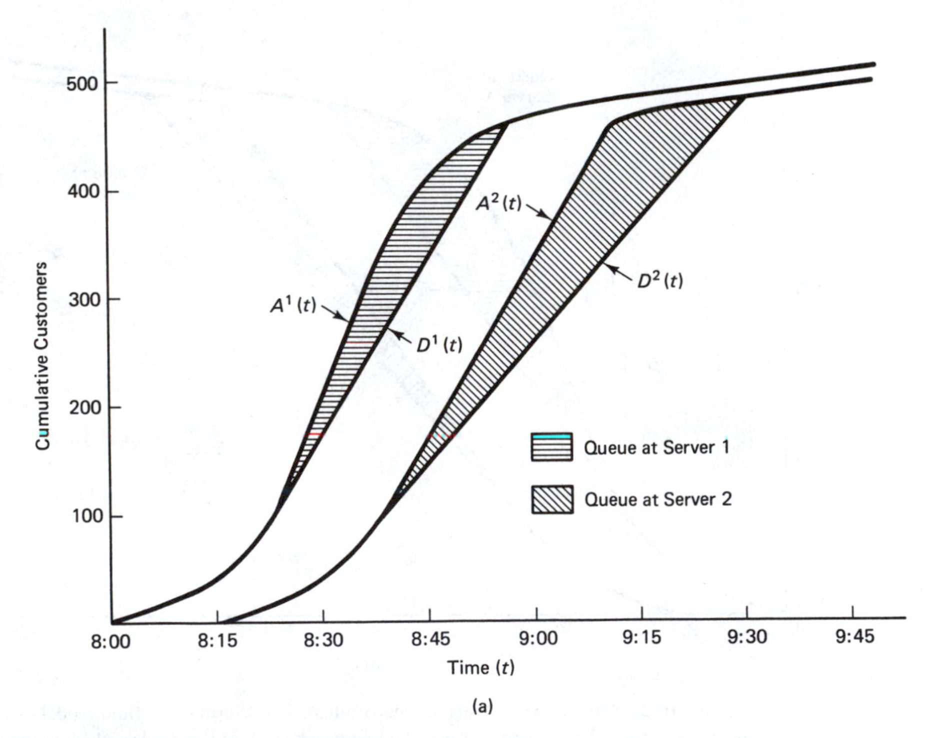

Figure 10.3a Increasing the capacity of the nonbottleneck server (server 1) shifts delay to the second queue, but does not affect total waiting time.

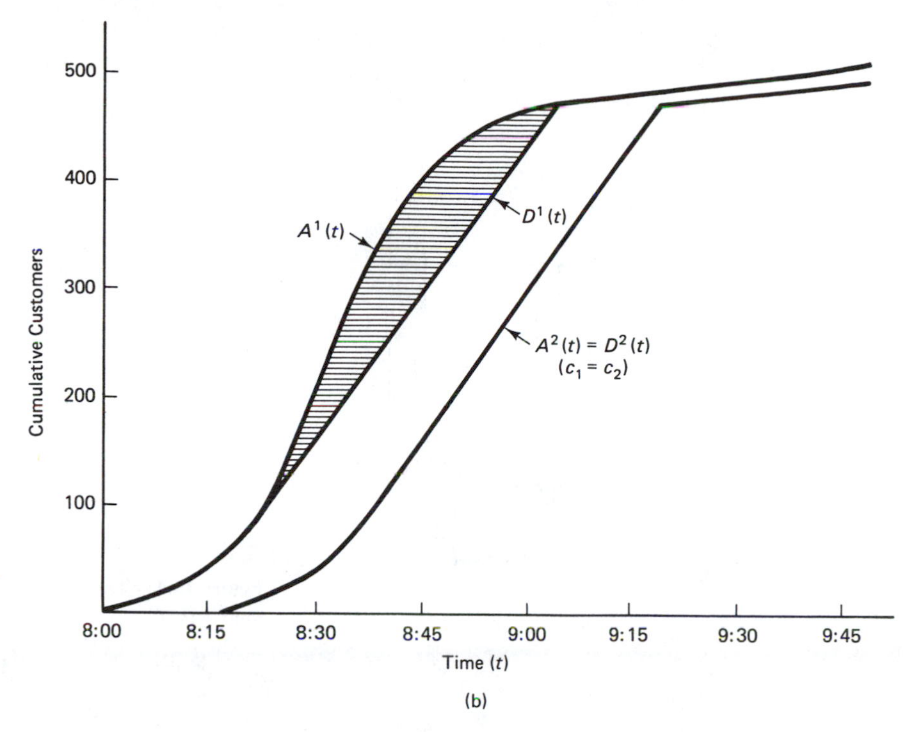

Figure 10.3b Increasing capacity of bottleneck to that of the nonbottleneck server eliminates second queue and reduces total waiting time.

(With two or more bottlenecks, the *first* of the bottlenecks will be the last place where the customer encounters a queue.) So there are two ways to detect that server 2 was the bottleneck in Fig. 10.2. It had the smallest service capacity (that is, smallest slope in its departure curve) and it was the last place that a queue was encountered.

The significance of the bottleneck in a deterministic/fluid environment, with queues in series, is that it alone determines how much time customers spend in the system. Specifically

Time in system can only be reduced by increasing the service capacity of a bottleneck server. If there is more than one bottleneck in a series queue, *all* of the bottleneck capacities must be increased. Increasing the capacity of nonbottleneck servers will not decrease time in system.

Though the deterministic assumption may seem restrictive, this is certainly one of the most important results in the queueing literature. It provides a simple way to attack problems in a series of queues: Identify the bottleneck (or bottlenecks) and improve its performance. Moreover, as will be seen later, the concept can be extended to more complicated systems.

Figure 10.4a is a good way to visualize the situation. Customers are represented by a fluid, servers by pipes, and queues by storage reservoirs. Whereas the arrival rate from

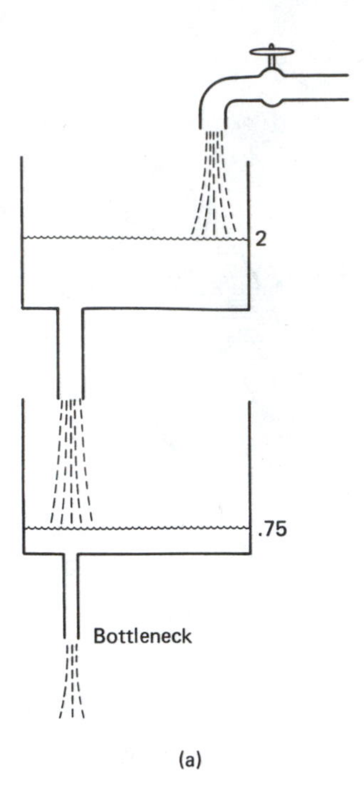

Figure 10.4a Queue size and waiting time depend on the server with the smallest capacity, as can be conceptualized with this fluid model.

outside the system may be highly variable, the arrival rate at the second server is much more steady—the first server smooths out the flow, not to exceed its capacity. The amount of time that the fluid spends in the system depends on the smallest capacity of the two drains. That is, the bottleneck is the narrowest drain, much as the neck is the narrowest part of a bottle. The time in system does not depend on the sequence of the two drains. Putting the bottleneck first pushes the queue upstream, but does not affect the total size, as shown in Fig. 10.4b.

10.1.2 Balanced Systems

The bottleneck results were first identified in the work by Oliver and Samuel (1967) studying letter delays in post offices. But Oliver and Samuel went further. They examined a large post office in which employees could be dynamically shifted from one sorting station to another to obtain variable server capacities. At any time, the sum of the capacities must be a constant (determined by the number of employees available), but the constant can be apportioned in any manner between the servers. Their remarkable result for closely spaced servers was that time in system is minimized when the capacities of the servers are *identical, even when the total capacity varies over time*. That is, there is no point in providing excess capacity at any one stage because delay will increase somewhere else. This principle contradicted prior practices in which workers were shifted from stage to stage as the day's mail progressed through the system.

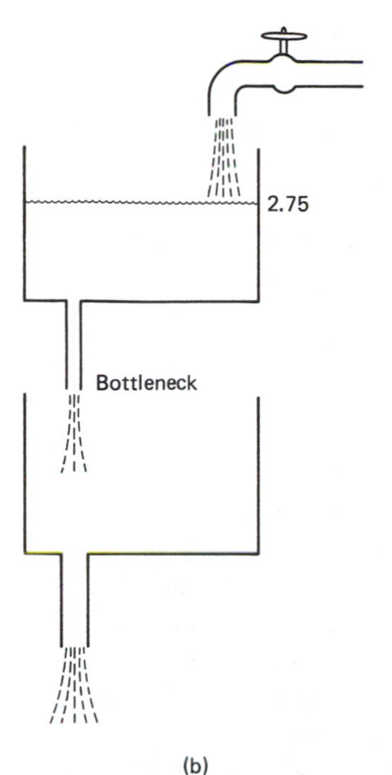

Figure 10.4b Switching server order moves the queue but does not reduce total queue size.

The implication of Oliver and Samuel's results is that series queueing systems should be designed so that the server capacities are the same. That is, any investment in server capacity should be allocated in a fashion that balances server capacity. Following this rule, major queues should only be encountered at the first server. After passing through this server, the customer should move through the system relatively unimpeded. In production systems, such a design has an added advantage. Customers enter the shop in the form of work orders, which, by their nature, are easy to store in queue. Once the work begins, customers (now called *work-in-process*, or *WIP*) become difficult, and expensive, to store. A balanced shop allows orders to queue but moves the WIP through the shop as rapidly as possible.

There are exceptions to the rule. Most importantly, if capacity can only be provided in discrete increments (for example, one can only operate an integer number of machines), it may be impossible to provide identical capacities at all servers. Suppose that

μ_n = capacity per machine at station n

N_n = number of machines operating at station n

p_n = cost per machine at station n (\$/unit time)

c = system capacity

$$= \min_n \{N_n \mu_n\}$$

The total operating cost (not counting waiting) can be expressed as the following function of system capacity:

$$C(c) = \text{operating cost per unit time, given capacity } c$$
$$= \Sigma\, p_n[c/\mu_n]^+ \qquad (10.3)$$

The symbol $[\ ]^+$ denotes the smallest integer larger than the closed quantity. That is, the number of machines at each station is the minimum required to achieve a capacity of c.

$C(c)$ is a step function, with each step representing the addition of a machine (or machines). For example, suppose that

$$p_1 = \$200/\text{hr} \quad p_2 = \$50/\text{hr} \quad p_3 = \$150/\text{hr}$$
$$\mu_1 = 200/\text{hr} \quad \mu_2 = 500/\text{hr} \quad \mu_3 = 100/\text{hr}$$

$C(c)$ should be plotted, as in Fig. 10.5, to identify design targets. For values of c between 0 and 100/hour, one of each machine type is needed, for a total cost of \$400/hour; to achieve a capacity of 100 to 200/hour, a second machine of type 3 is required, raising the cost to \$550/hour, and so on. Over the range shown, no capacity provides a perfect balance, achieving 100 percent utilization of all machines. However, system capacities of 200/hr and 400/hr provide 100 percent utilization of the most expensive machines and come close to minimizing the operating cost per unit capacity (it would take a capacity of 1000/hr to achieve a perfect balance). The actual design might be selected by substituting the arrival rate for c in Eq. (10.3). Or a more detailed analysis could be undertaken to assess the relationship between profit and the price for the service and consequent demand.

In addition to discrete capacity increments, different capacities might be used at different servers to compensate for random fluctuations in service times, but one must

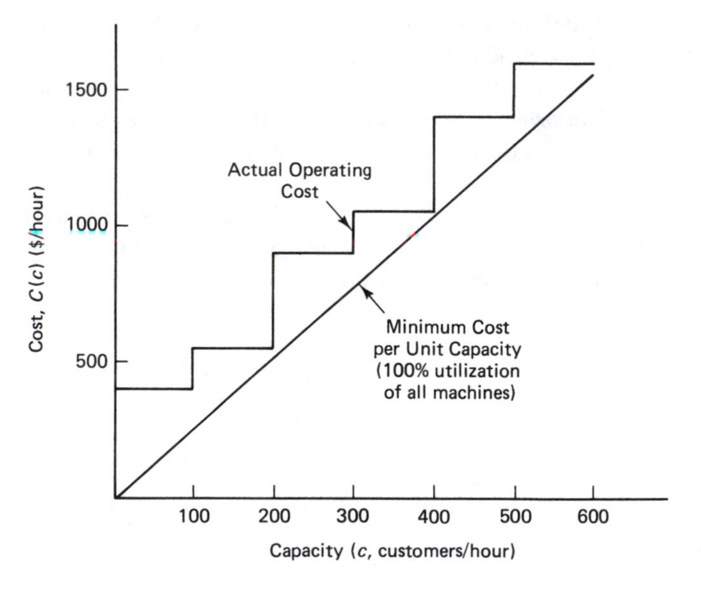

Figure 10.5 Operating cost versus capacity with discrete capacity increments. System should be designed so that the cost per unit capacity is small.

be careful here. A naive approach would be to provide extra capacity (on average) at the server with the greatest fluctuations. But one must remember that randomness in the service process at one server creates randomness in the arrival process at downstream servers. Hence, little may be gained in adding capacity at the server suffering random service fluctuations unless similar capacity increases are provided downstream.

10.1.3 Bulk Service

In Chap. 7, we saw that the motivation for serving similar customers in bulk is to reduce the amount of time the server is engaged in setups. Bulk service is used in transportation (traffic signals, freight transport) and in production (stamping presses). The same motivation applies to queueing networks, though the issues surrounding bulk service are more complex. Bulk service creates bulk arrivals at downstream queues. Instead of smoothing out arrivals (as was the case in the previous section), bulk service can create added variability downstream.

Despite the differences, the principle of the bottleneck is still applicable. A simple illustration will demonstrate why. Suppose that a series of traffic lights is spread along a one-way street, as in Fig. 10.6. All of the traffic on the one-way street enters at its start and exists at its end. Traffic on the cross streets pass without entering the one-way street. The signal system alternates between red and green phases, and is defined by the following parameters:

$\lambda(t)$ = arrival rate at entrance to one-way street at time t

c_n^g = service capacity at intersection n when signal is green

g_n = length of green phase at signal n, per cycle (not counting any loss time)

T_n = cycle length of signal n

τ_n = free-flow travel time from signal $n - 1$ to signal n

10.1.3.1 Synchronized Signals. Anyone who has driven in city traffic knows that signal systems are most effective when the green phases are synchronized at the intersections. If service rates are the same, signals should be timed so that a vehicle that receives a green light at the first intersection receives a green light at all subsequent intersections. Sychronization can be planned through the use of a ***time-space diagram***, as

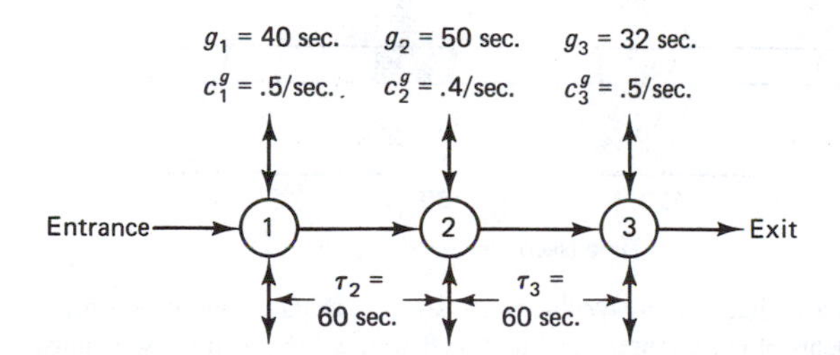

Figure 10.6 Diagram of traffic signals along a one-way street.

shown in Fig. 10.7. The diagram shows the progression of traffic from signal to signal, along with the required offsets between the green phases of the signals. Note that perfect synchronization demands that the cycle lengths be identical, and that the green phase at any signal should begin at a time τ_n after the start of the green phase at the signal immediately upstream.

Example

Three traffic signals are placed in series along a one-way street, all with a 60-second cycle time. The green phases currently have lengths $g_1 = 40$ sec., $g_2 = 50$ sec., and $g_3 = 32$ sec.; the capacities during the green phases are $c_1^g = .5/sec.$, $c_2^g = .4$ sec., and $c_3^g = .5$/sec. Cumulative diagrams in Fig. 10.8 show the system under congestion, with synchronization.

Even with synchronization, differences in capacity and green phase cause queues to form at all three servers. Queues at the second server are short-lived, vanishing before the end of each green phase. Queues at the first and third are not short-lived and grow over time. Which server is the bottleneck? As before, it is the server with the smallest capacity. But now it is the server with the smallest *effective capacity*, which can be calculated as follows (whether or not servers are synchronized):

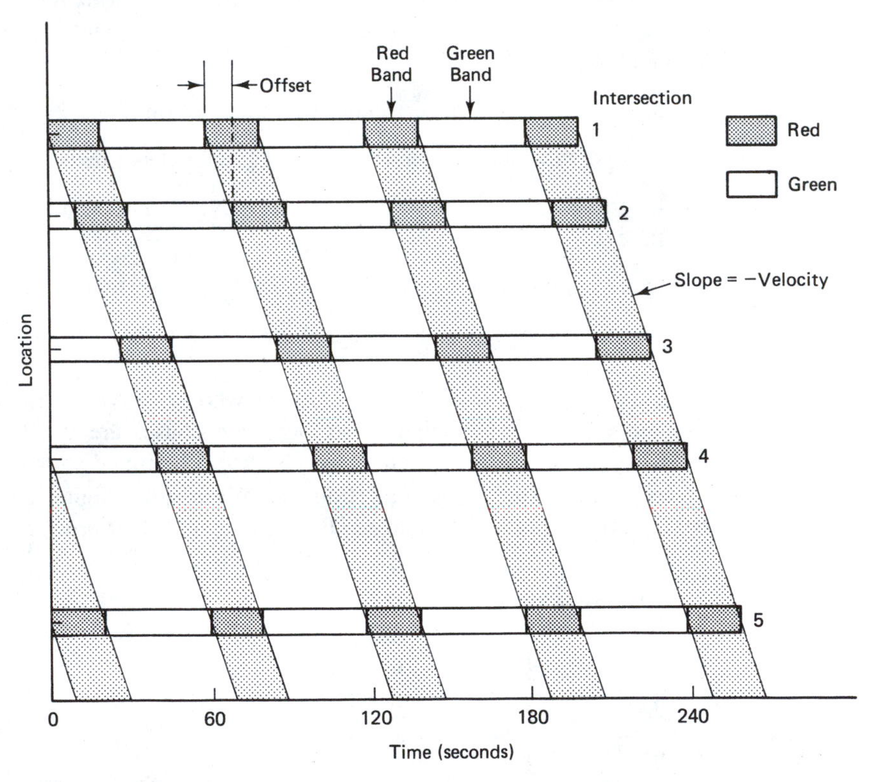

Figure 10.7 Time-space diagram of synchronized signals along a one-way street. Signals should have identical cycle lengths and be timed so that the green phase begins just as vehicles arrive.

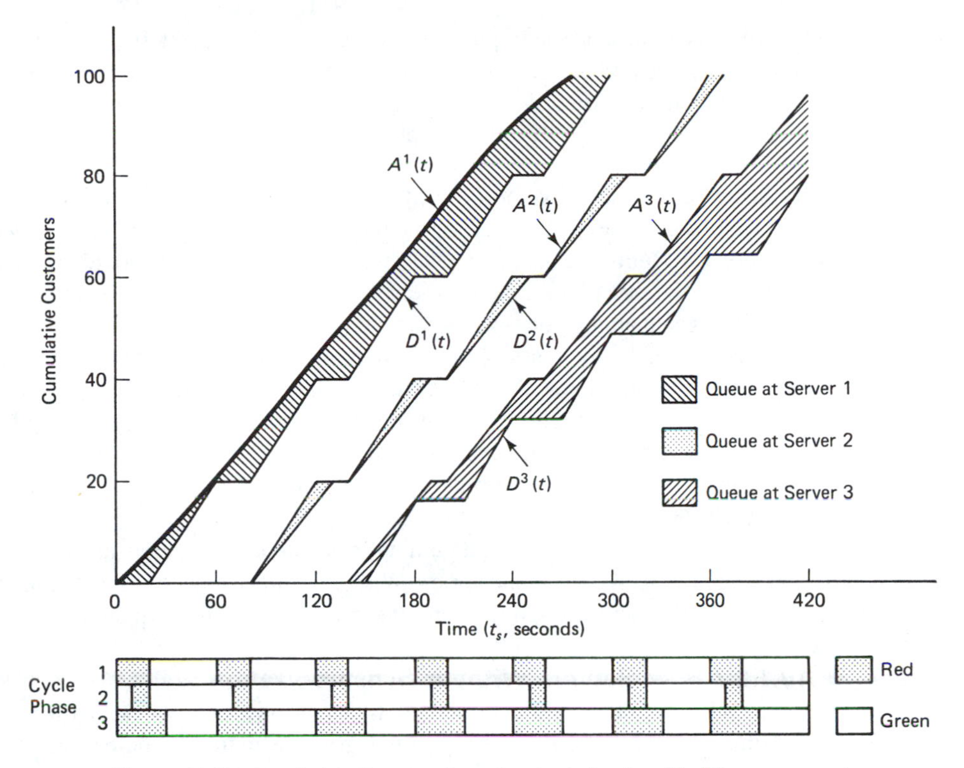

Figure 10.8 Cumulative diagram of synchronized signals with different green phase lengths. Server 3 is the bottleneck because it is the last place where a noncyclic queue is encountered.

Definition 10.3

$$
\begin{aligned}
c_n &= \textit{effective capacity} \text{ of bulk server } n \\
&= c_n^g(g_n/T_n) \qquad (10.4)
\end{aligned}
$$

In the example, server 3 has the smallest effective capacity, 16/60, and is therefore the bottleneck. The effective capacity is smaller at server 3 than at 2, not because it is not capable of serving at as high a rate but because it is in the green phase for a smaller percentage of the time. Less of the server's total capacity is allocated to serving the one-way street and more is allocated to serving cross traffic.

With bulk servers, *the bottleneck is the last server experiencing a noncyclic queue*. Or, put another way, it is the last server for which the queue does not vanish before the end of the green phase. As with the continuous server, the bottleneck can be identified by looking at capacity or looking at the queues.

The most effective way to reduce delay, as before, is to increase the capacity of the bottleneck server. But now the goal can be accomplished in any of three ways: by extending the green phase, by extending the cycle length *and* the green phase, or by serving customers at a faster rate (for example, by adding lanes to the road). Of course, extending the green phase alone implies that capacity is taken away from the cross traffic, which may not be desirable. Extending both cycle length and green phase can increase effective capacity in both directions by reducing capacity losses during changes in phase. To find the optimal cycle length and green phase at the bottleneck, the techniques covered in Chap. 7 can be applied.

If the effective capacity of the bottleneck cannot be increased to match other servers, it may be that the green phase of the nonbottleneck servers on the one-way street should be *reduced*. Reducing the green phase for the one-way street should not increase its delay. It may, on the other hand, reduce the delay for the cross streets. In a progression of equal capacity signals along a one-way street, all intersections should have identical green phases (as determined by the bottleneck). With modern equipment, the length of the green phase and cycle can be changed with traffic conditions, provided that the changes are synchronized among the signals. As the overall traffic increases, the cycle lengths should increase; as traffic decreases, the cycle should decrease. The length of the green phase should vary in accordance with the relative amounts of traffic on the one-way street and the cross streets.

As with the continuous server, a guiding principle is that major queues should only exist at the entrance to the queueing series. With a balanced system, the customer should proceed through the network unimpeded after passing the initial queue.

10.1.3.2 Bulk Servers Without Synchronization. Even without synchronization, efforts for reducing delay should focus on the bottleneck. However, if the system is uncongested (that is, queues do not grow over time), increasing capacity of nonbottleneck servers might also reduce delay somewhat. Figure 10.9 (drawn with $\tau_{n+1} = 0$) shows that intermediate queues will appear when two bulk servers, of equal capacity, are not synchronized. Unlike the queue behind server 2 in Fig. 10.8, this queue does not vanish at the end of each cycle. If the process is deterministic, the average wait per customer amounts to one-half the length of the red phase at each server (this can be seen by inspecting the figure):

Average time between server n and $n + 1$ =

$$\frac{T_n - g_n}{2} + \frac{T_{n+1} - g_{n+1}}{2} + \tau_{n+1} \tag{10.5}$$

A proportionate reduction in both cycle length and green phase at either server will reduce the queue size between the servers. This reduction is due solely to eliminating delay caused by the *lack* of synchronization. Therefore, a preferred strategy would be to institute synchronization. But this may not always be feasible. In traffic, for instance, it is difficult to synchronize signals along two-way streets. Vehicles coming from two different directions do not necessarily arrive at the signal at the same time.

Though explained in the context of traffic signals, the results are easily transferable to production. A signal operates in the same manner as a machine with setups, alternating between on and off phases. Once a lot of items is created, it should not be split, and should move through the network without delay.

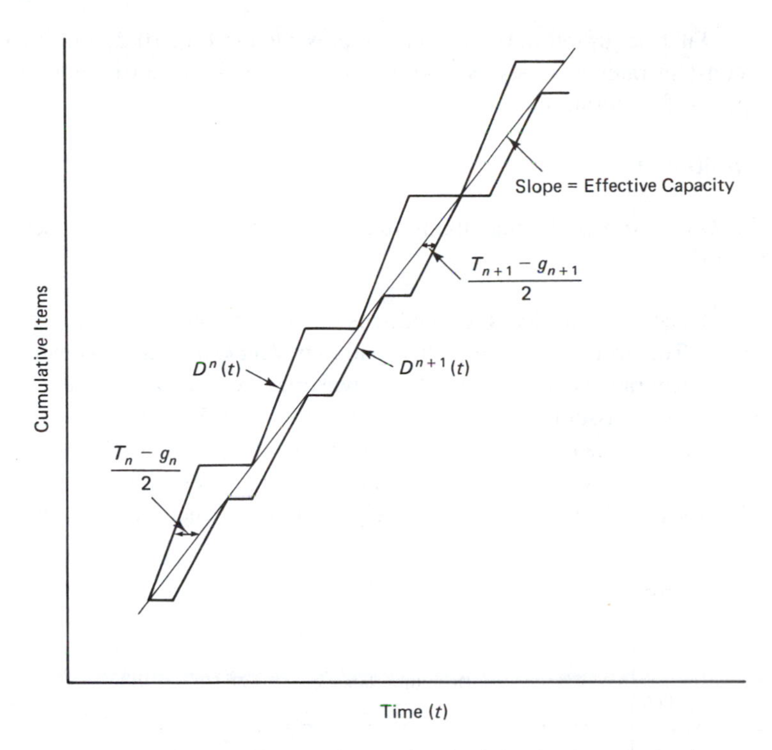

Figure 10.9 Cumulative diagram of unsynchronized signals. Area of triangles represents added delay due to lack of coordination.

In both contexts, production and traffic, bulk service typically operates in networks that are more complex than a series of servers. While much can be learned by examining components of such networks, the effects may extend well beyond a small set of servers. For instance, it may be necessary to coordinate signals over an entire grid of streets. There exist no ways to optimize signal timings over large networks. A logical heuristic (nonoptimal) is to set signal timings one street at a time, in decreasing order of total traffic. Other heuristics, such as the TRANSYT system, are described in Skabardonis (1986), Skabardonis and Kleiber (1983a, 1983b), and U.S. DOT (1981).

10.1.4 Buffers, Blocking, Starvation, and Gridlock

In recent years, the study of queues has been something of a popular pastime, particularly in congested cities such as New York. The term ***gridlock*** is one of the by-products of this craze. As originally intended, the term was meant to describe a traffic condition endemic to Midtown Manhattan, in which streets became so congested that they reached a standstill. It has since been incorrectly applied to all sorts of queues, from air traffic congestion to court delays, for gridlock is actually the product of specific type of queueing phenomenon called "blocking," which will be explained now.

First, let us return to the scenario presented in Fig. 10.2, with two queues operating at constant rates in series. Suppose that the second queue (the bottleneck) has a limited capacity for storing customers.

Definition 10.4

The ***buffer size*** is the maximum size of the queue between two servers in a queueing network.

Once this capacity is reached, customers can enter the queue no faster than they are served. The upstream server is then said to be ***blocked*** by the downstream queue, because its service rate is limited by the downstream capacity. Figure 10.10 illustrates this phenomenon, assuming that $\tau = 0$. Between time 8:50 and time 9:10, server 1 is blocked by server 2, because queue size has reached its capacity of 50. While blocked, server 1 and server 2 serve customers at identical rates. Blocking does not end until the queue behind server 1 vanishes, immediately after which the queue behind server 2 begins to

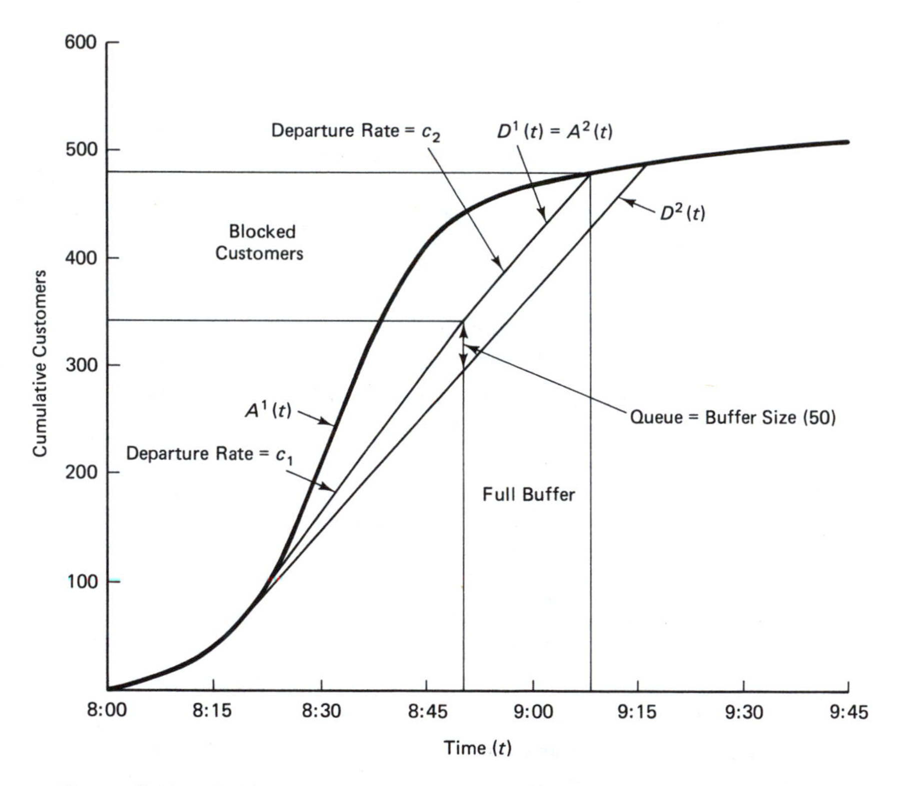

Figure 10.10 Blocking at a continuous server may not increase queueing. In the example, server 1 is blocked when the buffer at server 2 becomes full. This prevents server 1 from operating at full capacity, but because it is not a bottleneck, total waiting time is not affected.

dissipate. Blocking can also occur when τ is greater than 0. However, a lag might occur between the time the buffer is full and the time server 1 slows down.

Within the deterministic/continuous environment, buffer size actually has little impact on system performance. Blocking reduces service rate at nonbottleneck servers, but this only shifts delay upstream, not adding to total delay. Buffers are primarily needed to cushion against stochastic variations (in service and arrivals) and variations in queue size due to bulk service. There are two guiding principles here:

The bottleneck server should not operate at less than full capacity for lack of customers to serve (called ***starvation***).

The bottleneck server should not operate at less than full capacity due to being blocked by a downstream server.

To a lesser extent, these principles apply to nonbottleneck servers that operate close to capacity.

Now, let us consider the case of Midtown Manhattan. Suppose that two traffic signals control traffic at intersections located very close to each other, say, 100 meters apart. The buffer size is the number of vehicles that can fit between the intersections and therefore is proportional to the separation. When the upstream server turns green, the buffer begins to fill with vehicles and does not empty until the downstream signal turns green. But what if the downstream signal is out of phase with the upstream signal? In Fig. 10.11, two signals have identical cycle lengths (60 seconds) and green phase lengths (30 seconds), but are completely out of phase. In fact, the buffer (40 vehicles) fills completely before the green phase at the upstream signal ends. The upstream signal is blocked and cannot operate at full capacity. The downstream signal also cannot operate at full capacity. It starves when the queue empties before its green phase finishes. The maximum number of vehicles that can be processed in a cycle only equals the buffer size and a capacity of ten vehicles is lost in each cycle, a loss in capacity that compounds from cycle to cycle. Blocking can occur even when signals are not completely out of phase, particularly when a time lag exists between the start of a green phase and movement of customers from the *tail* of the bottleneck (as is common in traffic).

Gridlock in Manhattan is the extension of blocking to an entire grid of queues. When one intersection loses capacity due to blocking (especially when the bottleneck loses capacity), the effect can easily propagate upstream to other signals. Blocking grows because at each intersection there may be an additional drop in capacity. A compounding effect, when the buffer fills, is that vehicles may extend into the intersection after the end of the green phase, blocking traffic for the cross street. Blocking can even cascade around a block, so that a signal effectively blocks itself.

The solution to gridlock is fundamental. Provide an adequate buffer, at least as large as the number of vehicles that can be served in the green phase of a traffic signal. How does one increase the buffer size? In traffic, it can be increased by increasing the spacing between intersections or by widening the street. In Manhattan, where spacing is notoriously small, this may mean closing some of the cross streets. Enforcing regulations prohibiting vehicles from blocking intersections when the cycle turns red (the so-called

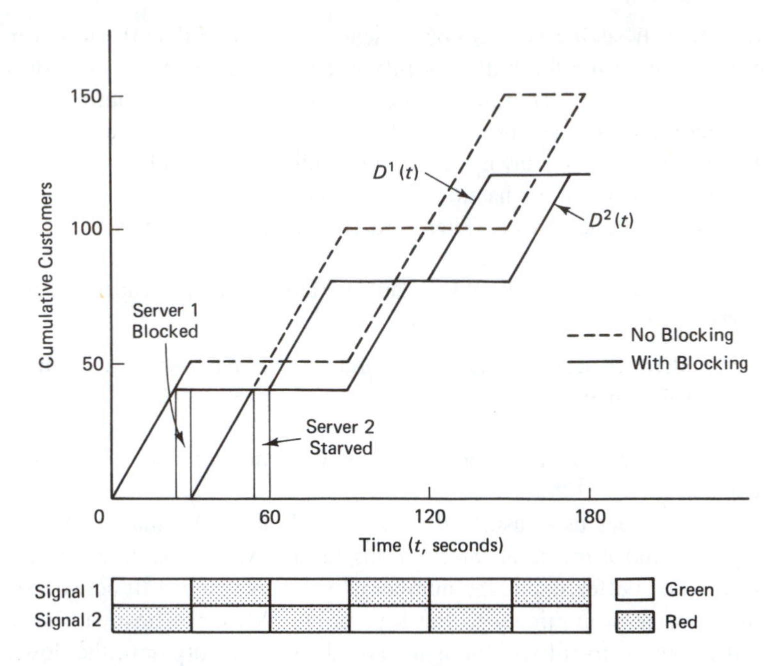

Figure 10.11 Blocking at bulk servers increases queueing. In the example, server 1 is blocked 24 seconds after the start of each cycle, and server 2 starves 54 seconds after the start of each cycle. Service capacity is lost at both servers.

''don't block the box'' regulation) is a second step. But it is mostly effective at keeping already bad situations from turning miserable. Better signal synchronization may also be a solution, but it can be difficult. A perfect synchronization in the absence of congestion may not be perfect once the buffer becomes full. Signal offsets must be adjusted dynamically as conditions change, a very difficult process.

The thing to recognize here is that providing adequate buffers does not prevent queueing. Buffers are there to ensure that servers (particularly bottleneck servers) can operate at full capacity. They help prevent blocking and starvation.

10.1.5 Sequencing Servers

The last issue to be considered in this section on series queues is the server sequence. In most cases this sequence is dictated by the flow of work. That is, whichever task must be performed first dictates the server that is visited first. Furthermore, even if the sequence can be changed, it often has little impact on system performance. We have already seen in the deterministic fluid model that server sequence has no impact on time in system. So why consider it?

Suppose that a queueing system consists of one bulk server and one continuous server. If any sequence is acceptable, which server should be put first? As a rule of thumb

The first server should be the one whose service process most closely matches the system's arrival process.

Invariably, this means that the continuous server should be placed first.

Example

> The U.S. Customs Service would like to process people traveling from Canada to the United States via airplane. Travelers can be processed in either Canada, before they board the aircraft, or in the United States, after exiting the aircraft.
>
> The process can be viewed as two servers in series, Customs and transportation. The Customs Service operates continuously, while the airplane trip operates as a bulk server. Because passengers arrive at a relatively even rate before the flight, they can be processed with less delay if the Customs operation is placed at point of departure.

As a general rule, if servers can be placed in any sequence, all of the continuous servers should be visited before the bulk servers. This is because arrivals at the entrance to a queueing network usually resemble a (possibly nonstationary) Poisson process. Bulk servers should, as usual, be synchronized and balanced, when possible. If this plan is followed, major queues should only form at the entrance to the system, and cyclic queues should only form prior to the first bulk server.

10.2 PARALLEL SERVICE QUEUEING SYSTEMS

Not all tasks have to be performed consecutively. Tasks performed on different types of servers (called ***multiple resources***), without precedence constraints, may be performed concurrently. Suppose that you are given the job of analyzing a data set with a sophisticated computer program. The job entails inputting data, then running the program multiple times and summarizing the results. One way to complete this job would be to perform the tasks serially, first inputting, then running the program, changing the data, running the program again, and eventually writing the summary. But a more efficient procedure would be to perform tasks simultaneously. If the summary is written while the program is running, the entire job can be completed faster.

It should be obvious that parallel processing leads to reductions in the service time. It can also, under certain circumstances, increase the server capacity through elimination of idle time, thus reducing the time spent in queue. In this section, we examine two forms of parallel service: parallel service in a workstation and parallel service with a set of machines and a human operator.

10.2.1 Workstation: All Tasks Parallel

In the simplest system, N servers operate in parallel at a single workstation, each performing a different task. A job is completed when the *last* of the N servers is finished. The next job enters service only after this last task is completed, so the station serves just one customer at a time.

A workstation like this can be analyzed as a single server queue, with a service time defined by the maximum of N random variables, each representing the time to complete one of the N tasks. Let

S_n = the service time random variable for server n

$F_n(s)$ = the probability distribution function for the service time at n

$F(s)$ = the probability distribution function for the workstation's service time

Then the station's service time is

$$S = \max_n\{S_n\} \tag{10.6}$$

If service times are independent, then

$$F(s) = P(\max_n\{S_n\} \leq s) = P(S_1 \leq s)P(S_2 \leq s) \cdots P(S_n \leq s) \tag{10.7}$$

$$= \prod_n F_n(s)$$

Π represents the product of the $F_n(s)$.

The expectation of a non-negative random variable equals the integral of its distribution function complement over $[0,\infty)$. Because a service time must always be non-negative

$$E(S) = \int_0^\infty [1 - F(s)]ds = \int_0^\infty [1 - \prod_n F_n(s)]ds \tag{10.8}$$

This expectation will always be less than $\Sigma E(S_n)$, so parallel service must always yield a reduction in service time. The calculations are illustrated in Fig. 10.12 for two hypothetical distribution functions. Another example is provided below.

Example

Two tasks are performed simultaneously at a workstation, one with mean 10 minutes and the other with mean 13 minutes. Both tasks have exponential distributions. The expectation of the maximum service time is determined as follows:

$$F(s) = [1 - e^{-s/10}][1 - e^{-s/13}] = 1 - e^{-s/10} - e^{-s/13} + e^{-s(1/10+1/13)}$$

$$E(S) = \int_0^\infty [e^{-s/10} + e^{-s/13} - e^{-s(1/10+1/13)}]ds = [10 + 13 - 1/(1/10 + 1/13)]$$

$$= 17.3 \text{ minutes}$$

If the service times are normally distributed, the distribution of S can be approximated with a recursive equation developed by Clark (1961). (The calculations are complicated, and some readers may choose to proceed directly to Sec. 10.2.2.) Because a normal random variable can be negative, the normal approximation is only applicable to service times when the probability of a negative outcome is small. At a minimum, this

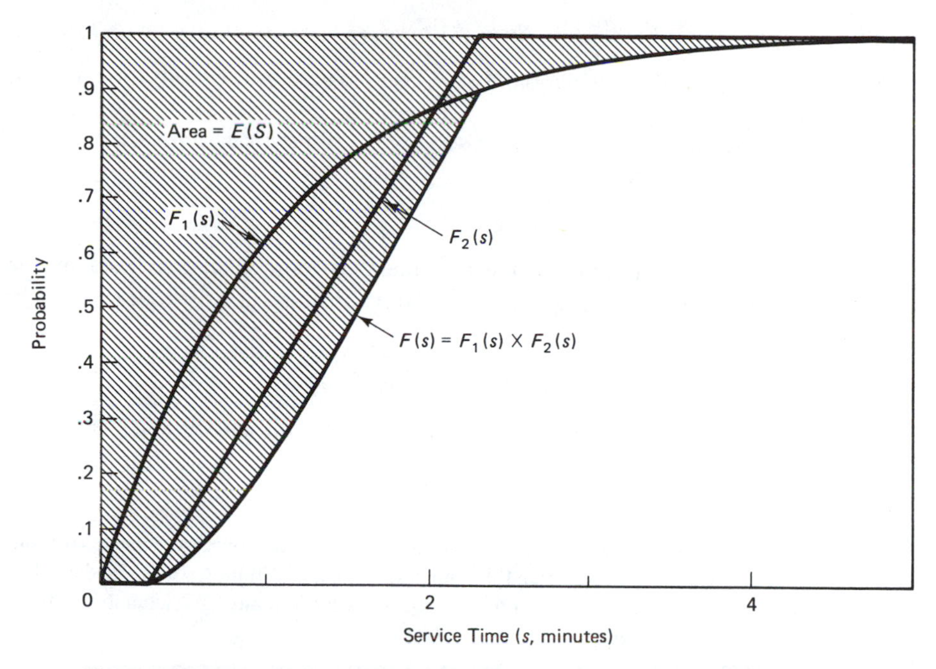

Figure 10.12 Probability distribution function for the maximum of two independent random variables is found by taking the product of their distribution functions.

requires that the mean should be at least twice as large as the standard deviation.

Unlike the previous calculation, Clark's approximation can be used even when the service times are not independent. Let

m_n = mean of service time at server n

m_n^2 = expectation of squared service time at n

σ_n^2 = variance of service time at server n

σ_{ij}^2 = covariance between the service times of server i and server j

Clark's approximation states that the maximum of any two random variables with normal distributions is also normal, with mean and variance defined as follows:

$$E[\max\{S_n, S_{n+1}\}] = m_{n+1} + (m_n - m_{n+1})\Phi(\alpha) + a\phi(\alpha) \quad (10.9)$$

$$E[\max\{S_n, S_{n+1}\}^2] = m_{n+1}^2 + \sigma_{n+1}^2 + [m_n^2 + \sigma_n^2 - m_{n+1}^2 - \sigma_{n+1}^2]\Phi(\alpha) + [m_n + m_{n+1}]a\phi(\alpha) \quad (10.10)$$

$$V[\max\{S_n, S_{n+1}\}] = E[\max\{S_n, S_{n+1}\}^2] - E^2[\max\{S_n, S_{n+1}\}] \quad (10.11)$$

where

Φ = the standardized normal distribution function (Table A.2)

ϕ = the standardized normal density function (Table A.1)

$$a = [\sigma_n^2 + \sigma_{n+1}^2 - 2\sigma_{n,n+1}^2]^{1/2} \qquad \alpha = (m_n - m_{n+1})/a \tag{10.12}$$

The term a is the standard deviation of $S_n - S_{n+1}$. Therefore, $\phi(\alpha)$ is the probability that S_n is greater than S_{n+1}.

The covariance with any other normally distributed random variable can also be approximated as follows:

$$\mathrm{COV}[\max\{S_n, S_{n+1}\}, S_j] = \sigma_{n+1,j}^2 + (\sigma_{n,j}^2 - \sigma_{n+1,j}^2)\phi(\alpha) \tag{10.13}$$

To find the distribution of the maximum of more than two random variables, Clark's approximation can be applied recursively, as illustrated for four random variables below:

$$\max\{S_1,S_2,S_3,S_4\} = \max\{\max[\max(S_1,S_2),S_3],S_4\} \tag{10.14}$$

Thus, the distribution for the maximum of N normal random variables can be found in $N - 1$ steps.

Example

Two tasks are performed simultaneously at a workstation, one with mean 10 minutes and standard deviation 2 minutes and the other with mean 13 minutes and standard deviation 1 minute. The covariance between the service times is .5 minutes^2. Calculations for the Clark approximation are performed as follows:

$$a = (2^2 + 1^2 - 2 \cdot .5)^{1/2} = 2 \text{ minutes}$$

$$\alpha = (10 - 13)/2 = -1.5$$

$$\phi(-1.5) = .067 \qquad \varphi(-1.5) = .130 \qquad \text{(Tables A.1, A.2)}$$

$$m_1^2 = (10^2 + 2^2) = 104 \text{ min}^2 \qquad m_2^2 = (13^2 + 1^2) = 170 \text{ min}^2$$

$$E[\max\{S_1, S_2\}] = 13 + (-3)(.067) + 2(.13) = 13.06 \text{ minutes}$$

$$E[\max\{S_1, S_2\}^2] = 170 + 1 + (104 + 4 - 170 - 1)(.067) + (10 + 13)(2)(.13) = 173$$

$$V[\max\{S_1, S_2\}] = 173 - 13.06^2 = 2.44 \text{ minutes}^2$$

As one might expect, the mean of the maximum is greater than the maximum of the means. However, the difference between these two values is small in the example, due to the relatively small standard deviations of the random variables. Also as one would expect, the mean of the maximum is much less than the sum of the means (as would apply to serial service).

Because of the small standard deviations, $E[\max\{S_1,S_2\}]$ is much smaller for the example normal distributions than it was for the previous example, which used exponential distributions.

10.2.2 Serial and Parallel Tasks at a Workstation

A somewhat more complicated situation is where a workstation processes some tasks serially and some in parallel, but, as in the previous section, a workstation still processes just one customer at a time.

In a deterministic environment, the ***critical path method (CPM)*** is an effective technique for determining the minimum possible service time in the workstation. The critical path method determines the sequence of tasks that has the greatest impact on the length of the entire service. Naturally, this sequence is called the *critical path*. CPM can also be used to schedule the tasks, achieving the optimal level of concurrency between tasks, given the precedence constraints. CPM is used extensively in the planning and management of major construction and design projects, but also applies to the evaluation of shorter repetitive tasks.

Figure 10.13 gives an example. Each task is represented by a node, and each precedence constraint is represented by an arrow (called an *arc*). Different tasks may or may not be completed concurrently, depending on the arcs. For instance, the arc from node 1 to node 2 indicates that these two tasks must be performed serially, in the prescribed order. Serial processing might be dictated by two tasks demanding the same resource, or it might be due to the fact that a task depends on the completion of a related task at another server. If two tasks are connected by an arc, or by a directed path, then they *cannot* be performed simultaneously (nodes 1 and 4, for example); they must be sequential (a directed path is a sequence of arcs pointing in the same direction). The path length is the sum of the service times for the tasks in the path. And the critical path is the path from start to finish with the longest length (path 3–2–4 in the example).

While in simple networks, such as Fig. 10.13, the critical path can be found by inspection, complicated networks require an algorithmic approach.

Definitions 10.5

$\mathscr{P}_n$ = set of tasks preceding task n in the critical path network

ES_n = *earliest* feasible *start time* for task n

EF_n = *earliest* feasible *finish time* for task n

S_n = service time of task n

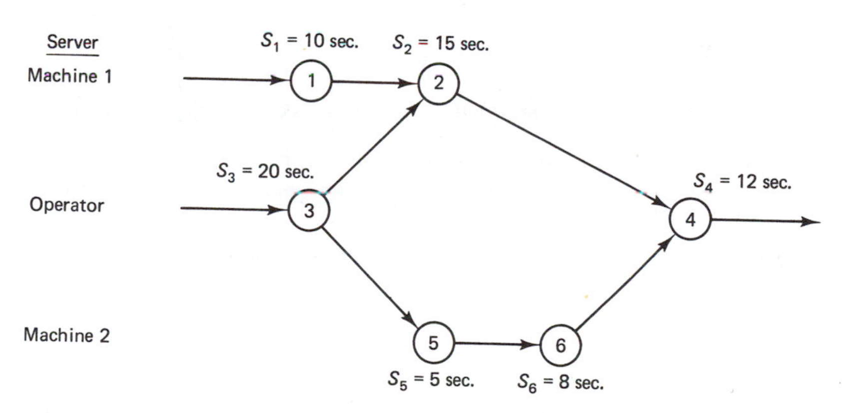

Figure 10.13 Example critical path network for workstation. An arrow indicates precedence, and a node denotes an activity.

(Script letters denote sets of nodes.) If service begins at time 0, then

$$ES_n = \max_{j \in \mathscr{P}_n} \{EF_j\} \tag{10.15}$$

$$EF_n = ES_n + S_n \tag{10.16}$$

The earliest start and finish times can be evaluated sequentially in a ***forward pass*** through the network. Initially, the earliest start times of all tasks without predecessors are set to zero, and their earliest finish times are set to equal their service times. At each subsequent stage a task can be evaluated if all of its predecessors have already been evaluated. If the critical path has no directed cycles (a directed path that begins and ends at the same node), there must always be at least one task with this property.

Example

For the network in Fig. 10.13, the earliest start and finish times are calculated in the following order:

$$ES_1 = 0 \qquad EF_1 = 10$$

$$ES_3 = 0 \qquad EF_3 = 20$$

$$ES_2 = \max\{10,20\} = 20 \qquad EF_2 = 20 + 15 = 35$$

$$ES_5 = 20 \qquad EF_5 = 20 + 5 = 25$$

$$ES_6 = 25 \qquad EF_6 = 25 + 8 = 33$$

$$ES_4 = \max\{35,33\} = 35 \qquad EF_4 = 35 + 12 = 47$$

The entire service duration is simply the maximum of the earliest finish times (47 seconds in the example).

To find the critical path, a ***backward pass*** is carried out. This determines the latest times that a task can start and finish *without delaying the completion of the entire service*. Let

Definitions 10.6

LS_n = *latest* feasible *starting time* for task n

LF_n = *latest* feasible *finish time* for task n

$\mathscr{S}_n$ = set of tasks succeeding task n

The latest start and finish times are defined as follows:

$$LF_n = \min_{j \in \mathscr{S}_n} \{LS_j\} \tag{10.17}$$

$$LS_n = LF_n - S_n \tag{10.18}$$

A task is critical if its latest finish time and earliest finish time (as well as its latest and earliest start) are the same. This can also be expressed in terms of the *slack time*.

Definition 10.7

$$SS_n = \textbf{\textit{slack time}} \text{ for task } n$$
$$= LF_n - EF_n = LS_n - ES_n \quad (10.19)$$

The slack time indicates how much flexibility is available in scheduling a task. If the slack is zero, the task is critical and there is no flexibility. If the slack is greater than zero, it may be possible to delay the start of the task without delaying completion of service. The set of tasks with zero slack defines a critical path (or critical paths).

The calculations are carried out in the same manner as for the earliest start and finish, except reversed. At each stage, a task can be evaluated if all of its *successors* have already been evaluated. The first task evaluated must have no successors, and its latest finish time is set equal to the service time for the entire network.

Example

Again returning to the example, the latest start and finish times and the slack times are calculated as follows:

$LF_4 = 47$	$LS_4 = 47 - 12 = 35$	$SS_4 = LF_4 - EF_4 = 0$
$LF_2 = 35$	$LS_2 = 35 - 15 = 20$	$SS_2 = LF_2 - EF_2 = 0$
$LF_6 = 35$	$LS_6 = 35 - 8 = 27$	$SS_6 = LF_6 - EF_6 = 2$
$LF_1 = 20$	$LS_1 = 20 - 10 = 10$	$SS_1 = LF_1 - EF_1 = 10$
$LF_5 = 27$	$LS_5 = 27 - 5 = 22$	$SS_5 = LF_5 - EF_5 = 2$
$LF_3 = \min\{20,22\} = 20$	$LS_3 = 20 - 20 = 0$	$SS_3 = LF_3 - EF_3 = 0$

As expected, tasks 2, 3, and 4 are critical, as indicated by their slack times.

Identifying the critical path at a workstation is analogous to identifying the bottleneck for serial tasks. It indicates which sequence of tasks has the biggest effect on delay. Any effort to reduce service time and time in system should focus on these tasks. Reducing the service time of any one task on a critical path will reduce the total service time. If there is more than one critical path, reducing the service time on one task from each critical path will reduce the total service time. Reducing the service time of noncritical tasks will have a lesser (or possibly no) impact on time in system.

Critical paths can also be analyzed probabilistically. The service time distribution for a set of serial tasks is the distribution for a sum of random variables. The mean service time is then the sum of the mean task times, and the variance of the service time is the sum of the variances of the task times plus twice the sum of the covariances of the task times. The distribution for a set of parallel tasks is the distribution for the maximum of a set of random variables. If normally distributed, Clark's approximation can be applied. Taking a building-block approach, the entire service time for the example in Fig. 10.13 can be calculated as follows:

$$S = \max\{S_4 + \max\{S_1 + S_2, S_2 + S_3, S_3 + S_5 + S_6\}\} \qquad (10.20)$$

If, for example, the coefficient of variation is .25 for all tasks, and task times are independent, carrying through Clark's approximation would indicate that the service time distribution is normal with mean 48.7 and standard deviation 6.4. Owing to randomness, this mean is slightly larger than the mean length of the critical path (47 seconds).

Though probabilistic analysis provides a more accurate picture of queue behavior, added complexity makes it less effective at identifying critical tasks than CPM. It is more of a supplement to CPM than a replacement.

10.2.3 Shared Resource: An Operator for Several Machines

Workstation networks are relatively easy to analyze because only one customer is allowed at a station at a time. But networks do not always impose this restriction. Many customers can enter the network, and queues can form at any and all servers. Efficiency is gained in the process, for a server does not have to wait for the other servers to complete processing before starting on the next customer. In this more general network, identifying the critical path is no longer relevant.

Except for determining which servers are bottlenecks (see Sec. 10.3), general parallel service networks are difficult to evaluate. However, one queueing model will be presented for a special case. The model is based on the finite calling population model of Chap. 5. Suppose that m operators and N machines constitute a service team. Each service entails both a setup of the machine (performed by the operator) and a production run (performed by the machine). When a machine finishes a job, it must wait for an operator to set up the next run. At any given time, either an operator will be idle, waiting to set up a machine, or a machine will be idle, waiting for an operator.

The capacity of the system can be determined from Eq. (5.60) to (5.63) by viewing the machine as a customer (drawn from a finite population) and the operator as a server (the roles can be reversed). As a simple illustration, suppose that run times are independent, exponentially distributed random variables (mean $1/\lambda$). And suppose that the setup times are also independent, exponential random variables (mean $1/\mu$). The arrival rate of customers (machines) is a decreasing function of the number of customers already in service or waiting for service. The service rate is an increasing function of the number of customers in service. Let the state be the number of customers in the system (that is, the number of machines being set up or waiting to be set up). The steady-state transition rates are defined as follows:

$$\lambda_n = \text{transition rate from state } n \text{ to state } n+1$$

$$= \begin{cases} (N-n)\lambda & n = 0, 1, 2, \ldots, N \\ 0 & n = N+1, N+2, \ldots \end{cases} \qquad (10.21a)$$

$$\mu_n = \text{transition rate from state } n \text{ to state } n-1$$

$$= \begin{cases} n\mu & n = 1, 2, \ldots, m \\ m\mu & n = m + 1, m + 2, \ldots \end{cases} \qquad (10.21b)$$

The only difference between these equations and Eq. (5.61) and (5.62) is that $(N - n)\lambda$ replaces $(N - n)\lambda/N$. In Chap. 5, λ was defined to equal the total arrival rate among all customers; here, λ is the arrival rate per machine.

For a single server (that is, single operator, $m = 1$), the following results are obtained (the multiple operator result is more complicated and is omitted):

L_s = expected number of machines waiting for an operator or being served by an operator

$$= N - \frac{\mu}{\lambda}(1 - P_0) \qquad (10.22)$$

where

$$P_0 = \frac{1}{\sum_{n=0}^{N} \frac{N!}{(N-n)!} \left(\frac{\lambda}{\mu}\right)^n} \qquad (10.23)$$

The expected number of operating machines is simply $N - L_s$, a value that is plotted as a function of N in Fig. 10.14. Thus, the capacity of the system is

$$\text{Capacity} = \lambda(N - L_s) = \mu(1 - P_0) \qquad (10.24)$$

Or the capacity is the operator's service rate multiplied by the probability that the operator is busy.

As might be expected, increasing the number of machines tended per operator leads to higher output but lower machine utilization. Further, after a point, increasing N results in only a small increase in the system's capacity. This should be expected because the output of the system cannot exceed the minimum of λN and μ (the machine capacity and the server capacity).

Example

An operator is responsible for the control of three machines. Jobs are completed in FCFS order, and have running times that are exponentially distributed, with mean 10 minutes. Each job also requires a setup, which is also exponentially distributed with mean 5 minutes.

P_0, calculated from Eq. (10.23), equals .211, and L_s, calculated from Eq. (10.22), equals 1.42. The three machines are running 53 percent of the time and the operator is busy 79 percent of the time. The capacity of the system is .16 job per minute.

As with the workstation models, Eq. (10.22) to (10.24) are directed at determining the mean service time, not the queueing delay, for the work. To ascertain queueing delay, one would also have to determine the mean and standard deviation of the service time (both

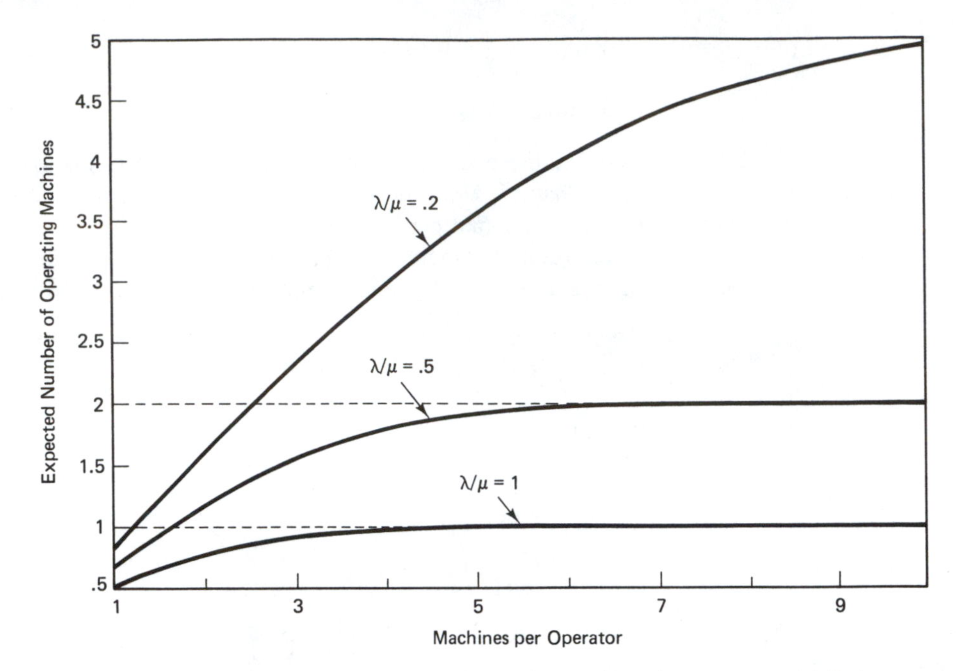

Figure 10.14 The expected number of operating machines increases toward a limit (μ/λ) as the number of machines per operator increases. (Exponential service times and a single operator are assumed here.)

setup and run time) and substitute the values in one on the general service time models in Chap. 5.

The principle that operator productivity increases and machine productivity decreases as the number of machines per operator increases applies to nonexponential distributions, though effective scheduling can provide higher productivity than indicated. Analytical evaluation of nonexponential distributions is beyond the scope of this book. However, simulation might be used to evaluate different configurations and select the most cost-effective solution, balancing operator against machine costs.

10.3 MAXIMIZING THROUGHPUT IN QUEUEING NETWORKS

In a large queueing network, the kind of network common to factories, communications, and transportation, it may be impossible to model systemwide queueing analytically or graphically. Alternatives are needed. One candidate is ***linear programming (LP)***. Within the context of queueing, linear programs can be used to determine the maximum system ***throughput*** (the rate at which customers depart from the system), or to optimize system revenue, subject to server capacity constraints. Linear programming can also identify system bottlenecks.

A linear program is a mathematical model. The components of the model are an objective function (for example, the objective of maximizing throughput), a set of constraints (server capacities), and a set of decision variables (the departure rates from the servers). The linear program is solved with an algorithm, such as the Simplex method, usually by computer. The ***primal solution*** to the LP specifies the optimal value of the objective function along with the optimal values of the decision variables. Computer software also provide a ***dual solution***. The dual solution indicates that optimal values of dual decision variables, which are associated with the constraints. The value of a dual variable equals the marginal change in the objective function due to a small change in the value of a constraint. In the case of server capacity, the dual variable would indicate the change in the objective function for each unit of capacity increase.

In queueing, the dual solution is significant because it identifies network bottlenecks. Suppose that the objective is to maximize throughput. Then

If the dual variable for a server equals zero
1. Increasing its capacity will not increase throughput.
2. The server is not a bottleneck (unless there are multiple bottlenecks).

If the dual variable for a server is greater than zero
1. Increasing its capacity may increase throughput.
2. The server is a bottleneck.

Computer software for solving linear programs is widely available, as are textbooks on the theory of linear programming (Bradley et al. 1977; Hillier and Lieberman 1986). Anyone interested in evaluating large queueing networks should be familiar with this subject. Applications are described in the following two sections.

10.3.1 Communication/Transportation Network

The two fundamental ways to transmit information across a communications network are ***circuit switching*** and ***store-and-forward switching***. In the former, common in telephone networks, a fixed path and capacity are allocated to each transmission at the start of a session (for example, when a phone call is dialed). The capacity is held constant even when the transmission rate is not constant. In the latter, common in data communications, transmissions are not allocated a fixed capacity and path. Instead, messages are transmitted one at a time. Whenever the transmission rate exceeds the capacity of a link in the network, the messages will form a queue at the start of the link. Store-and-forward networks achieve greater capacity utilization because no fixed capacity is allocated to each transmission.

Store-and-forward networks bear a strong resemblance to traffic networks. A message is like a vehicle, and a communications link is like a road. The techniques for analyzing throughput are the same for communications and transportation. In either case, the service process for each customer is *serial* (even though queues are not in series). As with all serial processes, customers are served sequentially, one server at a time.

Capacity limitations on communication links, switches, or, in the case of transportation, roads may lead to queues in the system. Linear programming cannot analyze the stochastic behavior of these queues, but it can identify the maximum throughput. The throughput is constrained to be the minimum of the network capacity and the rate at which customers arrive. If the arrival rate exceeds the network capacity, some of the arriving customers must be turned away or queued at the entrance, both undesirable consequences.

To create the LP, the communications network is fomulated in terms of ***nodes*** and ***arcs***. Arcs represent servers, where a server can be either a communication link or a switching unit. Nodes represent connections between servers. Origins and destinations are also represented by nodes. In words, the decision variables, objective function, and constraints can be described as follows:

Decision Variables

Customer flows across arcs (that is, departure rates of customers from the servers).

Objective Function

Maximize customer flow exiting the network (that is, maximize throughput), *or*
Maximize revenue (that is, maximize product of flow and revenue per customer).

Constraints

Arc flow does not exceed arc (server) capacity.
Flow into each node equals flow out of each node (flow conservation).
Flow into network does not exceed arrival rate.
Arc flow is greater than or equal to zero.

Figure 10.15 shows a simple communications network. Capacities are indicated in the figure, and the arrival rate is 140 messages per hour. The LP formulation for maximizing throughput is provided in Table 10.1. Table 10.2 provides the LP solution. In the primal solution, the maximum throughput is shown to equal 120 messages per hour. Because it is less than the arrival rate, throughput is limited by the network capacity. But not all of the arcs in the network are limiting; some have excess capacity.

Just as in the series queue, the concept of the bottleneck is critical. But unlike the series queue, *the bottleneck is not necessarily a single server; it may be a set of servers*. In the example, the dual variables for constraints 17, 24, and 25 equal one, meaning arcs (3,10), (7,10), and (8,9) form the bottleneck.

The set of bottleneck arcs has an important property. They form a ***cut*** between the message origin and destination. A cut is a set of arcs that disconnects the origin from the destination. That is, every path from the origin to the destination must contain at least one arc in the cut.

Definition 10.8

The **cut value** is the sum of the capacities of the arcs in a cut.

For networks with a single origin and destination, as in Fig. 10.15, the following theorem applies:

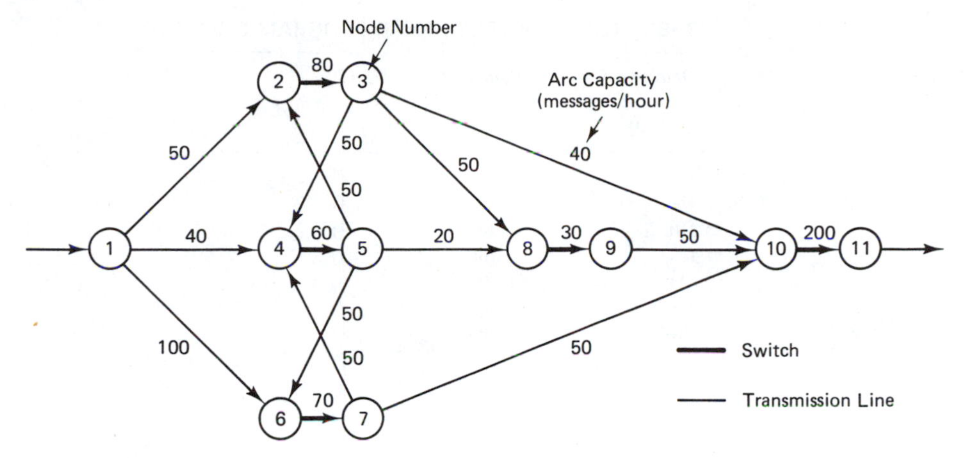

Figure 10.15 Representation of a communications network, to be used for throughput maximization.

TABLE 10.1 EXAMPLE LINEAR PROGRAM FORMULATION

1)	MAX	f1011	
	SUBJECT TO		
2)		f12 + f52 − f23 = 0	
3)		f23 − f34 − f38 − f310 = 0	
4)		f34 + f14 + f74 − f45 = 0	
5)		f45 − f52 − f56 − f58 = 0	
6)		f16 + f56 − f67 = 0	Conservation
7)		f67 − f74 − f710 = 0	
8)		f38 + f58 − f89 = 0	
9)		f89 − f910 = 0	
10)		f310 + f710 + f910 − f1011 = 0	
11)		f12 ≤ 50	
12)		f14 ≤ 40	
13)		f16 ≤ 100	
14)		f23 ≤ 80	
15)		f34 ≤ 50	
16)		f38 ≤ 50	
17)		f310 ≤ 40	
18)		f45 ≤ 60	
19)		f52 ≤ 50	
20)		f56 ≤ 50	Arc Capacity
21)		f58 ≤ 20	
22)		f67 ≤ 70	
23)		f74 ≤ 50	
24)		f710 ≤ 50	
25)		f89 ≤ 30	
26)		f910 ≤ 50	
27)		f1011 ≤ 200	
28)		f12 + f14 + f16 ≤ 140	Arrival Rate

Note: Program assumes that decision variables are non-negative.

TABLE 10.2 EXAMPLE LINEAR PROGRAM SOLUTION

Objective Function Value: 120

Variable	Value	Row	Dual variable
f12	50	2)	0
f14	40	3)	0
f16	30	4)	0
f23	70	5)	0
f34	0	6)	0
f38	30	7)	0
f310	40	8)	0
f45	40	9)	−1†
f52	20	10)	−1†
f56	20	11)	0
f58	0	12)	0
f67	50	13)	0
f74	0	14)	0
f710	50	15)	0
f89	30	16)	0
f910	30	17)	1*
f1011	120	18)	0
		19)	0
		20)	0
		21)	0
		22)	0
		23)	0
		24)	1*
		25)	1*
		26)	0
		27)	0
		28)	0

*Bottleneck arc.
†Nodes immediately following bottleneck arcs.

Max Flow/Min-Cut Theorem. The maximum feasible flow from origin to destination equals the minimum cut value among all cuts in the network.

Because all messages must traverse all cuts, the min cut is a natural limit on throughput. This fact is illustrated in Fig. 10.16. In terms of linear programming, the primal problem amounts to finding the maximum flow through the network, and the dual problem amounts to finding the minimum cut. The max flow/min-cut theorem states that these two values are the same.

In interpreting the solution, one must also be careful that there are not *two or more* min cuts. The dual solution will only identify one of these cuts. Thus, just because a dual variable equals one does not mean that increasing its capacity will increase throughput (however, *decreasing* its capacity will decrease throughput). The min cut can also change as capacities change. Therefore, if a capacity increase is contemplated, it is prudent to rerun the LP with the increase in place, to ascertain the true change in throughput.

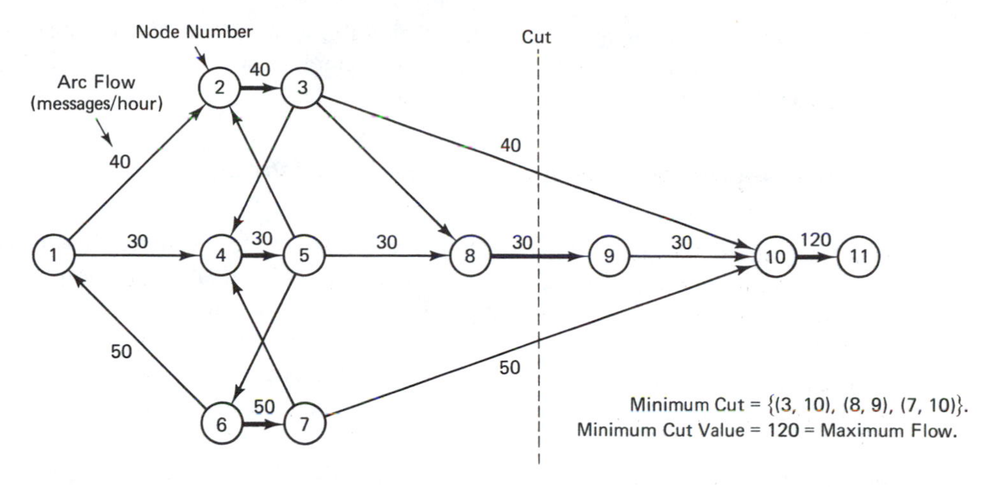

Figure 10.16 The maximum flow through the network equals the minimum cut capacity.

The example network only serves one type of customer, but it is easy to generalize a linear program to multiple customer types. For each arc in the network, there would be one flow variable for each customer type, and the arc capacity would apply to the combined flow among all types. One can also associate a different price with each type of customer and set the objective as maximizing system revenue.

10.3.2 Production Based Formulation

Unlike in communications, customers served in a factory are not allowed to follow any path from origin to destination. Instead, customers (jobs) are usually restricted to a set route, based on the type of processing needed and precedence constraints in manufacture. This leads to a slightly different, and simpler, throughput formulation that can be used with either serial or parallel processing. However, if parallel processing is used within a workstation, that workstation would have to be treated as a single entity, with a capacity determined by the techniques discussed in Sec. 10.2.

The basic difference in the formulation is that it is unnecessary for the LP to determine customer routings. This means that decision variables are not written in terms of arc flows, and conservation constraints are not needed.

Decision variables
  Departure rate of each job type at each machine.
Objective
  Maximize rate at which jobs depart from the system, *or*
  Maximize revenue (or profit) of jobs departing from the system.
Constraints
  Total departure rate among all jobs using server does not exceed capacity.
  Job departure rate does not exceed job arrival rate.
  Job departure rate is non-negative.

As with the communications/transportation formulation, the dual variables indicate which servers constitute the bottleneck. For example formulations, consult Hax and Candea (1984) and Johnson and Montgomery (1974).

10.3.3 Interpretation and Adaptation of Linear Programs

A linear program is capable of determining how much work to accept from each customer type in order to maximize throughput, revenue, or profit. The dual of the linear program is capable of identifying the system bottlenecks. In either case, the solution is premised on the system being congested. That is, results are only meaningful when the arrival rate exceeds system capacity. In factories, this is not unrealistic, because they strive to operate at or near capacity, adjusting prices and providing incentives in order to achieve this balance. Communication networks, on the other hand, strive to provide a high level of service, and consequently most queues are stochastic. The models presented in the following sections partially address this concern.

Linear programs can be modified in a variety of ways. For example, a server might operate in either of two modes, regular time or overtime. Each mode can be represented by an arc, with different costs for the different types of service. Linear programs also can be used to study systems with arrival rates that vary in a predictable fashion. The methodology is based on the concept of a "time-expanded network," where each node and each arc are replaced by a set of duplicate nodes and duplicate arcs. Each arc in the set would then represent a different time period, such as an hour or a day. The decision variables would denote the rate at which each customer type is served in the corresponding time period. In a time-expanded network, it is possible for the bottlenecks to change from time to time, depending on the prevailing arrival pattern. The LP will identify these changes.

Though time-expanded networks provide added realism, the large increase in effort (both human and computer) entailed in their creation often is not justified, unless the demand patterns for different customer types are substantially out of synchronization with each other. Readers interested in the subject can consult Maxwell and Wilson (1981), which discusses the application of time-expanded networks to materials-handling systems.

10.4 ROUTING THROUGH STOCHASTIC NETWORKS

The simplifying assumption of linear programming is that congestion is a simple on/off process. So long as the arrival rate is below the capacity of a server, there is no delay; as soon as the arrival rate exceeds the capacity of a server, there is infinite delay. A more realistic relationship is that congestion and delay grow at an increasing rate as the capacity is approached, as was shown in Chap. 5 (Fig. 5.3, for example). This relationship is often important in communications systems, which typically have excess capacity on average, but are susceptible to random queueing.

In a stochastic environment, the routing objective is changed from a linear function to a non-linear function, of the following type:

Objective $$\min \sum_{(i,j)} g_{ij}(f) \tag{10.25}$$

where $g_{ij}(f)$ is the total waiting time incurred on arc (i,j) as a function of the flow, f, on the arc.

The optimal network flows are ordinarily solved with convex programming methods, such as the *Frank-Wolfe algorithm* (Zangwill 1969). While the details of the algorithm are too involved to cover here, it is not difficult to characterize an optimal solution:

A solution is optimal if and only if total waiting time cannot be reduced by moving a customer from its assigned path to an alternative path.

Optimality is represented by the fact that no small change in the solution can reduce the total waiting time. This is something like saying the optimum of a convex function occurs where its derivative is zero.

As already seen in Chap. 8, system optimality does not imply that customers cannot selfishly reduce their own waiting times by changing paths. For instance, it may be possible in an optimal solution for a customer to reduce its own delay by shifting from a circuitous, uncongested path to a direct, congested path. But the customer would impose additional delay on customers already using the congested path, which more than compensates for the customer's personal savings. The net result is an overall increase in delay.

That personal interests (individual optimality) do not always coincide with societal interests (system optimality) is an unfortunate fact of life that applies to queueing networks like anything else. In communications networks, routing is taken out of the hands of the customer, so it is possible for the system operator to dictate an optimal solution. Unfortunately (or, perhaps, fortunately) traffic engineers have not yet been successful in doing the same for highway networks. Nevertheless, there are mechanisms to encourage customers to make optimal decisions, most notably pricing.

Even more sophisticated methods than Frank-Wolfe can be applied to communications networks. Routing might be performed dynamically, in response to changing queue lengths, and adaptively, as the customer progresses through the network. For a review of these approaches, consult Bertsekas and Gallager (1987).

10.5 QUEUE DISCIPLINES IN NETWORKS

Queue discipline, a topic first explored in Chap. 9, can take on new forms in the context of networks. The characteristics used to select customers need not be limited to attributes of the local server. It is desirable to adopt some form of ***global control***, which incorporates information taken from other servers. This information may include current queue sizes, expected utilization, and so on. If downstream queues are smaller for one job than another, it may then be reasonable to give it priority.

10.5.1 Bottleneck Dependent Disciplines

As with other topics in this chapter, it is important to account for the presence and location of bottlenecks. Consider the example presented in Fig. 10.17. Four types of customers are processed at three servers. Each customer has a different route: Customer 1 is only processed at server 1, customer 2 only at server 2, and customer 3 only at server 3, but customer 4 is processed at all three servers. The arrival rates for the customers are 40, 50, 60, and 70 customers/hour, and the service rates are 100/hour at each server.

First, suppose that customers are processed according to a *localized* queue discipline, FCFS at each server. Then the service rate for each customer type would be proportional to its arrival rate at the server. At the first server, customer 1 would be served at the rate of $100(40/110) = 36$ and customer 4 would be served at the rate of $100(70/110) = 64$. At the second server, the process is repeated, this time with an arrival rate of 50/hour for customer 2 and 64/hour for customer 4. Therefore, customer 2 is served at the rate $100(50/114) = 44$, and customer 4 is served at the rate of $100(64/114) = 56$. At the last server, the rates are 52 for customer 3 and 48 for customer 4.

The defect in local control should now be apparent. Customer 4 is served at the first server at the rate of 64, even though it leaves the system at server 3 at the rate of 48. Capacity at server 1 is wasted on customer 4, and customer 1 is unnecessarily delayed. The situation repeats at server 2. In fact, if customer 4 were served at the rate of 48 at all three servers, delay would decline for both customers 1 and 2 without increasing appreciably for customer 4.

If the queueing network were formulated as a linear program with the objective of maximizing throughput, an optimal service rate for customer 4 would be 40/hour. The same service rate would be used at all three servers, and the remaining capacity would be

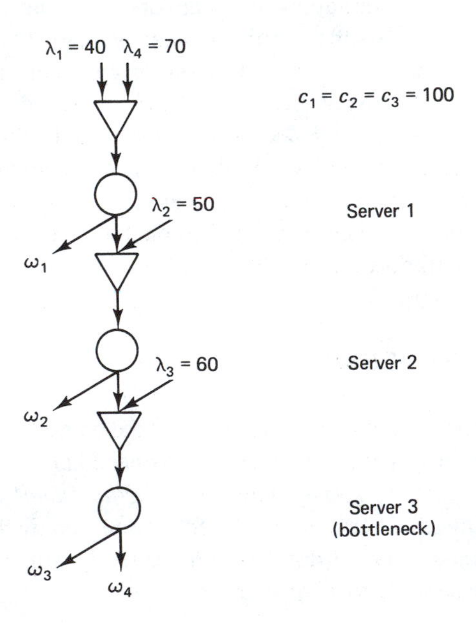

Figure 10.17 Diagram of a queueing network where not all customers pass through the bottleneck. Priority should be given to nonbottleneck customers.

allocated to the other customers. This would achieve a throughput of 190 customers/hour and eliminate delay for customers 1–3. With this plan, servers 1 and 2 would be operating below capacity, and the bottleneck would be server 3. Increasing capacity at any other server would not increase throughput.

FCFS does not optimize the throughput because it gives too high a priority to customer 4 at servers 1 and 2. Because customer 4 will inevitably be delayed downstream at the bottleneck, it should receive lower priority than customers 1 and 2. The example is indicative of the following general principle:

Customers that do not visit bottlenecks should not wait in queues.

Unless stochastic or cyclic queues are significant, there is not much sense in delaying a customer if it only visits underutilized servers. But executing this goal is not simple. As in the example, it is possible for queues to exist *upstream* from the bottleneck. Therefore, priority must be assigned to customers based on their subsequent route.

Definition 10.9

Bottleneck Priority Discipline (BPD) Assign highest priority to customers that will not visit *any* downstream bottleneck.

The above is really only a partial discipline, for it does not state how to break ties. A secondary criterion should then be employed, as will be discussed in a subsequent section.

Though the bottleneck priority discipline is premised on deterministic operation, it can also be effective in a stochastic environment. The key characteristics are prominent bottlenecks exist in the network and not all customers visit bottlenecks. Without these characteristics, it is impossible to prioritize customers by BPD.

10.5.1.1 Entry Control. The most effective place to control queue discipline is often at the entrance to the network. With the BPD discipline, *major queues should exist only at the entrance to the system*. Once the customer has passed through this queue, it should move unimpeded until completion. Further, *only customers that pass through bottleneck servers should wait at the entrance*. Other customers should bypass the queue at the system entrance and move through the network without delay. This is how the system should operate in a deterministic environment.

Execution of this principle requires a special form of queue discipline that will be named ***entry control***. This is particularly important in communications networks (for which the name ***flow control*** is used) because queues can lead to retransmission of data, putting extra strain on congested servers. In one type of flow control, called the ***isarithmic method***, a fixed number of messages is allowed to circulate within a network. Each message carries a permit, which it releases to a new message upon exiting the network. Flow control might also be affected by restricting the number of terminals logged onto the network to a fixed number. In production networks, entry control is accomplished by restricting the ***release*** of jobs into a shop. In freeway networks, entry control is accomplished through ''ramp metering,'' a traffic signal controlling the flow at the freeway entrance. In all three applications, the objective is to keep internal queues from interfering with the smooth operation of the network. Ideally, these entrance rates should be specific to customer type and regulated to match the results of a linear programming model, as in Sec. 10.3. But, in the case of ramp metering and many types of flow control, customers cannot be distinguished on the basis of their subsequent route. Then entry rates must be based on the average mix of customer types at the entrance point, certainly a less than optimal situation.

10.5.1.2 Networks with Cyclic Routings. The principle that major queues should exist only at the system entrance can be extended to queueing networks with cyclic routing. In many types of production processes, jobs are cycled through a piece of equipment several times before completion. This is particularly common in semiconductor manufacturing. Queues for the machines include jobs at various stages of manufacture, each stage defining a different job type. Following the principle that major queues should exist at the system entrance, a reasonable queue discipline would be the following.

Definition 10.10

Nearest Completion Discipline (NCD) At a server in a cyclic network, give priority to jobs that are nearest to completion.

This again reinforces the concept that a customer should move through the network unimpeded after entrance.

10.5.1.3 Impacts of Stochastic Variations. There are limits to the bottleneck disciplines. It would certainly be unwise to let the bottleneck go idle due to ''starvation'' because too much priority was given to nonbottleneck customers. In any time period, a sufficient amount of work should be released into the network to achieve 100 percent utilization at the bottlenecks:

Bottlenecks should never be idle for lack of work (except when there are no queues in the system).

Only the remaining capacity should be allocated to nonbottleneck customers. Further, a buffer of customers should be maintained in back of each of the bottlenecks to cushion against stochastic variations in arrival and service rate. This buffer should be sufficiently large that the number of customers in the buffer rarely drops to zero. If, historically, the queue size has frequently dropped to zero, then the buffer should be gradually increased. This can be accomplished by releasing more customers into the network than the bottleneck can serve over some period of time. Once the buffer has reached an adequate level, the release rate can be returned to a level matching the bottleneck capacity. If, historically, the queue has never dropped to zero, it may be worthwhile to reduce the buffer, by lowering the release rate below the service rate for some time period.

Adjustments such as these can also be made from time to time due to changes in arrival patterns and changes in the service process.

10.5.2 Bottleneck Independent Queue Disciplines

So far, we have seen that queueing can be reduced by giving priority to customers that do not pass through bottlenecks. We have also seen that queueing can be regulated by controlling the rate at which customers enter the system. And we have seen that major queues should only exist at the entrance to the network.

Even with the above, considerable discretion remains for the queue discipline. Because customers are divided into just two classes, bottleneck and nonbottleneck, rules are needed to select among equal priority customers. Further, if the network is well balanced, or if all customers pass through a bottleneck, the above provides no way to select between customers. And just because major queues only exist at the entrance does not mean that stochastic queues will not exist elsewhere, effective queue disciplines being needed at each server in the network. Finally, if the system capacity exceeds the arrival rate, the system will act like it has no bottlenecks, also mandating other disciplines.

There is a large literature on queue disciplines in balanced job shops, most of which relies on simulation. The references provided in Chap. 9 provide numerous examples, and the reader is especially encouraged to consult Conway, et al. (1967). In addition to the local disciplines cited in Chap. 9, some of the most prominent network disciplines are the following:

Definition 10.11

Shortest Imminent Service Time (SIST) Select the customer with the shortest service time at the *current server*.

Shortest Total Remaining Service Time (STST) Select the customer with the shortest total service time among all subsequent operations.

Shortest Dynamic Slack (SS) Select the job with the smallest difference between remaining time until due (due date minus date now) and total remaining service time.

Dynamic Slack/Remaining Operations (DS/RO) Select the job with the smallest ratio of dynamic slack to number of operations remaining.

Work in Next Queue (WINQ) Select the job with the smallest amount of work currently waiting at the next queue the job will visit (also might be based on shortest expected amount of work).

First Arrived, First Served (FAFS) Select the customer that arrived at the *network* first.

Critical Ratio (CR) Select the customer with the smallest ratio of remaining time until due to total remaining service time.

Simulation studies, such as the ones by Conway et al., have found SIST to be most effective at minimizing time in system. This should come as no surprise, given the advantages of the SST discipline for single servers (Chap. 9). As with a single server, however, minimizing time in system does not necessarily minimize number of jobs late or mean tardiness. Conway et al.'s work found the DS/RO rule to be most effective in this regard.

It may be surprising that the look-ahead rule WINQ did not perform as well as SIST. One reason for this may be that the rule does not distinguish between a bottleneck queue and a nonbottleneck queue. The most important queues may even be far downstream. Further, it may not be important whether the subsequent queue is small or large, so long as there is a queue. Therefore, the message is not so much that queue discipline should not account for downstream queues but rather that one must be careful to see which downstream queues are the bottlenecks.

COVERT is a somewhat more complicated discipline (Carroll 1965) that has performed well relative to due date oriented performance measures (Russell et al. 1987). It is based on the idea that a deadline can be created for each job at each step in its service process. This deadline equals the due date, minus remaining service time, *minus an expected remaining waiting time*. If the job is proceeding on schedule, then it should begin service before this deadline.

With COVERT, a job's priority at any server depends on the difference between current time and the deadline at that stage, and is computed as follows:

Definition 10.12

COVERT Discipline. Select the customer with the *highest* ratio of

$$\frac{\textbf{``Delay cost for customer''}}{\textbf{Imminent service time}}$$

The "delay cost" is calculated as follows:

$$\text{Delay cost} = \begin{cases} 1 & SS < 0 \\ 0 & SS > kW \\ \dfrac{kW - SS}{kW} & \text{otherwise} \end{cases}$$

where

SS = the dynamic slack

W = the remaining waiting time

k = a user-specified parameter (approximately one)

Following this rule, jobs receive lowest priority when their slack is greater than the expected remaining waiting time. These jobs would be considered ahead of schedule. A job would receive highest priority if its slack is less than zero. This job would be so far behind schedule that it could not possibly be completed on time. Giving it priority would minimize its tardiness. For jobs between these extremes, varying levels of priority are provided.

Example

The COVERT discipline, with $k = 1$, is used in a large queue network. Servers operate eight hours per day (8:00 A.M. to 4:00 P.M.). Currently, the time is 10:00 A.M. on May 20, and server 5 has two jobs waiting:

Job 1 Deadline: 11:30 A.M. 5/22

Server	Remaining Service Time	Remaining Waiting Time
5	1 hour	
6	4 hours	5 hours
7	4 hours	6 hours

Job 2 Deadline: 2:00 P.M. 5/21

Server	Remaining Service Time	Remaining Waiting Time
5	.5 hour	—
8	1 hour	1 hour
9	.5 hour	2 hours
10	1.5 hours	.5 hour

Then currently

Job 1 Slack = 17 hours − 9 hours = 8 hours
Delay cost = (11 − 8)/11 = .27
Delay cost/imminent service time = .27/1 = .27

Job 2 Slack = 12 hours − 3.5 hours = 8.5 hours
Delay cost = 0 [$SS > W$]
Delay cost/imminent service time = 0/.5 = 0

Job 1 currently has higher priority than job 2.

In the example COVERT assigns different priorities than the DS/RO or SIST rules, both of which would give higher priority to job 2. This is because COVERT accounts for job 1's longer waiting times at subsequent queues.

A major assumption of COVERT is that the subsequent waits can be predicted with reasonable accuracy. This task is not easy. One might estimate the waits based on current waiting times, but these are bound to change by the time a job arrives at the subsequent queue. With effective entry control, the internal queues should largely be stochastic, making them especially hard to predict. For more on ways to estimate waits, consult Russell et al. (1987).

10.5.3 Queue Disciplines in Parallel Processing Networks

In a network with parallel processing, the time in system will depend on the maximum of various service and waiting times, and not just on their sum. This leads to rather different, and more complicated, queue disciplines. It is desirable to finish dependent steps in the service process at the same time. Little is to be gained by completing a few parts early, only to wait for a critical component.

Consider, as an example, a process in which parts are assembled in stages to produce one final product. As depicted in Fig. 10.18, the assembly network has a tree structure, as the various parts are joined to form components, subassemblies, and so on.

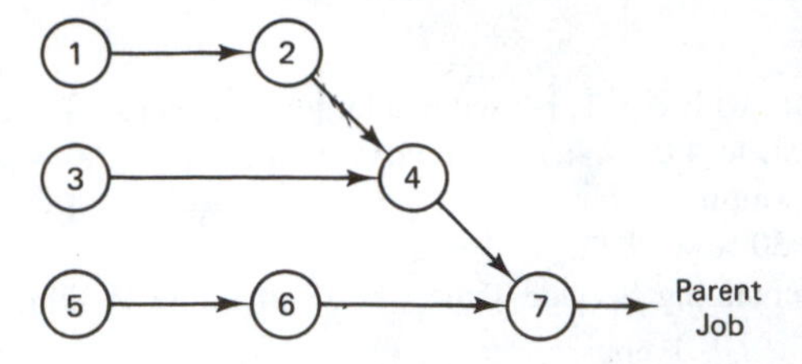

Figure 10.18 Network representation of an assembly process involving seven steps.

Branches of the tree converge when two or more of the components are merged to form a larger component. The final task in the network produces a finished product, which is called the ***parent job***. The time that the parent job is completed defines the service time. For an assembly operation, the service time is also called a ***makespan***. Following the network structure, two operations can be performed simultaneously if no directed path exists between the operations. The existence of a directed path would imply serial precedence. With perfect synchronization, the final tasks on each branch joining at a node would be completed at the same time.

Parallel queue disciplines have not been studied as extensively as serial queue disciplines. Nevertheless, Sculli (1987) has identified several possibilities:

Definitions 10.13

Shortest Network Slack (NSL) Perform a critical path analysis of the tasks not yet completed for each parent job and determine the slack time for each task. Select the job with the smallest slack.

Remaining Job Time (RJT) Select the job whose parent has the smallest total service time remaining.

Number of Parts (NOP) Select the job whose parent has the smallest number of uncompleted parts.

Dynamic Slack/Remaining Operations (DS/RO) Select the job with the smallest slack per remaining successor operation (a successor being linked to the current operation by a directed path).

As far as which discipline is best, there are few conclusive answers. Simulation studies, such as the one by Sculli (1987), have cast doubt on whether the simple SIST discipline is effective in a parallel environment because it does not adequately account for interdependencies. More complicated rules, such as NSL, which do account for interdependencies, are more adept at minimizing time in system and tardiness. However, even here there is room for improvement, for the NSL rule is based on the unrealistic scenario of perfect synchronization. Nevertheless, the NSL rule requires information on the progress of all job tasks, a potentially unreasonable demand in many factories.

10.6 EVALUATION OF STOCHASTIC QUEUEING NETWORKS

The stochastic behavior of queueing networks is extremely difficult to evaluate. In a well-balanced system, random fluctuations in arrival rate and service time (especially machine breakdowns) can cause the bottleneck to shift from server to server. This will inevitably decrease system throughput and increase queueing compared to what a deterministic model would predict.

Before analyzing stochastic behavior, it is a good idea to examine first the deterministic behavior, as discussed in previous sections. This will at least provide an indication of which servers are most critical and where changes should be made. Before the changes are implemented, however, it is wise to test their effectiveness with a stochastic model. Two basic tools are available for this purpose: queueing network analyzers and simulation.

10.6.1. Queueing Network Analyzers

Network analyzers are designed to evaluate the approximate steady-state performance of queueing systems. At the simplest level, that of a single server in the network, these tools might use something like the Allen-Cunneen approximation for estimating the expected number of customers in queue. For instance, American Telephone and Telegraph's (AT&T) Queueing Network Analyzer (QNA; Whitt 1983) approximates time in queue from the mean and coefficient of variation of the interarrival time and the mean and coefficient of variation of the service time. That is, the result depends on the moments of the probability distributions, not on their exact shapes.

At a higher level, network analyzers must determine how the arrival pattern at a server is influenced by the performance of other servers. The most difficult aspect is calculating the coefficient of variation of the interarrival time, which first requires determining the coefficient of variation of a departure process and, second, may involve determining the coefficient of variation of a superposition of arrival processes (associated with the merger of customers from various sources). In both cases, approximations are used.

The attractive feature of network analyzers is that they quickly provide information on the expected behavior of a queueing network. Like the steady-state models in Chap. 5, the performance measures are not susceptible to random fluctuations. The predictions are precise. On the other hand, because they employ approximations, network analyzers are not always accurate. In addition, network analyzers are much more restrictive than simulation. They cannot handle all queueing behavior, especially nonstationary behavior. Hence, they often cannot be used without compromising model realism. Finally, because model assumptions are often hidden from view, it is difficult for users to assess the validity of model predictions.

Analysis of queueing networks is an area of active research. For instance, at the University of Michigan, an approximation called "SIMP" has recently been developed to account for blocking caused by restricted buffer sizes (Lee and Pollock, 1989). New commercial products are also being introduced, such as MANUPLAN, a manufacturing oriented system which accounts for lot sizes, and TRAFFEN, which is oriented toward communication systems and can account for both delay and loss of customers.

10.6.2 Simulation Programs

Simulation is the most robust, and often most realistic, way to evaluate queueing performance. Virtually any probability distribution can be accommodated. Complicated queue disciplines can be incorporated. Networks with both parallel and serial processing, dynamic routing, and so on, can be represented. In fact, virtually any queueing phenomenon can be tackled through simulation.

Ever since IBM's introduction of GPSS in 1961, we have been blessed by the creation of many powerful computer simulation languages. In its 1987 survey, *Simulation* magazine listed over 150 commercially available languages. All of these languages include built-in features to simplify the creation and interpretation of simulation models, features including random number generation, list processing, and output formatting. Many of the languages also have special capabilities for analyzing queueing systems and manufacturing queues in particular.

10.6.2.1 The Limitations of Simulation. All of these features sound attractive, and indeed they are, but there are caveats. Simulation can be expensive and time consuming. Some of the better known languages cost more than $10,000, and this does not even include the cost of developing the simulation model, which can take years and hundreds of thousands of dollars. Once the model is completed, computational costs can also be enormous. There are famous examples of poorly structured models that took longer to run than the system being simulated!

Simulation is also dangerous. It is all too easy to put faith in a poorly specified model and not realize it is incorrect. Particularly for the untrained, good simulations are difficult to distinguish from poor simulations. There is even a tendency to forget that *simulation models require empirical data*. The data created by simulations look so natural that they can be mistaken for the real thing. In fact, it has been said that simulations can replace data collection. But simulation (like a network analyzer) can only be accurate as the data used to create it—data that must be obtained through observation. Whereas a simulation can be used to experiment with a model, the model must be properly specified in the first place.

10.6.2.2 What Simulation Can Do. Linear programming quickly estimates, and optimizes, the deterministic behavior of queueing networks. Network analyzers quickly estimate the *approximate* behavior of stochastic queueing networks. Simulation slowly (and imprecisely) estimates the exact behavior of stochastic networks.

Each technique can be used at a different stage of analysis, as one gets a better idea of what changes are most promising. Simulation's place comes at the end of analysis. It is an effective tool for evaluating specific changes in a queueing system. Simulation does not identify these changes. It merely checks to see how well they perform. This is fundamentally different from linear programming, which both identifies changes and measures their performance (though less accurately than simulation).

Simulation is useful when it is important to test a change before implementation.

Example 1

An industrial engineer employed by a major manufacturer has identified a machining bottleneck through use of a linear program. He is considering purchasing an additional piece of equipment. However, because the network is complex with many interrelated servers, he is not sure just how much delay will decline and just how much throughput will increase. Without this information, he cannot tell whether the investment is economical.

Example 2

In another factory, the engineer has identified a labor-intensive assembly stage as the bottleneck. Here, she is considering increasing capacity by reassigning an employee from another stage which currently has excess capacity. Again she is unsure how much delay will change and whether the action is justified.

The first example is an appropriate application of simulation. Because the change requires a major investment, the engineer should be confident that it will work. The second example is *not* an appropriate application of simulation. It is not because the system change is easily tested by implementation. If the evidence points in favor of a change, and the change is both easy to institute and to reverse, then just go ahead and try it out. If the idea turns out to be a failure, then try something else.

10.6.2.3 Simulation Languages. There are two distinct types of simulation: discrete simulation and continuous simulation. In ***discrete simulation*** the state of the system at any point of time is measured with a discrete variable. (This does not imply that the state of the system is only measured at discrete times; time can be a continuous variable.) In ***continuous simulation*** the state of the system is measured with a continuous variable. Applications of continuous simulation include rocketry and chemical processing, phenomena that are inherently continuous. Discrete simulation is most often applied to queueing systems; customers are discrete entities. Discrete simulation can also be used to approximate continuous phenomena, such as movements of robot arms, and continuous simulation can approximate discrete phenomena, such as population growth. Because the error introduced by rounding off a continuous value to a discrete number, and vice versa, is usually slight, the choice of simulation type is often dictated by modeling convenience, and not by whether the true system exactly meets model specifications.

It should be no surprise that the logic of continuous simulation is based on calculus. Continuous simulation languages usually allow the user to describe the system in the form of differential equations, which the computer solves numerically. Alternatively, *difference equations* (the discrete analog to differential equations) might be used, in which case time is decomposed into finite periods, and the state of the system is calculated one period at a time.

The logic underlying discrete simulation has already been described in Chap. 4 and will not be repeated. However, within the framework of discrete simulation, there are alternative ways to orient a simulation model. With the ***event orientation***, the system is defined by the changes that occur at isolated points in time. The system is only evaluated at event times, and not between. Within the context of queueing, an event might be an arrival, departure from queue, departure from service, or possibly an interruption or renege. Event-oriented simulation languages, such as SIMSCRIPT, maintain a list of all future events that have not yet occurred and process the events in chronological sequence. As this happens, new events might be added to the list. For instance, the event "customer enters service" will generate the future event of departure from service.

Process-oriented simulation languages, such as GPSS, combine sequences of events into single subroutines called blocks. For example, the GENERATE block in the

GPSS language creates a transaction, which might be an arrival or some other entity, and the SEIZE block causes a transaction to seize a system resource (for example, enter service). Conceptually, the user arranges the blocks in the form of a system diagram, which is executed by the GPSS processor. The process orientation often makes it easier to write a simulation program, though the user loses some power to customize the program to a given application.

Activity scanning orientation is an option available on many simulation languages. With this approach, the user specifies the conditions under which an activity (for example, wait for server) will begin or end. At each advance of the simulation clock, all activities are scanned to identify which (if any) conditions are met and which actions should be taken. The activity orientation is needed when the duration of an activity is undefined at its start and is affected by future events. For example, if servers operate at a faster rate as queue length increases, then the length of a service will be affected by future arrivals. Other complex phenomena, such as reneging, may also require an activity scanning orientation. Unfortunately, the added capabilities of activity scanning do not come without a price. Scanning all activities at each time advance consumes added computation time and slows the simulation.

The trend in simulation languages is to support a wider variety of modeling options. This is the case of SLAM, a language that allows all three types of discrete simulation, as well as continuous simulation. Some languages also contain application-specific features. SIMAN, SIMFACTORY, and XCELL, for instance, include special commands for modeling factories. In recent years simulation languages have become widely available for personal computers, creating new options, such as interactive simulation and computer animation. One such system is Cinema, which is based on the SIMAN language. While these features may be no more than glitz to the experienced analyst, managers often appreciate the capability of watching a system evolve. The model is taken out of the "black box" realm, and results become more meaningful.

In addition to its features for modeling factories, the SIMAN language has adopted a framework that divides the simulation program into a system model and an experimental frame. The ***system model*** defines the behavior of the system when exposed to external conditions. For instance, for a given set of customer arrival times and types, the system model might define the departure process. The ***experimental frame*** defines the process for creating the external conditions (for example, the simulation of random arrival times). The advantage of SIMAN's approach is that the system can be simulated under alternative conditions merely by substituting one experimental frame for another, without altering the system model.

An important thing to keep in mind is that different languages use different statistical methods for analyzing simulation results. One should be especially careful to make sure that correlations between dependent random variables are adequately accounted for in the estimation of standard errors. If correlations are not considered, then model accuracy will be greatly overestimated. Simulation models also do not necessarily present results in the most useful form: as cumulative graphs. Finally, keep in mind that confidence intervals and other estimates of accuracy are meaningless if the model is poorly specified, which is another way of saying "garbage in, garbage out."

Readers interested in using a simulation language should consult Kiviat et al. (1969), Pegden (1986), Pritsker (1986), and Schriber (1974), as well as the magazine *Simulation*. The IBM case study at the end of this chapter gives a good overview of how to use simulation in combination with a network analyzer in designing a factory. Another good case study is the article by Welch and Gussow (1986) on expansion of the Canadian National Railway.

10.7 CHAPTER SUMMARY

A queueing network is a collection of servers acting together to serve customers. Queueing networks are used when it is more efficient to divide the service process into separate tasks than to have a single server complete the entire service.

In a queueing network, servers do not operate independently of each other. In serial processing, the arrival curve at a downstream server is dictated by the service process at the upstream server. In parallel processing, the service time depends on the maximum service time among a set of servers. Despite the interdependencies, the message of this chapter is indeed simple:

To improve the performance of a queueing network, identify the bottleneck servers and improve their performance.

In a series queueing network, the bottleneck is the server (or servers) with the smallest capacity. It is also the last server where the customer encounters a nonstochastic/noncyclic queue. The bottleneck is not necessarily the server with the largest queue. The largest queue may be upstream from the bottleneck. In a workstation network, the bottleneck is defined by a critical path of tasks. And in a general network, the bottleneck is defined by a set of tasks with nonzero dual variables, as identified by a linear program.

Queues are accentuated when blocking and starvation occur. When a buffer fills to capacity, the upstream server is prevented from operating at full capacity; it can operate no faster than its downstream server. Starvation occurs when a buffer becomes empty. Then a server can operate no faster than its *upstream* server. In any queueing network, it is important that buffers be sufficiently large to keep bottlenecks from starving or being blocked. This is especially important in networks with bulk service and in networks with significant stochastic queueing.

Linear programming is an effective tool for identifying bottlenecks in large networks. The primal solution determines the maximum throughput and the departure rates. The dual solution determines the bottlenecks. Linear programming can also be used to analyze routing through congested communication, transportation, and production networks. In networks where stochastic queues dominate, the Frank-Wolfe method can be applied.

Queue discipline should account for the presence of bottlenecks. Ideally, customers should only queue if they pass through bottlenecks; nonbottleneck customers should move through the network unimpeded. Further, queueing should be controlled at the entrance to the network. Once a customer has passed through the initial queue, it should move through the network without major delay. Finally queue disciplines should account for other customer characteristics, such as service times and due dates. A variety of disciplines were proposed with this in mind (Sec. 10.5).

As a final step, stochastic queueing behavior can be evaluated with a queueing network analyzer or with simulation. Simulation is a robust tool that can accurately accommodate a wide range of queueing phenomena. It is effective at testing changes in a queueing system when it is difficult and/or expensive to test the changes on the real system. Simulation, however, may lack precision because it is susceptible to random fluctuations. As an alternative, or in addition, a queueing network analyzer can be used to evaluate expected steady-state performance. Network analyzers are not susceptible to random fluctuations and offer much shorter computation times. However, because they employ approximations, and because they cannot handle all queueing behavior, accuracy is not guaranteed. Neither tool can identify courses of action. This must come from one's understanding of queueing systems or, perhaps, through optimization techniques (such as linear programming).

Within the context of systems analysis, as introduced in Chap. 1, linear programming, network analyzers, and simulation all fall within the evaluation step. Before any of these techniques can be used, one must first understand the system (through observation and measurement), define the performance measures, and create a model (or models). Linear programming, along with one's insights, falls into the first phase of the evaluation step—using the model to generate alternatives. Simulation and network analyzers fall in the last phase of evaluation—measuring the performance of alternatives. The final step in systems analysis is deciding. Here, there is no formula. It is up to an educated individual to decide, based on the information that the models provide.

FURTHER READING

Arthurs, E., and B. W. Stuck. 1983. "Upper and Lower Bounds on Mean Throughput Rate and Mean Delay in Memory-Constrained Queueing Networks," *Bell System Technical Journal*, 62, 541–581.

Bertsekas, D., and R. Gallager. 1987. *Data Networks*. Englewood Cliffs, N.J.: Prentice Hall.

Bradley, S. P., A. C. Hax, and T. L. Magnanti. 1977. *Applied Mathematical Programming*. Reading, Mass.: Addison-Wesley.

Bronson, R. 1984. "Computer Simulation, What It Is and How It's Done," *Byte*, 9, 95–102.

Carroll, D. C. 1965. "Heuristic Sequencing of Single and Multiple Component Jobs," Ph.D. dissertation, Sloan School of Management.

"Catalog of Simulation Software," *Simulation*, 49, 165–181 (annual).

Clark, C. E. 1961. "The Greatest of a Finite Set of Random Variables," *Operations Research*, 9, 145–162.

Conway, R. W., W. L. Maxwell, and L. W. Miller. 1967. *Theory of Scheduling*. Reading, Mass.: Addison-Wesley.

Disney, R. L., and D. König. 1985. "Queueing Networks: A Survey of Their Random Processes," *SIAM Review*, 27, 335–403.

Garzia, R. F., M. R. Garzia, and B. P. Zeigler. 1986. "Discrete-Event Simulation," *IEEE Spectrum*, 23, 12, 32–36.

Haight, F. A. 1963. *Mathematical Theories of Traffic Flow*. New York: Academic Press.

Hax, A. C., and D. Candea. 1984. *Production and Inventory Management*. Englewood Cliffs, N.J.: Prentice Hall.

Hillier, F. S., and G. J. Lieberman. 1986. *Introduction to Operations Research*. Oakland, Calif.: Holden-Day.

Jackson, J. R. 1957. "Networks of Waiting Lines," *Operations Research*, 5, 518–521.

Johnson, L. A., and D. C. Montgomery. 1974. *Operations Research in Production Planning, Scheduling, and Inventory Control*. New York: John Wiley.

Kiviat, P. J., R. Villanueva, and H. Markowitz. 1969. *The SIMSCRIPT II Programming Language*. Englewood Cliffs, N.J.: Prentice Hall.

Kleinrock, L. 1976. *Queueing Networks, Volume 2: Computer Applications*. New York: John Wiley.

Koenigsberg, E. 1982. "Twenty-five Years of Cyclic Queues and Closed Queue Networks," *Operational Research Society Journal*, 33, 605–619.

Lee, H. S., and S. M. Pollock. 1989. "Approximate Analysis for the Merge Configuration of an Open Queueing Network with Blocking," *IIE Transactions*, 21, 122–129.

Maxwell, W. L., and R. C. Wilson. 1981. "Dynamic Network Flow Modeling of Fixed Path Material Handling Systems," *AIIE Transactions*, 13, 12–21.

Newell, G. F. 1982. *Applications of Queueing Theory*. London: Chapman and Hall.

Oliver, R. M., and A. H. Samuel. 1967. "Reducing Letter Delays in Post Offices," *Operations Research*, 10, 839–892.

Pegden, C. D. 1986. *Introduction to Siman*. State College, Pa.: Systems Modeling Corp.

Pritsker, A. A. B. 1986. *Introduction to Simulation and SLAM II*. New York: John Wiley.

Rubenstein, R. Y. 1986. *Monte Carlo Optimization, Simulation and Sensitivity of Queueing Networks*. New York: John Wiley.

Russell, R. S., E. M. Dar-El, and B. W. Taylor III. 1987. "A Comparative Analysis of the COVERT Job Sequencing Rule Using Various Shop Performance Measures," *International Journal of Production Research*, 25, 1523–1540.

Schriber, T. 1974. *Simulation Using GPSS*. New York: John Wiley.

Sculli, D. 1987. "Priority Dispatching Rules in an Assembly Shop," *Omega*, 15, 49–57.

Skabardonis, A. 1986. "Guidebook for Improving Traffic Signal Timing," UCB-ITS-RR-86-10. Berkeley, Calif;: Institute of Transportation Studies.

———, and M. C. Kleiber. 1983a. "Signal Timing Optimization: A Bibliography," UCB-ITS-LR-83-4. Berkeley, Calif.: Institute of Transportation Studies.

———. 1983b. "Traffic Signal Timing—Before and After Studies: A Bibliography," UCB-ITS-LR-83-5. Berkeley Calif.: Institute of Transportation Studies.

U.S. Department of Transportation. 1981. *The TRANSYT Signal Timing Reference Book*. Washington, D.C.

Walrand, J. 1988. *An Introduction to Queueing Networks*. Englewood Cliffs, N.J.: Prentice Hall.

WELCH, N., and J. GUSSOW. 1986. "Expansion of Canadian National Railway's Line Capacity," *Interfaces*, 16, 51–64.

WHITT, W. 1983. "The Queueing Network Analyzer," *Bell System Technical Journal*, 62, 2779–2815.

ZANGWILL, W. 1969. *Nonlinear Programming: A Unified Approach*. Englewood Cliffs, N.J.: Prentice Hall.

PROBLEMS

1. Give an example of a system that uses serial service and an example of a system that uses parallel service. Would it be possible to convert the serial system into a parallel system? If so, what would be the impact of this change? What would be the impact of converting the parallel service into serial service?

2. Commuters using a highway bridge have become irate over the queues that build up behind the toll plaza, which they must pass through before reaching the bridge. They have demanded that the toll plaza be removed so that delay can be eliminated. Are the commuters right? Discuss.

3. Suppose that Fig. 7.13 represents the arrivals of people at an employment development office. Each person has to pass through a series of three queues, first filing a form, then obtaining approval from an inspector, and finally receiving a check from a cashier. The service time at each stage is short, but queues still exist because the capacity falls below the arrival rate. The capacities are 4900 customers/hour at the first stage, 4500 customers/hour at the second stage, and 5000 customers/hour at the third stage. The walking time equals 5 minutes between each pair of stages.

(a) Draw the cumulative arrival and departure curves on a piece of graph paper. Which stage is the bottleneck, and what is the maximum throughput?

(b) Management has noticed that queues are large at the first stage but nonexistent at the third. Therefore, they would like to move a worker from the third to the first stage. If the worker can serve customers at the rate of 1000 per hour at either stage, should management make the change?

(c) Management is considering a new process in which the first two stages are combined. However, they are concerned because the combined stage will have a capacity of just 4700 customers per hour, and they fear queues there will become intolerable. The total walking time between stages will remain at 10 minutes. Do you believe that the change will be an improvement?

4. Repeat Prob. 3 with the following arrival data:

Time	6:00–6:30	6:30–7:00	7:00–7:30	7:30–8:00	8:00–8:30	8:30–9:00
Arrivals	2000	3000	4500	2000	1000	500

5. An automated manufacturing process requires four serial stages. The cost of purchasing equipment and the associated capacities are shown below:

Stage

A:	Machine 1:	Capacity 100/day	Cost $300/day
	Machine 2:	Capacity 200/day	Cost $500/day

B:		Capacity 500/day	Cost $200/day
C:	Machine 1:	Capacity 25/day	Cost $100/day
	Machine 2:	Capacity 700/day	Cost $400/day
D:		Capacity 200/day	Cost $100/day

In stages A and C, there is a choice of two machines (it is not necessary to buy both).

(a) Plot cost/day versus system capacity, for a capacity range of 0/day to 600/day. For what capacity is the cost/unit the smallest?

(b) Suppose that work arrives at the constant rate of 225/day, and that the company earns a profit of $15/unit (before subtracting costs above). How many machines, of each type, should the company buy, and what should be the plant capacity?

(c) A reduction in selling price by $5/unit would increase sales to 500/day. Should the company take this action?

6. The stages in Prob. 5 have been realigned so that they operate concurrently rather than serially. Would any of your answers change and, if so, how? Also, would there be any change in service time?

***7.** Tax forms are processed at an IRS facility in three stages. The processing rates, per employee, are shown below:

Stage 1:	20 forms/hour
Stage 2:	15 forms/hour
Stage 3:	30 forms/hour

The facility currently has 21 employees trained to work in all three stages, 10 employees trained only for stage 1, 20 employees trained only for stage 2, and 2 employees trained only for stage 3.

(a) How would you assign employees to stages in order to minimize the queueing? What is the maximum system capacity?

(b) Would the capacity change if half of the employees currently trained for stage 1 are also trained for stage 2?

(c) Because the arrival rate for forms is highly variable from week to week, the IRS is considering using the 21 flexible employees as ''roving'' employees. That is, at the start of each day, these employees will be shifted to whichever stage has the largest queue. Is this a good idea? Explain.

***8.** Customers at an order center are currently processed in one stage by either of two parallel servers. The service times are independent and exponentially distributed (mean 5 minutes) and customers arrive by a Poisson process, at the rate of 15/hour.

(a) Estimate expected time in system for the $M/M/2/\infty$ queue.

(b) The service process can be divided into two stages and performed by two separate servers. The service time at each stage will then have a mean of 2.25 minutes (exponential and independent). Assuming that the servers are immediately adjacent, calculate expected time in system. (Hint: The arrival process at the second server is Poisson.) Would you want to follow a two-stage process?

(c) Suppose that the arrival process is nonstationary, and that the peak arrival rate exceeds 40/hour. Would you want to follow a two-stage process? Explain.

(d) Would your answer to part b be different if the service times were identical for every customer at both stages? Explain in words.

*Difficult problem.

9. A metal part is processed in batch at three successive machines. The first machine devotes 1 hour/day to the part, the second devotes 3 hours/day, and the third devotes 2 hours/day. Included in these times is a setup time (8 minutes, 1 hour, and 5 minutes, respectively) and a running time (rates of 600/hour, 150/hour, and 225/hour).

(a) Given current operations, which machine is the bottleneck for the part and what is the current capacity of the system?

(b) If the current demand rate for the part is 250/day, would you want to change the amount of time devoted to the part on any machine?

(c) Suppose that the current demand rate is 350/hour. How could demand be met without changing the proportion of any machine's time devoted to the part?

***10.** Two adjacent traffic signals on a one-way street have the following cycle times and service rates:

Signal 1 $c_1^g = 54/\text{minute}$ $g_1 = 20$ seconds $T_1 = 60$ seconds

Signal 2 $c_2^g = 36/\text{minute}$ $g_2 = 20$ seconds $T_2 = 40$ seconds

The free-flow travel time between each pair of signals is 15 seconds. Vehicles arrive at signal 1 at a constant rate of 18/minute. (assume that the green time does not include a loss time.)

(a) Suppose that vehicles begin to arrive at 8:00, that signal 2 first turns green at 8:00.20 and signal 1 first turns green at 8:00.40. Draw the cumulative arrival and departure curves for the first 5 minutes.

(b) Given the phase lengths, cycle times, and capacities, what is the largest queue size that would be experienced at the second server?

(c) If there were only space for ten vehicles between the signals, would blocking or starvation exist? If so, what would happen to the queues at the first and second server?

(d) What could be done to the cycle times and phase lengths to improve the system performance?

11. The term *gridlock* has been used to describe congestion in the U.S. air traffic system. Is this really an example of gridlock, or is it something else? Discuss in words.

12. Work orders arrive at a two-stage machine shop at the following rates:

8:00–10:00	100/hour
10:00–12:00	200/hour
12:00– 2:00	300/hour
2:00– 5:00	75/hour

The first stage has 25 machines working in parallel, each with a service time of 6 minutes. The second, immediately adjacent, stage has 19 machines working in parallel, also with a service time of 6 minutes each. As an approximation, treat the two servers as though the service time is very short.

(a) The buffer between the two servers can store 100 jobs, and the buffer in front of the first server has unlimited capacity. Draw the cumulative arrival and departure curves.

(b) What would be the advantage of increasing the buffer size between the servers?

(c) What would be the advantage of switching the order of the servers?

***13.** Two machines process jobs in serial order. The first machine can process jobs at a rate of 210 per hour, and the second can process jobs at the rate of 225 per hour. However, both machines suffer from frequent breakdowns. The time between breakdowns on the first machine is

*Difficult problem

exponentially distributed with mean of 4 hours, and the time between breakdowns on the second machine is exponentially distributed with mean of 1 hour. In both cases, a repair always takes 20 minutes.

(a) Which machine is the bottleneck and what is the system capacity if the buffer has unlimited size?

(b) Suppose that the buffer between the two servers has a capacity of 25 jobs. Is there a large risk that the second machine will starve? Is there a large risk of the first machine being blocked? Would you expect the actual capacity to be much less than part a?

(c) Would an increase in buffer size to 70 change the risks from part b appreciably?

***14.** Using a computer, simulate the system described in Prob. 13 over a period of 40 hours. Assume that the queue of arriving jobs is unlimited. Perform your simulation for buffers of size 25, 70, and 100. Calculate the system throughput, the proportion of time that server 2 starves, and the proportion of time that server 1 is blocked.

15. After vacationers arrive at a resort community, they must go through two operations. In one operation, vacationers are given a group orientation (for about 30 people) that lasts 10 minutes. In the other operation, vacationers register at a desk (two servers, with a service time of one minute per person).

(a) Suppose that vacationers arrive by buses, carrying 30 people each. To minimize waiting time, in which order should the servers be sequenced? Explain.

(b) Suppose that vacationers arrive individually, at the rate of about 2 people per minute. To minimize waiting time, in which order should the servers be sequenced? Explain.

16. Each job that enters a workstation must undergo two tasks. The task times are independent and have $U[1 \text{ min}, 5 \text{ min}]$ and $U[1 \text{ min}, 10 \text{ min}]$ distributions.

(a) Suppose that the tasks are performed sequentially. What would be the mean service time?

(b) Suppose that the tasks are performed concurrently. What would be the mean service time?

***17.** In Prob. 16, suppose that customers arrive by a Poisson process, with rate of 8/hour. What is the expected number of customers in queue for concurrent processing? (Hint: Determine the coefficient of variation for the service time and use the Pollaczek-Khintchine formula.)

18. Suppose that the jobs in Prob. 16 have independent normal distributions, with means of 3 minutes and 6 minutes and standard deviations of 1 minute and 2 minutes. Using the Clark approximation, estimate the mean service time.

19. Suppose that a new task, task 7, is added to Fig. 10.13. Task 2 must precede task 7, and task 7 must precede task 6. Task 7 requires 3 seconds to complete.

(a) Determine the minimum time required to complete the entire set of tasks. Which sequence of tasks is critical?

(b) Suppose that a new process is developed whereby tasks 7 and 6 can be performed concurrently. However, task 7 must now precede task 4. What is the new time required to complete the entire set of tasks?

20. Nine separate tasks are performed in a workstation. Tasks A–D can be performed concurrently, as can tasks E and F and H and I. Tasks E and F must follow task C, task G must follow task E, and tasks H and I must follow task G.

(a) Draw the critical path network.

(b) If each task requires 15 seconds, what is the total service time?

*Difficult problem

***21.** Vehicles enter and exit a freeway from four different locations. During a period of particularly heavy demand, the arrival rates, by origin and destination, are the following:

Vehicles/hour

	To			
From	**1**	**2**	**3**	**4**
1	—	500	500	2000
2	—	—	750	3000
3	—	—	—	1500

The freeway is capable of serving 5000 vehicles/hour in each segment.

(a) Suppose vehicles entering each segment are served FCFS. Would total delay decrease if the capacity of segment (2,3) is increased to 6000 vehicles/hour?

(b) Suppose that vehicles with destinations at exit 3 have priority when entering segment (2,3). Then would total delay decrease if the capacity of segment (2,3) is increased to 6000 vehicles/hour? Would total delay change for vehicles traveling from exit 3 to exit 4?

(c) Describe a practical scheme that would give priority to vehicles that do not pass through the bottleneck.

22. The following jobs are waiting to be processed at server 5 at time 12:00, January 13. Servers operate ten hours per day, seven days per week.

Job 1: Deadline: 12:00 1/20

Server	Remaining Service Time	Remaining Waiting Time (estimated)
5	1 hour	—
7	8 hours	3 hours

Job 2: Deadline: 12:00 1/17

Server	Remaining Service Time	Remaining Waiting Time
5	3 hours	—
7	6 hours	3 hours
8	5 hours	1 hour
9	4 hours	0 hours
10	2 hours	2 hours

Job 3: Deadline: 12:00 1/17

Server	Remaining Service Time	Remaining Waiting Time
5	6 hours	—
6	5 hours	11 hours
11	6 hours	7 hours
12	5 hours	12 hours

(a) Calculate dynamic slack for each job. According to the DS/RO rule, which job currently has highest priority?

(b) According to the SIST rule, which job has highest priority?

*Difficult problem

(c) According to the critical ratio rule, which job has highest priority?
(d) According to the COVERT rule ($k = 1$), which job has highest priority?

23. Construct an example to illustrate why it is desirable to give priority to jobs that are closer to completion in cyclic queueing systems.

24. Construct an example to illustrate why it may be desirable in a queueing network to give priority to jobs that are performed sequentially over other jobs that depend on parallel tasks.

CASE STUDY

CAUTIOUS CONFIDENCE AT I.R.S.

Less Chaos in Philadelphia

William K. Stevens

Last year at this time a bureaucratic horror story was unfolding inside the long, low building in northeast Philadelphia where the Internal Revenue Service processes Federal income tax returns from 14 million taxpayers in the Middle Atlantic region.

By summer a jumble of errors, bottlenecks, delayed refunds and lost information, outraging hundreds of thousands of taxpayers, had made the Philadelphia processing center the worst example ever known of a breakdown of the I.R.S. system.

This year, with an air of cautious confidence, the center's new director predicts that nothing like what happened in 1985 is in store for 1986.

Joseph Cloonan, who took over as head of the Philadelphia center last August, said it is far ahead of its 1985 pace in processing returns. As of last Friday, nearly one-third of the 1,534,000 returns received at the center had been processed. At the same time last year it had worked through less than 10 percent of the 1,845,000 returns that had come in by then.

"That is not to say that there won't be problems or that some taxpayers won't complain," said Mr. Cloonan. "But for the majority of taxpayers there will be a return to the kind of service they're accustomed to, and our early performance indicates that."

The center is now about one-fourth of the way through the filing season. At this early stage, Mr. Cloonan noted, some taxpayers get refunds as soon as four weeks after mailing their returns. At the peak of the season, it usually takes seven weeks.

The official I.R.S. goal is to send out all refunds within 45 days after the April 15 filing deadline. This would mean by the end of May, and the Government would thus avoid paying interest to taxpayers. For a lot of reasons, the goal will never be 100 percent achieved, but the I.R.S. intends to come as close as it can.

I.R.S. officials attribute last year's chaos to a new, untested computer system, which was still in its shakedown period and prone to malfunctions, aggravated by widespread human error and inexperience.

The result was a tremendous log-jam of unprocessed tax returns. Before 1985 was over, hundreds of thousands of returns were backed up, delaying some people's refunds for many months. The delays have cost the Government more than $15 million in interest paid to the taxpayers.

On top of that, data from thousands of tax returns were lost. Some taxpayers were required to file returns twice or even three times. Some received refunds they were not supposed to get, while others were dunned for taxes they did not owe. Letters from furious citizens piled up. The telephone exchange serving the I.R.S. office in downtown Philadelphia became so jammed that at one point officials feared it might collapse from the overload.

And amid all this, it was reported—and later substantiated—that returns and checks for payment of taxes had been found in a trash barrel. In a separate incident, I.R.S. officials also learned a clerk habitually threw taxpayers' forms and checks into a wastepaper basket.

The debacle prompted Congress, the General Accounting Office and the I.R.S. itself to investigate.

Now officials are waiting to see whether the Philadelphia center can sustain its improved pace all year. If it does, Mr. Cloonan said, the improvement should be credited to a number of moves.

In one of them, the center's Sperry Univac computer system, which was new last year, had its programs tested and refined.

"It has run predictably and well since last spring, and it continues to run well," Mr. Cloonan said. The computer capacity has also been increased by 50 percent with the addition of two more central processing units, making a total of six.

Last year, he said, the computer system was scheduled to operate 65 hours a week. This year it is scheduled for 85 hours. Moreover, according to Mr. Cloonan, the system was out of order and not working for half its scheduled working time during much of 1985, causing serious bottlenecks. This year the system has been working properly for 95 percent of its scheduled time, he said.

An information-retrieval unit that had been particularly troublesome is now operating well, he added.

Some of the Improvements

Other improvements cited by Mr. Cloonan include these:

Expanded training, retraining and experience for employees dealing with the computer system. "We now have 13 or 14 months under our belt," he said. "We anticipate problems and jump on them faster. We've done a lot of rebuilding of the work force."

A 12 percent increase in the center's staff, to about 1,500 permanent employees and 2,000 seasonal ones, for a total of 3,500 at the peak of the tax season in April and May.

The addition of more data-base managers.

A new emphasis on accuracy as well as quantity, in rewarding the center's workers and incentive pay.

The introduction of error-checkers. These people review the data that other employees type into the computer directly from tax returns. In this way, it is hoped any error-prone employees can be identified.

Although Mr. Cloonan did not make a point of it, a new corps of senior managers has also been assigned to the center.

A Big Proportion of Woes

This year the Philadelphia center is expected to receive and process 8.3 million returns from individuals and 5.7 million returns from businesses in Pennsylvania, Delaware, Maryland and the District of Columbia. This center also handles all returns from Americans living abroad.

A study of the center's 1985 deficiencies by the General Accounting Office found that a disproportionate share of last year's problems—there were some problems at all 10 I.R.S. processing centers—occurred here.

The Philadelphia center handles about 8 percent of all individual tax returns in the country, but it was responsible for 23 percent of delayed refunds last year, or 523,137 in all.

It was also responsible for 30 percent of the nationwide total of almost $50 million in interest paid on delayed refunds by the end of 1985. No other center came close to Philadelphia's $15.5 million of interest paid. (The center still has not issued refunds on 664 of the returns, that were filed by last April 15—the end of last year's enormous backlog.)

The center's main job is to process returns, but that task often requires correspondence or telephone discussions with taxpayers whenever problems with the returns arise.

When Mail Trucks Arrive

Processing begins soon after mail trucks unload envelopes full of returns. A sorting machine then slices open the envelopes, scans the bar codes on the outside and deposits them in trays according to broad categories. The sorting machine also "looks" inside the envelope to see if a check is enclosed.

On the next floor, in a space the size of six football fields, the returns go to employees seated at special tables. Each table contains 15 "In" boxes arranged in semicircular tiers.

These "Tingle tables" (so-called after James Tingle, the I.R.S. employee who designed them) enable returns to be classified by type of return or document and by whether or not a check is enclosed. A mechanism on each table strips the envelopes of their contents.

For purposes of control, the returns are next gathered into blocks of 100 documents. Clerks stamp a locator number on each document and each remittance check. The thump-thump of their stamps is heard above other sounds in the room.

Then the 100-document blocks are grouped into batches of 2,000, placed on cars and taken to the next section, where employees examine each return to make sure that it is legible, properly filled out, signed and accompanied by wage statements.

Computer Stage is Next

The next operation—and the biggest involving tasks performed by employees—is entering the data from the returns into the center's computer system. For this purpose, 336 computer terminals are arranged in rows of 14 by 24. Many of last year's errors occurred at this stage, according to Mr. Cloonan, when people made typing mistakes.

The computer checks the accuracy of all calculations. Employees review any returns with arithmetic errors, correct them and re-enter them in the computer system.

With that, the job of the return itself is done. It is kept in the center's files for up to six months, then turned over to the Federal Records Centers. Returns are held for usually seven years and then destroyed.

Magnetic tapes containing the information gathered from returns are sent to the National Computer Center in Martinsburg, W. Va., for posting in a master file. From there, the details necessary for mailing refunds to taxpayers are sent to the Government check-disbursing centers in various parts of the country.

When the checks are mailed out, the annual tax cycle is completed.

Source: W. K. Stevens, "Less Chaos in Philadelphia," *The New York Times*, Feb. 20, 1986, pp. D1, D6.

CASE STUDY

IBM COMBINES RAPID MODELING TECHNIQUE AND SIMULATION TO DESIGN PCB FACTORY-OF-THE-FUTURE

Evelyn Brown
Departments Editor

New factories are reaching ever increasing levels of sophistication, and as their levels ascend, so must the level of the tools used in the design process.

Using simulation models that are powerful, flexible and quick to develop and analyze are still years away. However, a methodology combining a rapid modeling technique and a general purpose simulation model is bringing that time closer. The rapid modeling technique enables one to substantially reduce the number of design alternatives in the analysis stage. Selected alternatives can then be analyzed on the simulation model to use analysis time more effectively.

IBM decided to design a factory of the future to manufacture printed circuit boards, which are used in almost all the company's products from mainframes to PCs. The factory producing the circuit boards had to be flexible enough to produce the various types needed and adapt to produce changeovers quickly.

The company wanted to design the factory using its continuous flow manufacturing philosophy which is similar to the just-in-time manufacturing concept. Both share the goal of achieving a lot size of one unit. In this environment, automated material handling systems would move work between the process stations (which would use state-of-the-art process technology), minimizing work-in-process levels.

The high degree of computer integration necessary for this to work plus the expense of the automated equipment led management to bring in an analysis team early on.

The design team was faced with making the following decisions:

- Number and types of equipment required.
- Number and types of material handling and storage devices needed.
- Desired equipment layout.
- Buffer requirements.
- Desired part volumes and mixes.
- Alternative material routings.
- Operating policy alternatives and their implications.

The three-person analysis team from IBM (IIE senior member S. Wali Haider and IIE member David G. Noller worked with Thomas B. Robey, who was the interface person for the design team) had as their objective developing information to evaluate alternatives at each stage of the design. Their objectives in order of increasing information requirements were to determine:

- Initial assessment of the equipment's reliability requirements.
- Preliminary line balancing.
- Appropriate lot sizes.
- Size of a storage device under different material flow requirements.

- Choices for the transportation of materials for each sector. The choices included conveyors, monorails, material handling crews and AGVS.
- The impact of processing equipment and material handling reliabilities on workstation buffers, product cycle times and total production throughput.

Based on the analysis results, capital and floor space requirements were determined for each alternative.

The analysis team would be designing a simulation model, but analysts are often frustrated, according to Haider, because of the time it takes to develop and use simulation models. At early stages, the design team makes rapid changes in the overall design concept. It is difficult, using simulation, for the analysis team to keep up. And as they are trying to keep up, they can only generate a few alternatives for the design team to consider.

"Quite often, looking at the limited number of alternatives at high levels, you may have from the very beginning missed an alternative that could have been right on. You'll never know," he said, explaining, "and if you look at two bad alternatives and decide that one's better than the other one, you really haven't gained a whole lot."

The analysis team looked for a tool that would be able to meet the changing demands of the design team at a high level, quickly enough to look at a large number of alternatives. Once the alternatives were narrowed down, they would use a simulation package to perform a more detailed analysis.

For the initial analysis the following features were desired:

- All technical details masked from the user.
- The input and output schemes with easily understandable manufacturing terms.
- Rapid development and modifications.
- Model execution times under one minute.
- The capability to model multiple products, equipment breakdowns, process set-up, lot sizing, equipment yield and feedback flow loops.

Manuplan, a queueing network analysis tool from Network Dynamics Inc., was chosen because it met the team's specifications. They used the mainframe version on an IBM 43xx Series machine. A version that runs on the IBM PC AT with a Lotus 1–2–3 interface was not available then.

Manuplan uses node decomposition and approximation techniques which are based on mean value analysis. It also incorporates reliability modeling, and it was designed to model manufacturing environments.

The analysis team also considered using a special purpose manufacturing simulator. "They certainly cut down the time of model development because they incorporate features that allow you to put together a model much more rapidly," Haider explained. "So that's an advantage. The problem is that they are still slow in execution or in running. Especially if they are PC-based. A simple model may even take a half hour to 45 minutes or even longer for a single run."

Also, with the special purpose manufacturing simulator, there are still significant statistical issues facing the correct interpretation of the output data.

"That to us, was still a major negative, whereas in the queueing network approach, none of the simulation output analysis issues have to be addressed. For example, it gives you steady state values, which means the issue of how long to run a model or how many times to replicate the runs is not relevant," Haider said.

The analysis team also liked the speed of Manuplan. "When we were running on the mainframe, we were getting our output in less than a minute," Haider said. "There is an advantage to keeping your thought process going while you're generating alternatives and evaluating those alternatives.

"Your thought process is not broken," Haider explained, "in the sense that you run a scenario, look at the output, see what you don't like, and automatically say, 'Okay, let's try this alternative.' " If a simulation model takes 45 minutes to run, then it takes you time to get back into what you were doing.

The Simulation Model

For the detailed analysis, the analysis team wanted a program which would allow the model to be flexible, powerful and simple to use because the design team was interested in working with the models. Complexities such as material flow and resource assignment logic and entry of data using the language syntax, were to be masked from the user.

The simulation model needed to accommodate:

- Data entry using a template fill-in scheme.
- Multiple products.
- Several types of material handling devices.
- Different types of process equipment (batch, flow-through, etc.)
- Output provided in clear, uncluttered language, using terminology familiar to the user.

The team chose to use SLAM II simulation language from Pritsker and Associates, Inc. Other general purpose simulation languages could have been used, Haider explained, like Siman from Systems Modeling Corp. or GPSS from IBM or Wolverine Software Corp., but the analysis team was familiar with SLAM II and had expertise in Fortran, the language SLAM II uses.

The analysis team carefully developed the SLAM II simulation model so that it was flexible enough for the design team to modify it themselves. They spent the time because they wanted the design team to be able to make a variety of changes without writing a line of code.

"Typically, in most general purpose simulations you don't do that," Haider explained. "You almost have to go and mess with the model itself to make any changes, but we knew we couldn't do that because the design team wasn't intimately involved with the modeling activities or the SLAM II package itself. So we paid the price at the beginning in terms of designing and developing the model, but it paid off later on because the design team could still make reasonably significant changes to the model on their own."

Simulation Model Structure

There are three components to the custom-designed model: a preprocessor, model and postprocessor.

The preprocessor takes the input data, defined by the design team using a template with labels that make sense to them. This is passed on to the simulation model and creates all of the logic within the model. The postprocessor takes the output data from the simulation model and then translates it into a format that makes sense to a non-modeler. Instead of using Device 1, Device 2, etc., the preprocessor and postprocessor use the names of the material handling devices and equipment.

"They were able to take the model and do pretty much what they wanted to do at the detail level, which was quite gratifying for us," Haider explained.

The Analysis

In the initial analysis stage, the choices for the number and types of equipment, lot sizes and storage requirements were narrowed down using the queueing network analysis (Manuplan). Other decisions, like the impact of constraining the buffer sizes of equipment and the selection of the appropriate material handling configuration, required further investigation on the simulation model (SLAM II). Once the material handling logistics and buffer constraints issues were determined, the conclusions from the initial analysis stage needed to be verified on the simulation model.

Case Studies

The following two examples illustrate how the modeling techniques were used in the design of the printed circuit board facility.

Initial Analysis

The printed circuit board factory was to operate using IBM's continuous flow manufacturing philosophy. Since several of the pieces of equipment in the process require substantial set-up time for each product change, a lot size of one board was not realistic. The analysis team was to:

1. Determine whether the attainment of unit lot size was feasible given the constraints of set-up times.
2. If the unit lot size was not attainable, identify the optimal lot size in terms of equipment utilization and work-in-process (W-I-P).

Identifying the probable lot size early would help in determining the design of the material handling system. Then, the development of the simulation model and the design of the material handling system could proceed simultaneously.

A large number of runs were made on the queueing network model to identify the effect of lot size on system W-I-P and process equipment processes. These runs showed that a lot size of one panel was not attainable without adding much more equipment.

The queueing network model established that to minimize W-I-P, the optimal lot size is approximately 20 boards (See Figure 1). Larger lot sizes increase W-I-P and smaller ones require too much set-up.

Establishing unit size of 20 boards per tote provided valuable insight to the material handling control logic. Haider said this initial information decreased the simulation development time "at least three to five fold."

Using the queueing network analysis done earlier, Haider said "you can tailor your simulation model much better this way and decide where to add the details, where not to, or even, when not to model certain segments of the line." If one segment of the line were to be used only 20% of the time, it might not be necessary to include this in the model, he explained.

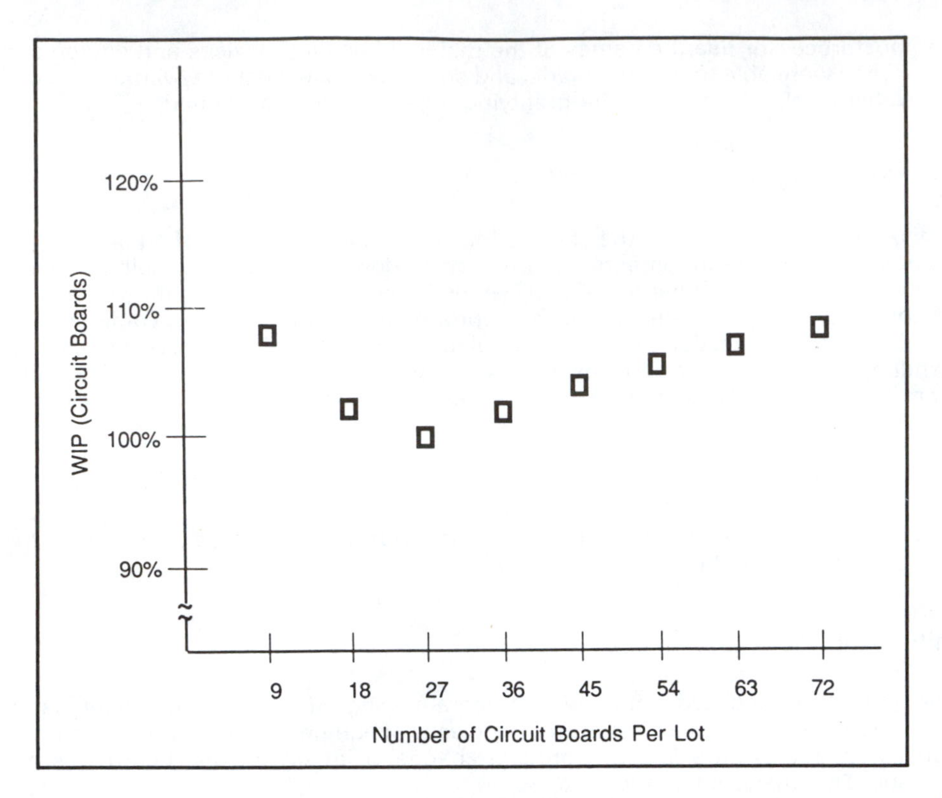

Figure 1 Determining Lot Size with Queueing Network Model

"We did not realize that this would be one of the by-product benefits of going with this approach," Haider noted.

Detailed Analysis

The queueing network model used in the initial analysis was not capable of modeling constrained buffers. The implications of equipment integration were studied with the simulation model. Using the same data set as the queueing network model, the simulation model was used to determine the minimum buffer sizes required between adjacent pieces of equipment.

The simulation showed that the buffer constraints tended to increase the use of the process equipment. This was attributed to blocking effects due to equipment and material handling device breakdown and operator unavailability. Figure 2 illustrates the effect of equipment utilization for a segment of the manufacturing process. Despite having constrained buffers, equipment utilization was maintained below the desired target of 80%.

Verification

Model verification is an important part of the analysis. Another unexpected by-product of this two-phased approach was the ability of the high-level analysis to verify the simulation results. "Results from the high level model provide you with some of the benchmarks to look at in your simulation model," Haider found.

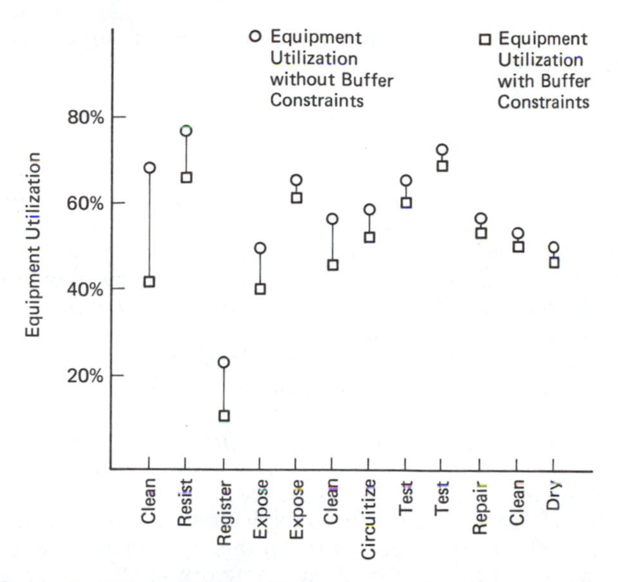

Figure 2 Effect of Buffer Constraints on Equipment Utilization

The scenario tested was a limiting case. Since the queueing network model cannot accommodate constrained buffers, both models were run with infinite buffers between processes. Through a comparison of the two, some errors in the simulation model were identified and corrected, and the two models were brought into close agreement. Figure 3 shows the results from the different analysis methods for equipment utilization.

The two models generally showed agreement within 2%. The utilization values from the simulation model are generally higher than the values from the queueing network model. Haider attributes this to the fact that the queueing network model did

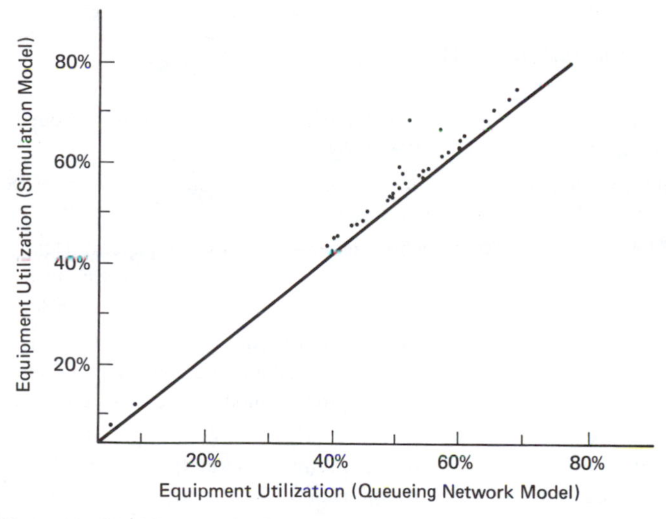

Figure 3 Verification of Simulation Model Using Queueing Network Model

not capture any blocking effect on the equipment due to operator and material handling device/crew unavailability.

Conclusion

Haider feels that the queueing network and simulation models complement each other and provide an effective analysis methodology for manufacturing system design. The queueing network model saved the analysis team time by decreasing the complexity of the general purpose simulation model and by allowing the analysis team to make certain decisions at an early stage. Additionally, it provided an effective means of verifying the simulation model.

Karl Steigele, a manufacturing engineer at IBM, was a key member of the design team and is now involved as a part of the planning team for the printed circuit board factory.

He said that the plans derived from this two-pronged approach "gave us numbers that satisfy us," and added that in comparison to those developed for other sites, the numbers this team developed were better. "We came up with the capacity of one optimized line initially, and we were able to replicate that," he explained. "This gave us an advantage over the other sites not using this approach because of the sensitivity." Steigele also noted that the queueing network analysis was very helpful in determining capital-to-output ratio and optimum line size and in setting strategic goals for lot sizes.

This two-phased analysis technique is currently in use at five or six IBM manufacturing sites, according to Haider.

Source: E. Brown, "IBM Combines Rapid Modeling Technique and Simulation to Design PCB Factory-of-the-Future." Reprinted from *Industrial Engineering*, June 1988, 90, pp. 23–26. Copyright Institute of Industrial Engineers, 25 Technology Park/Atlanta, GA 30092.

CASE STUDY QUESTIONS

Cautious Confidence at IRS

1. Diagram the Philadelphia forms processing center as a multiserver queue. Which server (or servers) was the bottleneck in 1985?
2. What actions have been taken to increase capacity during periods of peak demand? What actions should be taken to increase capacity?
3. The article refers to a clerk who "habitually threw taxpayers' forms and checks into a wastepaper basket." What features of the system might cause this to happen? What actions have been taken to improve accuracy?
4. Consider this change in operations: Instead of removing checks as a first stage in processing, remove checks at the time that data are entered into the computer. In this way, the amount on the check can be verified against the proper amount owed. Would this be a better way to operate?
5. The article does not mention a queue discipline. Would it be a good idea to use a discipline other than FCFS? If so, what discipline should be used?

6. The form processing system has been automated through use of computer terminals for entering data. What advantages and disadvantages have come from automation? How has automation affected the queueing system design?
7. Is minimizing total time in system a proper objective for the IRS? Should other concerns take priority?

IBM Combines Rapid Modeling Technique and Simulation

1. IBM's objective was to "design the factory using its continuous flow manufacturing philosophy." Is IBM's goal of reducing lot size to one unit sound and, if so, what should IBM do to achieve this goal? Judging by the analysis, was IBM successful?
2. Why is it that a circuit board factory can be modeled as a queueing system? What are the servers and what are the customers? What is the interpretation of work-in-process?
3. Haider is quoted as saying, "If you look at two bad alternatives and decide that one's better than the other one, you really haven't gained a whole lot." Why does Haider believe this is a problem with simulation?
4. What are the advantages and disadvantages of the "queueing network analysis tool" (Manuplan) relative to simulation?
5. IBM's methodology was to use Manuplan to narrow down options and use simulation to analyze options in detail. How might linear programming or lot-size models be used in place of Manuplan?
6. How do buffer sizes enter into the analysis? Why is buffer size an important consideration in system design?

chapter 11

Queue Design

To conceive, to invent, to form a plan—this is the role of design. It is one thing to analyze a queueing system in the abstract; it is quite another to create the design that transforms the results into physical reality. Design is the concern of this chapter, a concern that is manifest in the following ways:

1. Creating a pleasant waiting environment
2. Implementing effective and appropriate queue disciplines
3. Planning a queueing layout that promotes ease of movement and avoids crowding
4. Locating servers so that they are convenient to customers, while minimizing waiting
5. Providing sufficient space to accommodate ordinary queue sizes

Unlike in previous chapters, most of the ideas presented here are qualitative. Though much is known about the features of good and bad queue design, it is often impossible, or undesirable, to quantify these features. Queueing and waiting are at once personal and emotional. For even if the service is being performed on an object, as in manufacturing, there is invariably some human waiting for the service to be completed. Here the human aspects of waiting are emphasized. With this in mind, the term *customer* will be used in the conventional sense, that of designating the person waiting for the service to be completed. When appropriate, the customer will be distinguished from the job receiving the service.

What emotions does queueing elicit? In a poorly designed system, the emotions may well be boredom, anxiety, confusion, or even anger. Are customers clock-watching? Are they pacing the floor? Are their facial muscles tense, and are they speaking in elevated voices? In a well-designed system, customers should be productive, relaxed, and even happy. Are customers occupied while waiting? Are they sitting or standing still in a relaxed posture? Are they smiling or quietly conversing with other customers? These qualitative signs are just as important to look for as the quantitative signs—queue length, waiting time, and so on. And they are at least as important in assessing the effectiveness of a queueing system.

Divided into five sections, corresponding to the five issues cited above, the chapter begins with the most qualitative material and ends with the most quantitative. For this reason, the beginning sections can be read with full comprehension without first having read Chaps. 1 through 10.

11.1 CREATING A PLEASANT WAITING ENVIRONMENT

A premise of this book is that it is the server's responsibility to find ways to reduce or eliminate customer waiting. A customer's time is valuable, usually at least as valuable as the server's, and should not be taken unless essential to achieving service efficiency. Nevertheless, some waiting is unavoidable. But all waiting time is not the same. Few customers would not prefer spending ten minutes seated, in a quiet, air-conditioned waiting room to even one minute in a sweltering, noisy, smoky lobby. The server's responsibility goes beyond minimizing the wait; the server's responsibility is also to make the wait as pleasant as possible, however short it may be.

11.1.1 Psychological Aspects

With a bit of creativity, the wait can be enjoyable, not unpleasant. Here is a piece of queueing folklore:

> The owner of a large office building had a dilemma. Ever since her building was fully occupied, she had received frequent, and vociferous, complaints about the waiting times for the elevators. She looked into adding another elevator, but the high cost was out of the question. So she hired a pricey consultant. Two weeks later, the consultant came back with the answer: install mirrors by the elevator doors. Sure enough, the solution worked. People had to wait just as long, but no longer did anyone complain.

What is it about the environment that makes waiting pleasant, uncomfortable, or downright annoying? First, there is the psychological side of waiting. One of the major causes of stress in everyday life is lack of control (Bateson 1985). Because queueing takes away people's control, it is an inherently stressful experience. Giving back customer control, and giving customers something to do while waiting, will minimize this stress. For example, Maister (1985) suggested that entertainment, information, and equity can affect the customer's perception of time. That is, waits will seem shorter if customers are entertained and well informed, and if the queue discipline is fair. Further, Larson (1987) states that ''social justice'' and eliminating ''empty time'' are keys to making waits tolerable. Following are some ideas for achieving these ends. (For a more complete discussion of the psychological aspects of time perception, consult Fraisse, 1984.)

Entertainment. Waiting time will not be wasted time if the customer has something to do. Either the server can entertain the customer or the server can create an environment where the customer can entertain him- or herself.

If waits are long (more than a few minutes), the waiting environment should encourage customers to sit down and read, write, or otherwise relax. Reading materials, such as magazines and newspapers, should be provided and perhaps food and beverages too. Customers will be especially appreciative if these are available free of charge, a just compensation for their inconvenience.

For shorter waits, customers might be entertained in a variety of ways. In San Francisco, the cable car terminal at Ghirardelli Square is famous for attracting musicians, jugglers, sword swallowers—you name it—whose shows provide welcome relief to their captive audiences. Though something of a mixed blessing, the music broadcast while waiting on telephone lines can also make the wait more pleasant. The calming influence of music might also remove some of the tension created in waiting. Even a well-decorated waiting room with unusual artwork, aquariums, or informative displays can provide entertainment.

Alternatively, or additionally, an environment might be created to encourage customers to wait in groups, with acquaintances or possibly even with strangers. As Maister states, a solo wait feels longer than a group wait, a fact that is especially true when the service will be unpleasant (as at a doctor or dentist). Some or all of the seating in a waiting room might be arranged to promote conversation rather than privacy. Deasy and Lasswell (1985) suggest flexible seating, which the customer can move to suit individual preferences, or if this is not possible, seating that allows people to sit at 90 degree angles relative to each other. They also state that ''lighting should be arranged to illuminate the faces of people'' and ''the color of light should be appropriate, such that people's flesh tones are rendered correctly.'' In addition, the server might even take the effort to introduce customers to each other. (Students have sometimes commented that the best time to meet other students is while waiting in line to talk to their professor.)

Better yet, for long waits, the customer should have the freedom to drop off work and return later. Or perhaps the service can be performed over the phone, in which case the customer has the benefit of waiting in his or her own home or office.

Information. Customers should be reminded when the system is working properly and told when it is not. But providing good information begins at time of arrival, when the customer should be given a reasonable estimate of his or her waiting time. Reasonable does not mean accurate, for it is usually best to overestimate waits (and certainly never to underestimate waits). Using restaurants as an example, Sasser et al. (1979, p. 89) write: ''If people are willing to agree to wait this length of time, they are quite pleased to be seated earlier, thus starting the meal with a more positive feeling.'' On the other hand, it is annoying to the customer when the wait turns out to be longer than predicted. This can only aggravate the customer's sense that he or she lacks control.

Waiting information can be conveyed orally. A receptionist might tell each customer how long his or her wait will be, or perhaps announcements can be broadcast over a public address system. Information can be presented in writing: signs might be placed at various positions of the queue indicating how long the wait will be from the point (a practice used extensively at Disneyland). And waiting information can be presented nonverbally. The customer's own observations of the queue length and queue progression provide an indication of the wait. Unfortunately, nonverbal information sometimes gives customers the wrong message. A queue of a given length may translate into either a short or a long wait, depending on the service rate. If the service rate is very fast, customers may overestimate the wait. It would then be desirable to make the queue appear shorter than it really is, perhaps by snaking the queue back and forth rather than stretching it out in one long line. When possible, the service process should also be designed to promote a smooth even flow, so as to provide reassurance that the system is working and reduce the uncertainty in the wait. This works against bulk service. Unless customers are aware of why service repeatedly stops and starts, anxiety will be created.

When a problem does occur, customers should be told what is happening. If a machine breaks down, they should be told why service is interrupted and how long it will take to fix the problem. If a plane will be delayed leaving an airport, passengers should be given frequent and realistic updates as to when the plane will depart. Servers should refuse the temptation to be overly optimistic, for little is gained in the long run through deception. As Maister states, "Waiting in ignorance creates a feeling of powerlessness."

Customers should also be reminded that they have not been forgotten. On telephone lines, statements that calls are served in the order received can help make long waits tolerable. In almost all cases, it is worthwhile for the server to acknowledge the customer's presence soon after arrival, perhaps through use of a receptionist. The time spent waiting for the first human contact is usually the most painful. Maister argues that servers should try to "convey the sense that service has started" soon after the customer arrives, perhaps by collecting some information from the customer, or otherwise beginning a portion of the service.

Equity. A queue's discipline should be equitable, both in the eyes of the server and in the eyes of the customer. The most widely accepted queue discipline is FCFS (first-come, first-served). This is how people naturally queue up in most countries, so serving customers by any other discipline means breaking the norm. This holds true even when the system might perform better under other disciplines (see Chap. 9).

Larson (1987) explains inequitable disciplines in terms of "slips" and "skips." A slip is when another customer, arriving later, slips ahead of you, and a skip is when the server passes over you to serve someone farther back in the queue. As an example, Larson describes an airline that changed the landing gate for one of its flights in order to *increase* the walking time to the baggage carousel. With this change, customers arrived at the carousel nearer to the time that bags arrived, which also meant that all passengers were served at nearly the same time. This eliminated the *perception* of slow service and the aggravation of watching other passengers depart earlier (an example of a skip). Despite the fact that average time in system increased, the problem of customer complaints was solved.

More generally, to reduce average waiting time, it is often better to serve customers according to the SST (shortest service time) discipline rather than FCFS. Yet, despite the attractive properties of SST, customers with long service times may not accept being served after later arrivals (after all, they arrived first). Even harder to justify in egalitarian societies is giving preferential treatment to favorite customers—allowing wealthy patrons to bypass a line, for instance. If a business hopes to keep its regular customers, it better keep such special treatment well out of sight. As a matter of principle, such preferential service ought to be available to all customers and should be attained through paying an extra price, not special favors. Customers should have the option of whether to pay or to wait.

If the discipline is FCFS, customers should be reminded that they are being served accordingly. People inherently view an orderly FCFS queue as fair. If the discipline is not FCFS, a good policy is to serve high priority customers and low priority customers in separate queues. In supermarkets, for instance, it is better to have separate express lanes than to constantly rearrange the queueing sequence as higher priority customers arrive (at least, if one wishes to avoid a riot). In cases where different priority customers are mixed, as in a hospital emergency room, the queue discipline should be clearly stated and justified. Further, customers should be discouraged from physically waiting in a line and encouraged to sit down. The purpose here is to inhibit people's natural tendency to follow the FCFS discipline.

Equity is less of an issue when customers drop off work. Unless customers actually witness differential treatment, it is not likely to be a concern. It may then be easier to invoke a high-performance non-FCFS discipline.

11.1.2 Physiological Aspects

Entertaining the customer, providing information, and ensuring equity all help put the customer in the right state of mind while waiting. But what about the physical environment? This too can make the wait pleasant or annoying. There is no reason why the physical environment should not be just as pleasant as a customer's own home. The waiting area should have an attractive architectural design and be well decorated. Further, the waiting area should be well designed from the standpoints of noise, illumination, ventilation and climate, and crowding.

Noise. A noisy environment, especially an environment punctuated by intermittent sounds, high frequency sounds, or very low frequency sounds, is not only annoying but can have damaging health consequences. Certainly no waiting area should be so loud that the customer risks hearing loss. But what are the consequences of intermediate noise levels? Studies have shown that noise both reduces helpfulness and increases aggressive behavior (see Jones and Broadbent 1987). Sudden bursts of noise can also induce a startle response, causing facial muscle contraction and head jerk. And voices make it difficult for people to read or work—even if the voices are in a foreign language (other types of noises have to be very loud before reading is affected). Studies have also shown that a person's feeling of control over a noise influences the degree of annoyance. Because customers usually do not wait out of choice, it is especially important to minimize disturbing noises in a queueing environment.

Noise levels can be controlled by modifying or eliminating the noise source or by dampening the sound. Public address systems, machinery, ventilation equipment, and even loud employees are all potential sources of excessive noise. For instance, one remedy to excess noise is to have servers walk over to customers to call them to service rather than have them shout out names. It might also be less distracting to readers if verbal messages are replaced by simple sounds, such as bells, even if the noise level is the same. Alternatively, visual displays might be used (see Fig. 11.1). Wall insulation and selective use of floor, ceiling, and wall materials can help dampen loud noises. Concrete floors and brick walls perform especially poorly with respect to sound absorption (a carpeted floor can absorb from 5 to 30 times as much noise as a linoleum on concrete floor, depending on noise frequency; Hill 1973).

Besides being too noisy, an environment might be too quiet. A certain amount of steady noise will mask distracting intermittent sounds. By necessity, most waiting rooms require some talking on the part of servers and customers. Low level background noise—a quiet ventilation system, music, or even a white noise generator—can keep these sounds from irritating customers who are reading or working while waiting. Low level noise can also stimulate customer interaction. When a room is dead quiet, few will venture to say a word, for fear of being heard by everyone in the room. In most waiting environments, 40

Figure 11.1 Lighted display indicates when customers can be served.

to 48 decibels is a good target noise level, approximating a library at the lower end and a medium-sized office at the upper end.

Illumination. Lighting can affect the mood of the customer and affect his or her ability to use waiting time constructively. An area that is well lighted, particular through use of natural light, will be more pleasant than a dark room. And a well-lighted room will make it easier to read.

In a study by Flynn et al. (1973), lighting was found to influence people's sense of spaciousness, relaxation, privacy, pleasantness, and perceptual clarity. These senses were enhanced in the following ways:

Spaciousness: uniform wall lighting

Relaxation: nonuniform lighting with wall emphasis

Privacy: low intensity lighting near user/higher intensity elsewhere

Pleasantness: nonuniform lighting, some wall lighting or high reflectance walls

Perceptual clarity: bright, uniform lighting, some wall or high wall lighting or high reflectance walls

In a waiting environment, it is especially important to enhance the customer's sense of relaxation and pleasantness, which would favor nonuniform wall lighting. However, if the waiting room is small, uniform wall lighting might be used instead to enhance the sense of spaciousness. If customers are able to sit and read while waiting, low intensity table lamps can be installed to give them a sense of privacy, though this might discourage customer interaction.

Other factors to consider are the amount of light provided, or *illuminance*, and glare. The Illuminating Engineering Society (IES 1983) states that for visual tasks of medium contrast, the illuminance should be in the range of 500 to 1000 *lux* (one lux equals one *lumen* per square meter, and one lumen equals 1/683 watts of light of 555 nm wavelength), depending on people's age, task background (for example, paper) reflectance, and the importance of speed and accuracy. Combining these factors, lighting toward the lower end of the range (500 to 600 lux) should be adequate for reading while waiting. IES recommends much lower lighting levels in working spaces where "visual tasks are only occasionally performed"—100 to 200 lux. For cost efficiency, a good policy is to provide greater illumination in areas where people are reading and less illumination elsewhere.

Glare is created by light that is overly bright relative to its background. Lighting should be positioned so that it does not reflect or shine directly into people's eyes, especially while reading. It is also helpful to provide lighting that customers can adjust to suit their preferences.

Ventilation and climate. The customer's comfort depends on temperature, humidity, and air ventilation rates. When customers are sedentary, as is most often the case while waiting, a temperature range of approximately 68° F to 75° F is considered comfortable in the winter and a range from 72° F to 80° F in the summer, the two ranges attributable to seasonal differences in clothing (ASHRAE 1985). The exact comfort range depends on a combination of temperature and humidity. Toward the high end of the summer scale, relative humidities between 20 and 30 percent can be tolerated, while at the low end of the winter scale, humidities between 50 and 80 percent are comfortable. In addition, high temperatures can be compensated somewhat through increased air circulation, perhaps through the use of fans.

In addition to controlling the temperature and humidity, the ventilation system should also provide fresh air from outside sources. This is important both in terms of removing smoke, odors, and fumes and in eliminating the general stuffiness of a room. Because waiting often occurs within confined spaces, with people close together, it is best not to allow smoking at all, or at least make the majority of the waiting area nonsmoking. As always, ambiguity is to be avoided, and these areas should be clearly and prominently marked.

Crowding. Crowding is almost surely the least pleasant aspect of waiting in a queue. Fruin (1971) notes that people waiting in line for buses typically space themselves only 19 to 20 inches apart "with very little variation." This puts people within what Hall (1982, p. 116) calls an "intimate distance," an area where "the presence of the other person is unmistakable and may at times be overwhelming because of the greatly stepped-up sensory inputs. Sight (often distorted), olfaction, heat from the other person's body, sound, smell, and feel of the breath all combine to signal unmistakable involvement with another body." While intimate distance may be desirable in the contexts of comforting, protecting, and lovemaking, most would find 19 to 20 inches too close in a social setting (though distance standards vary among countries).

If such close contact is uncomfortable, why do people wait so close together? The answer is largely to protect their place in the queue. Interpersonal spacing below 2 feet impedes the passage of other people through the line. (The same phenomenon can be witnessed on highways; drivers tailgate to prevent other vehicles from cutting into line.) But in a well-designed queue (see Sec. 11.3), customers should not have to be closely spaced to defend their position.

Fruin (1971) developed standards for judging the level of service of a queue of standing people based on crowding. His classification is as follows

A. Free circulation zone: spacing 4 feet or more, 13 ft^2/person
"free circulation through the queueing area"

B. Restricted circulation zone: spacing 3½–4 feet, 10–13 ft^2/person
"restricted circulation through the queue without disturbing others"

C. Personal comfort zone: spacing 3–3½ feet, 7–10 ft^2/person
"space is provided for standing and restricted circulation through the queueing area by disturbing others"

D. No-touch zone: spacing 2–3 feet, 3–7 ft^2/person
"space is provided for standing without personal contact with others but circulation through the queueing area is severely restricted, and forward movement is only possible as a group"

E. Touch zone: spacing 2 feet or less, 2–3 ft^2/person
"personal contact with others is unavoidable"

F. The body ellipse—2 square feet per person
"space is approximately equivalent to the area of the human body"

At level F, Fruin states that "no movement is possible, and in large crowds potential for panic exists." Though level F does occasionally occur, it should always be avoided, through the provision of buffers separating people.

Fruin (1981) writes: "The combined pressures of massed pedestrians and shock-wave effects that run through crowds at critical density levels produce forces which are impossible for individuals or even small groups of individuals to resist." It is this type of condition that has caused crowd disasters, such as stampedes at the entrance to soccer contests. Fruin sees the only recommended application for level E as elevators, while he sets level C as the standard for "ordered-queue ticket selling areas." This level falls in what Hall calls the far phase of personal distance, a distance at which "a person cannot easily 'get his hands on' someone else."

Unfortunately, people space themselves too closely in queues. This demands action on the part of the server. At a minimum, the queueing area should allow 7 square feet per standing customer, as specified by level C. And if waits are longer than a few minutes, seating should ordinarily be provided, which will require even greater amounts of space. But merely providing space does not necessarily eliminate crowding, for crowding is often spontaneous. Crowding is a sign that the queue discipline is not being adequately enforced. Ideas for resolving this problem are provided in the following section.

A summary of psychological and physiological factors affecting the queueing environment is presented in Table 11.1.

11.2 MAINTAINING QUEUE DISCIPLINE

It is one thing to select a queue discipline and quite another to enforce it. Who has not patiently waited in line, only to see others cut in front, or signed in with a receptionist and subsequently found your name was lost and eight later arrivals served first. Implementing a queue discipline requires an effective means for tracking and selecting customers. This can be accomplished by either passive or active means.

11.2.1 Passive Enforcement

Passive enforcement entails the provision of information, either verbal or nonverbal, which directs the movement of customers to and through queues. This information should allow the queue to "self-organize." That is, servers merely select customers from the front of the queue (or queues).

The information should leave no ambiguity as to what customers are expected to do. Not only will this eliminate confusion, it will also discourage queue jumpers who are merely looking for excuses for cutting in line. Passive enforcement depends on the

TABLE 11.1 WAYS TO IMPROVE THE WAITING ENVIRONMENT

Psychological	*Entertainment*
	Encourage customers to sit and read
	Provide reading material, food, and drink
	Hire performers
	Install aquariums or informative displays
	Encourage customers to wait in groups
	Allow customers to drop off work and return later
	Information
	Give reasonable and frequent waiting time estimates
	Promptly tell customers when there is a problem
	Remind customers that the system is working and that they have not been forgotten
	Equity
	Use FCFS discipline where appropriate
	Do not provide special favors
	Separate high priority from low priority customers
	Do not try to physically rearrange the queue order
Physiological	*Noise*
	Eliminate loud and unpleasant noises
	Use building materials that absorb sound
	Create low level background noise to mask voices
	Illumination
	Provide sufficient light for reading
	Use nonuniform wall lighting
	Remove bright lights or other causes of glare
	Ventilation and Climate
	Control summer temperatures in the 72° F–80° F range and winter temperatures in the 68° F–75° F range
	Circulate air from outside at a fast rate
	Prohibit smoking
	Crowding
	Provide at least 7 to 10 square feet of space per standing person
	Install queue barriers to protect customers' places in the queue

understanding and goodwill of the customers. Hence, it is most effective when the queue discipline is simple, such as FCFS or other head-of-line disciplines. It is also most effective when waits are short, for it usually requires that customers physically wait in a line.

In a head-of-line discipline, each customer is placed at the end of a queue at the time of arrival. Within each queue, customers are processed according to FCFS, but different queues might receive different priority or be processed by different servers. Physically, the general scheme is for customers to enter the queue at its end, move forward through the queue while waiting, and exit the queue from its front. This contrasts with the undesirable LCFS discipline (last-come, first-served), in which the customer can be viewed as exiting and entering the queue from the same end (as in a stack of work orders resting on a desk).

Head-of-line disciplines require that customers be provided two key pieces of information—which queue to join and where the end of the queue is. Customers should also be made aware of the discipline so that they will know that the order of customers in the queue is identical to the service order. Communicating this information usually does not require a great deal of sophistication. Simple, well-placed signs suffice. Nevertheless, some information should always be provided. It is both unfair and unwise to rely entirely on customers to self-regulate the queue discipline. Customers should never have to answer the question "Who's next." Let's look at some common problem areas:

Example 1

> A busy four-lane street has experienced large queues at the intersection with a two-lane road due to insufficient capacity on the two-lane road. Queues have been known to back up a half mile in the right lane, as drivers attempt to make the turn onto the two-lane road. Queues in the left lane, for traffic proceeding straight, have not been significant. However, a problem has occurred in this lane. A large percentage of drivers turning right do not join the end of the queue. Instead, they stay in the left lane until almost reaching the intersection, where they finally switch into the right-hand lane.

Example 2

> A heavily patronized library employs two people to check out books. About half the time, customers form two separate queues behind the two servers, each of which is served FCFS by its own server. The other half of the time, customers form a single line served by both of the servers (again FCFS). However, when this line is short, there is a tendency for new arrivals to form a second line, behind one or the other of the servers.

These two examples illustrate consequences of not providing customers with adequate direction. In the traffic example, drivers who stay in the left lane effectively cut in line in front of those in the right. Yet they are doing nothing illegal or, perhaps, even wrong. They might argue that it is ambiguous why vehicles in the right lane are waiting in a queue (they might be backed up behind a slow truck or the queue might be for a turn preceding the congested intersection). Thus, drivers have a valid excuse for cutting in line. In the library example, there is no societal norm as to whether people should form one queue for all servers or separate queues for each server. This ambiguity leaves the person who tries to create a single queue in an inferior position to the person who arrives later and begins a separate line.

Here is a rather different problem:

Example 3

> To relieve congestion, traffic engineers have widened a road from two lanes to three lanes, before and after a busy intersection. To their surprise, delay has not gone down. Queues have persisted in the original lanes, and drivers are not using the added lane.

Sometimes people are too courteous. In this example, driving in the added lane has the appearance of cutting in line. In fact, driving in the added lane amounts to choosing an underutilized server (each lane is a server at the intersection). Choosing this server not only reduces the driver's delay but helps relieve congestion for other drivers. Drivers need to be informed that this is the proper way to act (a sign should state "through traffic, use all three lanes").

11.2.1.1 Resolving Information Problems. Problems might be resolved through provision of adequate information. Here are some general ideas:

Mark the front position of the queue.

Post the sign "form single line" at the beginning and end of the queue.

Post signs giving the purpose of each queue (for example, signs stating "use right lane for right turns on Greer Road").

Install maps that indicate where one should wait for each type of service.

Create a line on the ground, with arrows, indicating the proper position and direction of the queue.

The information should be positive. It is better to tell customers what they should do than what they should not do, and it is better not to emphasize the penalties for doing something wrong (for example, $100 fine for doing this or that). If customers are not following the proper discipline, it is usually not so much because they are flagrantly violating the rules as it is that the rules are ambiguous or unknown.

What if information is being provided, but customers still do not follow the discipline? Then they should be surveyed to see whether they are aware of the proper discipline and aware of the information. This should help identify trouble spots, such as the following:

Potential Information Problems

Messages are not sufficiently brief to be comprehended.

Messages are illogical, poorly written, or inconsistent.

Sign lettering is not sufficiently large, or there is insufficient contrast between lettering color and background color.

Signs are not adequately illuminated, or glare makes them invisible.

Signs are not posted in prominent locations.

Audio messages are not loud enough to be heard.

Customers cannot read or speak the language in which information is provided.

Verbal messages are not consistent with the architectural design of the waiting area.

Information is provided either too early or too late.

Sometimes it takes several reminders to get the information across, and customers might have to be told when they are doing something wrong. This responsibility should not fall on other customers, but on the servers.

11.2.1.2 Barriers and Queue Markers. Beyond disseminating information, many queue operators believe that queue discipline should be enforced through use of physical barriers, most often ropes. This approach has major strengths and weaknesses. On the negative side, physical barriers may make customers feel that they are being treated like children or, worse yet, like cattle. Barriers can also restrict customers' freedom to move about, and may lead to a feeling of confinement or entrapment. On the positive side, physical barriers are very effective at preventing cutting in line; they remove virtually all ambiguity as to the queue discipline. Less obvious is that barriers can help *relieve* crowding and ease tension. Customers do not feel compelled to stand as close to the person in front merely to protect their position in line. Barriers also allow customers to divert their attention from the queue, without fear that a gap in front will be filled by some interloper. This can make the wait much more relaxing.

Barriers do not have to be overly obtrusive to be effective. They can even be attractive. Planters and trellises will do the job. Barriers also do not have to be overly tall; they merely have to be visible (a planter should not be so low that customers trip over it). Short barriers will reduce the feeling of confinement.

Even less obtrusive is the idea of marking a line on the ground, tracing the position and direction of the queue. The line might simply be painted or perhaps created through the use of colored tiles, bricks, or other materials that contrast with the surrounding surface. Rather than create a line, per se, circles or triangles might be spaced every 2½ to 3 feet to signify a customer's position in the queue (these might even be numbered).

11.2.1.3 Take a Number. On the boundary between active and passive enforcement is the common take-a-number system. With this approach, customers are told to pull a tag from a sequenced roll upon arrival, and servers call out the numbers in the order in which the numbers were taken. Typically, the last number to enter service will also be displayed on a sign. If customers are waiting for a service to be completed, as in the preparation of a prescription at a pharmacy, information might also be displayed on a lighted board, as in Fig. 11.1.

Take-a-number systems are most useful where the customers can use their waiting time more productively when not restricted to standing in a set line. They are especially common in bakeries and delicatessens, where waiting time can be spent inspecting the food and deciding what to order. Besides giving the customer something to do, this acts to reduce the service time, for the customer will be better prepared to place an order once his or her number is called. Though less common, take a number is also effective in camera, jewelry, and other retail stores where merchandise is kept behind a glass counter; waiting time can be spent inspecting the merchandise. All too often, unfortunately, the queue discipline followed by these stores is haphazard—customers are served at random and no one is really sure who arrived first.

Unless numbers are computer generated, they will have to be distributed at a single location by a single machine. Care should be taken to ensure that the machine is in a prominent location, that customers see it immediately upon entering the service area, and that customers know when it is in force. A variant of take a number that is used by the Kaiser Health Plan is to have customers drop their health cards in a box, from which a nurse selects customers in the FCFS order (pulling from a slit at the bottom of the box). This is more personal than take a number, for the nurse can call customers by name rather than by number. It also does not require the customer to remember an unfamiliar number.

As with other passive systems, take a number can be used only with FCFS or other head-of-line disciplines. More complicated disciplines demand active enforcement, as discussed in Sec. 11.2.2.

11.2.1.4 Stacking Systems. A frequent problem in processing work orders or objects is that items are stacked vertically. The natural order for removing items from a vertical stack is the undesirable LCFS—the first item to come off the stack will be the last item put on. Figure 11.2 shows an alternative. The device, called a flow-through bin,

Figure 11.2 Flow-through bin operates FCFS.
Source: Courtesy of Kinston-Warren, Newfields, New Hampshire.

operates by gravity. Items are fed from the back and flow through to the front. A second alternative (not shown) is to stack items horizontally on a conveyor or rollers and allow the conveyor to push items from the back of the queue to the front. In either case, the work enters the queue from one end and leaves from the other; unlike a vertical stack, it does not enter and leave from the same end.

11.2.2 Active Enforcement

Active enforcement directly involves the server in the selection and sequencing of customers, beyond merely taking the next customer in line. Virtually all queueing systems need active enforcement from time to time, such as when a new server is opened, and the first customer in an existing queue is allowed to be served first. Active enforcement is essential for complicated queue disciplines, especially dynamic disciplines. Active enforcement also frees customers from having physically to stay in line. They can sit down and wait for their names to be called. Or, in some cases, work can be dropped off and picked up later.

Active enforcement requires that the server maintain a data file of all the customers waiting in queue, either on a sheet of paper or in a computer, or possibly in the server's head (the latter, being the most prone to error, is not advised except for queues of two or three customers). Each time a customer arrives, a record of the customer is entered in the queue data file. Each time a service is completed, the file is scanned and the customer with highest priority is selected to enter service next. If interruption is allowed, the queue file might also be scanned when a customer arrives, at periodic intervals or even continuously.

The two common ways to store the customer information are in list format and card format.

11.2.2.1 Waiting Lists. With the ***list format***, the server maintains a list of all customers waiting in the queue. The list should contain all information pertinent to the queue discipline. At a minimum, this will include the arrival time and customer name, and it may also include information on job size, expected service time, due date, and priority. Whenever a service is completed, the server scans the list and determines each customer's current priority. The customer with the highest priority is selected next.

Provided that a strict regimen is followed in entering customer data, computerized systems are not likely to make errors in selecting customers. Bar code systems are particularly effective in this regard. Not only do they offer speed, but they can help eliminate keying mistakes.

When the waiting list is manually kept, complicated queue disciplines should be avoided, and extra care should be taken to ensure that customers are not forgotten. For instance, in restaurants, the order in which customers are served partly depends on the size of the dinner party, not just on arrival time. Depending on what size tables are available, certain dinner parties might be skipped over. Unfortunately, once a party is skipped, it is more likely to be overlooked. Another potential problem is that a customer might not be available when his or her name is called (perhaps the customer is in the rest room). These customers too might be forgotten.

There is really no magic to maintaining good records. Waiting lists should be well organized and legible. To this end, the server, not the customer, should write the customer's name on a lined form (not a blank piece of paper), and the server should verify the pronunciation. When customers enter service, their names should not be crossed off, which would destroy the record. Instead, the time should be recorded (this may provide valuable information for analysis). As a double check, the time that each customer completes service might also be recorded. Finally, waiting customers should be surveyed periodically to see who might have reneged and to determine whether someone was overlooked.

A drawback of all forms of active enforcement is that many of the nonverbal information cues are removed. Customers can see how many people are sitting and standing in the waiting area, but cannot ascertain their relative position in the queue. Customers also cannot tell whether higher priority customers might have temporarily left the waiting area, only to return later. The absence of nonverbal information puts greater demands on the server to provide feedback. Ideally, the current waiting list should be prominently displayed on a CRT, a large sheet of paper, or a large chalkboard, perhaps along with waiting time estimates. The waiting list should not be kept secret, and customers should not have to ask to see it. Failing this, the server should individually update customers as to their status at periodic intervals.

11.2.2.2 Cards. The ***card format*** is a method of manual record keeping in which customer information is kept on individual cards rather than a central list. The advantage of cards is that it allows the server to rearrange the customer sequence without rewriting any of the data. This can be useful when customer priorities change over time, in which case the cards can be moved to new positions reflecting current priorities. To select a customer for service, the server need only pick the first card from the stack; the entire stack does not have to be scanned. Unfortunately, the ancillary risk is that cards might be lost or misfiled. Thus extra care is required.

A typical implementation, common in manufacturing, is to attach a card to the customer, whether the customer be a person or piece of work. When the work arrives at the server, the card is detached, a notation is made of the arrival time, and the card is entered in a queue file from which work is selected for service. Once service is completed, the card is removed from the queue file, a notation is made of the time service is completed, and the card is reattached to the work. Often, the card will contain routing information directing the work to another server, where the card may be detached a second time and entered into a new queue file, after which the whole process might repeat. When the work finally emerges from the system, the card can be detached and filed with historical records. The approach is attractive from the position that the information travels with the work, from station to station. The system is also simple, effective, and widely used. It is fundamental to the Toyota Kanban system and the common bin-card reorder system.

11.3 QUEUE LAYOUT

The layout of a queueing system can have a profound effect on crowding, congestion, and server efficiency. A good layout will promote smooth flow of customers and noncustomers through the waiting area. And it will make servers readily accessible and visible to customers.

The following discussion of queue layout presents a variety of alternative queue designs, which can be applied to two major types of service systems:

Definitions 11.1

Turn-back service Service occurs at a counter, with the customer on one side and the server on the other. Upon completion, the customer exits by turning back, away from the counter (Fig. 11.3a).

Flow-through system Service is completed at a station, with the customer adjacent to the server. Upon completion of service, the customer proceeds forward, away from the queue (Fig. 11.3b).

Turn-back systems are used when servers need to access shared and/or large pieces of equipment or files, as in a bank or post office. Flow-through systems are used when the server only requires a small station, as at a supermarket checkstand. Flow-through is also used when it is difficult for the "customer" to turn around, as when vehicles are served (in a gas station, for example).

In this section, queue layouts are further classified according to whether a single queue feeds all servers or whether separate queues feed each server. Special-purpose

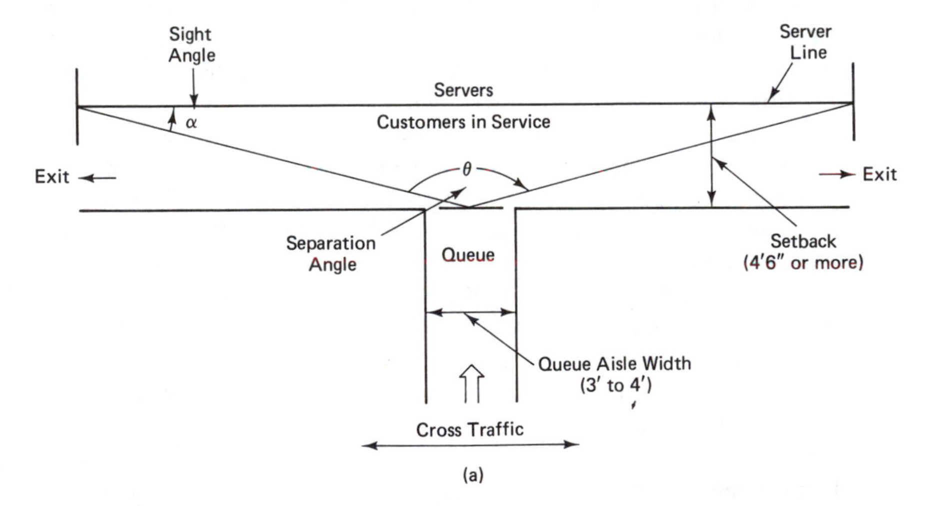

Figure 11.3a Central queue position shortens customer to server distance, but increases separation angle. Perpendicular queue impedes crosstraffic (Turn-back arrangement shown.)

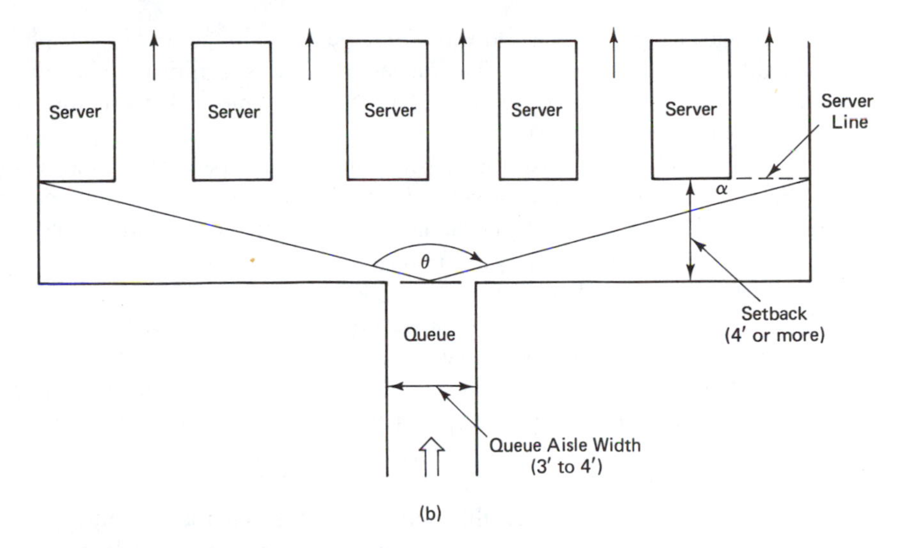

Figure 11.3b Central queue position with flow-through arrangement. Cross traffic conflicts still exist.

designs, for situations where lateral space is limited, are also provided. These designs are primarily applicable to flow-through systems.

The designs are oriented toward queues of standing people, though some are transferable to other types of customers. Readers specifically interested in manufacturing applications should consult the facility design texts by Tompkins and White (1984) and Sule (1988), as well as the *Handbook of Materials Handling* by Lindkvist (1985), all three of which include lengthy discussions of materials storage devices. Also not covered is seating layout of waiting rooms. The books by Deasy and Lasswell (1985), Woodson (1981), and Panero and Zelnik (1979) should be consulted. Layout of the service area, another factor in the design process, is also not covered.

11.3.1 Single Queue versus Separate Queues

In Chap. 9's discussion of queue discipline, the merits of a single queue feeding all servers were evident. A single queue allows differences in arrival rate and service time to be averaged out and reduces variations in waiting time. This advantage is relative to separate server queues with jockeying (customers allowed to change their queue choices). Compared to separate queues without jockeying, a single queue will also reduce expected waiting time. Without jockeying, it is possible for some servers to be idle while others have queues, an obviously inefficient situation. A single queue has another advantage. It eliminates customer anxiety as to which queue to join (one of Murphy's laws is that the other queue always looks faster), making waiting more relaxing.

Based on these merits, it would appear that it is never desirable to have separate queues. But this is not the case (Rothkopf and Rech 1987). Separate queues can sometimes improve the service time and waiting time by reducing the *movement time*, the time it takes for the customer to walk from the queue to the server. The physical distance is smaller for separate queues. The movement time is significant when the service times are short and the number of servers is large (for example, ticket windows at sports arenas or tollbooths on highways).

A second reason for using separate queues is to personalize the service. The best example is health care. People generally prefer to wait for their own personal physician rather than see whichever physician happens to be free at the moment. While establishing a separate queue for each physician must increase average waiting time, it should also improve the quality of the service.

A third reason for using separate queues might be that the service capacity approaches or exceeds the speed that customers can move in a line. Pedestrians walking single file at 2 miles per hour and spaced 3 feet apart move at the rate of 3500 customers per hour. If the service capacity approaches or exceeds this figure, then the system will likely be "queue constrained" if a single queue feeds all servers. But even if the service capacity is much less than this figure, it would still be wise to use more than one queue. Customers are annoyed when they are marched through a long line at a rapid clip.

A final reason for using separate queues is that little benefit may be gained from using a single queue. This would be true if the standard deviation of the service time is very small. But even when the standard deviation is not small, little is lost if servers are divided into groups of four or five, each with its own queue. Beyond four or five servers per queue, reductions in waiting time variability are usually small.

There are strengths and weaknesses to both alternatives. These should be kept in mind while reading about the layouts that follow.

11.3.2 Single Queue for All Servers

Figure 11.3 shows probably the most common, and the simplest, designs for a single queue feeding multiple servers. Servers are arranged in a line, with the queue projecting perpendicularly away.

Definitions 11.2

The **server line** is defined by the front edges of the service counters.

The **setback** is the distance from the front of the queue to the server line.

To provide ample space for customers to exit upon completion of service, the setback should be at least 5 feet for the turn-back design (Fig. 11.3a); 4 feet is adequate for flow-through (Fig. 11.3b).

The design has the appeal of simplicity, but suffers from the following:

1. The queue blocks traffic crossing the center of the room.
2. Customers may have to walk a long distance from the front of the queue to the server.
3. Distant servers are difficult to see, so customers might not know when they are free.

A better design is provided in Fig. 11.4. By directing the queue parallel to the server line, cross traffic is not impeded. Further, by snaking the queue back and forth, customers are allowed to wait in closer proximity to the counter, thus removing the sense of distance between customer and server.

If barriers are used to mark the queue boundary, the queue aisle should always be at least 3 feet wide, both to alleviate crowding and to allow sufficient space for wheelchair movement (Access America 1980). Where turns are needed, at least 3.5 feet should be used, again to accommodate wheelchairs. A 4-foot offset is needed at the end of any 180 degree turn (see Fig. 11.4). For queues where customers tend to wait alone, aisles should not be so wide that customers can pass each other without touching. As a rule of thumb, this means that width should be no more than 4 feet. If customers wait in groups, a width of 4 to 5 feet might be adopted, to allow customers to comfortably wait side by side. In all situations, crowding can be further alleviated through use of wide queue barriers, such as planter boxes, which increases the space between adjacent aisles.

11.3.2.1 Adjustments to Improve Visibility. The modified design of Fig. 11.4 does not improve visibility, an issue that depends on several factors:

- Distance from customer to server
- Quality of displays for informing customers when a server is ready
- Separation angle between most distant servers
- Angle of server line relative to the customer's line of sight

Figure 11.5 is the image a customer might see when looking down a long service counter. Not only do the distant servers appear smaller but, more importantly, they are hidden behind obstructions. This makes it difficult to detect when servers are free.

Good visibility is important for two reasons. First, it provides for a more pleasant waiting experience—customers do not have to strain to see servers—and removes anxiety caused by not being able to see when servers are free. Second, good visibility reduces

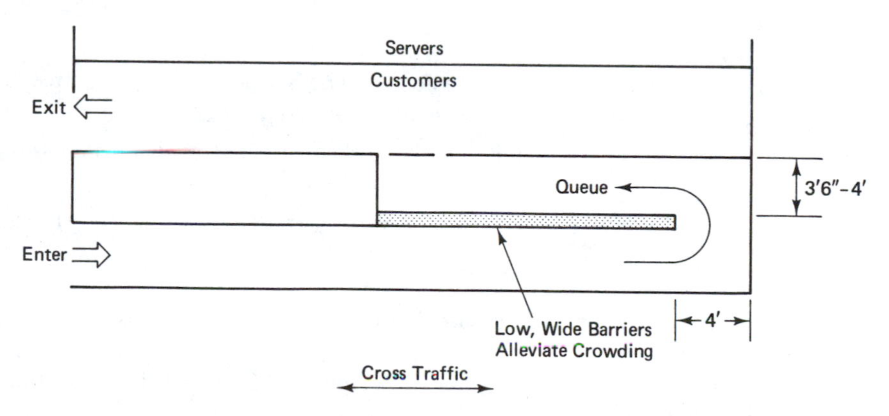

Figure 11.4 Snaked queue arrangement eliminates cross traffic conflict.

Figure 11.5 If setback is too small, it is difficult to see when servers are free.

customer reaction time—the time from the moment when the server becomes free until the moment the customer detects that the server is free. The reaction time is one of two factors influencing the idle time between the completion of one service and the beginning of the next. The other factor, customer movement time (from front of queue to server), is also influenced by design, and will be discussed later.

Problems with poor visibility can be mitigated through use of appropriate information displays:

Post a light at each counter indicating whether or not a server is free. These should be installed at staggered heights to promote visibility.

Put a bell in the vicinity of the server, to be sounded whenever a service is completed.

Place a display directly in front of the customer, showing the positions of available servers.

Care should also be taken to ensure that no obstructions, such as high counters or glass walls, impair visibility.

In addition, changes in the queue design might be considered, such as the circular counter in Fig. 11.6.

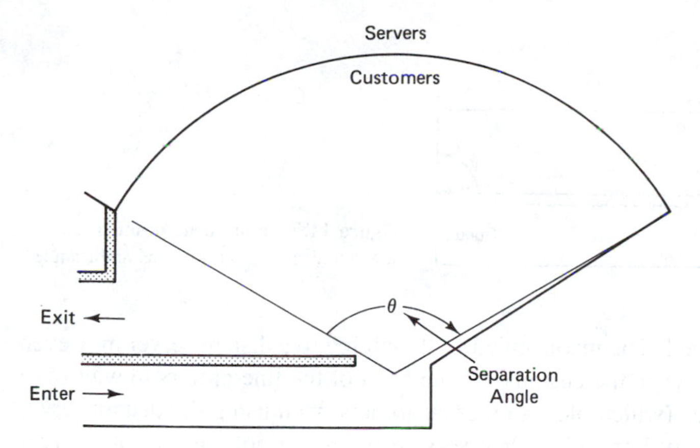

Figure 11.6 Circular arrangement provides optimal 90 degree sight angles, but increases customer to server distance. Extra space is also required between queue and servers.

Definitions 11.3

The **separation angle, θ**, is the angle formed by the lines connecting the customer to the most distant server to the left and the most distant server to the right.

The **sight angle, α**, is the angle formed by the customer's line of sight and the server line.

At the expense of increased customer to server distance, the circular counter provides optimal 90 degree sight angles. Another possibility is to increase the setback for the linear arrangement. Again, average distance from customer to server increases, but sight lines to the more distant servers are improved.

Both alternatives reduce the separation angle. The separation angle is important because it influences the effort required of the customer to scan the servers, as well as the customer's reaction time once a server becomes free. When the separation angle is less than 30 degrees, customers can scan the entire counter through use of their peripheral vision alone (Sanders, 1970, calls this the *stationary field*). Angles between 30 and 80 degrees can be scanned through eye movements (the *eye field*) and angles in excess of 80 degrees require both eye and head movements (the *head field*). Each change in field causes the reaction time to jump, from a few tenths of a second to one or more seconds. The jump can be especially large when the angle increases above 80 degrees and customers are not vigilant in watching for available servers.

To reduce the separation angle, the front of the queue can be offset from the center of the server line. With an end position, as in Fig. 11.7, the separation angle can even be reduced below 80 degrees. Unfortunately, an end position worsens the sight angle. So while customers will not have to move their heads, it will be more difficult for them to detect when distant servers are free.

11.3.2.2 Movement Time and Distance. The improvements in sight angle and the associated reaction time must be weighed against changes in the movement time and in the distance from the front of queue to servers. Both increase when the basic linear

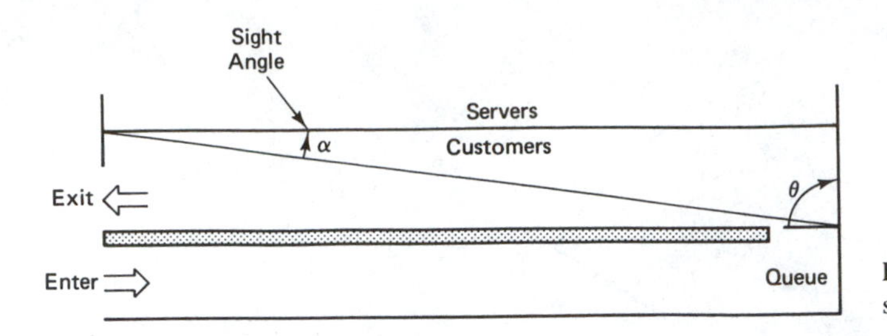

Figure 11.7 End queue position reduces separation angle but worsens sight angle.

design (Fig. 11.3) is modified. The inconvenience of walking to a distant server may even lead to underutilization, as when the customer at the head of the line prefers to wait for a closer server to become free (which blocks other customers from using the distant server too). This factor is especially important when service times are not significantly larger than the movement time.

The relationship between time and distance is determined by the speed of the customer. Studies of pedestrian movement, such as those reported in Fruin (1971), have found that the average walking speed of pedestrians is about 3 miles/hour (4.40 feet/second), with a range of 2 to 4 miles/hour (2.93 to 5.87 feet/second). These results are confirmed in a study of teller queues conducted at a major bank branch in Berkeley, California, shown in Fig. 11.8. In a data set of 604 customers, with 11 server positions,

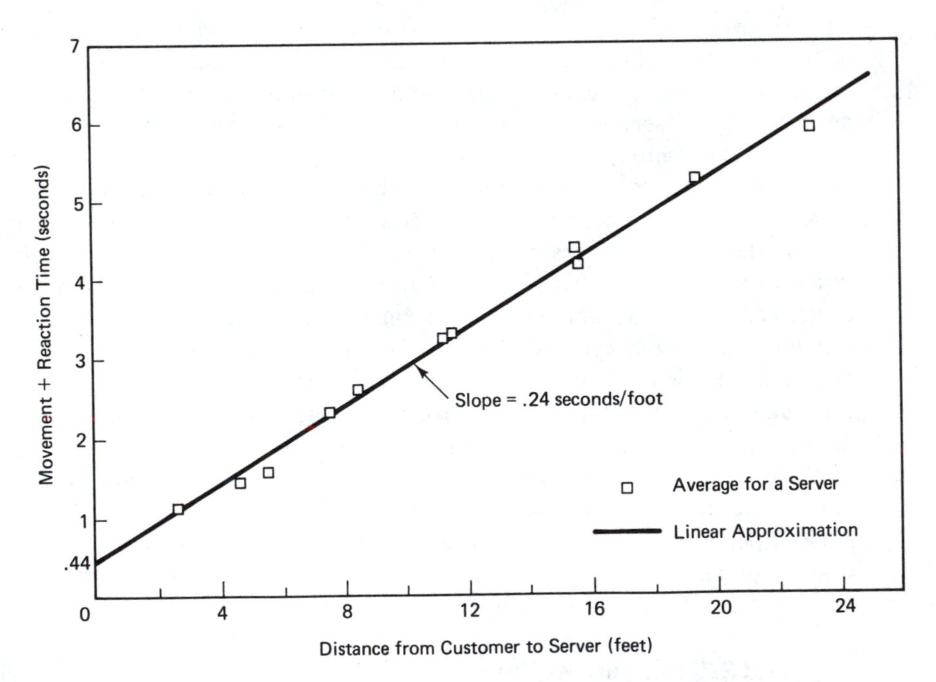

Figure 11.8 Increasing the distance from queue to server increases movement time. (Relationship based on measurement at a Berkeley bank branch.)

the time from completion of one service to start of the next was closely approximated by the equation

$$t = .44 + (1/4.09)(\text{walking distance, in feet}) \qquad \text{(seconds)} \tag{11.1}$$

The number .44 represents the average reaction time, in seconds, and the number 4.09 represents the average walking speed, in feet per second (2.79 miles/hour).

The average walking distance can be approximated for the different configurations through mathematics. Let

x = length of service counter

θ = separation angle (degrees)

y = distance from front of queue to server line for centered linear design

$= (x/2)/\tan(\theta/2)$

For the circular design, servers are equidistant from the customer:

d_c = average customer to server distance with circular design

$$= x\left(\frac{1}{\theta} \cdot \frac{360}{2\pi}\right) \tag{11.2}$$

For the linear design, an average distance can be calculated through calculus. The following assumes that all servers are equally likely to be used:

d_ℓ = average customer to server distance with linear design

$$= \frac{1}{x} \int_{-x/2}^{x/2} \sqrt{X^2 + y^2}\, dX$$

$$= \sqrt{(x/2)^2 + y^2} + \frac{y^2}{2}\, ln\left(\frac{y + \sqrt{(x/2)^2 + y^2}}{x}\right) \tag{11.3}$$

Equations (11.2) and (11.3) are plotted as a function of θ in Fig. 11.9. For both, average distance is expressed as a proportion of the counter length, x. As might be expected, distance is smaller for the linear design than the circular design for all values of θ by as much as 21 percent. Note also that small reductions in separation angle can be attained without increasing distance enormously. However, to make θ small enough to fall in the eye field, distance will have to increase by a factor of more than 2 above the minimum. A better alternative may then be to use an end position (Fig. 11.7).

The figures have been translated into movement times for differing counter lengths and counter distances, in Table 11.2. Note that for all counter lengths, increasing the setback from 4 to 7 feet increases the movement time by no more than about .5 second, while reducing the separation angle substantially.

As far as selecting between different designs, many considerations come into play. The improved sight angles of the circular counter provide shorter reaction times. More importantly, it should be more pleasing to the customer's eye (servers would not be

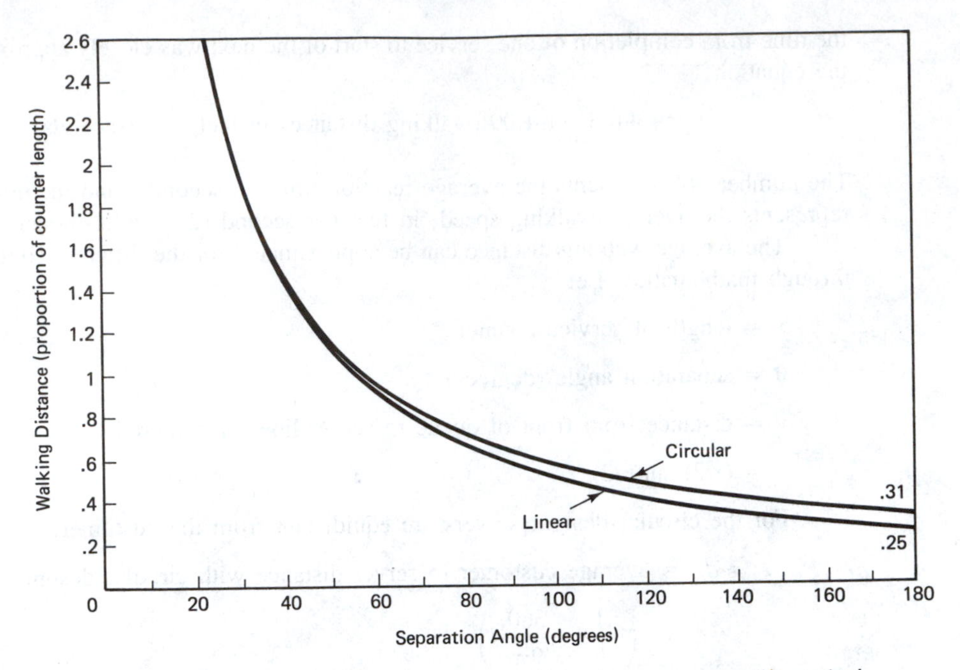

Figure 11.9 Increasing the separation angle provides a shorter movement distance (and movement time). However, reaction time may increase.

TABLE 11.2A AVERAGE WALKING TIME VERSUS SETBACK (SECONDS)

Linear counter—center position*

	Setback				
	4 Feet	5 Feet	7 Feet	10 Feet	15 Feet
Counter Length					
50	3.3	3.4	3.6	4.1	5.0
40	2.7	2.8	3.1	3.6	4.6
30	2.2	2.3	2.6	3.2	4.2
20	1.6	1.8	2.2	2.8	4.0
10	1.2	1.4	1.8	2.6	3.8

Linear counter—end position*

	Setback				
	4 Feet	5 Feet	7 Feet	10 Feet	15 Feet
Counter Length					
50	6.3	6.4	6.5	6.8	7.5
40	5.0	5.1	5.3	5.7	6.4
30	3.9	4.0	4.2	4.6	5.5
20	2.7	2.8	3.1	3.6	4.6
10	1.6	1.8	2.2	2.8	4.0

*For parallel counters, divide counter separation by 2 to obtain setback. Counter length is the length of each counter. Does not include reaction time.

TABLE 11.2B SEPARATION ANGLE VERSUS SETBACK (DEGREES)

Linear counter—center position*

	Setback				
	4 Feet	5 Feet	7 Feet	10 Feet	15 Feet
Counter Length					
50	162	157	149	136	118
40	157	152	141	127	106
30	150	143	130	113	90
20	136	127	110	90	67
10	103	90	71	53	37

Linear counter—end position*

	Setback				
	4 Feet	5 Feet	7 Feet	10 Feet	15 Feet
Counter Length					
50	85	84	82	79	73
40	84	83	80	76	69
30	82	81	77	72	63
20	79	76	71	63	53
10	68	63	54	45	34

*Results do not apply to parallel counters. In general, separation angle for parallel counters is at least twice the indicated value.

hidden from view), enhancing the waiting atmosphere. The advantage of a linear counter with end queue position is reduced separation angle. Finally, a linear counter with a central queue position offers a shorter movement distance, a factor that will dominate all others when the server line is especially long.

A final design, the parallel linear arrangement, is shown in Fig. 11.10a and b. The design is something like the layout of elevators in large buildings. A minimum separation of 7½ to 12 feet is needed between the counters, both for visibility of distant servers and ease of movement. The parallel design is capable of providing the smallest average movement distance (see Table 11.2), but only if the separation angle is extremely large (more than 180 degrees), providing new meaning to the expression ''having eyes in the back of one's head.'' Certainly, the parallel system would not make sense without a supplemental information system, such as bells that sound whenever servers are free.

11.3.3 Separate Server Queues

Separate server queues are simpler to arrange than a single queue and do not require that customers move as far from the front of the queue to the server. The reduced walking time can provide increased server efficiency through elimination of idle time. This factor is crucial when customer movement is slow, as is the case of grocery shoppers pushing heavy carts. On the other hand, separate queues increase the variation in waiting times. Further, server efficiency might be lost due to the simultaneous presence of queues and idle servers, as when customers are not aware that a server is free.

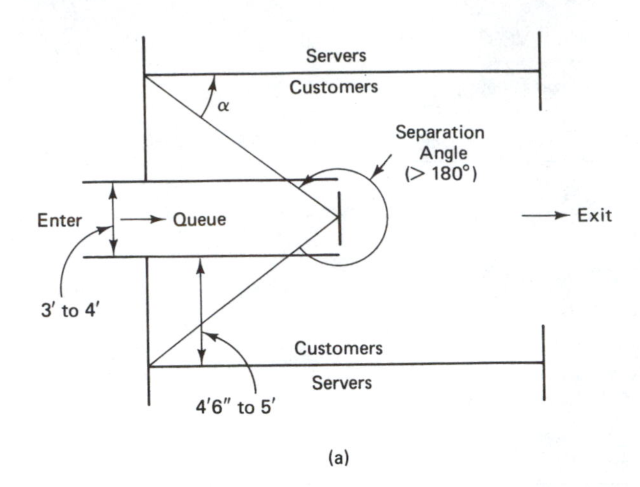

Figure 11.10a Parallel counters reduce distance from front of queue to servers. Separation angle is much worse and sight angles improve with middle position.

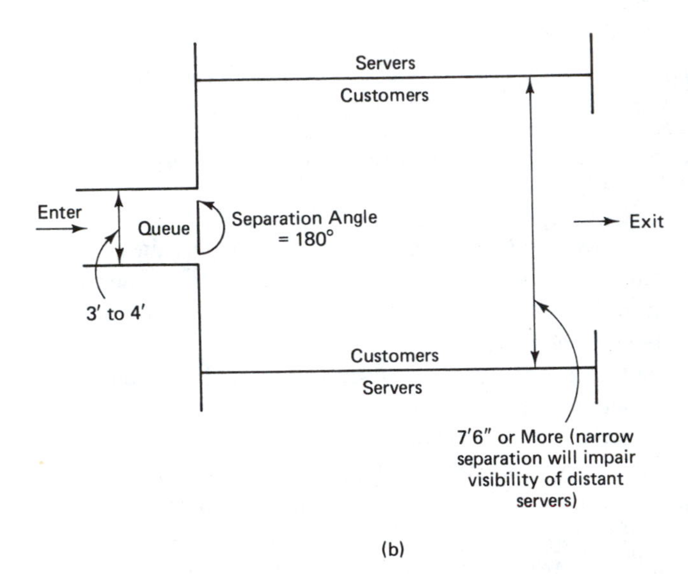

Figure 11.10b Parallel counters with end position reduce separation angle but increase distance from front of queue to servers. End position allows the separation between server lines to be reduced.

The most common arrangement, a single line of servers with perpendicular lines of customers, is shown in Fig. 11.11. In the turn-back designs, sufficient space must be allocated for the customer to turn around and exit upon completing service. This can be accomplished by providing an aisle for lateral movement (side exit, Fig. 11.11a) or, when space permits, by providing exits immediately adjacent to the server (Fig. 11.11b). The guidelines for aisle width presented in Sec. 11.3.1 apply here. However, keep in mind that the spacing between servers and aisles must be at least as wide as the server workstation. A further difference is that the setback with adjacent exits can be reduced to 3½ feet, because the lateral space is not used as a passage for exiting customers.

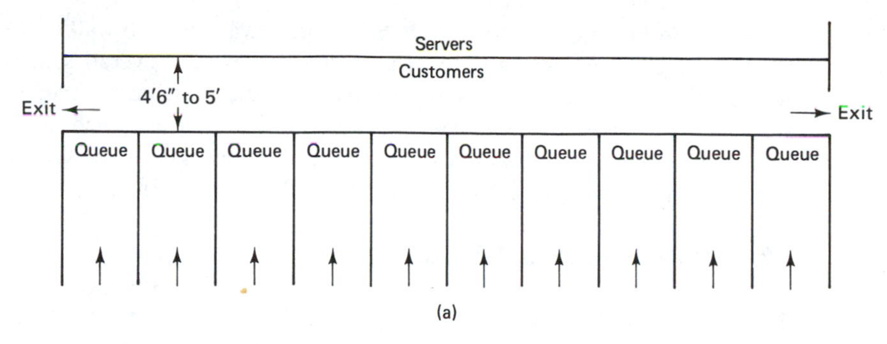

Figure 11.11a Separate queues with turn-back arrangement. Side exits allow more servers to be fit within a given lateral space.

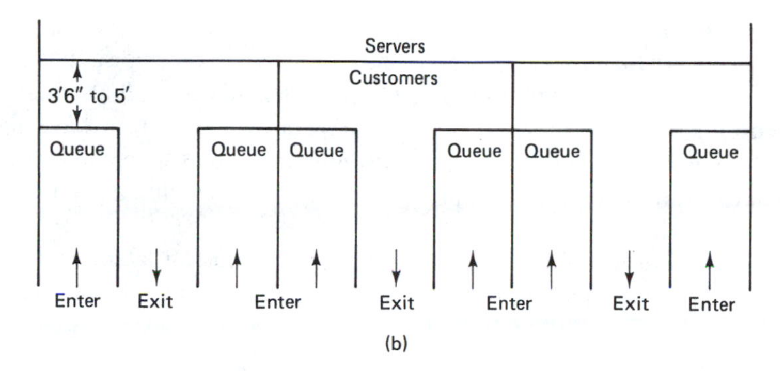

Figure 11.11b Adjacent exists provide easier egress, but require more lateral space.

Crowding around the server is a more pervasive problem for separate queues than for a single queue. Therefore, extra care is needed to enforce setbacks. Signs and markers should prominently indicate the front position of the queue. When the service process permits it, a flow-through design is preferred, for this prevents interference between customers who have completed service and customers who are waiting for service.

With separate queues, it is important to make sure that all servers are fully utilized. Customers have a tendency to join the first queue they see. If all customers arrive from the same direction, the first servers might perpetually have queues, while the last are perpetually idle. As a precaution, all servers should be fully visible to customers as they arrive. Usually this means that the server line should be perpendicular to the direction from which customers arrive and that an adequate setback be provided to keep long queues from obstructing views of idle servers.

In most situations the linear arrangement is perfectly adequate. Once the customer joins a queue, visibility is not a significant issue, for the customer only has to watch the server immediately in front. However, if the queue is located in a congested building with considerable cross traffic (such as an airline terminal), an angled system might be adopted, as in Fig. 11.12. This directs the queue toward the top wall of the building and provides more space for passersby. But it does not reduce overall space requirements.

If the service time is very short relative to the movement time, it may even be advantageous to have two queues feed each server, one on each side. This design, common in cafeterias, allows the server to work on one side while the customer moves into service on the other, effectively reducing the service time.

11.3.4 Restricted Lateral Space

It is almost axiomatic that service enterprises will grow over time. In the United States, it is not unusual to see supermarkets with 30 or more cashiers spread over a wide expanse at the front of the store, or highways with 20 tollbooths spread across the horizon. Large queueing systems like these have enormous appetites for space, appetites that demand alternative designs.

Of particular concern is the demand for lateral space, the overall width of the system. As the number of servers increases, the server line simply gets wider and wider. A large width adds to the customer's movement distance. It also makes it difficult for the customer to see which servers are free or which servers have the shortest queues. Moreover, sufficient lateral space just might not be available to accommodate the queues and servers. One can only squeeze so many servers within a given space before crowding becomes intolerable.

For turn-back systems, the options for reducing lateral space are limited. Where separate server queues are used, a side-exit design might be adopted instead of adjacent

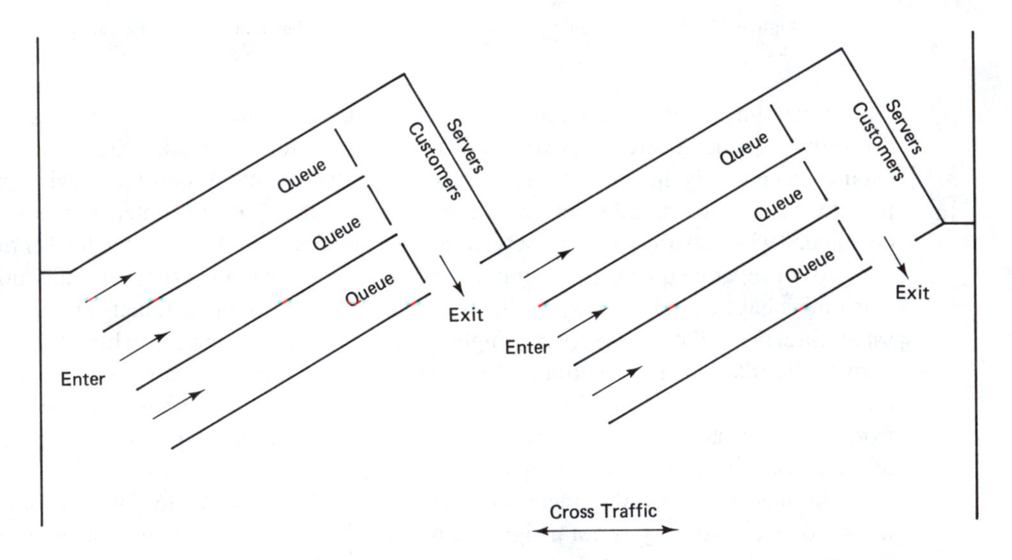

Figure 11.12 Angled server line can reduce conflicts with cross traffic when separate queues are used.

exits. Otherwise, the only solution is to add servers at an alternative location, or reorient the counter direction, as in Fig. 11.13, in which case a separate queue might serve each group of servers.

For flow-through systems, there are several additional options. In Fig. 11.14, servers are placed in ***tandem***, one processing customers to the left and the other processing customers to the right. Let

$$w_s = \text{width of server counter}$$

$$w_c = \text{width of customer aisle}$$

Then a basic flow-through design, as in Fig. 11.3b, has a server density of

$$\gamma^b = \text{servers/unit lateral distance for basic design}$$

$$= \frac{1}{w_s + w_c} \tag{11.4}$$

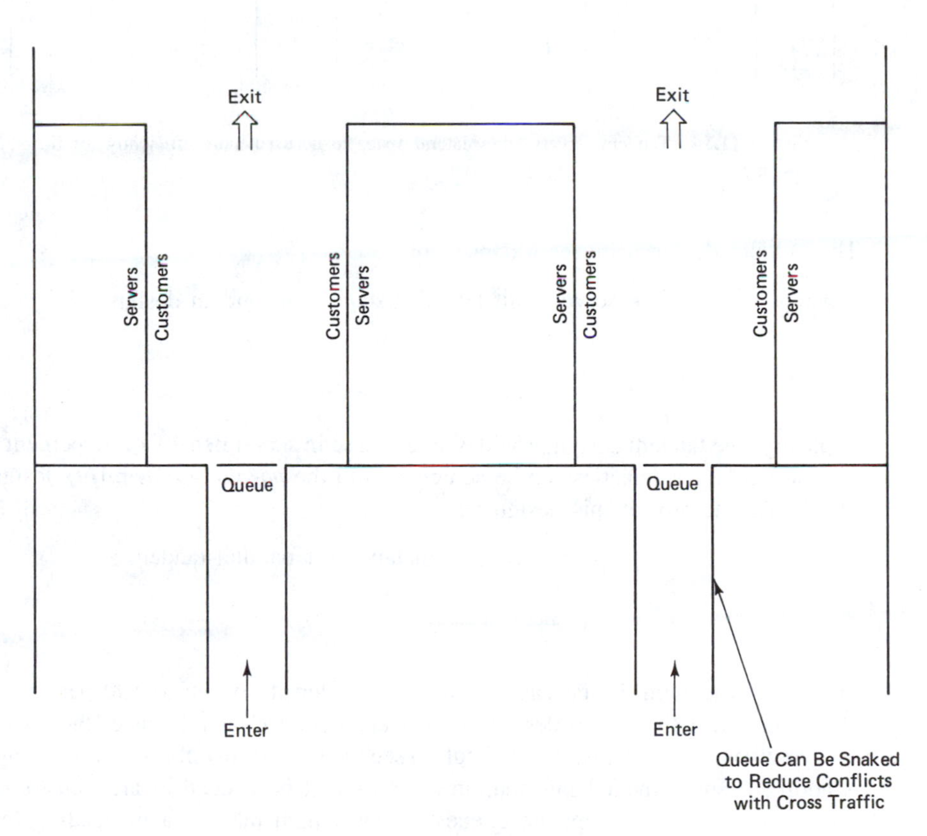

Figure 11.13 Reoriented service counters allow more servers to be fit within a given lateral space. Some confusion may occur when customers select queues.

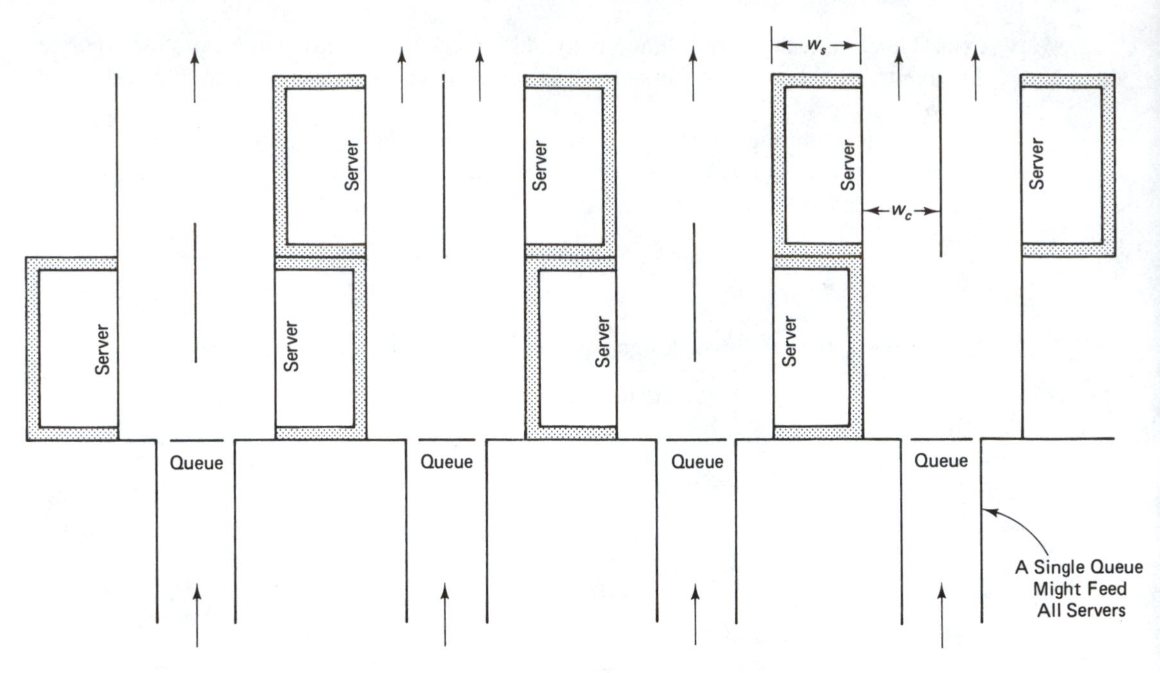

Figure 11.14 Tandem servers allow lateral space to be used more efficiently for flow-through systems.

The tandem design achieves a density of

$$\gamma^t = \text{servers/unit lateral distance for tandem design}$$

$$= \frac{2}{w_s + 2w_c} \tag{11.5}$$

Typically, the tandem design provides an increase in server density of 30 percent or more.

Even larger densities can be achieved with the ***parallel-tandem*** (*pt*) design in Fig. 11.15. The density of this design is:

$$\gamma^{pt} = \text{servers/unit distance for parallel tandem}$$

$$= \frac{4}{w_s + 3w_c} \tag{11.6}$$

The parallel-tandem design can increase server density by up to 100 percent.

Though both tandem designs significantly reduce lateral space, they also present special visibility problems: It is difficult to see the second row of servers. If a single queue is adopted, supplemental information systems will be needed to alert customers when servers are free. With separate queues, queue length may be a misleading indictor of waiting time because queues begin at different locations relative to the server line. These are just two reasons why the tandem designs should only be used when lateral space is limited.

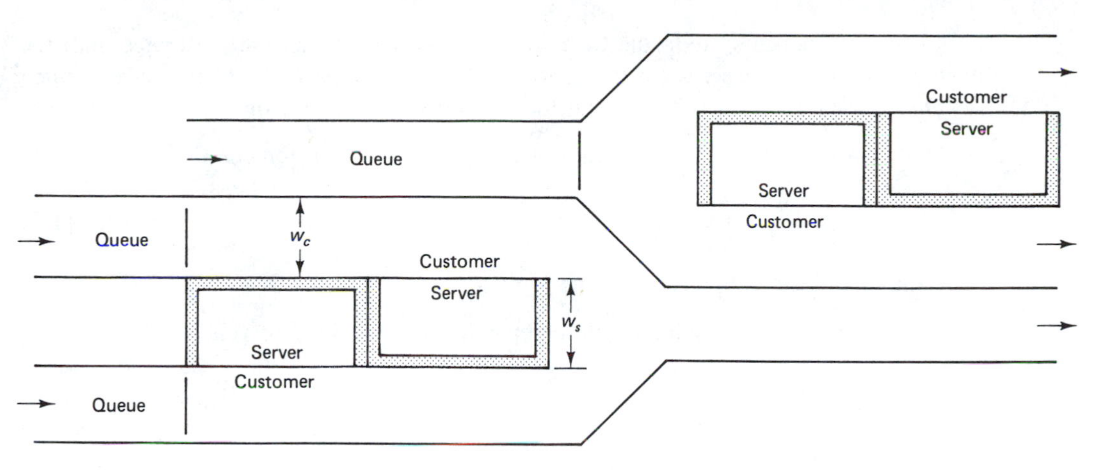

Figure 11.15 The parallel-tandem arrangement further increases utilization of lateral space, but may confuse customers.

Increased server density can also be achieved by reorienting counter direction, as was shown for turn-back systems in Fig. 11.13. Several customers are now served in the same lane. Unfortunately, this arrangement may make it impossible for a customer to depart from service until all of the customers ahead of it have departed. This is the case when customers move on fixed tracks or conveyors, as illustrated in Fig. 11.16. Customers cannot pass each other once they enter service; instead, they move through the system in platoons, of size equal to the number of servers in the lane.

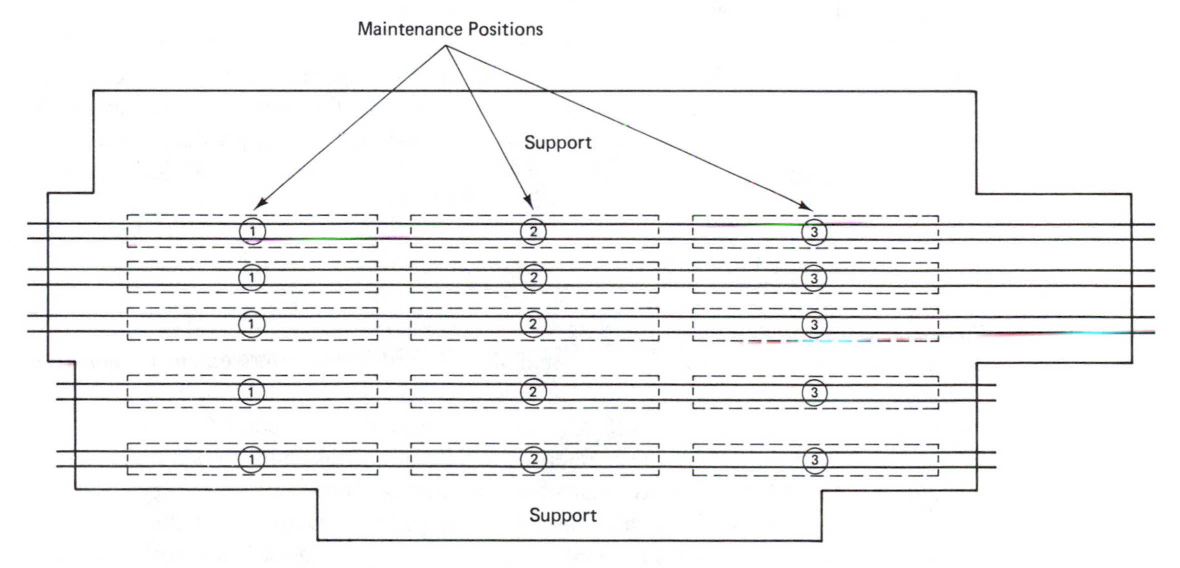

Figure 11.16 In rail maintenance garages, vehicles are served in lanes.

Source: Courtesy of San Francisco Public Utilities Commission and Manna Constultants, San Francisco.

In a platooned system, the time from when the customer enters service until the customer departs from service is the same for all customers in a platoon, and is determined by the *maximum* service time among the customers in the platoon:

$$S_l = \text{service time for a platoon with } l \text{ customers}$$

$$= \max_{i=1,\ldots,l} \{S_i\} \tag{11.7}$$

The maximum throughput of the lane is then

$$c_l = \text{maximum throughput for a lane with } l \text{ servers}$$

$$= \frac{l}{E(S_l)} \tag{11.8}$$

Techniques for calculating $E(S_l)$ were presented in Chap. 10. Suppose, for instance, that service times are independent, identically distributed and can be approximated by normal random variables. Then Clark's approximation yields the relationship in Fig. 11.17. Note that increasing l, the number of servers per lane, provides a less than proportional increase in throughput, especially when the coefficient of variation is large. In fact, with a CV of .5 and 10 servers, the throughput is barely half of what it would be if passing were allowed. The conclusion is that serving multiple customers in a lane is most reasonable when there is little variation in service times and the number of servers per lane is small. A further conclusion is that when multiple lanes are provided, effort should be taken to ensure that the customers assigned to each lane have similar service times. This can be accomplished by grouping customers demanding similar amounts or types of service on the same lane.

The reason why platooned systems are used at all is because they provide a more compact service area. A wide and skinny maintenance garage, for instance, would be both difficult to construct and difficult to supervise. Placing multiple servers in the same lane reduces width and increases length. As usual, the lateral space is of greater concern than longitudinal space.

11.3.5 Managing Flows between Servers

Once the customer emerges from a server, he or she may have to travel on to another queue. The layout and information system should facilitate this movement by ensuring that customers do not lose their way and that customer paths do not interfere with each other. Directional signs and markings should be prominent and frequent. A line pointing in the correct direction might be marked on the floor, for instance. In addition, to help customers who have gone astray, maps and information booths might also be installed.

Layouts can be divided into those that are ***product focused*** and those that are ***process focused***. In a product-focused design, servers are arranged in *serial* order, the order followed by all customers. The servers then work as a team to produce an identical service for all customers. In a process-focused design, service is not identical. Different

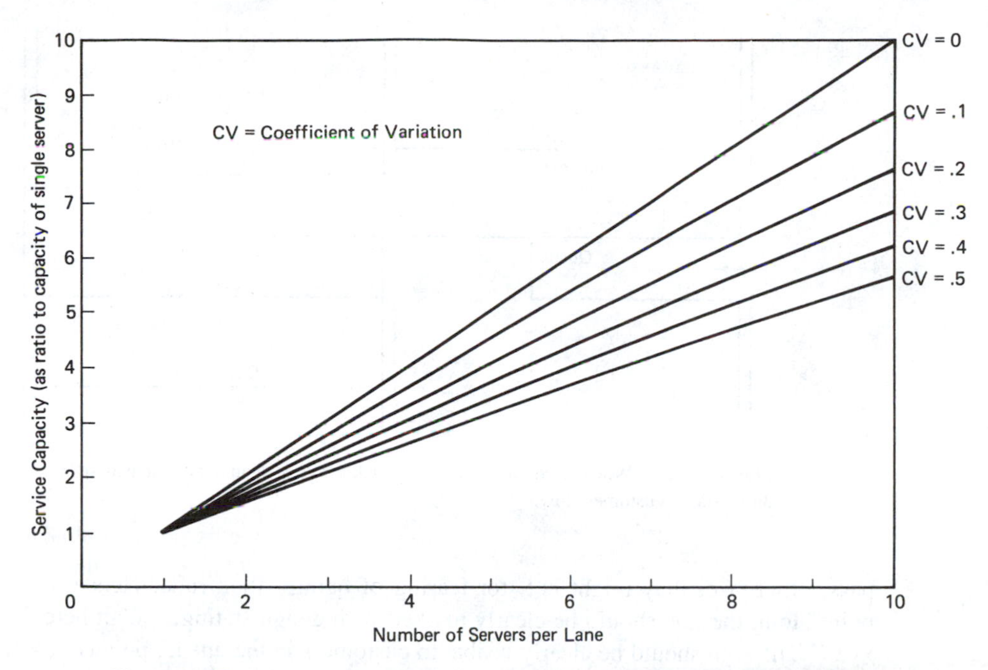

Figure 11.17 Increasing the number of servers per lane provides a less than proportional increase in lane capacity, especially when the coefficient of variation of service times is large. (Based on normal service times and Clark's approximation.)

customers visit different servers, in different orders (many machine shops operate this way).

For the product-focused system, the main principle is that the next server visited should be in the direction that customers would naturally head, and away from the direction customers just came. For flow-through systems, that direction is straight ahead, which means that serial servers should be oriented along a line. For turn-back systems with side exits, the natural direction to head is toward the side, leading to designs like Fig. 11.18. Adjacent exits are to be avoided for serial queues, for customers will emerge from different places and head different ways.

For process-focused systems, the layout follows from an analysis of the "flows" of customers between pairs of servers. Generally speaking, if there is a large flow between a pair, the servers should be adjacent, or at least close together. Accomplishing this end is not always easy and requires fairly sophisticated techniques. Interested readers should consult the texts on facility layout presented at the end of this chapter.

Falling between product focused and process focused are "cafeteria systems." In a cafeteria, all servers are located along a line, but different customers use different services. The greatest concern here is that customers may be forced to wait in queues for servers that they have no desire to use (perhaps I have no interest in eating the pot roast, but I have to wait behind the person in front of me anyway). Clearly, this is inequitable and inefficient. Sufficient space should be provided to allow customers to bypass queues. Barriers should not interfere with movement. Customers should even be encouraged to

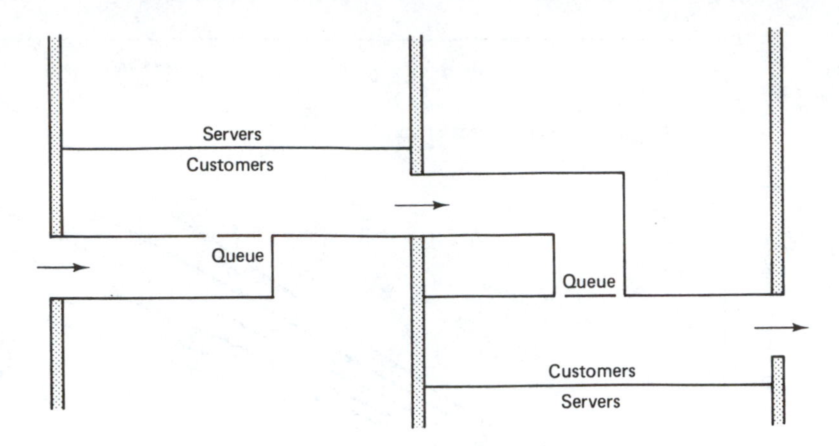

Figure 11.18 When used in series, turn-back servers should be oriented in natural direction of customer flow.

pass; otherwise, they might not, for fearing of being called rude. Hence, each service point along the line should be clearly marked with a sign stating, ''Wait here for service xyz.'' The sign should be clearly visible to customers in the queue, pointed toward them, not posted flat against the service counter.

As a final comment, the designs should be modified for bulk service queues. Instead of creating single-file lines, it might be better to partition the queue into waiting areas, each of sufficient size to comfortably accommodate the number of customers served in one batch. Active enforcement will be needed to ensure that the waiting area is not filled beyond intended capacity.

11.4 SERVER LOCATION

In this section we step back from the question of queueing layout for a cluster of servers and look at the large-scale issue of where these clusters should be located. That clusters should be placed where they are convenient to customers is an obvious fact. Less obvious are *how many different locations* should be used and how the number of locations affects level of service. Here are some examples:

Example 1

A large insurance company would like to plan the acquisition of office copy machines. As part of this plan, it will determine where copy machines should be placed, and how many should be placed at each location. In addition to the cost of acquiring the machines, the company would like to consider the time spent by employees walking to and from the machines and the time spent waiting in queue at the machines.

Example 2

A West Coast city known for its radical politics and good restaurants is creating a plan for the deployment of fire engine crews. Specifically, it would like to know how many fire stations should be open, and how many crews should be housed at each station. The objective is to minimize the response time to fires.

Example 3

The same West Coast city is creating a plan for the deployment of police cars. Again, the objective is to minimize response time.

The first system is an example of a ***customer-to-server*** system (Larson and Odoni 1981). The customer physically moves to the server to obtain the service, but it might also be that service is *completed* over the phone. In either case, the server is stationary. The second two systems are examples of ***server-to-customer*** systems, for the obvious reason that the server moves to the customer. Examples 2 and 3 can be further classified according to whether the server has a ***fixed base*** (as is ordinarily the case of fire crews) or a ***roving base*** (as is ordinarily the case of police cars).

11.4.1 Customer-to-Server System

The ordinary scenario is that when a service is desired, the customer travels to the nearest server cluster and waits for the first available server in a single queue. A somewhat more complicated scenario is that if the queue is sufficiently large, the customer travels on to another cluster, in hopes of finding a shorter queue. The latter is uncommon, due to only minimal improvements in waiting time, so the focus here will be on the former.

The heart of this question is a trade-off between accessibility and waiting time:

The more clusters that are used, the shorter will be the average travel distance from customer to server.

The more servers placed at each cluster, the shorter will be the average waiting time.

For a fixed number of servers, N, the product of the number of clusters used, and the number of servers per cluster, must be a constant. Hence, increasing the servers per location increases travel distance, and increasing the number of clusters increases waiting time. The trade-off can be expressed in terms of either the number of clusters or the number of servers per clusters. A secondary trade-off is that increasing the total number of servers can decrease both the waiting time and the travel distance, at the expense of higher service cost. The focus here is on the first trade-off.

11.4.1.1 Average Distance. Calculating average travel distance from customer to server amounts to calculating the expectation of a random variable. There is even a field of research called ***geometric probability***.

Finding the expected distance between a set of discrete locations requires calculating the expectation of a discrete random variable. Let

f_i = number of trips made by customer i per unit time

x_i, y_i = x and y coordinates of customer i's location

$\bar{x},\bar{y}$ = x and y coordinates of server located closest to customer i

$\bar{D}$ = expected travel distance

If customers follow straight-line (Euclidean) paths, expected distance is the following:

$$\bar{D} = \frac{\Sigma f_i\sqrt{(x_i - \bar{x})^2 + (y_i - \bar{y})^2}}{\Sigma f_i} = \frac{\text{total distance}}{\text{total trips}} \tag{11.9}$$

Paths over road networks tend to be 10 to 20 percent longer than straight line. So, if the exact distance is not known between a pair of locations, a good estimate is the Euclidean distance multiplied by 1.15.

An alternative approach is to approximate customer locations with a ***continuous space model***. This approach involves calculating the expectation of a continuous random variable. Instead of thinking of customers as discrete entities, one looks at them in terms of the number of trips generated per unit area:

$\rho(x,y)$ = trip density at location x,y (trips per unit time, per unit area)

Example

Within a rectangular region of size 5 miles × 7 miles, 140 trips are generated per week. $\rho(x,y) = 140/35 = 4$ trips per week, per square mile, for locations within the rectangle.

As can be seen, a major advantage of the continuous space approach is that less data are needed to represent customer demand.

From the trip density function, expected distance is calculated with an integral instead of a summation:

$$\bar{D} = \frac{\int\rho(x,y)\sqrt{(x - \bar{x})^2 + (y - \bar{y})^2}dxdy}{\int\rho(x,y)dxdy} = \frac{\text{total distance}}{\text{total trips}} \tag{11.10}$$

The appeal of the continuous space approach is that when demand is distributed uniformly over space, $\bar{D}$ reduces to very simple expressions. For example, if clusters are placed at the centers of square service regions, then

Square Service Region, Servers in Center

$$\bar{D} = .383\sqrt{A} \tag{11.11}$$

where A is the area of each square. As is always the case with a fixed region shape, average distance is the square root of the region size multiplied by a constant, a less than linear relationship. If the region is circular, the constant equals .376. Other constants can be found in Larson and Odoni (1981).

11.4.1.2 Travel Time versus Waiting Time Trade-Off.

For a *given arrival rate per server*, expected waiting time decreases as the arrival rate increases, because random variations in service time and arrival rate are averaged out over servers. This implies that expected waiting time will decrease as the number of servers per location increases (and as the area per server location increases). Expected travel distance (and expected travel time) is affected in the opposite way.

This trade-off can be evaluated by measuring the expected number of customers in queue and traveling, $L(m)$:

$$L(m) = L_q(m) + \sqrt{m \cdot k} \tag{11.12}$$

where

m = number of servers per location

$L_q(m)$ = expected customers in queue, given m servers per location

A_0 = total service area

N_0 = total number of servers

v = customer movement velocity

λ = total arrival rate among all servers

k = $.383\sqrt{A_0/N_0}\,(\lambda/v)$

The trade-off is depicted in Fig. 11.19 for a busy $M/M/m$ queueing system, with $\lambda/c = .9$ and $k = 1$. In the example, five to 10 servers per cluster provide nearly optimal values of $L(m)$. Figure 11.20 gives $L(m)$ for different values of k, from .25 to 16.

It is impossible to write a general analytical expression to capture the trade-off. Nevertheless, the important conclusion is that when k is small (below 4) and λ/c is large (.9), there is a strong incentive for placing two or more servers at each location. On the other hand, for large values of k, Eq. (11.12) is dominated by travel time, and just one server should be placed at each location. Not revealed in the figure is the fact that server clusters should not be as large when λ/c is small as when λ/c is large. For small values of λ/c (under .5), cluster sizes larger than one are seldom warranted.

Example

In a 100,000-square-foot, single-story office building, the number of trips to copy machines has been calculated to be 3000/day. The building has 20 machines, and the machines are busy 90 percent of the time. Workers walk at the rate of 3 miles/hour, and work 8 hours/day.

Solution k is calculated as follows:

$$k = .383\sqrt{100{,}000/20}\,(3000/8)/(3 \cdot 5280) = .64$$

Referring to Fig. 11.20, approximately ten copiers should be placed at each location, or about 50,000 square feet per location.

Of course, a preferred solution may be to purchase more copiers, an option that might be studied in greater detail.

11.4.2 Server-to-Customer Systems

For server-to-customer systems, the customer will ordinarily phone the company providing the service, and the company will dispatch an appropriate server to the customer's location. That server is often just the nearest available server to the customer. However,

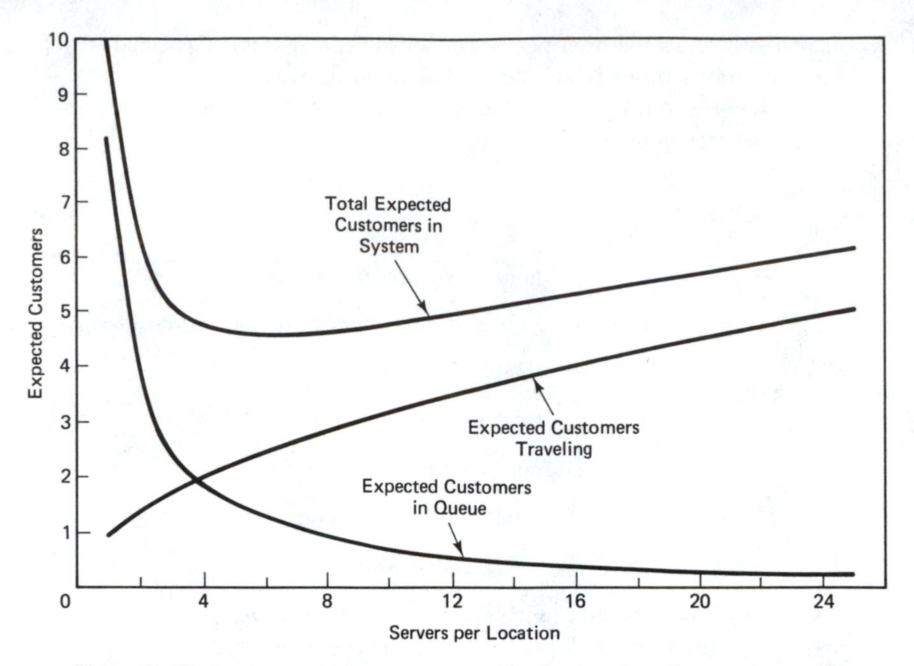

Figure 11.19 In a customer-to-server system with a fixed number of servers, increasing the number of servers per location will increase travel time but decrease queueing time. (Figure assumes that $\lambda/c = .9$, $k = 1$, and $M/M/m/\infty$ queues.)

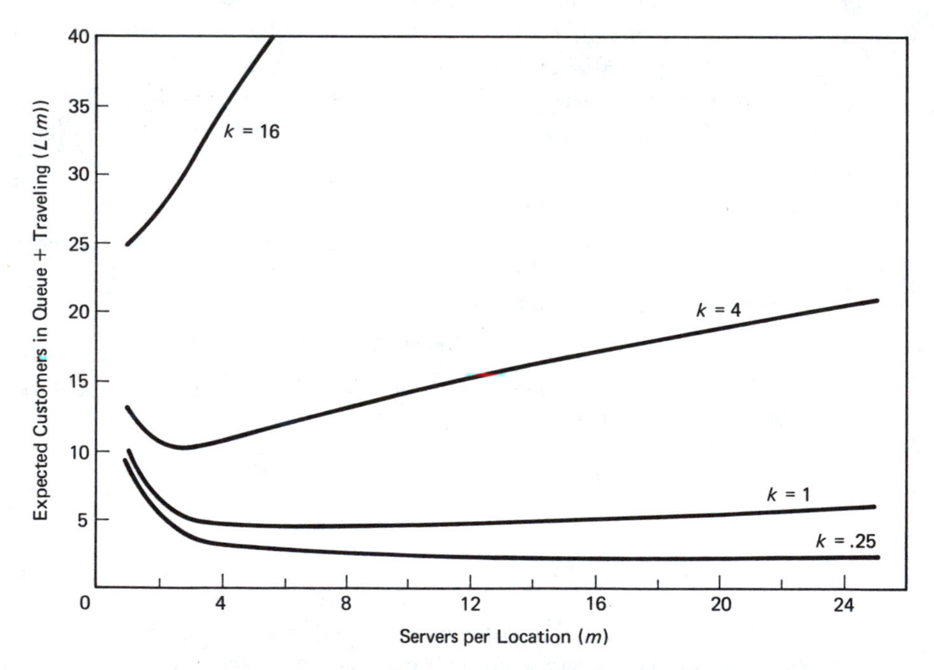

Figure 11.20 The optimal number of servers per location increases as the parameter k decreases ($\lambda/c = .9$, $M/M/m/\infty$).

alternative schemes are possible, such as customers are divided into districts, and the nearest server assigned to the customer's district is dispatched, or servers travel on set routes, and customers are not visited until the server reaches their portion of the route.

The unique characteristic of server-to-customer routes is that the customer does not have to travel to the server to see whether or not it is free; the dispatcher has this information available. This removes the waiting time incentive for concentrating several servers at the same location. In fact, waiting time and travel time combine to form a single value: response time. In the case of emergency services, the total number of servers should be sufficiently large that the probability of all servers being busy at the same time is very small. For nearest server dispatching, this means that the expected response time is simply the expected travel time from the nearest available server (plus any time spent by the dispatcher and server processing the call), a value that can be calculated as follows:

$$E(RT) = E(T_1) + P(\text{Serv 1 busy}) \cdot [E(T_2) - E(T_1) + P(\text{Serv 2 busy} \mid \text{Serv 1 busy}) \cdot \{E(T_3) - E(T_2) + [\ldots \quad (11.13)$$

where $E(T_i)$ is expected time from ith closest server.

Kolesar (1975) studied this relationship for fire stations in New York City (also see Ignall et al. 1975). He found that the expected distance is accurately approximated by an expression of the following type:

$$E(D) = k\sqrt{\frac{A_0}{N_0(1 - \tilde{\rho})}} \quad (11.14)$$

where

A_0 = area of the entire service region

N_0 = total number of stations

$\tilde{\rho}$ = proportional utilization = busy probability (including response time)

Dividing $E(D)$ by the velocity gives the expected response time. As with the customer-to-server model, the quantity under the square root sign represents the area per server, except now it is the area per *available* server. For New York City, the constant was found to be .55. This is slightly less than it would be if stations were completely random (by Poisson process) and travel occurred over rectangular paths, .63. While the constant .55 is not transferable to all cities, the general relationship of Eq. (11.14) is widely applicable. In particular, the constant for *roving* servers ought to be slightly larger than the constant for fixed base servers, due to greater randomness in their locations. For any given application, the constant can be estimated through regression techniques from empirical data.

The effect of heavy system utilization is to increase the chance that a server is busy. This increases the area per available server and increases the chance that a more distant server will be needed to respond to the call. In Eq. (11.14), the increased area is reflected by the fact that the denominator of the expression includes $1 - \tilde{\rho}$, the probability that a server will not be busy.

The literature on server-to-customer systems is quite extensive. It is well documented by Larson and Odoni (1981). Also noteworthy is Larson (1972), which studies police departments.

11.5 QUEUE CAPACITY

The final design issue is the queue capacity, also called the *buffer size*. Ordinarily, the buffer should have sufficient capacity to accommodate maximum queue sizes. That is, customers should not be turned away due to inadequate queue capacity. Especially when people are involved, providing adequate space for the largest queues is essential, both for customer comfort and customer safety. The following are some simple guidelines:

Predictable deterministic queues Maximum queue size equals maximum difference between $A(t)$ and $D_q(t)$, which occurs at a time when the arrival rate equals capacity, as described in Chap. 6.

Bulk service queues, effective capacity exceeds arrival rate Maximum queue size occurs immediately prior to the start of the service phase and equals the number of customers that arrived during the preceding "red phase."

Bulk service queues, arrival rate exceeds effective capacity Maximum queue size equals maximum separation between $A(t)$ and $D_q(t)$, which occurs at the start of a service phase near a time when effective capacity equals the arrival rate.

Stochastic queues, steady state The state probability distribution gives the 95th, 99th, or higher percentile for customers in the system, which can be a design target for queue capacity, as described in Chap. 5.

Stochastic, nonstationary systems are more difficult to evaluate and may demand simulation. To convert queue capacity into space for standing people, the level C standard discussed in Sec. 11.1.2 can be applied. However, keep in mind that some customers will be accompanied, in which case space requirements must be factored up accordingly. Other human conversion factors can be found in FAA (1975).

There are instances where a buffer might be designed for less than the maximum queue size. This is true of telephone information queues. The customer may rather be blocked from joining the queue than be allowed to enter the queue and pay telephone toll charges while waiting. In such instances, the queue capacity should approximate the maximum wait a customer is willing to endure.

11.6 CHAPTER SUMMARY

It is not by accident that this book has ended with the subject of queue design. The thesis of queueing methods is that by understanding queue behavior, queueing problems can be eliminated—by changes in the service process (reducing the service time or varying the service capacity), changes in the arrival process (congestion pricing and reservations), and changes in the queue discipline—and they can be eliminated at little or no cost to the server.

The topic of queue capacity, in fact, the entire chapter on queue design, strays from this thesis. Instead of showing how to eliminate queues, this chapter shows how to *accommodate* queues. Creating a pleasant waiting environment is one such concern. On the psychological side, customers should be entertained and well informed, and the queue discipline should be equitable. On the physiological side, the environment should not be overly noisy (or overly quiet), it should be well illuminated, and temperature should be controlled at comfortable levels. More importantly, sufficient space should be provided so that customers are not crowded.

A second concern is implementing queue disciplines. This can be accomplished by passive means—providing information so that customers will self-organize—or by active means—direct involvement of the server in selecting customers.

The third concern is queue layout (Table 11.3). The layout should allow for cross traffic, minimize the distance from the front of the queue to the servers, and provide good server visibility from the front of the queue. In addition, the choice of whether to use a single queue feeding all servers or separate queues feeding each server depends on factors such as the service time standard deviation, average service time, number of servers, and the need to personalize the service.

A fourth concern is selecting good server locations. In the case of customer-to-server queues, this amounts to balancing a trade-off between waiting time and travel time, the former decreasing with cluster size and the latter increasing with cluster size. In the case of server-to-customer queues, equations were presented for response time, which can be used in calculating the optimal number of servers.

The message of this chapter is that actions can be taken to alleviate the consequences of queueing. Queueing need not be unpleasant. Queueing need not be chaotic. But no matter what, queueing should still be prevented. It should be prevented because it takes away the customer's freedom to do as he or she chooses. Nevertheless, after all avenues for eliminating queues have been exhausted, occasional queueing might still remain. The last step is then to design the queue—to create a pleasant environment capable of accommodating ordinary queue sizes.

TABLE 11.3 ALTERNATIVE QUEUE DESIGNS

	Strengths	Weaknesses
Separate Queues	Personalized service Short movement time	Simultaneous queueing and idle servers Larger waiting time variation
Single Queue	Small waiting time variability Balanced server utilization	Large movement time Poor server visibility
Linear: Central Position	Short movement distance	Large separation angle
Linear: End Position	Small separation angle	Large movement distance Poor sight angles
Circular	Optimal sight angles	Large movement distance
Parallel Counters	Short movement distance	Poor sight angles Large separation angle (center position)

FURTHER READING

Access America. 1980. *Designing for Everyone*. Washington, D.C.: U.S. Government Printing Office.

ASHRAE. 1985. *ASHRAE Handbook, 1985 Fundamentals*. Atlanta: American Society of Heating, Refrigerating, and Air Conditioning Engineers.

Bateson, J. E. G. 1985. "Perceived Control and the Service Encounter," in *The Service Encounter, Managing Employee/Customer Interaction in Service Businesses*, ed. J. A. Czepiel, M. R. Solomon, and C. F. Surprenant. Lexington, Mass.: Lexington Books.

Bender, P. S. 1976. *Design and Operation of Customer Service Systems*. New York: Amacom.

Deasy, C. M., and T. E. Lasswell. 1985. *Designing Places for People: A Handbook on Human Behavior for Architects, Designers, and Facility Managers*. New York: Watson-Guptill.

Federal Aviation Administration. 1975. *The Apron and Terminal Building Planning Report*, Report FAA-RD-75-191. Washington, D.C.

Flynn, J. E., T. J. Spencer, O. Martyniuck, and C. Hendrick. 1973. "Interim Study of Procedures for Investigating the Effect of Light on Impression and Behaviour," *Journal of the Illuminating Engineers*, 3, 87–94.

Fraisse, P. 1984. "Perception and Estimation of Time," *Annual Review of Psychology*, 35, 1–36.

Fruin, J. J. 1971. *Pedestrian Planning and Design*. New York: Metropolitan Association of Urban Designers and Environmental Planners.

______. 1981. "Causes and Prevention of Crowd Disasters," *Student Activities Programming*, 14, 48–53.

Hall, E. T. 1982. *The Hidden Dimension*. Garden City, N.Y.: Doubleday.

Hill, V. 1973. "Control of Noise Exposure," in *The Industrial Environment: Its Evaluation and Control*, 37. Washington, D.C., Superintendent of Documents.

IES Industrial Lighting Committee. 1983. "Proposed American National Standard Practice for Industrial Lighting," *Lighting Design and Application*, 13, 7, 29–68.

Ignall, E. J., P. Kolesar, A. J. Swersey, W. E. Walker, E. H. Blum, and G. Carter. 1975. "Improving the Deployment of New York City Fire Companies," *Interfaces*, 5, 48–61.

Jones, D. M., and D. E. Broadbent. 1987. "Noise," in *Handbook of Human Factors*, ed. G. Salvendy. New York: John Wiley.

Kolesar, P. 1975. "A Model for Predicting Average Fire Engine Travel Times," *Operations Research*, 23, 603–613.

Larson, R. C. 1972. *Urban Police Patrol Analysis*. Cambridge, Mass.: MIT Press.

______. 1987. "Perspectives on Queues: Social Justice and Psychology of Queueing," *Operations Research*, 35, 895–905.

______, and A. R. Odoni. 1981. *Urban Operations Research*. Englewood Cliffs, N.J.: Prentice Hall.

Lindkvist, R. G.T. 1985. *Handbook of Materials Handling*, Chichester, U.K.: Ellis Horwood Ltd.

Maister, D. H. 1985. "The Psychology of Waiting Lines," in *The Service Encounter, Managing Employee/Customer Interaction in Service Businesses*, ed. J.A. Czepiel, M. R. Solomon, and C. F. Surprenant. Lexington, Mass.: Lexington Books.

Newell, G. F. 1982. *Applications of Queueing Theory*. London: Chapman and Hall.

PANERO, J., and M. ZELNIK. 1979. *Human Dimension and Interior Space*. New York: Whitney Library of Design.

ROTHKOPF, M. H., and P. RECH. 1987. "Perspectives on Queues: Combining Queues Is Not Always Beneficial," *Operations Research*, 35, 906–909.

SANDERS, A. F. 1970. "Some Aspects of the Selection Process in the Functional Visual Field," *Ergonomics*, 13, 101–117.

SASSER, W. E., J. OLSEN, and D. D. WYCKOFF. 1979. *Management of Service Operations: Text, Cases and Readings*, New York: Allyn Bacon.

SULE, D. R. 1988. "*Manufacturing Facilities, Location, Planning, and Design*," Boston: PWS-Kent Publishing Company.

TOMPKINS, J. A., and J. A. WHITE. 1984. *Facilities Planning*, New York: John Wiley.

VOSS, C., C. ARMISTEAD, B. JOHNSTON, and B. MORRIS. 1985. *Operations Management in Service and the Public Sector*. Chichester, U.K.: John Wiley.

WOODSON, W. E. 1981. *Human Factors Design Handbook*. New York: McGraw-Hill.

PROBLEMS

1. With the help of your instructor, select a queueing system for observation. Based on what you see, describe how you can improve the waiting environment by each of the following means:
   (a) Providing entertainment
   (b) Improving customer information
   (c) Making the queue discipline more equitable
   (d) Adjusting the noise level
   (e) Adjusting the illumination
   (f) Improving ventilation and climate control
   (g) Reducing crowding

2. Give examples of three queueing systems for which customers are not provided with adequate information and/or queue discipline is not properly controlled. For each, prescribe a remedy.

3. Describe three examples of good passive systems for controlling queue discipline.

4. Describe three examples of good active systems for controlling queue discipline.

5. Customarily, ticket counters at airport terminals are designed as turn-back systems. Show how a ticket counter could be redesigned as a flow-through system. Do you believe the flow-through design is an improvement?

6. Customarily, supermarket checkout counters are designed as flow-through systems. Show how a checkout counter could be redesigned as a turn-back system. Do you believe the turn-back system is an improvement?

7. Under what conditions would you use each of the following layouts (describe in words)?
   (a) Linear turn-back
   (b) Linear flow-through
   (c) Circular turn-back
   (d) Parallel linear counters
   (e) Tandem servers
   (f) Parallel-tandem servers
   (g) Platooned service

8. A large proportion of the clients at a social service agency are elderly. How would this client population affect the layout and design of the queueing system?
9. The entrance to a large auditorium can process customers at a very high rate. Nevertheless, large queues sometimes exist prior to major events. How would these considerations affect the layout and design of the queueing system?
10. Discuss the advantages and disadvantages of the following queue designs, relative to a system where all servers are located together and a single queue feeds all servers:
    (a) All servers located together, with a separate queue at each server. Customers jockey between servers.
    (b) Each server located at a separate location and fed by a separate queue. Customers do not jockey.
11. A server line is located 5 feet from a 30-foot-long service counter. For each of the following, calculate average movement time, maximum distance to server, separation angle, and minimum sight angle.
    (a) End position queue
    (b) Center position queue
12. The toll plaza on a heavily traveled turnpike has been experiencing long delays. Nevertheless, some of the tollbooths (especially those on the edge of the plaza) are underutilized, even when queues are long. Prescribe a remedy.
13. A queueing system requires 10 servers, which, due to back-room activities, must be oriented in a turn-back layout. Each server requires a counter width of 5 feet. The maximum queue ordinarily experienced amounts to 50 customers. For each of the following service times, suggest a queue layout.
    (a) 15 seconds
    (b) 1 minute
    (c) 5 minutes
14. Would you change your answer to Prob. 13, part a, if each server only required a counter width of 3 feet? If so, suggest a new design. If not, explain why your previous design is still valid.
15. A building is being designed for the maintenance of railcars. The building has multiple tracks, and each track has multiple maintenance positions. The tracks are one-directional, so if a track has positions to repair three cars, all three cars must be cleared simultaneously before the next three cars are brought onto the track. The service time is normally distributed and independent, with a mean of 24 hours and a standard deviation of 8 hours.
    (a) If each track has two maintenance positions, what is the capacity per track (approximate from Clark's formula)?
    (b) If each track has three maintenance positions, what is the capacity per track (approximate from Fig. 11.17)?
    (c) Given that your answer to part a achieves a higher productivity per position, is there any reason for preferring three positions? Explain.
16. Describe a multiple server queueing system that has a product-focused layout. Describe a multiple server queueing system that has a process-focused layout. Explain why the systems use different layouts.
17. Suppose that 12 clusters of servers are spaced over a rectangular building of size 600 feet by 800 feet.
    (a) Assuming that people follow Euclidean paths, and customer locations are uniformly distributed throughout the building, what is the average walking distance from customer to nearest server?

(b) Suppose that there are only six clusters in the building. Estimate the average walking distance (approximate distance with the equation for a square region).

(c) Suppose that the arrival process is Poisson with rate 100 per hour, and that service times are exponential with mean 4 minutes. Customers walk 3 miles per hour. Also, suppose that the total number of servers equals 12. For both 6 clusters and 12 clusters, calculate the expected number of customers in queue and the expected number of customers walking. How many clusters do you prefer?

(d) Based on your intuition, would you prefer a different number of clusters if the service time increased to 7 minutes? Explain.

18. A small city would like to determine how many patrol cars to have on duty during the evening hours. Calls are received at the rate of 10/hour, and calls are served at the rate of 4/hour per car (counting response/travel time). In an emergency, patrol cars average 30 miles/hour. In addition to the travel time, it takes 2 minutes, on average, before a car begins moving to the call. The city covers 15 square miles.

(a) The city would like to use four patrol cars. Estimate the average time from when a call is received until a patrol car arrives (use $k = .5$ in Eq. (11.14)).

(b) Using the Erlang loss formula, calculate the probability that all cars will be busy when a call arrives.

(c) It would be undesirable for a critical call to go unanswered. Describe a practical procedure for responding to emergencies when all patrol cars are busy.

DESIGN PROJECT

Under your instructor's direction, visit and observe a queueing system, then study how to improve all aspects of its performance, including reduction in service time and/or change in capacity, scheduling temporary capacity increases, altering the arrival process, queue discipline, queueing environment, and queue layout. Prepare a report that covers these elements and suggests an improved queue design.

Appendix

Statistical Tables

TABLE A.1 ORDINATES OF THE NORMAL DENSITY FUNCTION

$$\phi(x) = \frac{1}{\sqrt{2\pi}} e^{-x^2/2}$$

x	.00	.01	.02	.03	.04	.05	.06	.07	.08	.09
.0	.3989	.3989	.3989	.3988	.3986	.3984	.3982	.3980	.3977	.3973
.1	.3970	.3965	.3961	.3956	.3951	.3945	.3939	.3932	.3925	.3918
.2	.3910	.3902	.3894	.3885	.3876	.3867	.3857	.3847	.3836	.3825
.3	.3814	.3802	.3790	.3778	.3765	.3752	.3739	.3725	.3712	.3697
.4	.3683	.3668	.3653	.3637	.3621	.3605	.3589	.3572	.3555	.3538
.5	.3521	.3503	.3485	.3467	.3448	.3429	.3410	.3391	.3372	.3352
.6	.3332	.3312	.3292	.3271	.3251	.3230	.3209	.3187	.3166	.3144
.7	.3123	.3101	.3079	.3056	.3034	.3011	.2989	.2966	.2943	.2920
.8	.2897	.2874	.2850	.2827	.2803	.2780	.2756	.2732	.2709	.2685
.9	.2661	.2637	.2613	.2589	.2565	.2541	.2516	.2492	.2468	.2444
1.0	.2420	.2396	.2371	.2347	.2323	.2299	.2275	.2251	.2227	.2203
1.1	.2179	.2155	.2131	.2107	.2083	.2059	.2036	.2012	.1989	.1965
1.2	.1942	.1919	.1895	.1872	.1849	.1826	.1804	.1781	.1758	.1736
1.3	.1714	.1691	.1669	.1647	.1626	.1604	.1582	.1561	.1539	.1518
1.4	.1497	.1476	.1456	.1435	.1415	.1394	.1374	.1354	.1334	.1315
1.5	.1295	.1276	.1257	.1238	.1219	.1200	.1182	.1163	.1145	.1127
1.6	.1109	.1092	.1074	.1057	.1040	.1023	.1006	.0989	.0973	.0957
1.7	.0940	.0925	.0909	.0893	.0878	.0863	.0848	.0833	.0818	.0804
1.8	.0790	.0775	.0761	.0748	.0734	.0721	.0707	.0694	.0681	.0669
1.9	.0656	.0644	.0632	.0620	.0608	.0596	.0584	.0573	.0562	.0551
2.0	.0540	.0529	.0519	.0508	.0498	.0488	.0478	.0468	.0459	.0449
2.1	.0440	.0431	.0422	.0413	.0404	.0396	.0387	.0379	.0371	.0363
2.2	.0355	.0347	.0339	.0332	.0325	.0317	.0310	.0303	.0297	.0290
2.3	.0283	.0277	.0270	.0264	.0258	.0252	.0246	.0241	.0235	.0229
2.4	.0224	.0219	.0213	.0208	.0203	.0198	.0194	.0189	.0184	.0180
2.5	.0175	.0171	.0167	.0163	.0158	.0154	.0151	.0147	.0143	.0139
2.6	.0136	.0132	.0129	.0126	.0122	.0119	.0116	.0113	.0110	.0107
2.7	.0104	.0101	.0099	.0096	.0093	.0091	.0088	.0086	.0084	.0081
2.8	.0079	.0077	.0075	.0073	.0071	.0069	.0067	.0065	.0063	.0061
2.9	.0060	.0058	.0056	.0055	.0053	.0051	.0050	.0048	.0047	.0046
3.0	.0044	.0043	.0042	.0040	.0039	.0038	.0037	.0036	.0035	.0034
3.1	.0033	.0032	.0031	.0030	.0029	.0028	.0027	.0026	.0025	.0025
3.2	.0024	.0023	.0022	.0022	.0021	.0020	.0020	.0019	.0018	.0018
3.3	.0017	.0017	.0016	.0016	.0015	.0015	.0014	.0014	.0013	.0013
3.4	.0012	.0012	.0012	.0011	.0011	.0010	.0010	.0010	.0009	.0009
3.5	.0009	.0008	.0008	.0008	.0008	.0007	.0007	.0007	.0007	.0006
3.6	.0006	.0006	.0006	.0005	.0005	.0005	.0005	.0005	.0005	.0004
3.7	.0004	.0004	.0004	.0004	.0004	.0004	.0003	.0003	.0003	.0003
3.8	.0003	.0003	.0003	.0003	.0003	.0002	.0002	.0002	.0002	.0002
3.9	.0002	.0002	.0002	.0002	.0002	.0002	.0002	.0002	.0001	.0001

(*Source: A. M. Mood, F. A. Graybill, and D. C. Boes,* Introduction to Theory of Statistics, *p. 551. New York: McGraw-Hill. Copyright © by McGraw-Hill. Reprinted by permission of the publisher.)*

TABLE A.2 CUMULATIVE NORMAL DISTRIBUTION

$$\phi(x) = \int_{-\infty}^{x} \frac{1}{\sqrt{2\pi}} e^{-t^2/2}\, dt$$

x	.00	.01	.02	.03	.04	.05	.06	.07	.08	.09
.0	.5000	.5040	.5080	.5120	.5160	.5199	.5239	.5279	.5319	.5359
.1	.5398	.5438	.5478	.5517	.5557	.5596	.5636	.5675	.5714	.5753
.2	.5793	.5832	.5871	.5910	.5948	.5987	.6026	.6064	.6103	.6141
.3	.6179	.6217	.6255	.6293	.6331	.6368	.6406	.6443	.6480	.6517
.4	.6554	.6591	.6628	.6664	.6700	.6736	.6772	.6808	.6844	.6879
.5	.6915	.6950	.6985	.7019	.7054	.7088	.7123	.7157	.7190	.7224
.6	.7257	.7291	.7324	.7357	.7389	.7422	.7454	.7486	.7517	.7549
.7	.7580	.7611	.7642	.7673	.7704	.7734	.7764	.7794	.7823	.7852
.8	.7881	.7910	.7939	.7967	.7995	.8023	.8051	.8078	.8106	.8133
.9	.8159	.8186	.8212	.8238	.8264	.8289	.8315	.8340	.8365	.8389
1.0	.8413	.8438	.8461	.8485	.8508	.8531	.8554	.8577	.8599	.8621
1.1	.8643	.8665	.8686	.8708	.8729	.8749	.8770	.8790	.8810	.8830
1.2	.8849	.8869	.8888	.8907	.8925	.8944	.8962	.8980	.8997	.9015
1.3	.9032	.9049	.9066	.9082	.9099	.9115	.9131	.9147	.9162	.9177
1.4	.9192	.9207	.9222	.9236	.9251	.9265	.9279	.9292	.9306	.9319
1.5	.9332	.9345	.9357	.9370	.9382	.9394	.9406	.9418	.9429	.9441
1.6	.9452	.9463	.9474	.9484	.9495	.9505	.9515	.9525	.9535	.9545
1.7	.9554	.9564	.9573	.9582	.9591	.9599	.9608	.9616	.9625	.9633
1.8	.9641	.9649	.9656	.9664	.9671	.9678	.9686	.9693	.9699	.9706
1.9	.9713	.9719	.9726	.9732	.9738	.9744	.9750	.9756	.9761	.9767
2.0	.9772	.9778	.9783	.9788	.9793	.9798	.9803	.9808	.9812	.9817
2.1	.9821	.9826	.9830	.9834	.9838	.9842	.9846	.9850	.9854	.9857
2.2	.9861	.9864	.9868	.9871	.9875	.9878	.9881	.9884	.9887	.9890
2.3	.9893	.9896	.9898	.9901	.9904	.9906	.9909	.9911	.9913	.9916
2.4	.9918	.9920	.9922	.9925	.9927	.9929	.9931	.9932	.9934	.9936
2.5	.9938	.9940	.9941	.9943	.9945	.9946	.9948	.9949	.9951	.9952
2.6	.9953	.9955	.9956	.9957	.9959	.9960	.9961	.9962	.9963	.9664
2.7	.9965	.9966	.9967	.9968	.9969	.9970	.9971	.9972	.9973	.9974
2.8	.9974	.9975	.9976	.9977	.9977	.9778	.9979	.9979	.9980	.9981
2.9	.9981	.9982	.9982	.9983	.9984	.9984	.9985	.9985	.9986	.9986
3.0	.9987	.9987	.9987	.9988	.9988	.9989	.9989	.9989	.9990	.9990
3.1	.9990	.9991	.9991	.9991	.9992	.9992	.9992	.9992	.9993	.9993
3.2	.9993	.9993	.9994	.9994	.9994	.9994	.9994	.9995	.9995	.9995
3.3	.9995	.9995	.9995	.9996	.9996	.9996	.9996	.9996	.9996	.9997
3.4	.9997	.9997	.9997	.9997	.9997	.9997	.9997	.9997	.9997	.9998

x	1.282	1.645	1.960	2.326	2.576	3.090	3.291	3.891	4.417
$\phi(x)$	.90	.95	.975	.99	.995	.999	.9995	.99995	.999995
$2[1 - \Phi(x)]$	.20	.10	.05	.02	.01	.002	.001	.0001	.00001

(*Source: A. M. Mood, F. A. Graybill, and D. C. Boes,* Introduction to Theory of Statistics, *p. 552. New York: McGraw-Hill. Copyright © by McGraw-Hill. Reprinted by permission of the publisher.*)

TABLE A.3 CUMULATIVE t DISTRIBUTION

$$F(x) = \int_{-\infty}^{x} f(t)dt$$

Degrees of Freedom \ F	.75	.90	.95	.975	.99	.995	.9995
1	1.000	3.078	6.314	12.706	31.821	63.657	636.619
2	.816	1.886	2.920	4.303	6.965	9.925	31.598
3	.765	1.638	2.353	3.182	4.541	5.841	12.941
4	.741	1.533	2.132	2.776	3.747	4.604	8.610
5	.727	1.476	2.015	2.571	3.365	4.032	6.859
6	.718	1.440	1.943	2.447	3.143	3.707	5.959
7	.711	1.415	1.895	2.365	2.998	3.499	5.405
8	.706	1.397	1.860	2.306	2.896	3.355	5.041
9	.703	1.383	1.833	2.262	2.821	3.250	4.781
10	.700	1.372	1.812	2.228	2.764	3.169	4.587
11	.697	1.363	1.796	2.201	2.718	3.106	4.437
12	.695	1.356	1.782	2.179	2.681	3.055	4.318
13	.694	1.350	1.771	2.160	2.650	3.012	4.221
14	.692	1.345	1.761	2.145	2.624	2.977	4.140
15	.691	1.341	1.753	2.131	2.602	2.947	4.073
16	.690	1.337	1.746	2.120	2.583	2.921	4.015
17	.689	1.333	1.740	2.110	2.567	2.898	3.965
18	.688	1.330	1.734	2.101	2.552	2.878	3.922
19	.688	1.328	1.729	2.093	2.539	2.861	3.883
20	.687	1.325	1.725	2.086	2.528	2.845	3.850
21	.686	1.323	1.721	2.080	2.518	2.831	3.819
22	.686	1.321	1.717	2.074	2.508	2.819	3.792
23	.685	1.319	1.714	2.069	2.500	2.807	3.767
24	.685	1.318	1.711	2.064	2.492	2.797	3.745
25	.684	1.316	1.708	2.060	2.485	2.787	3.725
26	.684	1.315	1.706	2.056	2.479	2.779	3.707
27	.684	1.314	1.703	2.052	2.473	2.771	3.690
28	.683	1.313	1.701	2.048	2.467	2.763	3.674
29	.683	1.311	1.699	2.045	2.462	2.756	3.659
30	.683	1.310	1.697	2.042	2.457	2.750	3.646
40	.681	1.303	1.684	2.021	2.423	2.704	3.551
60	.679	1.296	1.671	2.000	2.390	2.660	3.460
120	.677	1.289	1.658	1.980	2.358	2.617	3.373
∞	.674	1.282	1.645	1.960	2.326	2.576	3.291

(*Source: Taken from Fisher & Yates,* Statistical Tables for Biological, Agricultural, and Medical Research, *6/e, published by Longman Group UK Ltd., London (previously published by Oliver and Boyd Ltd., Edinburgh) and by permission of the authors and publishers.*

TABLE A.4 KOLMOGOROV-SMIRNOV DISTRIBUTION

Sample size (N)	Level of significance (α)		
	0.10	0.05	0.01
1	0.950	0.975	0.995
2	0.776	0.842	0.929
3	0.642	0.708	0.828
4	0.564	0.624	0.733
5	0.510	0.565	0.669
6	0.470	0.521	0.618
7	0.438	0.486	0.577
8	0.411	0.457	0.543
9	0.388	0.432	0.514
10	0.368	0.410	0.490
11	0.352	0.391	0.468
12	0.338	0.375	0.450
13	0.325	0.361	0.433
14	0.314	0.349	0.418
15	0.304	0.338	0.404
16	0.295	0.328	0.392
17	0.286	0.318	0.381
18	0.278	0.309	0.371
19	0.272	0.301	0.363
20	0.264	0.294	0.356
25	0.24	0.27	0.32
30	0.22	0.24	0.29
35	0.21	0.23	0.27
over 35	$\frac{1.22}{\sqrt{N}}$	$\frac{1.36}{\sqrt{N}}$	$\frac{1.63}{\sqrt{N}}$

(*Source: F. J. Massey, Jr.*, "The Kolmogorov-Smirnov Test for Goodness of Fit," *Journal of the American Statistical Society*, V. 46, p. 70, 1946.*) Reprinted by permission of the publisher.*

Answers to Selected Problems

CHAPTER 2

1a. $W_q = 19$ min, $\sigma_{W_q} = 15$ min
c. E (tardiness) $= 2.9$ min, SD (tardiness) $= 7.0$ min

2a. $L_q = .74$, $\sigma_{L_q} = .72$
b. $L_s = 1.64$, $\sigma_{L_s} = .87$

3a. $.74 = (7/180)(19)$
b. $1.64 \neq (7/180)(45) = 1.75$

5. $\rho = .90$, throughput $= 2$/hr

6a. 3 servers
d. $W_q = 2.4$ days, $L_q \approx .71$ customer
e. Only approximate; system begins and ends with customers in queue

7c. $W_s = 47$ min, $\sigma_{W_s} = 21$ min

9c. $E(S) = 52$ sec

10. $W_q = 11.7$ min, $\sigma_{W_q} = 16.5$ min (among non-reneges)
1/3 renege, average time to renege $= 3.7$ min
(other measures omitted)

12a. $W_q \approx .16$ min, $W_s \approx .65$ min
c. 7 servers

CHAPTER 3

1a. $P(n)$ = .35, .39, .19, .057, .014
b. $P(n)$ = .37, .37, .18, .061, .015
3a. $P(n)$ = .19, .31, .26, .15, .061, .020
b. .036
c. .84
d. .5
6. $P(0)$ = .51, $P(1)$ = .17
8a. .22
11a. Yes. (b). no, due to bulk arrivals
c. $\tilde{\lambda}$ = 9.4/hr, 95% → [7.1, 13.9]
d. K-S stat = .10 → do not reject at 5% sig level
e. $r = -.29$, $t = -1.82$ → do not reject at 5% sig level
f. K-S stat = .10 → do not reject at 5% sig level
12c. $\hat{\lambda}$ = 14.5/hr
d. K-S stat = .47
e. r = .023
f. K-S stat = .051
(some answers omitted)
15. 270 hr

CHAPTER 4

1a. 9507, 214, 1621
2. 1.2, 3.2
3a. 1, 3.5 (discrete)
4. a. 0. b. 0 c. 1
5a. 3.95, 47.5
b. 461, 69.3 [using ln(U)],
c. 69.6, 100
7. Using ln(U): .0597, .136, .574, .747, .767
9. 104.7, 106.4, 107.5, 109.0 (in minutes past 12:00)
12a. Service times: .194, .078, .166, .131, .133
b. $A^{-1}(n)$ = .0597, .136, .574, .747, .767
$D_q^{-1}(n)$ = .0597, .254, .574, .747, .878
D_s^{-1} = .254, .332, .740, .878, 1.01
14.

Customer	1	2	3	4	5
$A^{-1}(n)$	.0597	.136	.574	.747	.767
$D_q^{-1}(n)$	.0597		.574	.747	.878
$D_s^{-1}(n)$	.254		.740	.878	1.05
Renege?	—	Yes (U = .154)	—	—	No (U = .444)

15a. W_s = 1.58 min
b. .062 hr
c. Based on every seventh customer, 1.58 min ± .55 min

CHAPTER 5

2a. $L_s = 1.05$, $L_q = .45$
b. $W_s = .105$ hr, $W_q = .045$ hr
c. $\sigma_{L_s} = 1.12$, $\sigma_{L_q} = .805$
d. $1/\mu = .06$ hr, $\rho = .6$

4a. $L_s = 1.4$, $L_q = .2$
b. $W_s = .14$ hr, $W_q = .02$ hr
c. $\sigma_{L_s} = 1.02$, $\sigma_{L_q} = .4$
d. $1/\mu = .12$ hr, $\rho = 1.2$

5a. $P_n = .286 \cdot .714^n$
b. $L_s = 2.5$, $L_q = 1.8$
c. .903
d. .865

7a. $P_n = .138, .345, .345, .172$
b. $\bar{\lambda} = 2.4$/hr, $\bar{\mu} = 2.8$/hr
c. $L_q = .69$, $L_s = 1.55$, $W_q = .29$ hr, $W_s = .65$ hr

9. $L_q = .065$, $W_q = .013$ min

10a. $m = 1 \rightarrow W_s = \infty$. b. $m = 2 \rightarrow W_s \approx .57$ hr. c. $m = 3 \rightarrow W_s \approx .088$ hr. d. $m = 4 \rightarrow W_s \approx .071$ hr
b. 3 employees

12a. E (in service) = 3.3 = ρ
b. .88
c. P(7 or fewer) = .979, use 7 or 8 lines

14a. $L_q = .91$, $\bar{\lambda} = 4.1$ cust/hr
b. $L_q \approx 3$, $\bar{\lambda}$ will be between 4.1 and 5/hr.
c. $L_q = .035$, $\bar{\lambda} = 5.6$/hr
d. $10.20 added profit, not counting server wages

16. Employee 1: $L_q = .83$, $L_s = 1.5$
Employee 2: $L_q = .82$, $L_s = 1.52$

17. $L_q = .15$, $W_q = .03$ hr

18.

%	95	85	75	65	55	45	35	25	15	5
W_q	3.65	2.67	2.24	2.04	1.88	1.69	1.58	1.48	1.33	1.18

19a. $W_q = .033$ hr
b. $W_q = .24$ hr

20a. i .15 hr

21. $W_q \approx .15$ hr, inter. rate ≈ 1.3/hr, time bet. inter. ≈ .76 hr.
E(busy servers) = 2.1

22b. E(cabs) = 1.0
c. E(cab wait) = .087 hr (among those that join queue)

CHAPTER 6

1a.
$$\Lambda(t) = \begin{cases} 5(t - 8) & 8 \leq t < 12 \\ 20 + 20(t - 12), & 12 \leq t < 13 \\ 40 + 10(t - 13), & 13 \leq t \leq 15 \end{cases}$$

b. .018

c. 0: $.5^{20}$ 2: $\binom{20}{2}.5^{20}$

d. .097

3. between 7:00 and 7:40, $\approx$ 6,400/hr

5a.
$$L_q(t) \approx \frac{(\lambda(t)/.2)^2}{1 - \lambda(t)/.2}\left(\frac{1.16}{2}\right)$$

b. convex and increasing, largest at $t = 92$, $L_q(92) = .6$

6. max $\Delta = .0091$ at $t = 80$: approximation is valid

8. 8:00–12:00, 5/hr $\pm$ .5/hr
(other intervals omitted)

11b. max $L_q(t) \approx 1400$ at 7:50

c. between 7:00 and 7:40, $\approx$ 1600/hr

d. begins at 6:55, ends at 8:45

e. total wait $\approx$ 1000 pers-hr

13a. $E[L_q(t')] \approx 12$ customers

c. valid until 6:50, at which time $L_q(t) = 3.8$

15. at 6:55, approximately 1000 jobs.

CHAPTER 7

5a. .82 hr

b. 4 lanes

c. $24,000/day

7. from 6:20 to 6:55 and 7:45 to 8:15 (approximate)

9b. start at: 8:00, 8:30, 11:15, 12:00, 3:50, 4:00, 4:30, 4:50, 7:15

c. 9 employees

11a. 7 employees

b. no

c. yes—save 6% of cost

13a. 25 sec, 35 sec, 100 sec, 30 sec

b. approximately, 77 sec, and 23 sec (includes loss time)

c. 19 vehicles

16. weeks 0, 1.25, 3.5, 6.25

17b. \$300/day and \$40/day
c. 2000 parts, \$250/day
18a. 4.5 days

CHAPTER 8

6. Possible solution: 10 reservations at each of following times: 6:00, 6:30, 7:30, 8:00, 9:00, 9:30
18. 1049 tickets
10a. 6/hr and 9/hr, .23 hr
b. 7/hr and 8/hr, .22 hr
c. \$2.80 and \$3.10 (approximate)

CHAPTER 9

4a. .36 hr^2
b. 6.2 hr^2
c. .89 hr^2
5.

	Departure time from system						
Customer	**1**	**2**	**3**	**4**	**5**	**6**	**Sequence**
FCFS	16	22	37	58	60	70	(1, 2, 3, 4, 5, 6)
SST	16	22	37	70	39	49	(1, 2, 3, 5, 6, 4)
EDD	16	37	31	60	39	70	(1, 3, 2, 5, 4, 6)
MS	16	58	31	52	70	68	(1, 3, 4, 2, 6, 5)

7a. mean: 2, 2.5, 1, 2.3; max: 7, 8, 5, 10
b. number late: 2, 2, 2, 3
c. W_s: 24, 19, 26, 30

9a. $$\frac{2.5e^{-x/5}(1 + x/5) + 10e^{-x/20}(1 + x/20)}{.5(e^{-x/5} + e^{-x/20})} - x$$

12a. .20 sec, .98 sec, 2.0 sec, 3.9 sec, 9.8 sec
b. 1.14 sec
c. 1.43 sec
13a. 5.76 hr (same for both)
b. 1.9 hr, 7.7 hr, 3.8 hr (average)
c. .25 hr, 9 hr, 3.2 hr (average)
17b. 1083 person-hr
c. 833 person-hr

CHAPTER 10

3a. stage 2, 4500/hr

5a.

c	0–25	25–50	50–75	75–100	1–200	2–300	3–400	4–500	5–600
Cost ($/day)	700	800	900	1000	1200	1600	1800	2200	2600

Among options considered, capacity of 600 has cost of $4.33/unit capacity.

b. 200/day: A: mach, B: 1 machine, C: mach 2, D: 1 machine
profit = 3000 − 1200 = $1800/day (refuse work)
or 300/day: profit = $1775/day (do not refuse work)

c. profit increases to $2800/day

7a. stage 1: 8, stage 2: 3, stage 3: 10; capacity = 345/hr
(alternating one employee between stages raises capacity to 353/hr)

9a. machine 2, capacity = 300/day

10b. 18 vehicles

13a. machine 2, capacity = 150/hr

16a. 8.5 min

b. 5.8 min

18. 6.63 min

20. 60 sec

21a. yes

b. no, yes

22a. DS: 61 hr, 20 hr, 18 hr

b. Job 1

c. CR: 7.8, 2, 1.8 (choose 3)
(some answers omitted)

CHAPTER 11

11a. $\theta = 81^\circ$, $\alpha = 9^\circ$, ave. time = 4.0 sec, max dist = 30.4 ft

15a. 1.68/day

b. 2.3/day

17a. 77 ft

c. 12 clusters: 9.3 customers (walk + wait)

18a. 5.2 min

b. P(all busy) = .15

Index

A

B

C

H

I

J

K

L

M

N

O

P

Q

R

S

T

U

V

W

Y